Shaping the Future of Automation With Cloud-Enhanced Robotics

Rathishchandra Ramachandra Gatti
Sahyadri College of Engineering and Management, India

Chandra Singh
Sahyadri College of Engineering and Management, India

IGI Global
PUBLISHER of TIMELY KNOWLEDGE

A volume in the Advances in Computational Intelligence and Robotics (ACIR) Book Series

Published in the United States of America by
IGI Global
Engineering Science Reference (an imprint of IGI Global)
701 E. Chocolate Avenue
Hershey PA, USA 17033
Tel: 717-533-8845
Fax: 717-533-8661
E-mail: cust@igi-global.com
Web site: http://www.igi-global.com

Library of Congress Cataloging-in-Publication Data

CIP Pending

Shaping the Future of Automation With Cloud-Enhanced Robotics
Rathishchandra Gatti, Chandra Singh
ISBN: 979-8-3693-1914-7
eISBN: 979-8-3693-1915-4

This book is published in the IGI Global book series Advances in Computational Intelligence and Robotics (ACIR) (ISSN: 2327-0411; eISSN: 2327-042X)

British Cataloguing in Publication Data
A Cataloguing in Publication record for this book is available from the British Library.

For electronic access to this publication, please contact: eresources@igi-global.com.

Advances in Computational Intelligence and Robotics (ACIR) Book Series

Ivan Giannoccaro
University of Salento, Italy

ISSN:2327-0411
EISSN:2327-042X

Mission

While intelligence is traditionally a term applied to humans and human cognition, technology has progressed in such a way to allow for the development of intelligent systems able to simulate many human traits. With this new era of simulated and artificial intelligence, much research is needed in order to continue to advance the field and also to evaluate the ethical and societal concerns of the existence of artificial life and machine learning.

The **Advances in Computational Intelligence and Robotics (ACIR) Book Series** encourages scholarly discourse on all topics pertaining to evolutionary computing, artificial life, computational intelligence, machine learning, and robotics. ACIR presents the latest research being conducted on diverse topics in intelligence technologies with the goal of advancing knowledge and applications in this rapidly evolving field.

Coverage

- Algorithmic Learning
- Intelligent Control
- Agent technologies
- Evolutionary Computing
- Cognitive Informatics
- Artificial Intelligence
- Synthetic Emotions
- Natural Language Processing
- Fuzzy Systems
- Adaptive and Complex Systems

IGI Global is currently accepting manuscripts for publication within this series. To submit a proposal for a volume in this series, please contact our Acquisition Editors at Acquisitions@igi-global.com or visit: http://www.igi-global.com/publish/.

The Advances in Computational Intelligence and Robotics (ACIR) Book Series (ISSN 2327-0411) is published by IGI Global, 701 E. Chocolate Avenue, Hershey, PA 17033-1240, USA, www.igi-global.com. This series is composed of titles available for purchase individually; each title is edited to be contextually exclusive from any other title within the series. For pricing and ordering information please visit http://www.igi-global.com/book-series/advances-computational-intelligence-robotics/73674. Postmaster: Send all address changes to above address.

Titles in this Series

For a list of additional titles in this series, please visit: http://www.igi-global.com/book-series/advances-computational-intelligence-robotics/73674

Secure and Intelligent IoT-Enabled Smart Cities
Sushil Kumar Singh (Marwadi University, India) Sudeep Tanwar (Nirma University, India) Rajendrasinh Jadeja (Marwadi University, India) Saurabh Singh (Woosong University, South Korea) and Zdzislaw Polkowski (Wroclaw University of Economics, Poland)
Engineering Science Reference • copyright 2024 • 434pp • H/C (ISBN: 9798369323731) • US $325.00 (our price)

Using Real-Time Data and AI for Thrust Manufacturing
D. Satishkumar (Nehru Institute of Technology, India) and M. Sivaraja (Nehru Institute of Technology, India)
Engineering Science Reference • copyright 2024 • 343pp • H/C (ISBN: 9798369326152) • US $345.00 (our price)

AI Approaches to Smart and Sustainable Power Systems
L. Ashok Kumar (PSG College of Technology, India) S. Angalaeswari (Vellore Institute of Technology, India) K. Mohana Sundaram (KPR Institute of Engineering and Technology, India) Ramesh C. Bansal (University of Sharjah, UAE & University of Pretoria, South Africa) and Arunkumar Patil (Central University of Karnataka, India)
Engineering Science Reference • copyright 2024 • 432pp • H/C (ISBN: 9798369315866) • US $300.00 (our price)

Wearable Devices, Surveillance Systems, and AI for Women's Wellbeing
Sivaram Ponnusamy (Sandip University, Nashik, India) Vibha Bora (G. H. Raisoni College of Engineering, Nagpur, India) Prema M. Daigavane (G. H. Raisoni College of Engineering, Nagpur, India) and Sampada S. Wazalwar (G. H. Raisoni College of Engineering, Nagpur, India)
Engineering Science Reference • copyright 2024 • 321pp • H/C (ISBN: 9798369334065) • US $335.00 (our price)

Methodologies, Frameworks, and Applications of Machine Learning
Pramod Kumar Srivastava (Rajkiya Engineering College, Azamgarh, India) and Ashok Kumar Yadav (Rajkiya Engineering College, Azamgarh, India)
Engineering Science Reference • copyright 2024 • 296pp • H/C (ISBN: 9798369310625) • US $300.00 (our price)

Artificial Intelligence of Things (AIoT) for Productivity and Organizational Transition
Sajad Rezaei (University of Worcester, UK) and Amin Ansary (University of the Witwatersrand, South Africa)
Business Science Reference • copyright 2024 • 368pp • H/C (ISBN: 9798369309933) • US $275.00 (our price)

Internet of Things and AI for Natural Disaster Management and Prediction
D. Satishkumar (Nehru Institute of Technology, India) and M. Sivaraja (Nehru Institute of Technology, India)
Engineering Science Reference • copyright 2024 • 334pp • H/C (ISBN: 9798369342848) • US $345.00 (our price)

IGI Global
PUBLISHER of TIMELY KNOWLEDGE

701 East Chocolate Avenue, Hershey, PA 17033, USA
Tel: 717-533-8845 x100 • Fax: 717-533-8661
E-Mail: cust@igi-global.com • www.igi-global.com

Table of Contents

Section 2
Applications of Cloud Robotics

Detailed Table of Contents

In this digital age, every machine is full of data, telling stories about how it works, how much wear it's wearing, and what it's doing. In the middle of all this tech, the combination of advanced data analysis and predictive maintenance acts as a conductor, bringing machinery, data, and foresight together. Instead of just reacting to mechanical issues, industries can now anticipate them and turn potential downtime into organized maintenance periods. It's not just about fixing things before they break, it's a whole new way of understanding and working with industrial machines and robotics. Sensors, algorithms, and the power of data science are all part of this new era, where machines 'talk' to each other and let us know their health so we can act quickly. This combination of data analytics and proactive maintenance promises not only improved operational efficiency, but a jump towards safer, stronger, more resilient and more affordable industrial ecosystems. Read on to find out how this combination of data and foresight is transforming industries.

Cloud robotics is currently regarded as one of the cutting-edge research areas within the mainstream automation and artificial intelligence fields. The new machine-to-cloud communication will increase the productivity and efficiency of industrial robots. For decades, automated devices have been developing and affecting practically every element of life. Cloud computing and associated technologies have a great deal of potential to get around hardware limitations and boost performance. This study emphasises the developments in robotic technology with a particular focus on cloud-based automation as a new development. Unexpected effects of robotic technologies on human life have been seen. This chapter discusses the fundamental ideas, the creation process, and the general architecture of cloud robotics in order to explore the potential of clouds for improving robotics for industrial systems. Finally, it examines how useful cloud robotic systems might be in various real-world scenarios.

In the face of the development of robotic technology, integration with cloud technologies is a big opportunity to greatly improve robots. This chapter will synthetize the spirit which robotic applications lacks to refine and add depth to it. Also, this procedure not only gives the robotic device more computing power, but also ups its storage capacities and resource management, all without rendering robotical atrophy. In this chapter, the authors examine several different forms of application running away from home. In particular, with the growing public interest in robotics, it is vital that not only participants from industry but also researchers and developers who should realize that through cloud-based offloading they really have a chance to create a revolution.

In a landscape of technology that is changing at breakneck speed, the combination of the internet of things (IoT) with robotics represents an important step toward future automation and intelligent systems. In this chapter, the authors explain how such an integration will change things. This combination of IoT's connected devices and robotic dynamism brings unprecedented progress in many areas, such as industrial automation and healthcare. After cutting through the technical jumble and coming to terms with plausible applications, problems, and opportunities afforded by this match, readers will have a clear understanding of what this revolutionary field has today, as well as its promise for tomorrow.

Robotics needs to be applied in many different areas which requires the existence of better-quality platforms whose performance and function can be refined. The cloud and robotic cloning in this chapter offer an innovative idea. As a result, utilizing the immense capabilities and scalability of cloud computing can only make robotic systems better and more capable in multi-robot cooperative scenarios. The chapter offers an outline of the theoretical foundations, advantages, possible concurrent applications, and future

Cloud-enhanced robotics is reforming the healthcare sector through adventures. Cloud-enhanced healthcare robots aimed to provide potential support to medical professionals by accurately providing services like nursing, diagnosing, and executing critical surgeries. In the domain of healthcare, cloud robotics technology facilitates real-time data exchange, enabling professionals to deliver optimal services and care. Tasks ranging from laboratory testing to medication distribution can now be automated, and cloud-enhanced robotics have reached a level of proficiency that significantly lowers the risk of human faults in critical surgeries. This chapter explores the architecture and components of cloud-enhanced healthcare robots, provides various applications in the field, assesses the advantages and disadvantages of healthcare robots, and scrutinizes the security implications associated with cloud-enhanced robots.

The healthcare industry is currently experiencing a groundbreakingly revolution as cloud robotics and artificial intelligence (AI) come together. This research examines the synergistic relationship between these two advanced technologies and their significant influence on improving patient care. Through the utilisation of cloud-based computing power and the advanced capabilities of intelligent robotics, healthcare systems can attain unparalleled levels of efficiency, accessibility, and personalisation. Integrating AI algorithms with robotic systems enables enhanced diagnosis, treatment planning, and real-time monitoring, ultimately resulting in enhanced patient outcomes. This chapter examines the present condition of these technologies, explores instances where they have been effectively put into practice, and highlights the possible obstacles and ethical concerns. In this era of healthcare transformation, it is essential to recognise the collaborative capabilities of cloud robotics and AI in designing the future of patient-centred, data-driven healthcare systems.

There is an urgent need for innovative and efficient disaster response and recovery solutions. This chapter investigates the use of cloud technologies in conjunction with robotics to improve disaster management operations. Cloud-enhanced robotics (CER) appears as a significant innovation, offering a synergistic approach in which cloud-enabled robots provide a more dynamic, adaptable, and robust response to catastrophic occurrences. CER enables robots to access huge computational resources and data, improving operational efficiency, decision-making, and problem-solving abilities. This connection improves real-time data analysis, coordination, and deployment tactics, resulting in a more effective reaction to crisis scenarios unfolding. Furthermore, it encourages greater coordination and resource sharing among different robots and reaction teams, hence maximizing overall recovery operations.

This chapter provides a progressive thing classification and the most limited organising approach for robot-assisted night-time salvaging jobs. In reality, people are sorted using a system for determining skin tone. Boundaries were identified using the skin colour based (SCB), pixel count based (PCB), correlation coefficient based (CCB), and histogram draws near techniques in this study. The PCB, CCB, and SCB calculations are all utilised to snag distinct proof in order. The SCB calculation, on the other hand, gradually determines whether the observed entity is human or nonhuman. It also distinguishes between live and nonliving impediments, allowing for more efficient salvage efforts (except for flies, mosquitoes, and different creepy crawlies). According to continuous testing data, the accuracy of the CCB, PCB, and SCB calculations is 87.5%, 88.8%, and 90.9%, respectively, and the time length of the CCB, PCB, and SCB calculations is 6.27 seconds, 6.50 seconds, and 6.57 seconds, respectively.

Through case studies, the chapter illustrates how cloud-enhanced robotics positively influences education. It introduces popular platforms and their features conducive to educational settings, providing insights into hands-on programming exercises and projects. This equips educators and learners with valuable resources for cultivating programming skills in robotics. It emphasizes transformative impacts on learning experiences, emphasizing flexibility, inclusivity, and successful implementations in remote learning scenarios.

Education is continually evolving to adapt to the ever-changing technological landscape of the 21st century. The integration of robotics into classrooms has opened up new horizons for both educators and students. Robotics serves as a powerful tool for engaging students in hands-on learning, problem-solving, and critical thinking. However, as the world becomes increasingly interconnected, cloud technology has emerged as a game-changer in education, particularly in the realm of robotics. Cloud-enhanced robotics is transforming the educational landscape, offering students and educators unprecedented

efficiency, scalability, and intelligence to the agricultural landscape. Several case studies are presented to illustrate real-world applications of cloud robotics in diverse farming settings, highlighting their impact on resource management, crop yield optimization, and environmental sustainability. Furthermore, the chapter identifies the hurdles and barriers hindering the widespread adoption of cloud robotics in agriculture, including cost, connectivity, and data security concerns, and offers potential solutions to overcome these challenges.

Preface

Welcome to *Shaping the Future of Automation With Cloud-Enhanced Robotics*, a groundbreaking exploration of the intersection between cloud technology and robotics. This book delves into the transformative potential of cloud-enhanced robotics, offering insights into the cutting-edge advancements and challenges in this dynamic field.

Overview of the Subject Matter

The book delves into the fusion of cloud computing and robotics, exploring how this synergy revolutionizes automation across industries. From industrial applications to healthcare, education, and disaster response, cloud-enhanced robotics is reshaping how we perceive and utilize autonomous systems.

Relevance in Today's World

Integrating cloud computing and robotics is at the forefront of innovation in today's rapidly evolving technological landscape. This book provides a comprehensive understanding of how cloud-enhanced robotics drives efficiency, scalability, and intelligence in various domains, setting the stage for a future where automation is more interconnected and adaptable than ever.

Target Audience

This book is tailored for researchers, academics, industry professionals, and technology enthusiasts seeking to deepen their knowledge of cloud robotics. It is a valuable resource for anyone interested in exploring the potential and challenges of integrating cloud technologies with robotic systems.

Importance of Each Chapter

In Chapter 1, the author delves into the prospective developments of industrial automation through the utilization of AI and cloud robotics, thereby establishing the foundation for the paradigm-shifting consequences cloud technology will bring about in industrial environments. This chapter discusses the anticipated industrial automation movements of AI and cloud robotics. Instead, cloud robotics and artificial intelligence should be implemented to redefine industrial production processes in line with the global manufacturing industry's pursuit of greater flexibility, sustainability, and efficiency. The scalability and distributed computing capability of cloud robotics and AI's data-driven analysis overcome the greatest challenges in automating throughput, demonstrating that automation can and should be executed properly. The research paper elucidates how these technologies facilitate adaptive decision-making, predictive maintenance, and real-time data processing, thereby augmenting the productivity and ingenuity of the organization. The discourse also encompasses fundamental barriers to AI technology, including the transformation of the data security workforce and forthcoming trends. Therefore, by integrating current implementations, case studies, and expert forecasts, this research will analyze the impact of AI and cloud robotics on forthcoming manufacturing facilities and the subsequent evolution of the industry.

Cloud infrastructure for robotics is the subject of Chapter 2, which demonstrates how cloud-based deployment and development are transforming the field. Integrating cloud infrastructure and robotics revolutionizes the domain by furnishing an economically viable framework for creating, evaluating, and implementing robotic applications. The utilization of on-demand computing resources in the cloud expedites the process of innovation by enabling developers to test novel algorithms efficiently. By collaborating globally, it is possible to remotely control and test robots from any location, effectively surpassing the limitations imposed by geography. By transferring resource-intensive duties to the cloud, cloud infrastructure facilitates deployment by decreasing hardware demands and encouraging cost efficiency. Centralized fleet management ensures consistent updates and security upgrades to maintain system stability, thereby streamlining operations. Privacy concerns are mitigated by implementing safeguards such as encryption and authentication. Cloud robotics reduces the likelihood of security breaches by implementing peripheral computing for local data processing. Scalability facilitates the adjustment of businesses in response to expanding demands, thereby preventing big investments.

In Chapter 3, the convergence of data analytics and predictive maintenance is examined, emphasizing the significance of data-driven insights in improving the performance and dependability of robotic systems. Every machine we hear and see in the digital age has data revealing its operation, condition, and what it is doing. Amidst the plethora of technological advancements, the integration of predictive maintenance and advanced data analysis serves as a conduit, connecting machinery, data, and foresight. Industries can proactively identify and address mechanical issues, preventing delays through structured maintenance periods. It is a new method of comprehending and operating robotics and industrial machinery, not merely repairing items before they break. This new era comprises sensors, algorithms, and the power of data science; machines "communicate" with one another regarding their health to take prompt action. Integrating data analytics and proactive maintenance can enhance operational efficiency and propel industrial ecosystems towards enhanced safety, strength, resilience, and affordability. Continue reading to learn how this convergence of data and foresight revolutionizes various sectors.

Chapter 4 presents an all-encompassing examination of techniques, taxonomy, applications, and research issues in cloud robotics and automation, thereby furnishing a comprehensive perspective. Cloud robotics is considered one of the most advanced research domains in the conventional artificial intelligence and automation disciplines. The new machine-to-cloud communication will increase industrial automation productivity and efficiency. Automated devices have been advancing and influencing virtually every aspect of life for decades. The potential of cloud computing and related technologies to circumvent hardware limitations and enhance performance is substantial. This research paper centres on the advancements in autonomous technology, specifically emphasizing the emergence of cloud-based automation. The impact of robotic technologies on human existence has been observed to be unforeseen. Exploring the cloud's potential to enhance robotics for industrial systems, this article examines the fundamental concepts, development methodology, and overall architecture of cloud robotics. In conclusion, cloud autonomous systems' applicability to various real-world scenarios is evaluated.

The efficacy of application offloading strategies in cloud robotics is examined in Chapter 5, which illustrates how delegating tasks to the cloud can enhance the efficiency of robotic operations. Given the rapid advancements in robotic technology, incorporating cloud technologies presents a substantial opportunity to enhance robot performance significantly. To refine and augment the profundity of robotic applications, this chapter will synthesize the spirit that is absent from them. Furthermore, this process enhances not only the computational capabilities but also the storage capacities and resource management of the robotic device, all without causing it to deteriorate. Advantages and disadvantages of various application offloading forms: This chapter will examine various applications fleeing from home. Part I: Given the increasing public fascination with robotics, industry participants, researchers, and developers must recognize that cloud-based outsourcing represents an opportunity to instigate a paradigm shift.

Examining the Evolution of Robotics through the Integration of IoT and Robots, Chapter 6 demonstrates the synergies between robotic technologies and IoT devices. In an era of evolving technologies, integrating robotics and the Internet of Things (IoT) is a pivotal development towards intelligent systems and automation of the future. Chapter 4 delves into the notion of "Robotic Evolution Cumulating in the Internet of Things and Robots," elucidating the transformative implications of such an integration. The convergence of IoT-enabled devices and the dynamism of robotics enables unparalleled advancements in numerous domains, including healthcare and industrial automation. Upon deciphering the intricate technical complexities and confronting the practical implications, challenges, and prospects presented by this match, readers will comprehend this groundbreaking discipline's present state and future potential.

In Chapter 7, the author examines the utilization of cloud-based cloning platforms to scale autonomous intelligence, demonstrating how such platforms facilitate the replication and expansion of such intelligence. To accomplish this, robotics must be implemented in various domains, necessitating the availability of superior-quality platforms whose functionality and performance can be enhanced. This chapter discusses the innovative concepts of cloud computing and robotic replication. Utilizing the enormous capabilities and scalability of cloud computing can improve and expand the capabilities of robotic systems in multi-robot cooperative scenarios. The chapter delineates the theoretical underpinnings, benefits, potential concurrent applications, and prospective trajectory of cloud-based cloning platforms—a novel amalgamation of robotics and cloud computing. Through an in-depth examination of the topic, readers will understand how cloud-based replication platforms can serve as a significant enabling element, facilitating the expansion of intelligence and capability in robotics with the advent of multi-robot systems.

Communication protocols in cloud robotics are examined in Chapter 8, emphasizing the criticality of reliable communication to ensure uninterrupted operation. The emergence of cloud robotics has revolutionized conventional autonomous systems through their integration with the enormous computational capacity of cloud infrastructure. The present chapter delves into the critical significance of communication protocols in enabling smooth and uninterrupted exchanges between tangible robotics and the cloud. Cloud robotics is predicated on resolving particular obstacles, including but not limited to low latency, dependability, scalability, and security. Various communication protocols, such as WebSocket, MQTT, and RESTful APIs, are evaluated in terms of their appropriateness for various applications, including data offloading and real-time monitoring. Security concerns are of the utmost importance when transmitting sensitive data between robots and cloud servers; therefore, encryption, authentication, and authorization mechanisms are required. Additionally, the chapter explores the significance of interoperability and standardization in promoting cooperation among heterogeneous robotic systems.

Security and Privacy in Cloud Robotics, critical for ensuring data confidentiality and integrity in cloud-connected robotic systems, are discussed in Chapter 9—the amalgamation of robotics and cloud computation. Cloud robotics ushers in a period of profound change. Granting machines unparalleled access to the cloud facilitates complex operations but presents significant security concerns. Securing the transmission of sensitive data requires robust measures. Machine learning facilitates the detection of threats in real-time; best practices for data encryption, storage, authentication, and access control are investigated. In summary, the chapter guides stakeholders regarding the security of cloud robotics by examining crucial concerns, endorsing optimal methodologies, and predicting forthcoming developments. By providing data privacy and security analysis, this resource enables experts to utilize cloud robotics while minimizing potential hazards effectively. This succinct investigation, encompassing both the physical and digital domains, is a guiding light in the ever-evolving domain of cloud robotics privacy and security.

In Chapter 10, the author delves into the various applications of cloud computing in industrial robotics, illustrating the multifaceted way cloud technologies augment industrial automation processes. Over the last several decades, industrial and technological progress has been substantially influenced by robots. Industrial robotics integrated with the cloud provides a vast network of devices for various industries. The convergence of robotics and cloud computing presents many opportunities to leverage enhanced computational capabilities in industry, communication, healthcare, disaster management, defence systems, and space exploration. Furthermore, it ensures comprehensive and dependable accessibility throughout a network of devices. The present chapter explores the foundational principles of cloud computing and robotics. After introducing the fundamental concepts of cloud computing and robotics, we discuss their integration. We then delve into communication within cloud robotics, which includes both Machine-to-Machine (M2M) and Machine-to-Cloud (M2C) interactions. Then, we analyze the primary industrial robotics and other robot applications that use cloud robotics technology. In conclusion, we address security concerns and obstacles in cloud robotics.

Chapter 11 examines the utilization of swarm robotics in supply chain and logistics, emphasizing the increased flexibility and efficiency of swarm robotics. Implementing swarm robotics in the supply chain and logistics signifies a paradigmatic change in the operational procedures of these pivotal sectors. This chapter delves into the incorporation of swarm robotics into the domains of supply chain management and logistics, emphasizing the fundamental transformation from traditional approaches to automated systems that are more effective. Swarm robotics is defined at the outset of the chapter, emphasizing its attributes, including scalability, decentralization of control, and robustness. The subsequent section explores the particular implementations of these robotic systems across diverse facets of supply chain and logistics. The chapter demonstrates the transformative impact of swarm robotics on inventory management and warehousing operations by implementing automated storage, retrieval, and classification procedures. The chapter further explores how Artificial Intelligence (AI), Machine Learning (ML), and the Internet of Things (IoT) can enhance the functionalities of swarm robotic systems. The chapter discusses the obstacles and constraints of implementing swarm robotics in the logistics and supply chain industries.

Chapter 12 delves into the examination of contemporary healthcare through the lens of cloud-enhanced robotics, illustrating the profound impact that cloud robotics have on healthcare provision and patient care. With the incorporation of robotics and cloud computing, contemporary healthcare is enduring a profound overhaul. The convergence of these two technologies is the subject of this chapter, which defines and investigates the notion of cloud-enhanced robotics in the healthcare industry. The chapter illustrates the diverse uses of cloud-enhanced robotics, including telemedicine, diagnostics, and data analytics, and how they improve patient care utilizing real-world case studies. There are numerous advantages associated with cloud-enhanced robotics, including enhanced patient outcomes, improved precision, decreased costs, and expanded access to specialized knowledge. This chapter examines the benefits and obstacles associated with cloud-enhanced robotics in healthcare, providing valuable perspectives on the possible advancements in the field.

A comprehensive examination of the healthcare applications of cloud-enhanced robotics is presented in Chapter 13. Cloud-enhanced robotics is revolutionizing the healthcare industry. The objective of cloud-enhanced healthcare robots was to offer potential assistance to medical practitioners through the precise provision of nursing, diagnosis, and critical surgery execution. Real-time data exchange is facilitated by cloud robotics technology in the healthcare industry, allowing professionals to provide optimal care and services. Automation has become feasible for various responsibilities, including medication distribution and laboratory testing. The proficiency of cloud-enhanced robotics in this domain has substantially reduced the likelihood of human error during critical surgical procedures. This chapter examines the architecture and constituent parts of cloud-enhanced healthcare robots. It then presents a range of practical implementations within this domain, evaluates the merits and demerits of healthcare robots, and critically assesses the security ramifications linked to cloud-enhanced robots.

Chapter 14 examines the potential for a paradigm shift in healthcare through the integration of cloud robotics and AI to improve patient care. An unprecedented paradigm shift is occurring within the healthcare sector due to the convergence of Cloud Robotics and Artificial Intelligence (AI). The present study investigates the substantial impact of the synergistic relationship between these two cutting-edge technologies on enhancing patient care. By harnessing the computational prowess of the cloud and the sophisticated functionalities of intelligent robotics, healthcare systems have the potential to achieve unprecedented levels of effectiveness, availability, and customization. Integrating artificial intelligence algorithms with autonomous systems facilitates improved diagnosis, treatment strategizing, and real-time monitoring, enhancing patient outcomes. This article critically analyses the current state of these technologies, investigates cases in which they have been successfully implemented, and underscores potential challenges and ethical considerations. Amidst the current paradigm shift in healthcare, it is imperative to acknowledge the synergistic potential of AI and Cloud Robotics when devising healthcare systems that are data-driven and patient-centric.

The role of cloud-enhanced robotics in disaster response and recovery is the subject of Chapter 15. Innovative and efficient disaster response and recovery solutions are required immediately. This chapter examines the potential of robotics and cloud technologies to enhance disaster management operations. Cloud-enhanced robotics (CER) emerges as a noteworthy advancement, presenting a synergistic methodology whereby cloud-enabled robots deliver a more prompt, versatile, and resilient reaction to catastrophic events. CER enhances their problem-solving capabilities, decision-making processes, and overall operational efficiency by providing robotics with access to vast computational resources and data. By enhancing real-time data analysis, coordination, and deployment strategies, this connection facilitates a more efficient response to unfolding crisis scenarios. Moreover, it promotes enhanced collaboration and the exchange of resources between robots and response teams, thereby optimizing recovery operations as a whole.

Chapter 16 delves into using autonomous robots for night vision-based human identification and path planning in rescue operations. It presents inventive implementations of cloud-enhanced robotics in the context of emergencies. This chapter presents an incremental categorization and the most restricted method of organizing nighttime rescue operations aided by robots. In actuality, individuals are categorized according to their skin tone. In this investigation, the Skin Colour (SCB), Pixel count-based (PCB), Correlation Coefficient (CCB), and Histogram drawings methodologies were employed to ascertain boundaries. In that order, the PCB, CCB, and SCB calculations are all used to identify specific proofs. In contrast, the SCB calculation progressively ascertains the status of the observed entity as human or nonhuman. Additionally, it differentiates between living and nonliving obstructions, facilitating more effective salvage endeavours (except flies, mosquitoes, and various unpleasant crawlies). As determined by continuous testing data, the CCB, PCB, and SCB calculations have respective accuracies of 87.5%, 88.8%, and 90.9%. Furthermore, the computation times for the CCB, PCB, and SCB are 6.27 seconds, 6.50 seconds, and 6.57 seconds, respectively.

Chapter 17 delves into the application of cloud-enhanced robotics in education, specifically highlighting opportunities for remote learning, robotics programming platforms, and cloud-enhanced robotics. The initial dimension examines the integration of cloud technology in robotics education to improve accessibility, scalability, and collaborative learning experiences. The chapter uses case studies to demonstrate the positive impact of cloud-enhanced robotics on education. The subsequent facet examines pragmatic elements by investigating various cloud-based programming platforms customized specifically for robotics education. This text presents well-known platforms and their favourable attributes for educational environments while offering perspectives on practical programming exercises and projects. This provides educators and students with invaluable resources to foster their robotics programming abilities. The third dimension explores the transformative potential of cloud robotics in education by providing remote access to robotics resources. The focal points are successful implementations in remote learning scenarios, flexibility, inclusivity, and transformative impacts on learning experiences.

The educational applications of cloud robotics were discussed in Chapter 18. Education is in a constant state of evolution in order to accommodate the technological environment of the twenty-first century. Integrating robotics into classrooms has provided students and instructors with new opportunities. Robotics is a potent instrument to foster active student participation, problem-solving, and the development of critical thinking skills. However, with the growing interconnectivity of the world, cloud technology has emerged as a transformative force in education, specifically in the robotics domain. Cloud-enhanced robotics revolutionizes education by providing educators and students unparalleled access to collaborative programming platforms, remote control functionalities, and resources. This article explores the potential of cloud robotics in education, encompassing its various applications, advantages, and obstacles. Additionally, it offers insights into well-known robotics programming platforms and opportunities for remote learning.

Chapter 19 explores the utilization of cloud-enhanced robotics to augment security measures through the lens of IoT-enhanced spy robotics. The convergence of robotics and Internet of Things (IoT) technology has enabled the development of novel security and surveillance applications. This study describes the design and execution of an Internet of Things (IoT)-enabled spy robot intended to augment intelligence collection and surveillance in various environments. Through the use of IoT protocols, the robot's sensors, cameras, and actuators are all seamlessly interconnected. Spy Robot capabilities include autonomous navigation, remote control, and real-time data transmission, contributing to its efficacy in surveillance applications. This study investigates the obstacles, design factors, and possible uses of an Internet of Things (IoT)-enabled Spy Robot, thereby making a scholarly contribution to the dynamic domain of intelligent and interconnected robotic systems.

In Chapter 20, the author delves into the utilization of cloud robotics in the context of autonomous vehicles, illustrating how cloud technologies propel progress in autonomous vehicle systems. The emergence of cloud robotics as a revolutionary technology in autonomous vehicles has caused a paradigm shift in how these vehicles navigate, perceive, and make decisions. This chapter thoroughly examines the fundamental elements, benefits, and obstacles associated with integrating cloud robotics into autonomous vehicles. Cloud robotics capitalizes on scalability, machine learning, and real-time data processing capabilities. It improves the cost-effectiveness, efficacy, and safety of autonomous vehicles. Nevertheless, it presents obstacles of a different kind, including connectivity, security, and latency, that necessitate inventive resolutions. Additionally, the chapter addresses utilizing cutting-edge technologies in learning, adaptability, decision-making, and model training. Furthermore, it investigates autonomous vehicles' ethical and regulatory implications, including analyzing government regulations and the technological ramifications on society.

In Chapter 21, an examination is conducted into the utilization of cloud robotics for agricultural automation, demonstrating the transformative potential of cloud-enhanced robotics in augmenting productivity and efficiency in agriculture. Agriculture, a pivotal sector in the worldwide economy, encounters escalating obstacles, including the imperative for sustainability, the constraint of resources, and the escalating demand for food. Implementing cloud robotics and automation technologies in agriculture has emerged as a revolutionary approach to tackling these obstacles. This chapter introduces and explains how cloud computing and data analytics can enhance intelligence, scalability, and efficiency in the agricultural sector. Real-world implementations of cloud robotics in various agricultural contexts are exemplified through several case studies, which underscore their influence on the optimization of crop yields, resource management, and environmental sustainability. Additionally, the chapter delineates the obstacles and impediments that impede the extensive implementation of cloud robotics in the agricultural sector. These encompass financial implications, connectivity issues, and apprehensions regarding data security. It provides prospective resolutions to surmount these obstacles.

CONCLUSION

This book serves as a comprehensive guide to the evolving landscape of cloud-enhanced robotics, offering valuable insights, practical examples, and actionable strategies for leveraging the power of cloud technology in robotic systems. By bridging the gap between theory and practice, this book aims to empower readers to navigate the complexities of cloud robotics, driving innovation and advancement in automation across diverse industries. Join us on this journey as we shape the future of automation with cloud-enhanced robotics.

Rathishchandra Ramachandra Gatti
Sahyadri College of Engineering and Management, India

Chandra Singh
Sahyadri College of Engineering and Management, India

Acknowledgment

We, as the editors of *Shaping the Future of Automation With Cloud-Enhanced Robotics* would like to express our deepest gratitude to all those who have contributed to the creation and success of this book.

Acknowledgement to Parents and Family

First and foremost, we would like to thank our parents and family for their unwavering support, encouragement, and understanding throughout this journey. Their love and belief in us have been the foundation upon which we have built our careers.

Acknowledgement to Authors

To the authors of this book, we extend our heartfelt appreciation for your dedication, expertise, and passion in sharing your knowledge and insights on the transformative potential of cloud robotics in shaping the future of automation. Your contributions have enriched this work and will undoubtedly inspire readers in the field.

Acknowledgement to Reviewers

We are grateful to the reviewers who generously shared their time, expertise, and valuable feedback to ensure the quality and accuracy of the content presented in this book. Your meticulous review and constructive criticism have been instrumental in refining the ideas and concepts discussed within these pages.

Together, with the collective support and collaboration of our parents, family, authors, and reviewers, we have been able to bring *Shaping the Future of Automation With Cloud-Enhanced Robotics* to fruition. Thank you all for your invaluable contributions and unwavering support.

Section 1
Cloud Robotics Technologies

Chapter 1
Future of Industrial Automation With AI and Cloud Robotics

Mandeep Kaur
Department of Electronics and Communication Engineering, Punjabi University, Patiala, India

ABSTRACT

This chapter outlines the shifts induced by AI and cloud robotics to industrial automation in the near future. And, with the striving of manufacturing worldwide for more efficiency, flexibility, and environmentally sustainability, the integration of the AI and cloud robotics is pivotal in redefining the manufacturing production process. The synergetic relationship between AI's data-driven analysis and cloud robotics' scalable, distributed computing power solves the issue of greatest challenges in automating throughput by demonstrating that automation can and should be done properly. The study will thus bring together current implementations, case studies, and expert predictions in order to cast a glance on what AI and cloud robotics contribute to shape the factories of the future, and also to identify how they will continue to make the sector evolve in the future.

1. INTRODUCTION

The history of industrial automation is just another example of the human invention and the tireless quest for higher performance and the creation of the ultimate product in manufacturing. From water and steam utilized for operating the first manufacturing facilities by mechanical power in the late 18th century till the current era which is the threshold of the fourth industrial revolution, known conventionally as the industry 4.0 or Industry 4.0, the shape of manufacturing has been greatly influenced. Every step of industrial revolution is technically conjunct with new novelties for less manual labor, better product quality and faster processes (P. S. Kumar and A. Sharma, 2019). The spinning of the first stage of industrial revolution marked the dawn of this journey, which made people first mechanized machines and the factory system. The first commercialization of the production line was the result of the second industrial revolution, involving electricity use for goods production. This improved production immensely owing to the massive production and significantly lowering of costs and time allocation required in the process (S. El-Gendy, 2020).

DOI: 10.4018/979-8-3693-1914-7.ch001

Revolution three, the digital one, that started in 1950s replaced mechanical types of production by introducing computer science. The following period was in the development of programmable computers (PLCs) and robotics that gave rise to the latest version of automation(Kumar and R. Singhal, 2020). These innovations enabled activities that were more accurate, uniformity-based and complex in nature and together with other factors greatly boosted efficiency and productivity in industrial operations.

Now, the fourth industrial revolution, made possible by technology exploitation and ranking reality in terms of digital and physical twins, is going on. Seen at the heart of this new stage is de facto the joint use of artificial intelligence (AI) and cloud robotics with industrial automation systems, which would otherwise mean a big step in the very field's development. This integration is not about a marginal array of tools for improvement (K. N. V. Srinivas and P. K. Lakineni, 2023); it is a pivotal shift in industrial operations, allowing capabilities that redefine the boundaries of traditional industrial operations.

Beware that this is the area where AI stands heads and shoulders above all the others that allows the technology to scan, process (P. S. Kumar, S. Gangadharan and S. C. N, 2023), learn and deliver conclusions based on the data. Regarding the manufacturing automation in factories, Artificial Intelligence (AI) can foster production while it minimizes the prospects of downtime and ensures better quality by detecting defects that avoid our human inspectors (Nayyar and A. Nagpal, 2018). Besides, machine learning algorithms employed in automation systems to allow for the improvement thereafter providing for constant increasing of the efficiencies and effectiveness.

Having it on the one side, cloud robotics exploits cloud computing in its full capacity, with robots and automation making use of nearly limitless computational resources, storages, and special software algorithms (V. S. Prasad, Y. Mahajan and S. Suneetha, 2023). This gives the ability to companies to carry on heavy duty processing tasks by the cloud and thus they can create robots that are able to fulfill more complicated functions without having to possess powerful on board processing capabilities (A. Kumar, V. Kalpana and V. Hariram, 2023). Moreover, cloud robots uplift the cooperation among robots through the cloud, making it possible for them to share their data and they get better at each other's experience, thus improving their efficiency and ability of adapting to changing environments.

The joint of Artificial Intelligence together with cloud robotics in industrial automation facts, it means the factories can become more than merely automatic but intellective. These smart factories will have the systems that are capable of self-optimization, self-adaptation, automatic decision-making and coordination of these systems robustly in real-time(Hussain, N., Samal, A., 2024). The impact of this on the efficiency, productivity, and the flexibility ability of the manufacturing industry is large and offers a way to solve some of the current most relevant problems, including the need for customization, sustainability, and resistance to disruption.

On the other hand, the road to that foreseen future is full of opposing forces. The connection of AI and cloud system ecosystems with manufacturing creates a great deal of technical, economic, and social hurdles (M. Kranthi and R. C. Tanguturi, 2023). Key points are challenges with individual privacy and security, expensive initial systems installation and negative impact on jobs. Addressing this challenges needs to be done by all with different roles, namely, the authorities, the leaders of the industry and the work police to make sure that the transition to the new industrial automation goes smoothly.

The chapter ahead will explore that aspect along with other specifics pertaining to the manner in which AI and cloud robotics are reshaping industrial automation (A. Ranjan, K. K. Hegge and R. Agrawal, 2023). It will discuss the technologies that make it to happen, present case studies where they applied across different industries, and consequently see how this change could be in future manufacturing (Avinash, Kumaraswamy, Anand Kumar, 2023). To achieve that, the next part of the chapter should give an overview of where we stands today with regards to advanced automation technologies and prepares the young readers for the new era that will be brought by these technologies which they will probably live to see and witness. Finally, the path from mechanical looms to smart production lines is an image of merciless of human race to the evolution (Chinamuttevi, Veera Sivakumar, 2023). A whole new revolutionary era would therefore be ushered in by the advent of artificial intelligence and cloud robotics integration, which represents the final step in this adventure. It will promise to deliver a new dimension of efficiency, flexibility, and intelligence in the way industries operate. While manufacturers around the world are in the process of this transition, the possibilities for innovation and more growth undoubtedly remain unlimited; these possibilities even open the doors of a new future where automation not only changes the manufacturing but resolves the global economic issues.

2. UNDERSTANDING AI IN INDUSTRIAL AUTOMATION

Foundational Overview of AI

AI stands for various technologies allowing machines to imitate the human intelligence factors by means of learning, reasoning, and self-correcting. In this segment, we offer a bottom-up foundation of the most significant ideas, technologies, and methodology of AI engineering focusing on industrial settings.

Machine Learning (ML): ML, one of the AI components, gives the ability to systems to learn from data, detect patterns and still make decisions with little or no man power required(Shekhar, R. & Chaturvedi, Abhay, 2023. ML algorithms are divided into three main classes which are supervised and unsupervised learning as well as reinforcement learning; these types of algorithms are applied to different functions in industrial automation.

Neural Networks (NNs): ANNs, on the other hand, are modelled after the human brain which has layered network of nodes that mimic neurons. They are efficient when it comes to the processing of the patterns in data which could be complex(V. Srinivas, T. De, 2023). As a result they are the best option for tasks like image and speech recognition. Deep learning, which is a subfield of ML with neural networks that contain lots of layers, has been responsible for some notable gains in complex decision-making, as well as predictive analytics.

Data Analytics: The process of taking in some data sets to come up with a conclusion which they contain (K. K. Hegge and R. Agrawal, 2023). AI-enabled advanced data analytics is able to find out trends as well as patterns to improve the strategic and tactical decision-making relating to industrial automation thus improving operational effectiveness and efficiency.

AI Technologies in Industrial Automation

Predictive Maintenance: Applying AI in the prediction failure of the machinery and equipments in advance means that the downtime and the costs are reduced (K. . Prasad, and I. . Gugan, 2023). Sensors and logs that collect historical and real-time data can be analyzed by Artificial Intelligence algorithms to forecast possible failure modes before they happen.

Quality Control: AI driven visual inspection systems based on machine learning and neural networks, being more accurate and constant than human inspectors, can find manufacturing defects and/or quality problems (N. S. Reddy and B. Ashreetha, 2023). These systems use images from cameras and sensors along with the analysis to identify variations which guarantees efficient production with minimal waste and reworks.

Supply Chain Optimization: AI functionality in the supply chain is achieved by analyzing the large amount of data to forecast the demand, balance the inventory, and find the most effective routes to deliver (P. Pavankumar, N. K. Darwante, 2022). The adaptability of machine learning models can easily align to the moving market, which results in improved responsiveness and efficiency in the supply chain.

Use Cases of AI in Automation

Automotive Industry: AI solutions play a key part in predictive maintenance of equipment during production, quality control of automotive parts through visual inspection system, and the optimization of just-in-time manufacturing (S. Reddy P, P. S. Patwal, 2022), which reduces the inventory cost and increases the efficiency.

Pharmaceuticals: In pharmaceutical manufacturing, AI performs inspection processes according to the highest quality standards, schedules production based on the predictive analytics and updates research and development through analyzing the clinical data (A. Singla, N. Sharma, 2022).

Food and Beverage: AI applications that show up in the food and beverage industry can include utilizing machine learning algorithms to forecast demand, insuring consistent product quality by AI-powered visual inspection and optimizing supply chains to reduce waste and improve sustainability (K. S and M. R. Arun, 2022).

Electronics: The AI industry deploys AI, for example, for the exacting assembly of components in circuit boards, defect detection in the chips and displays by high-resolution image analysis and supply chain optimization which aims to deal with the difficulties of component sourcing and merchandising.

This fundamental grasp of AI's role in industrial automation highlights the transformative nature which this technology can have across different sectors (P. Ramesh Naidu and N. Guruprasad, 2021). Through the utilization of processing power leveraged by machine learning, neural networks, and data analytics, industries are able to realize unseen levels of efficiency, quality, and innovation propelling the fourth industrial revolution.

3. CLOUD ROBOTICS IN INDUSTRIAL AUTOMATION

Definition and Relevance of Cloud Robotics

The convergence of cloud computing and robotics produces a field called cloud robotics that makes robots take advantage of cloud computing capacities such as unlimited information storage, powerful processing units, and the software abilities provided through the cloud This coherence greatly improves the capability of robots (M. Shruthi and B. Naveen K, 2020), thereby making them wiser, likely and highly productive.

The significance of employing cloud robotics in industrial automation comes down to the ability of it to fill in the gaps in relation to computing power and data storage caused by on-boarding computing capacities (M. Ramesha, and G. N. Pai, 2020). For the robots it is a gift to be able to get more data through cloud in comparison to having data processed independently. This feature is of special importance for tasks based on the treating of real-time data and making sophisticated choices. Also, it is worth mentioning machine learning from the point of processing big data.

Enhancing Robotic Capabilities

Improved Data Processing: Cloud robotics provides the ability to transfer the computationally taxing work into the cloud, a place where it can be easily handled by strong computing power faster than the procedure that a robot's on-site computer will be able to perform within a short period of time(J. J. Jijesh, and Shivashankar, 2017). This is is very handy in real time processing of data sent from sensors, and it lets take response measures much faster, often with higher level of accuracy.

Expanded Data Storage: Now that cloud storage is available robots can exploit a heap of data, including archived data sources, to create permanent improvement latterly (S. M. Sali and K. R. Joy, 2023). This becomes vital for machine learning models that may only achieve good performing models after the training is complete with multitudes of datasets collected.

Enhanced Access to Software and Resources: A cloud-connected robot will be equipped with ample of software unwiring and libraries, allowing it to perform a more complex of jobs other than what its programmed for. Moreover (D. Pal and S. Joshi, 2023), robot performers are able to trade information and learn from each other, which is very important for collective learning and experiential knowledge accumulation among them.

Applications of Cloud Robotics in Industry

Remote Monitoring and Management: The cloud robotic technology provides for supervision and managing of the robotizing systems through a remote-operating method that makes it possible to monitor operations from any part of the world(H. M. Mishra, A. Qtaishat and M. Kumar, 2023). This is very important given that this function is one of the key maintenance activities that aims to keep production machines running smoothly and promptly identify and tackle problems.

Collaborative Robots (Cobots): Human-Cobot cooperation, or partnership, is intended to improve the effectiveness of work and safety conditions. Sensor cloud robotics provides cobots with opportunities to access valuable data from the cloud, rapidly change to new jobs, and gain experience of interacting with humans and other machines, which enhances their performance and versatility.

Automated Guided Vehicles (AGVs): AGVs have applications for material handling, and they used for transportation within facilities and warehouses alike (J. Khan, F. Ali, F. Ullah, J. Bacha and S. Lee, 2019). Clouding the robotics is the main technologies where AGVs can navigate more proficiently, can easily adapt to new circumstances and can do effective coordination with other robotes and systems.

Case Studies

Manufacturing: An automobile maker turns into a front line company, that has cloud-robotics for assembling the parts. These robots, which can adapt to various car models by downloading the specific assembly instructions from the internet (A. Krishnan, S. Swarna and B. H. S, 2020), without remaking robots for the new tasks, expand the range of their jobs and decrease the time and cost.

Logistics and Warehousing: E-commerce giants deploy cloud connected AGVs within their warehouses so that stocking can be done accordingly, picking of items and movement of products becomes more efficient (R. S. Varun, and T. Rajesh, 2018). The cloud based AAVs are coordinated running real-time adjustments to inventory change and order priorities. In other word s, they drastically optimizes the long cycle of fulfilment partly.

Agriculture: Through precision agriculture, drone and ground vehicles gaining connection with clouds handle the duty for data gathering - that concerns crop health, moisture and pests. This information together with the cloud computing is processed to farmers for the proper optimizing the irrigation, fertilization and pest control process.

Particularly cloud robotics is becoming more and more widespread as an essential advance in the industry sector, thanks to its data management and access tools that are of better quality (W. Liu, 2021). Thanks to the use cloud based robots, flexibility, intelligence and ability to do difficult tasks with optimal efficiency become the drivers of the upcoming wave of progress in the industrial operations.

4. SYNERGY OF AI AND CLOUD ROBOTICS

The merging of machine learning (ML) and cloud robotics technology opens a revolutionary door in the robotics and automation industry. The interface of different hardware and software technologies powers these flexible, intelligent, and productive automation systems that are changing the scene in factories and industrial plants of all sizes all over the globe. Combining the remains of AI's decision-making and learning functionality with cloud robotics' powerful processing abilities and high-capacity storage facilities allows industries to realize new unheard levels of productivity and novelty.

Synergistic Effects of AI and Cloud Robotics Are Multifold

Enhanced Decision-Making: AI algorithms which scan live data collected by the automated robots working in the cloud give the ability to the automatons to make accurate decisions while in action. This interactive way of decision making enables set-up of correct modes for the changing production environment which elevates the effectiveness of the whole system.

Scalability and Flexibility: The advantage of cloud is that it enables the deployment of AI models that works across different robotics solutions to permit the growth of scalable and distinctive types of automation. This means that the system being dynamic and able to change its degree of production can make this process very efficiently and there will be no big redundancy or reconfiguration in the system because of an adaptable nature.

Improved Learning and Adaptation: AI models, in access of the cloud, may be dynamically updated and increased with data generated by robot. This collective learning and obtainment results to a system that succeeds as the more its exposed to newer tasks and environments the better and sustains.

Predictive Maintenance and Optimization: Through the analysis of data trends and patterns, AI can forecast the times when machinery and robots are likely to become defunct or need maintenance services, which in return leads to the practical programmation of such interruptions for the purpose of preparing emergency actions in time to prevent any unexpected disruptions. Alongside this, AI is able to do good by optimizing the productions process by identifying inefficiencies and suggesting improvements.

IoT's Role in the Convergence of Different Areas of Human Life

The IoT is a smackdown vehicle in the integration of AI and the cloud with robotics. IoT concept comprises the use of the interconnection of the physical objects (Things) such as sensors, software, and other technologies aimed at collecting and exchanging data with other devices as well as systems on the Internet.

Real-Time Data Exchange: IoT-connected devices at the factory floor acquire and stream real-time data to the cloud that the latter extracts using the power of AI technologies and algorithms. This ease of data transfer is critical giving the necessity of real-time decision making and adjustability of AI-driven run-ins in the ubiquitous cloud robotics.

Enhanced Operational Visibility and Control: Through IoT, companies are given a proverbial hand with a lens, glaring its review lights on all aspects of production. This visibility grants better control and capability of automation process management. It will lead to more productive and precise operations.

Facilitating Collaborative Automation: IoT quasi-creates different ways of robots and automation systems interaction and cooperation. Having the system to interact and inform each other, these systems can operate more accurately on top of each other's back, which further boosts their efficiency by reducing the repetitive work.

Case Examples

Smart Manufacturing: AI along with cloud robotics enables intelligent factories where both lead to automation of smart manufacturing tasks. IoT sensors capture data on machine configurations, environmental conditions and production output, that AI algorithms subsequent base on to adjust working flows, anticipating maintenance and fitting changing production needs.

Precision Agriculture: Cloud accessed drones and ground robots, intelligent and internet embedded sensors provide health monitoring services and environmental condition checking. AI gets insight into this data which helps in giving advice to farmers related to irrigation, fertilization, and pest control making farms more efficient by optimizing resource usage.

Supply Chain Management: AI and cloud robotics which is supported by IoT makes forecasting suitable demand, inventory levelling and logistics management faster and more accurate. Therefore the whole system that provide service of waste processing in a unique way with less waste, improved efficiency and immediate response to changes in the market.

The whole process which "AI + cloud robotics + IoT" integrates gives a new meaning to "industrial automation". They offer smart, adaptable, and efficient ways to work, thus driving productivity and newness in industrial processes that set the bar of the unimaginable.

5. TRANSFORMATIVE IMPACTS ON INDUSTRIAL AUTOMATION

The implementation of automatic systems of artificial intelligence (AI) and cloud robots in industrial automation has paved the way for a revolutionary era that is characterized by immense productivity gains, risk minimization, cost reduction, and novel discoveries The section given below highlights the multidimensional effects of these technologies on the industrial level of production as well as provides some broader prospects for worker development, skill requirements and job vacancies.

AI along with cloud robotics immensely increase production efficiency by carrying out automated operations, minimizing the risk of human error, as well as acting as the end controller of production processes. AI algorithms are able to forecast and coordinate maintenance activities, thus fewer constraints on downtimes happen. Cloud computing facilitates the implementation of complex calculations immediately up to the production line, making it possible to change the process quickly and accurately. Through this tech factory, industrial operations can run more precisely and quickly as well as produce less waste and more products.

The development and implementation of AI and cloud robotics in the production environments led to considerable enhancement of security standards in industries. The robots can carry out activities which put human health at risk, for instance, pipe repair or removal of objects which pose a hazard to the humans, thus cutting down the time and risk of accidents. AI by taking data analytics from real world sensors can determine potential safety threats, alarms or even automatic correction can be applied to the system to avoid crashes.

Increasing the productivity of business processes, and reducing the necessity of human intervention, are the activities AI and cloud robotics make possible at a lower cost. The AI's predictive maintenance capabilities will help in reduction of high equipment failures and production stoppages which consequently will result in costs reduction. In addition, better performance and minimized operational costs come due to the improved efficiency and productivity, while cloud robotics implements theCloud resources, and consequently, leads to lower expenses.

Through AI and cloud robotics, automation systems integration gives way to the creation of things that were unthinkable before. Such technologies promote quick sampling, running tests, and revising, thus speeding up the exhibition of new things. Moreover, manufacturers are able to individualize products with larger volume, which leads to new market exploitation and product variety.

Not limited to mere technological progress, the application of AI and cloud robotics in industrial automation even influence the directions of workers and jobs. With the rise of automation routines jobs, there's a soaring need for relatively complicated skills, including system design, maintenance, and data analysis. Workers must be able to respond to such changes by upgrading their skills and to the areas of AI and robot management, not to mention optimization.

However, the growing trend of automatization in industrial sectors is not only a threat; it creates a possibility for new job roles as well. Firstly, the opportunities that revolve around artificial intelligence and data science, cloud computing, and robotics engineering are on high demand. Besides that, cybersecurity becomes to be another profession in demand. Due to cybersecurity, AI and robotics systems will be protected from being harmed.

While the usage of AI and cloud robotics bring numerous perks, they also give rise to job loss and competence mismatches among the workforce. Companies should take this issue to heart and invest in workplace training and education programmes, creating and giving employees a culture of constant learning and adjusting. It is a task of the government and educational institutions to create the curricula meeting demands of the labor market and to arrange for retraining and up-skilling people to keep pace with the industrial change.

As industrial automation gets revolutionized with both the advent of AI and cloud robotics, the improvements made in terms of productivity, safety, cost efficiency and innovation are simply huge. These technologies are not sitting idle but rather, they are remaking the industrial ecosystem, hence, a point has to be made to reflect on jobs and required skills. For AI and cloud robotics to deliver their benefits, there is a great need for industries, governments, and educational institutions to come together to provide for the workforce through education and training and ensure proper management for the more automated environment.

6. CHALLENGES AND CONSIDERATIONS

Integration of AI and cloud robotics in industrial automation though having a huge industrial application, carries many transformative capabilities but it is also loaded with a complex array of challenges and consideration that the industries have to consider. Among the vital cybersecurity risks of the digital age is the information systems' protection. As industrial sites become more and more integrated and customer-centric, cyber threats and attacks loom large and are potent. Such result can be accompanied by extensive disruption to the operational activities, loss of confidential information and manipulation of the security devices. As cybersecurity is increasingly imperative, the problem of privacy of data rises too. Significant data processed and stored by AI and robotics systems of cloud, some of them may be of sensitive nature, can lead to the imposition of security standards. As the compliance burden would be very large, organizations must confront the legal complexity.

Furthermore, the high preliminary costs of deploying AI and cloud robotic systems to replace the current operational technologies constitutes a tremendous entrance hurdle for most industries. The financial barriers to adopting new technology, such as those connected with the purchase of technology items, the upgrading of infrastructure and the training of workers, may be too high for small and medium businesses. This situation get even worse because of the fact that workforce must continually be trained and well-prepared for this field. Owing to that fast-paced rate of modern technology invention, companies are in need of workforces that not only have the skills to work with the available technology but also, can adopt to those innovations in the future. Skills, to maintain, optimize and manage AI and robotic systems, through ongoing programs - training and education, should be found in employees.

As well, the combination of new technology is also coming with the same problem of integration with old systems builds technical and operational obstacles. Asking for compatibility and continuity in a transition from one system to another may appear to be easier proposition, but it will require a great technetic expertise and strategic planning. Despite these challenges, the strategies that will beat the mentioned obstacles may entail adapting powerful cybersecurity measures, fighting against data privacy regulations, doing thorough cost-benefit analysis to know if one should invest or not, and using gradual approach for technology adaption. Through the smart application of AI and cloud robotics and workforce training, enterprise sectors will be able to make it through all such difficulties and gain the maximum from these technologies, putting them into their automation systems.

7. FUTURE TRENDS AND DIRECTIONS

Descending into the vista of industrial automation is an AI-cloud robotics integration that is capable to bring in the age of creativity, effectiveness, and ability. With the technological breakthroughs of today, the ongoing trends and future approaches in this field are being affected profoundly, which will have a significant impact on the future of all industries around the globe. One of the major development is the emergence of machine learning, in which the development of the algorithm is getting more and more complex. Such technological progresses are likely to further make the algorithms smarter as they will be capable of learning from smaller datasets, making more accurate predictions and may exhibit amazing adaptivity to new cases. This growth is aimed to grant systems with the newest automation capabilities, which result in higher adaptability, better control of the complex tasks, and enhanced precision.

Moreover, the concurrent development of 5G connectivity will provide the terrains for the dramatic shift in industrial automation through unparalleled speed and ultra-low latency of the connections. Such a dramatic uplift in the capability level of cloud robotics will be possible by bringing in new dimensions of real-time data exchange and communication among machines, they will perform at unprecedented high speeds. This will open new possibilities for real-time monitoring, control and autonomous decision making in industrial processes.

And now edge computing is gaining power as a partner to cloud type robotics. The edge computing capability enables the data processing process in the area that data is created for low-latency processing and therefore enhancing the response of robotics system. It turns out especially useful when real-time decision-making is pivotal in regard to suggesting the amalgam of cloud computing massive processing power and edge computing immediacy.

Eventually, with the arrival of these technologies, industrial manufacturing practices are likely to undergo unknown changes. On the other hand, AI and cloud robotics provide the means to make manufacturing more efficient and agile. This could result in the growth of the distributed manufacturing setup that uses local suppliers instead of global supply chains, which could potentially alter the structure of the global manufacturing scene. This transformation might have the substantial impact on sustainability as the conformable and adaptable production methods could be the cause of the less waste generation, energy saving, and the lower carbon footprint.

Complete Automation of Smart Factories as a possible milestone in this development based on application of AI and increased use of Robots that would control any single production process. Such facilities would certainly be in the vanguard of operation and tailoring, and celebrate a new age of industrial revolution that will be more adaptable, conservationist, and creative. In summary, the automation of industry including AI and cloud robotising, as a result, brings the promise of exciting perspectives. As these technologies continue – to advance and fuse, they will re-write the rules of what is possible in manufacturing, and break new grounds – for smart automation to be at best used in regular industrial operations.

8. CASE STUDIES

The impact of AI, robotics, and cloud technologies in a variety of sectors has demonstrated their capability to not only improve the quality of working processes but also to address complex issues. As an example, a well-known worldwide automotive producer had to apply efforts to increase production efficiency, quality of the car and to decrease downtime. Adding to this predicative maintenance and quality control systems with cloud robotics systems enabled the company to plan maintenance ahead of time, thus keeping production losses and increasing product quality index to a minimum. Through the use of this integration efficiency of production has been improved by 20%. A multinational e-retailer had to deal with the challenges that came with running a warehouse that was expected to accommodate a growing inventory assortment and order volumes. The deployment of the fleet of the cloud-connected automated guided vehicles (AGVs) and collaborative robots (cobots) which is controlled by AI for inventory management and order processing revolutionized business processes in the warehousing. This meant that order processing times were drastically reduced and practically no picking and packing errors occurred, as the company was being able to successfully handle the increased order volumes without high labor costs or big change in the warehouse space. For instance, agri-tech startups working on precision farming solutions have used cloud robotics and AI systems to enhance the growth of crops while cutting down the resource usage. The company implemented a drone and ground robots system organized with sensors. They collected data on soil and plants health, which was analyzed by AI algorithms and was used for the optimization of water, fertilizers and pesticides. This principle was not only a core factor of increasing crop yields but also responsible for more sustainable farming practices by preventing overuse of resources and hence improved pest and disease management. Such case studies reveal that the application of AI and cloud robotics is not limited to only a given areas. Instead, it enables processes to be streamlined, successfully, sustainably and productively, in general.

9. RESULTS AND DISCUSSIONS

Combination of artificial intelligence (AI) and cloud robotics as industrial automation tools has transformed manufacturing processes in various sectors by increasing productivity, throughput, safety and opening new dimension of innovation. The different practice already reflect the revolutionary possibility that these technologies have in facing up industrial processes. This part explores the major results of interplaying AI and the inter cloud robots, providing the data from the case studies and cloud trends too.

The AI cloud robot integration requires the most noticeable consequences of productivity and operation efficiency increase. Automated equipment involving machines can run all the time without the systems to be in shutdown mode which will reduce downtime to a significant level and at the same time increase speed and accuracy comparatively with manual works. The sentence could be: For instance, the continued progression of automation saw a 20% productivity boost for the automotive company due to the integration of AI as the key enabler. A tailspin position here reflects the value creation function of AI and cloud technology, promoting production efficiency while shaving off the decision-making time. Consequently, they deliver higher output, leading to better use of the available resources.

Planning Cloud robotics and AI use to create safety conditions for workers is significant. The robots substituting dangerous tasks, for example in the factory lines with their dangerous moving parts, can be hard for humans to do. Such a substitution contributes to a healthier environment, as there are injuries that occur due to the heavy machinery. Moreover, AI-solutions can forecast the areas of possible danger and subsequently mitigate the concerns towards the safety of the industrial processes. The transition together with the benefits of occupational health and safety outcomes does not only enable workers to be safe but also minimize the financial and reputational damages associated with workplace accidents.

Integrating AI and cloud robotization entails tyre-long cost savings as time goes on. Although the major expenditure of the beginning, the long-term reduction in labor costs, rejection of a waste, and the minimum downtime have resulted in the reasonable investment in return. The group from e-commerce has demonstrated how automation can effectively manage the volume at such painful cost as volumes do not bring more expensive, demonstrating the cost-effectiveness of these tools.

Ai and cloud robotics take innovation a step forward by increasing the launch of new products and services, and by optimizing production processes. Such technologies have provided a number of advantages to businesses by enabling them to quickly adjust to changing market conditions and clients needs. This would be, for example, the agri-tech startup use of precision farming which shows the way how AI and cloud robotics can be used as the source of novel solutions to the traditional problems, creating a completely new vent and pathways for business overall growth and achievement of sustainability.

The moving towards automated systems means that such workforce has to become skilled in new technologies. While profits job loss worries however is a real potential to provide employment in the field of Artificial Intelligence, Robotics, and data analysis. This particular trend creates a job opening in the workforce development championing the need for training sessions and education to update existing employees with the right skills.

The lasting effect on the industries such as change of global manufacturing system architecture as a result of together with domestically-oriented manufacture driven by additional automation efficiency. Moreover, the efficiencies provided by artificial intelligence (AI) and cloud robotics may take manufacturing into a new direction characterized by green and resource-conscious processes.

Incorporation of AI and cloud robotics in industrial robotics is a breakthrough in automation industry, as it raises efficiency and safety standards of industries, but also opens a new chapter in cybersecurity and data privacy matters, while adaptation of modern working conditions is also a big task. The present and future adoption of technologic advancements and implementation of face-to-face human-AI and cloud robotics across sectors implies that smart automated processes might one day take over and would require workers to always adapt and industries' plans to help tap the potential of such technologies.

Figure.1 annotates the production efficiency change of AI and cloud robotics application for 3 different companies (Company A, Company B and Company C). The graph depicts that the use of these technologies resulted in a visible increase in efficiency percentage, it can be assumed that the integration of these technologies helps in increasing efficiency of the industrial processes. The efficiency is calculated by considering the percentage of output being produced when the input used is taken into the consideration. Such enhanced workability underscores the technologies' proficiency to automate processes, slash down wasted resources, and ultimately yield better productivities.

The chart Figure 2 presents the cut of downtime in the machinery degree achieved prior to and following the implementation of AI and cloud robotics in the same range of enterprises. Downtime indicated in hours is shown in the bar chart which graphically denotes a marked lessening after implementation. This has been the result of two productivity improvement aspects in AI namely predictive maintenance and intelligent process monitoring that allow machines to operate at peak performance with crisis-less stops which in reality means creating more manufacturing cycles with reduced operating losses.

The graph (Figure 3) shows how AI as well as cloud robotics can be applied in several industries in order to ensure occupational safety, such as in manufacturing, logistics and the pharmaceuticals industry. By the line graph using as comparison the number of safety incidents before and after the technology integration, the graph shows almost nonexistent cases which are a great reduction compared to what used to be tendencies. This is made possible through technologies, which could help us forecast possible dangers ahead, take away dangerous tasks manually and assure solid operational standards are kept across the whole industrial process that turns industrial places into safer environments.

Figure.4 presents the operational cost savings approximately achieved over a five-year period when AI and cloud robotics (CR) are integrated into the industry automation process. This line graph shows the saving summing up to a net effect of the incremental benefit of efficiency gains, decreasing idle time and decreased labor costs. This graph reflects the financial advantages after some time of investing these kinds of technologies in line with their enormous cost reductions not only, but also the chance they can offer of a sustainable economic paradigm to the industry.

Figure 1. Increase in production efficiency

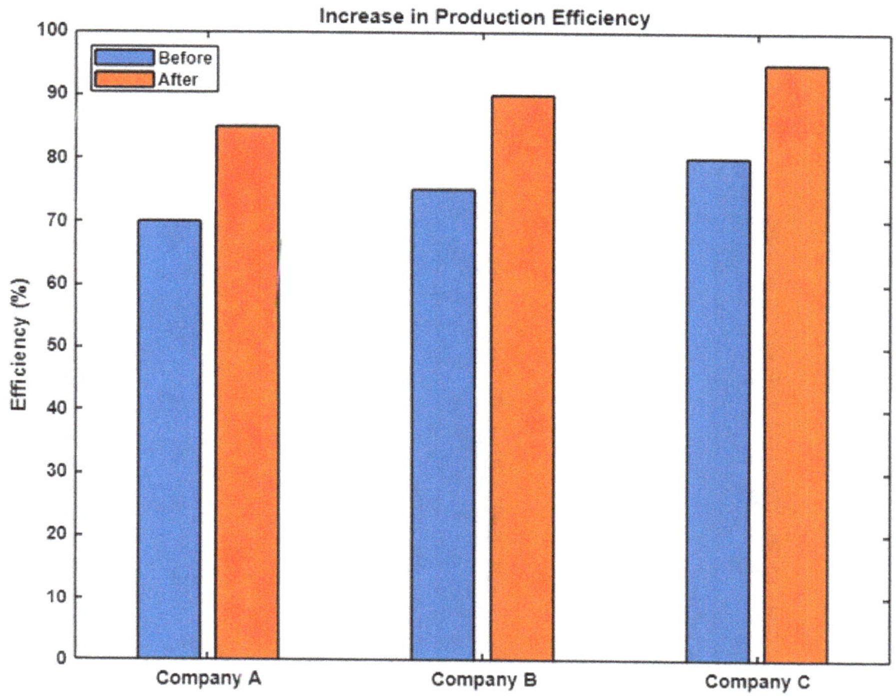

Figure 2. Reduction in downtime

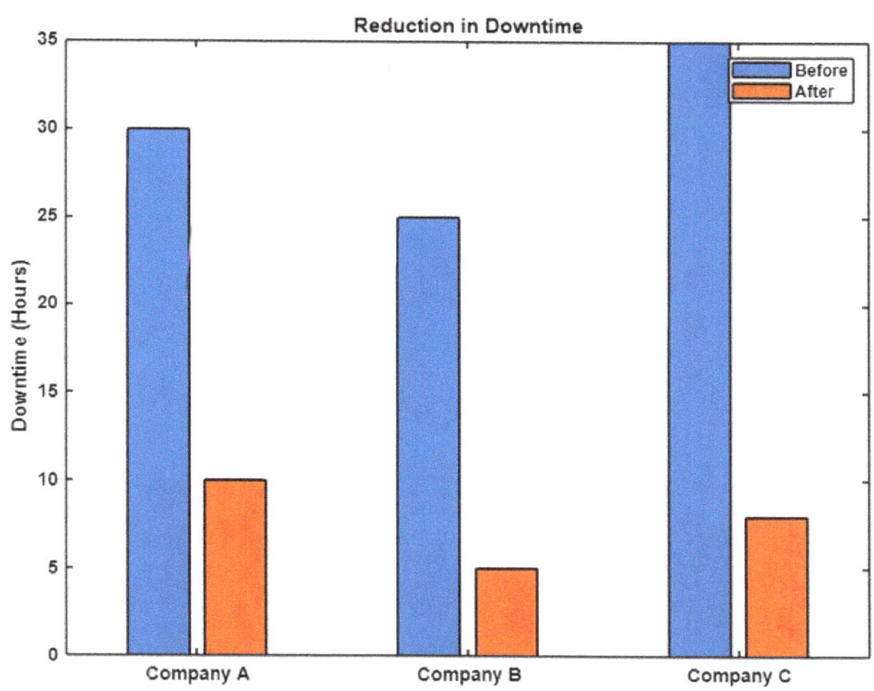

Figure 3. Safety incident reduction

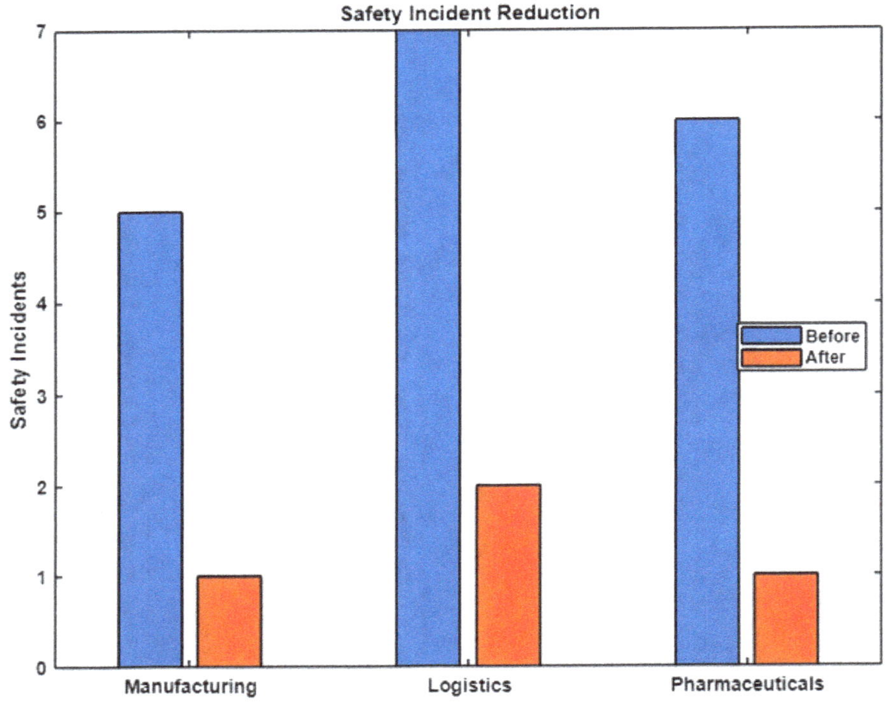

Figure 4. Cost savings over time

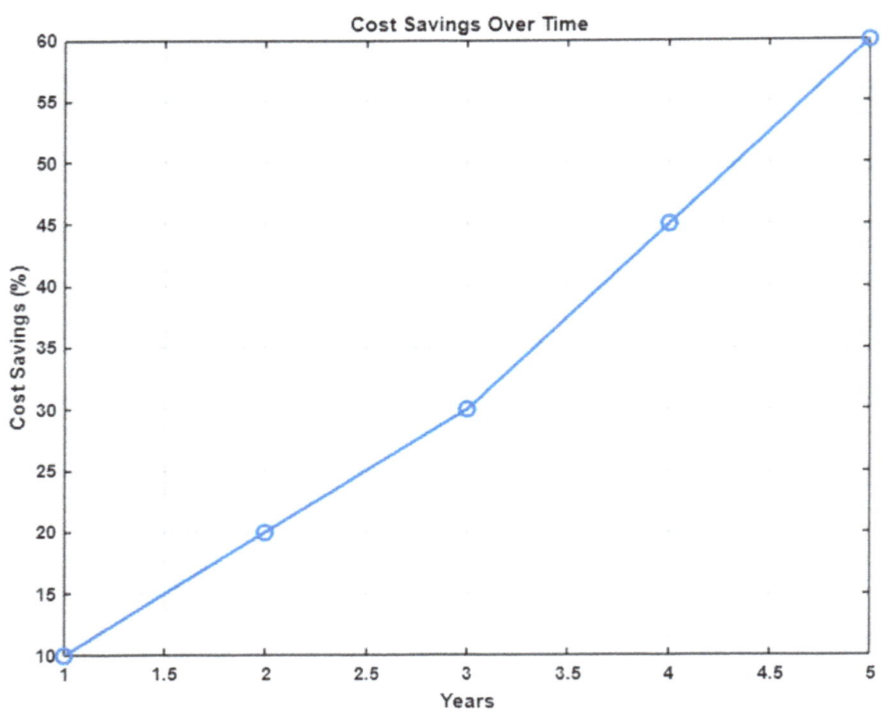

Figure 5. Workforce skill shift

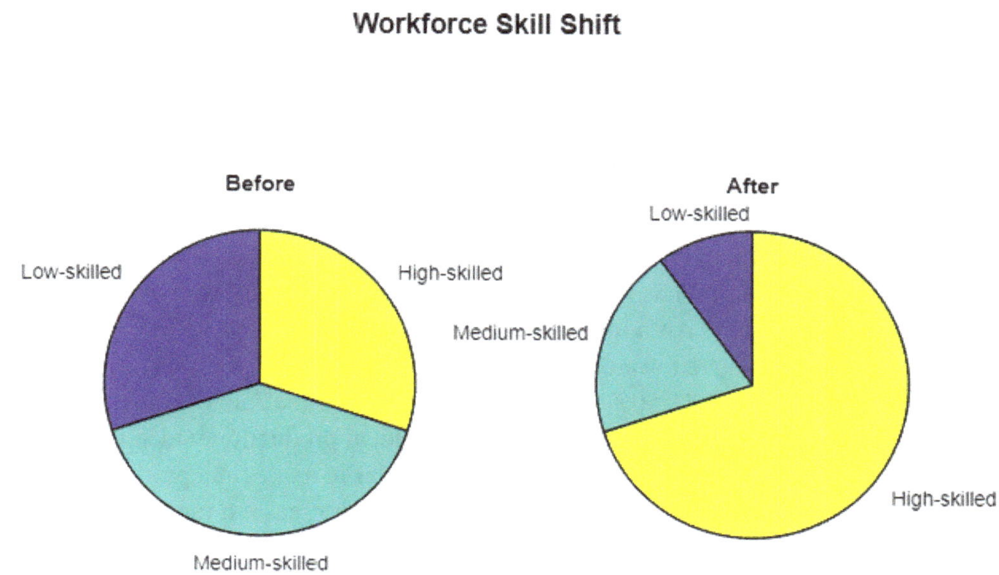

Figure 6. Global manufacturing shift

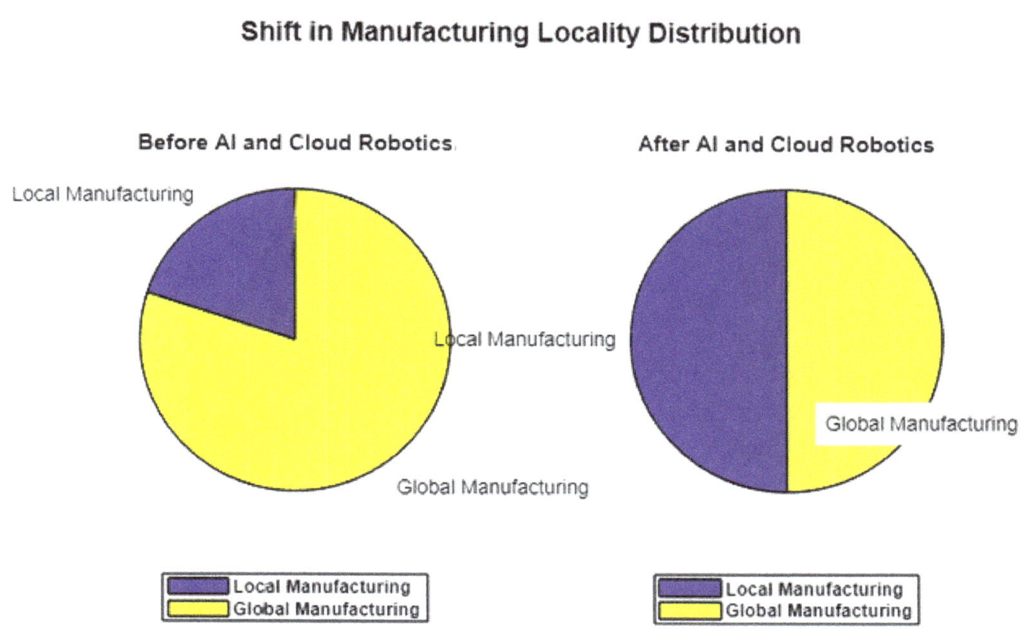

Figure.5 presents a two-tier comparison of tasks needed in the workforce, namely, the skill distribution distribution before and after AI and cloud-based robotics integration with the use of two pie charts. It provides evidence for the expansion of the labour market, in which the number of workers required to have a specific detail of the technical education as a requirement rises sharply after its implementation. This feedback showcases the new industrial work nature where the automated systems which are being managed, programmed, and keep in charge the technicians tasks which before they were considered to be manual and repetitive ones.

What figure.6 tries to do is to present a side by side comparison of local and global distribution of manufacturing activities before and after the transformation that was brought by AI and cloud robotics integration. The left pie chart named "Manufacturing before AI and Cloud Robotics," was the main illustration of worldwide practices, with a very large reliance upon them, among which 80% were globally outsourced and only 20% remained inside the country. A diverse distribution confirms the classic model of industrial production that has been thus far related to global supply channels for the purpose of scalability and cost efficiency. An appropriate pie chart, "With AI and Cloud Robotics," presents an illustration of the planned move toward a balanced dispersion, where the share of both the local as well as global manufacturing is 50%-50%. Therefore, it underlines the role of transformative effect of AI and cloud robotics in manufacturing as the means of the local production which is caused by higher automation efficiency, reduced production costs and possibilities to make final products quicker without concentrated production centers. The use of high-tech equipment creates a more distributed and agile manufacturing model, currently lessening reliance on offshore supply chains, and cushioning the effects of worldwide logistical and trade volatility. AI and cloud robotics automation highlights the strategic shift towards local manufacturing facilitated by the environmental, economic challenges, technology advancement, and the targeted market in a broader industry trends of sustainability, customization, and supply chain toughness. Through the AI integration for more accurate predictions and optimization purposes, along with the cloud robotics use, that enables the scale and flexibility, industries can develop their manufacturing into more productive, environmental, and adaptable ways. With this change there are not only economic and environmental sustainability repercussions but the rebuilding of the global manufacturing compartments, which in turn can be interrelated with things like new employment patterns, trade dynamics, and advantage positions in the digital age.

10. CONCLUSION

This journey demonstrates the evolution of AI and cloud robotics in industrial automation and is characterized by incredible improvement in efficiency, productivity, safety, and innovation. In this chapter, we have covered the main constructs of AI and cloud robotics, the manner in which these technologies can be applied across different industries, the complementariness of these technologies, the transformation that they bring, the trends that are emerging, as well as the challenges and considerations associated with them. What the world today knows as AI and cloud robotics is the leading edge of the fourth industrial revolution, offering new functionalities beyond mere automation and incorporating intelligent decisions, predictive analytics, and the unity of digital and physical systems. These technologies have proved to have the ability to change industries in the way that they make smart factories, facilitate the use of data, and bring the improvement of the era of efficiency and customization. The case studies discussed demonstrate the realistic implications and advantages achieved by integrating AI with the cloud robotics, thereby, manifesting remarkable augmentation of production efficiency, cost savings, safety, and ability to respond to dynamic market demands. Also, the outlook on the future going forward points the way, highlighting the advancements of machine learning algorithms, connectivity via 5G, and the inclusivity of edge computing as drivers for further development. With the industries adapting the interfusion of AI and cloud robotics, a change of paradigm in global manufacturing practices, sustainability, and the nature of work is expected. The trend towards domestic production is induced by the benefits of these technologies both in efficiency and flexibility, which outlines a different vision of global supply chains and probably a movement towards more sustainable and resilient types of manufacturing. To sum up, the introduction of AI and cloud robotics in industrial automation is not just a technological update but a strategic conversion which needs an all-encompassing strategy to be fully unleashed. Looking to the future, these technologies will keep evolving which means that they will improve operational efficiencies as well as change the contour of industrial automation. The road ahead is of transform, revolution, and partnership that forms a path to a future where smart automation supplants the industrial processes.

REFERENCES

El-Gendy, S. (2020). IoT Based AI and its Implementations in Industries. *2020 15th International Conference on Computer Engineering and Systems (ICCES)*, 1-6. 10.1109/ICCES51560.2020.9334627

Givehchi, O., Trsek, H., & Jasperneite, J. (2013). Cloud computing for industrial automation systems — A comprehensive overview. *2013 IEEE 18th Conference on Emerging Technologies & Factory Automation (ETFA)*, 1-4. 10.1109/ETFA.2013.6648080

Hegge, K. K., & Agrawal, R. (2023). Industrial Automated Multipurpose Robot Using WIFI. *2023 4th International Conference for Emerging Technology (INCET)*, 1-8.

K. S. & Arun. (2022). Priority Queueing Model-Based IoT Middleware for Load Balancing. *2022 6th International Conference on Intelligent Computing and Control Systems (ICICCS)*, 425-430.

Khan, J., Ali, F., Ullah, F., Bacha, J., & Lee, S. (2019). Artificial Intelligence and Internet of Things (AI-IoT) Technologies in Response to COVID-19 Pandemic: A Systematic Review. *IEEE Access : Practical Innovations, Open Solutions*, *10*, 62613–62660. doi:10.1109/ACCESS.2022.3181605

Krishnan, A., & Swarna, S. (2020). Robotics, IoT, and AI in the Automation of Agricultural Industry: A Review. *2020 IEEE Bangalore Humanitarian Technology Conference (B-HTC)*, 1-6. 10.1109/B-HTC50970.2020.9297856

Kumar, P. S., & Sharma, A. (2019). IoT Enabled Indoor Autonomous Mobile Robot using CNN and Q-Learning. *2019 IEEE International Conference on Industry 4.0, Artificial Intelligence, and Communications Technology (IAICT)*, 7-13.

Kumar & Singhal. (2020). Design and Analysis of IoT-Based Intelligent Robot for Real-Time Monitoring and Control. *2020 International Conference on Power Electronics & IoT Applications in Renewable Energy and its Control (PARC)*, 549-552.

Liu, W. (2021). Improvement of Navigation of Mobile Robotics Based on IoT System. *2021 IEEE International Conference on Robotics, Automation and Artificial Intelligence (RAAI)*, 69-72. 10.1109/RAAI52226.2021.9507826

Nayyar & Nagpal. (2018). Internet of Robotic Things: Driving Intelligent Robotics of Future - Concept, Architecture, Applications and Technologies. *2018 4th International Conference on Computing Sciences (ICCS)*, 151-160.

Pal, D., & Joshi, S. (2023). AI, IoT and Robotics in Smart Farming: Current Applications and Future Potentials. *2023 International Conference on Sustainable Computing and Data Communication Systems (ICSCDS)*, 1096-1101. 10.1109/ICSCDS56580.2023.10105101

Pavankumar, P., & Darwante, N. K. (2022). Performance Monitoring and Dynamic Scaling Algorithm for Queue Based Internet of Things. *2022 International Conference on Innovative Computing, Intelligent Communication and Smart Electrical Systems (ICSES)*, 1-7.

Ramesh Naidu, P., Guruprasad, N., & Dankan Gowda, V. (2021). A High-Availability and Integrity Layer for Cloud Storage, Cloud Computing Security: From Single to Multi-Clouds. *Journal of Physics: Conference Series*, *1921*(1), 012072. doi:10.1088/1742-6596/1921/1/012072

Ramesha, M., & Pai, G. N. (2020). Internet of things: Internet revolution, impact, technology road map and features. *Adv. Math. Sci. J.*, *9*(7), 4405–4414. doi:10.37418/amsj.9.7.11

Ranjan, A., & Agrawal, R. (2023). Wireless and Solar-Powered Multipurpose Robot for Industrial Automation: A Sustainable Solution. *2023 8th International Conference on Communication and Electronics Systems (ICCES)*, 146-151.

Ravi, R. V., Goyal, S. B., Neagu, B. C., Raboaca, M. S., & Verma, C. (2022). A Low-Cost Industrial Automation System Using IoT and Cloud Computing. *2022 International Conference and Exposition on Electrical And Power Engineering (EPE)*, 649-653. 10.1109/EPE56121.2022.9959772

Reddy, N. S., & Ashreetha, B. (2023). Technologies for Comprehensive Information Security in the IoT. *2023 International Conference for Advancement in Technology (ICONAT)*, 1-5.

Sali, S. M., & Joy, K. R. (2023). Intelligent Rover: An IoT Based Smart Surveillance Robotic Car for Military. *2023 2nd International Conference on Computational Systems and Communication (ICCSC)*, 1-6. 10.1109/ICCSC56913.2023.10143011

Saxena, A., & Chaturvedi, R. (2021). Advancement of Industrial Automation in Integration with Robotics. *2021 International Conference on Simulation, Automation & Smart Manufacturing (SASM)*, 1-6. 10.1109/SASM51857.2021.9841071

Shruthi & Naveen. (2020). Internet of Things: A Survey on Devices, Ecosystem, Components and Communication Protocols. *2020 4th International Conference on Electronics, Communication and Aerospace Technology (ICECA)*, 611-616.

Singh, G., Banga, V. K., & Yingthawornsuk, T. (2022). Artificial Intelligence and Industrial Robot. *2022 16th International Conference on Signal-Image Technology & Internet-Based Systems (SITIS)*, 622-625. 10.1109/SITIS57111.2022.00098

Singla, A., & Sharma, N. (2022). IoT Group Key Management using Incremental Gaussian Mixture Model. *2022 3rd International Conference on Electronics and Sustainable Communication Systems (ICESC)*, 469-474.

Varun, R. S., & Rajesh, T. (2018). Implementation of swarm intelligence in obstacle avoidance. *RTEICT 2017 - 2nd IEEE International Conference on Recent Trends in Electronics, Information and Communication Technology, Proceedings*, 525–528.

Chapter 2
Cloud Infrastructure for Robotics:
A Revolution in Robotics Development and Deployment

C. Harinakshi
https://orcid.org/0009-0003-0321-766X
Sahyadri College of Engineering and Management, India

A. Masooda
Sahyadri College of Engineering and Management, India

G. Suchetha
Sahyadri College of Engineering and Management, India

ABSTRACT

The integration of cloud infrastructure with robotics transforms the field by providing a cost-effective platform for developing, testing, and deploying robotic applications. Cloud's on-demand computing resources accelerate innovation, allowing developers to experiment with new algorithms efficiently. This global collaboration enables remote control and testing of robots worldwide, overcoming geographical constraints. Cloud infrastructure enhances deployment by offloading intensive tasks, reducing hardware requirements, and promoting cost-effectiveness. Centralized fleet management streamlines operations, ensuring uniform updates and security patches for system stability. Robust security measures, including encryption and authentication, address privacy concerns. Cloud robotics implement edge computing for local data processing, minimizing the risk of breaches. Scalability accommodates growing demands, enabling businesses to adapt without significant investments.

SECTION 1: INTRODUCTION

Cloud robotics is an emerging field at the intersection of robotics and cloud computing, poised to revolutionize the way robots operate and interact with the world. This innovative technology leverages

DOI: 10.4018/979-8-3693-1914-7.ch002

the power of the cloud to enhance the capabilities of robots, making them smarter, more versatile, and more connected than ever before. At its core, cloud robotics involves the integration of traditional robotics with cloud-based resources and services. Robots have traditionally operated with limited onboard computational power and storage capacity, which has restricted their ability to perform complex tasks and adapt to changing environments. However, cloud robotics aims to overcome these limitations by allowing robots to offload computation, storage, and data processing to the cloud. This means that robots can access vast amounts of data, powerful algorithms, and real-time updates from the cloud, enabling them to operate more efficiently and intelligently.

One of the fundamental components of cloud robotics is connectivity. Robots need a reliable and high-speed internet connection to access cloud resources seamlessly. This connectivity can be achieved through Wi-Fi, 4G/5G (Djellid, A., Benmeddour, F., Bahri, L., & Ghodbane, A, 2022) networks, or other communication technologies. With a stable connection, robots can send and receive data to and from the cloud in real-time, enabling them to make informed decisions and respond rapidly to changing situations.

Another critical element of cloud robotics is data storage and processing. Cloud platforms offer virtually unlimited storage capacity and powerful data processing capabilities. Robots can store and retrieve data in the cloud, allowing them to maintain a history of their interactions and learn from past experiences. Moreover, cloud-based machine learning algorithms can analyze data collected from multiple robots, enabling continuous improvement and optimization of robot behavior.

The benefits of cloud robotics extend beyond enhanced computing power and storage. Cloud-based services also provide robots with access to a wealth of information and knowledge. This knowledge can include maps, navigation data, object recognition models, and even human expertise. For example, a robot can leverage cloud-based maps and navigation services to autonomously navigate through complex environments. Additionally, robots can tap into human expertise through teleoperation, where human operators control the robot remotely and provide guidance in complex tasks.

Cloud robotics has a wide range of applications across various industries. In manufacturing, robots can use cloud-based quality control systems to detect defects in real-time, leading to higher production quality. In healthcare, telemedicine robots can connect patients with remote doctors, bringing medical expertise to underserved areas. In agriculture, cloud-connected drones and robots can monitor crops and optimize farming operations. In logistics, robots can work collaboratively in warehouses, coordinating their actions through the cloud to maximize efficiency.

One of the most promising applications of cloud robotics is in autonomous vehicles. Self-driving cars and drones can benefit immensely from the cloud's real-time data and computation capabilities. These vehicles can access up-to-date maps, traffic information, and weather data, allowing them to navigate safely and make split-second decisions. Moreover, cloud-connected autonomous vehicles can share information with each other, creating a network of vehicles that can cooperate and avoid accidents.

The potential impact of cloud robotics on society is profound. It can lead to increased automation across industries, improving productivity and reducing the risk of human error. It can also make robots more accessible and affordable, enabling smaller businesses and organizations to adopt robotics solutions. Furthermore, cloud robotics can enhance the quality of life for individuals with disabilities by providing them with robotic assistance and support.

The rest of the chapter content is organized as follows. In Section-II, Cloud computing and Robotics, Section-III, Cloud Technology Working Models, Section-IV: Challenges in networked Robotics, Section-V: IoT and Cloud Robotics integration, Section-VI: Cloud Robotics Technologies, Section-VII: Cloud Robotics Algorithms, Section-VIII: Applications.

SECTION 2: CLOUD COMPUTING AND ROBOTICS

Cloud computing and robotics represent two dynamic and rapidly evolving fields that, when integrated, open up new frontiers in technological innovation and automation. The convergence of these two domains is reshaping industries, offering unprecedented opportunities for efficiency, scalability, and remote capabilities.

Cloud computing, as a paradigm, revolutionizes the way computing resources are provisioned and utilized. It involves delivering a wide range of services over the internet, including storage, processing power, and applications. In the context of robotics, cloud computing becomes a powerful ally, providing a scalable infrastructure that enables robotic systems to transcend the limitations of onboard computing.

One of the fundamental contributions of cloud computing to robotics lies in its ability to handle intensive computational tasks. Robots, particularly those engaged in complex operations or requiring advanced decision-making capabilities, often face constraints in terms of onboard processing power. Cloud computing addresses this challenge by allowing robots to offload computational tasks to remote servers with substantial computing resources. This capability empowers robots to execute intricate algorithms and process vast datasets, enhancing their overall performance and functionality.

Remote monitoring and control constitute another key dimension of the cloud-robotics integration. Through cloud platforms, users can remotely monitor and control robotic systems from virtually anywhere with an internet connection. This has transformative implications for industries such as surveillance, where robots equipped with cameras and sensors can be deployed in remote or hazardous environments, while operators manage them from the safety of a control center.

The Software as a Service (SaaS)(Sun, L., Ma, J., Wang, H., Zhang, Y., & Yong, J, 2015) model in cloud computing further accelerates the development and deployment of robotic applications. Developers can leverage cloud-based tools and platforms to create, test, and deploy robotic software without the need to manage the underlying infrastructure. This not only streamlines the development process but also facilitates faster iterations and updates, a crucial aspect in the ever-evolving landscape of robotics.

Machine learning and artificial intelligence (AI) (Bogue, R, 2017).play a pivotal role in the capabilities of modern robotic systems. Cloud computing provides the necessary computational resources for training sophisticated machine learning models. Training such models requires substantial computational power, and the cloud offers a cost-effective and scalable solution for processing and analyzing vast amounts of data. As robotics increasingly relies on AI for tasks like object recognition, decision-making, and learning from experience, the collaboration with cloud computing becomes indispensable.

The communication and connectivity aspects of cloud computing contribute significantly to the collaborative functioning of robotic systems. In scenarios where multiple robots need to work together or where robots are part of a larger network, the cloud serves as a centralized hub for coordination. Real-time communication and data sharing between robots become seamless, enabling them to collaborate effectively in tasks such as search and rescue operations, environmental monitoring, or warehouse automation.

Scalability is a hallmark of cloud computing, and it addresses a critical need in the realm of robotics. The ability to scale resources based on demand is particularly valuable in dynamic environments where the workload may vary. Cloud services offer the flexibility to allocate more computing power or storage when needed, ensuring that robotic systems can adapt to changing circumstances and operate efficiently under varying workloads.

While cloud computing provides a centralized and powerful solution, edge computing has emerged as a complementary paradigm in the field of robotics. Edge computing involves processing data closer to the source, such as on the robot or nearby devices. This approach reduces latency and enhances real-time decision-making, critical factors in applications where split-second decisions are essential, like autonomous vehicles or drones navigating complex environments.

Security and privacy considerations (Fosch-Villaronga, E., & Millard, C, 2019). are paramount in both cloud computing and robotics. When data is transmitted between robots and the cloud, robust security measures must be in place to protect against unauthorized access and ensure the integrity of the data. Privacy concerns, especially in applications involving sensitive information, underscore the importance of secure and encrypted communication between robotic systems and cloud servers.

Cost efficiency is a driving force behind the adoption of cloud computing in robotics. The pay-as-you-go model offered by cloud services allows organizations to pay only for the resources they consume, avoiding upfront capital expenditures. This model aligns well with the variable resource needs of robotics projects, where computational demands may fluctuate based on the task at hand. As a result, cloud computing becomes an economically viable solution for organizations looking to deploy and scale robotic systems.

The integration of robotics with the Internet of Things (IoT)is a natural progression facilitated by cloud computing. Robotics often intersects with IoT devices, and cloud platforms provide a seamless integration point. This interconnectedness enhances the capabilities of both technologies, allowing robotic systems to interact with a myriad of IoT devices, sensors, and actuators. The collaborative synergy between cloud computing, robotics, and IoT opens up new possibilities in smart cities, industrial automation, and interconnected ecosystems.

SECTION 3: CLOUD TECHNOLOGY WORKING MODELS

Cloud technology operates based on several working models, each serving specific purposes and catering to diverse requirements. These models encompass different service types and deployment methods. Here's an overview of the primary working models in cloud technology (Sun, L., Ma, J., Wang, H., Zhang, Y., & Yong, J, 2015).

1. Service Models:
    a. Infrastructure as a Service (IaaS):
        - Definition: IaaS provides virtualized computing resources over the internet. It includes virtual machines, storage, and networks.
        - How it Works: Users can rent virtualized hardware resources on a pay-as-you-go basis. This model offers flexibility and scalability, allowing users to deploy and manage applications without investing in physical hardware.

    b. Platform as a Service (PaaS):
        - Definition: PaaS delivers a platform allowing customers to develop, run, and manage applications without dealing with the complexities of underlying infrastructure.
        - How it Works: Developers can focus on coding without managing the underlying infrastructure. PaaS provides tools and services for application development, such as databases, middleware, and development frameworks.

    c. Software as a Service (SaaS):

- Definition: SaaS delivers software applications over the internet on a subscription basis. Users access the software through a web browser.
- How it Works: Users don't need to install or maintain the software locally. Updates and maintenance are handled by the service provider. Examples include email services like Gmail and productivity suites like Microsoft 365.

2. Deployment Models:
   a. Public Cloud:
      - Definition: Public clouds are owned and operated by third-party service providers. Resources are shared among multiple users.
      - How it Works: Users access and manage resources through web browsers. Public clouds offer scalability, cost efficiency, and flexibility but might have limitations in customization.
   b. Private Cloud:
      - Definition: Private clouds are dedicated to a single organization. They can be hosted on-premises or by a third-party provider.
      - How it Works: Organizations have more control over resources, security, and customization. Private clouds are suitable for industries with strict compliance and security requirements.
   c. Hybrid Cloud:
      - Definition: Hybrid clouds combine public and private cloud models, allowing data and applications to be shared between them.
      - How it Works: Organizations can leverage the scalability of the public cloud for non-sensitive functions and maintain critical applications in a private cloud for enhanced security and control.
   d. Community Cloud:
      - Definition: Community clouds are shared by multiple organizations with common concerns, such as security, compliance, or industry-specific requirements.
      - How it Works: Organizations within a specific community collaborate to share resources and meet their collective needs.
3. Cloud Service Lifecycle Models:
   a. Cloud Service Creation:
      - Definition: This phase involves the development and creation of cloud services, including design, coding, and testing.
   b. Cloud Service Deployment:
      - Definition: The deployment phase involves making the service available to users. It includes activities like provisioning resources and configuring the infrastructure.
   c. Cloud Service Management:
      - Definition: Management involves monitoring, scaling, updating, and maintaining the cloud service throughout its lifecycle.
   d. Cloud Service Decommissioning:
      - Definition: The decommissioning phase involves retiring a cloud service when it's no longer needed. This includes data removal and resource deallocation.

Understanding these working models is crucial for organizations and individuals adopting cloud technology. The choice of service and deployment models depends on specific business requirements, including scalability needs, security considerations, and compliance standards. As cloud technology continues to evolve, new working models may emerge, providing even more options for users and organizations seeking efficient and scalable IT solutions.

SECTION 4: CHALLENGES IN NETWORKED ROBOTICS

Networked robotics, also known as swarm robotics or multi-robot systems, involves the coordination and collaboration of multiple robots to achieve a common goal. While networked robotics has the potential to revolutionize various fields, it also comes with its fair share of challenges (Arzo, S. T., Naiga, C., Granelli, F., Bassoli, R., Devetsikiotis, M., & Fitzek, F. H, 2021):

- Communication: Effective communication among robots is crucial for coordination. Challenges include dealing with communication delays, packet loss, and limited bandwidth. Ensuring reliable and low-latency communication is essential.
- Coordination: Coordinating the actions of multiple robots to achieve a common goal can be complex. Deciding which robot should perform which task, how to allocate resources, and how to avoid conflicts requires advanced algorithms and strategies.
- Scalability: As the number of robots in a networked system increases, scalability becomes a significant challenge. Scalable algorithms and communication protocols are needed to manage large robot teams effectively.
- Fault Tolerance: Robots can encounter hardware or software failures, and these failures can disrupt the networked system. Developing fault-tolerant mechanisms to handle failures gracefully is essential for robustness.

  Localization and Mapping: Accurate localization of robots and creating a map of the environment is crucial for navigation and coordination. In dynamic environments, maintaining up-to-date maps can be challenging.
- Energy Efficiency: Robots typically rely on batteries, and energy constraints can limit their operational time. Optimizing energy consumption while maintaining productivity is a constant challenge.
- Security: Networked robots are susceptible to cyberattacks, which can compromise their operation, data, or even safety. Implementing robust security measures is crucial to protect against threats.
- Heterogeneity: In many cases, robots in a networked system may be heterogeneous, with different capabilities, sensors, and communication hardware. Integrating and coordinating such diverse robots adds complexity.

  Task Allocation: Assigning tasks to robots optimally is a challenging problem. It involves considering factors like robot capabilities, task requirements, and the dynamic nature of the environment.
- Human-Robot Interaction: If humans are involved in the networked robotics system, ensuring effective interaction between humans and robots, including user-friendly interfaces and safety protocols, is essential.

- Ethical and Legal Issues: The deployment of networked robots in real-world scenarios raises ethical and legal questions, such as liability for robot actions, privacy concerns, and societal impacts.
- Adaptability and Learning: Building robots that can adapt to changing environments, learn from experience, and improve their performance over time is a challenging research area.

Standardization: The lack of standardized protocols and interfaces can hinder interoperability between different robot platforms, making it difficult to integrate robots from different manufacturers into a networked system.

SECTION 5: IOT AND CLOUD ROBOTICS INTEGRATION

IoT and Cloud Robotics integration represents a paradigm shift in the field of robotics, merging physical devices with cloud-based computing and connectivity. IoT devices, including sensors and actuators embedded in robots, collect a wealth of data about the environment, machine performance, and operational metrics. This data is transmitted to cloud platforms, where it undergoes analysis, processing, and storage. Cloud computing infrastructure provides the computational power and storage capacity necessary for handling large volumes of data and running sophisticated algorithms.

One of the key advantages of IoT and Cloud Robotics integration is the ability to enable remote monitoring and control of robotic systems (Kabir, H., Tham, M. L., & Chang, Y. C., 2023). Operators can access real-time data from anywhere with an internet connection, monitor robot operations, diagnose issues, and even perform remote maintenance or adjustments. This capability not only improves operational efficiency but also reduces downtime and enables proactive maintenance strategies. Additionally, cloud-based solutions facilitate scalability, allowing organizations to easily scale up or down their robotic deployments based on demand.

Let's consider a practical example of IoT and Cloud Robotics integration in the context of warehouse automation. Imagine a large e-commerce company that operates a massive warehouse with hundreds of robotic units for inventory management and order fulfillment. These robots are equipped with IoT sensors that monitor factors like inventory levels, item locations, environmental conditions, and equipment health. The data collected by these sensors is transmitted to a cloud-based platform for processing and analysis. In this scenario, the cloud platform uses machine learning algorithms to optimize warehouse operations. For instance, it can predict inventory demand based on historical data, recommend optimal storage locations for items, and dynamically adjust robot routes to minimize travel time and maximize efficiency. This real-time optimization helps the warehouse operate smoothly, reduce errors, and fulfill orders faster.

Moreover, remote monitoring and control capabilities enabled by IoT and Cloud Robotics integration allow warehouse managers to oversee operations from anywhere. They can access dashboards and reports that provide insights into inventory status, robot performance, and operational metrics. If any issues arise, such as a robot malfunction or a low inventory alert, managers can remotely troubleshoot, diagnose the problem, and take corrective actions without physically being present on-site. Additionally, the cloud-based platform facilitates scalability by allowing the company to easily add more robotic units or adjust workflows based on seasonal demand fluctuations or business growth. This flexibility ensures that the warehouse can efficiently handle varying workloads without compromising on performance or customer satisfaction.

SECTION 6: CLOUD ROBOTICS TECHNOLOGIES

A few advancements are being utilized including the Lego NXT robot, the Lego Java Operating System (leJOS), a Python mediator, the Raspberry Pi, Phidgets, and the Seattle distributed computing stage (Bogue, R, 2017).

1. Lego NXT Robot

The LEGO Mindstorms NXT (Not eXactly C) (Zygouris, N. C., Striftou, A., Dadaliaris, A. N., Stamoulis, G. I., Xenakis, A. C., & Vavougios, D, 2017) is a versatile robotics kit designed for education and hobbyist purposes. It was released by LEGO in 2006 as a successor to the original LEGO Mindstorms Robotics Invention System. The NXT kit allows users to build and program their own robots using LEGO Technic pieces, sensors, and motors.

Key Components of the LEGO NXT Robot Kit

Brick (NXT Intelligent Brick):

- The NXT Intelligent Brick serves as the brain of the robot. It has a programmable ARM7 microcontroller, a display, buttons, and ports for connecting sensors and motors. The brick can be programmed using the NXT-G graphical programming language or other programming languages such as Python or Java.

Motors:

- The kit typically includes two or three motors, each with built-in rotation sensors. These motors provide movement to the robot and can be programmed to control the robot's wheels or other moving parts.

Sensors:

- Various sensors are included in the kit to enable the robot to interact with its environment. Common sensors include:
- Touch Sensor: Detects physical contact.
- Ultrasonic Sensor: Measures distances using sound waves.
- Light Sensor: Measures ambient light levels.
- Sound Sensor: Detects sound intensity.
- Color Sensor: Identifies colors.

LEGO Technic Pieces:

- The kit includes a variety of LEGO Technic pieces, such as beams, axles, gears, and connectors. These pieces allow users to build a wide range of robot designs, from simple vehicles to more complex machines.

NXT-G Programming Software:

- The LEGO Mindstorms NXT-G software provides a graphical programming interface for creating programs for the NXT Intelligent Brick. Users can drag and drop blocks to create sequences of actions, control motors, read sensor values, and make decisions based on sensor input.

Bluetooth Connectivity:

- The NXT Intelligent Brick has built-in Bluetooth support, allowing wireless communication between the robot and other devices. This feature enables remote control and communication between multiple NXT robots.

Typical Steps in Building and Programming a LEGO NXT Robot

Assemble the Robot:
- Use the LEGO Technic pieces to build the physical structure of the robot according to a chosen design. Motors, sensors, and the NXT Intelligent Brick are attached to create a functional robot.

Connect Sensors and Motors:
- Attach sensors to the robot to enable it to interact with the environment. Connect motors to provide movement and control various parts of the robot.

Program the Robot:
- Use the NXT-G programming software to create programs for the robot. Drag and drop programming blocks to control motors, read sensor values, and make decisions based on the robot's surroundings.

Test and Iterate:
- Test the robot to ensure that it behaves as intended. If needed, iterate on the design and programming to improve the robot's performance.

Explore Advanced Programming:
- As users become more comfortable with the NXT platform, they can explore more advanced programming languages, such as RobotC, Python, or Java, to gain greater control over the robot and develop more sophisticated behaviors.

2. Lego Java Operating System

The Lego Java Operating System, commonly known as leJOS (Ewert, D., Schilberg, D., & Jeschke, S, 2013)., is an open-source firmware replacement for the Lego Mindstorms series of programmable bricks. leJOS allows users to program Lego Mindstorms robots using the Java programming language, providing a versatile and powerful environment for robotics development.

Key Features of leJOS

Java Programming Language:
- One of the primary features of leJOS is its support for the Java programming language. This allows developers to leverage the simplicity, readability, and object-oriented nature of Java for programming Mindstorms robots.

Firmware Replacement:
- leJOS replaces the original firmware on the Lego Mindstorms brick, offering an alternative operating system that extends the capabilities of the standard Lego software. This firmware replacement opens up new possibilities for advanced programming and control.

API for Robotics Programming:

- leJOS provides a rich set of libraries and APIs (Application Programming Interfaces) specifically designed for robotics programming. These APIs include functions for motor control, sensor input, Bluetooth communication, and more, simplifying the development process for users.

Remote Control and Monitoring:

- With leJOS, robots can be remotely controlled and monitored using Bluetooth or Wi-Fi connections. This feature is particularly useful for projects that require real-time interaction or for scenarios where physical access to the robot is limited.

Multi-Threading Support:

- leJOS supports multi-threading, allowing developers to run multiple tasks concurrently. This feature is essential for handling parallel processes, such as managing sensor inputs while controlling motor outputs simultaneously.

Integrated Development Environment (IDE):

- leJOS is compatible with popular Java development environments, including Eclipse and IntelliJ IDEA. This compatibility streamlines the development workflow, enabling developers to use familiar tools for coding, debugging, and testing.

Advanced Sensor Integration:

- leJOS facilitates the integration of advanced sensors, extending the capabilities of Mindstorms robots. This includes support for additional sensors beyond those provided in the standard Lego Mindstorms kit.

Community Support and Documentation:

- leJOS benefits from an active and supportive community of developers. This community contributes to the ongoing development of the firmware, provides assistance through forums, and shares code examples and projects. Comprehensive documentation is available to help users get started and explore advanced features.

Using leJOS

Installation:

- To use leJOS, users need to install the leJOS firmware on their Lego Mindstorms brick. This involves downloading the firmware, connecting the brick to a computer, and flashing the firmware using the leJOS installation tools.

Programming:

- Developers can then write Java programs for their Mindstorms robots using their preferred Java development environment. The leJOS API simplifies access to robot hardware, such as motors and sensors.

Uploading and Running Programs:

- Once the Java program is written, it can be uploaded to the Mindstorms brick using a USB cable or Bluetooth connection. The leJOS firmware then executes the Java program on the robot.

Debugging and Monitoring:

- leJOS supports debugging features, allowing developers to set breakpoints, inspect variables, and step through code. Additionally, the ability to monitor and control robots remotely enhances the debugging and testing process.

Community Interaction:

- Users can engage with the leJOS community through forums, where they can seek assistance, share experiences, and collaborate on projects. The community-driven nature of leJOS contributes to its continuous improvement and expansion of capabilities.

3. A Python Mediator

A Python mediator (Gundu, S. R., Panem, C. A., & Timmapuram, A, 2020) in cloud robotics acts as an intermediary layer that facilitates communication and coordination between robotic systems and cloud services. Python, as a versatile and widely-used programming language, provides a convenient and powerful platform for developing such mediators. Here's an exploration of the key aspects and functionalities of a Python mediator in the context of cloud robotics:

Communication Bridge:

A Python mediator serves as a communication bridge between robotic devices and the cloud. It handles data exchange, commands, and status updates between the robots and the cloud infrastructure. This involves leveraging Python libraries and frameworks for networking and communication protocols.

Cloud Service Integration:

The mediator integrates with cloud services to enable functionalities such as data storage, processing, and analytics. It interacts with cloud APIs (Application Programming Interfaces) using Python libraries, allowing seamless integration with platforms like AWS, Azure, or Google Cloud. This integration empowers the robotic systems with access to scalable computing resources and advanced cloud-based services.

Data Management:

Python's capabilities in data handling and processing make it well-suited for managing the flow of data between robots and the cloud. The mediator may preprocess sensor data from robots, perform data analytics, and store relevant information in the cloud for further analysis or long-term storage.

Real-time Communication:

Python mediators can facilitate real-time communication between robotic systems and the cloud, ensuring timely transmission of critical information. This is essential for applications such as remote monitoring, control, and collaboration among multiple robots.

Security Measures:

Security is paramount in cloud robotics, and Python's extensive set of libraries for encryption, authentication, and secure communication can be employed to implement robust security measures. This includes securing data transmission between robots and the cloud and implementing access controls to protect sensitive information.

Edge Computing Integration:

In scenarios where edge computing is employed, the Python mediator can coordinate the distribution of computational tasks between the edge devices and the cloud. This ensures efficient processing and reduces latency for time-sensitive applications.

Scalability and Flexibility:

Python's scalability and flexibility make it adaptable to various cloud robotics scenarios. The mediator can be designed to scale dynamically based on the number of connected robots, adjusting resource allocation as needed. Containerization technologies, such as Docker, can be leveraged for easy deployment and scaling.

Development and Testing:

Python's ease of development and extensive set of testing frameworks contribute to the efficient creation and testing of the mediator. This accelerates the development cycle, allowing for rapid prototyping and iteration.

Error Handling and Logging:

The mediator can incorporate robust error handling mechanisms and logging functionalities implemented in Python. This ensures that issues are identified and addressed promptly, contributing to the reliability and maintainability of the cloud robotics system.

Community Support and Libraries:

Python's vibrant community and rich ecosystem of libraries and frameworks provide a wealth of resources for building and maintaining a Python mediator in cloud robotics. Open-source contributions and community support contribute to the ongoing improvement and evolution of the mediator.

In summary, a Python mediator in cloud robotics acts as a crucial intermediary layer, facilitating communication, data management, and integration with cloud services. Its versatility, scalability, and extensive ecosystem make Python an excellent choice for developing mediators that enhance the capabilities and connectivity of robotic systems in cloud environments.

4. Phidgets Devices

Serve as pivotal components in the realm of cloud robotics, acting as programmable sensors, actuators, and controllers that seamlessly bridge the gap between physical hardware and cloud services. Offering a diverse array of devices, including sensors and motor controllers, Phidgets facilitate easy integration with cloud platforms through their built-in USB connectivity. Leveraging the Phidgets (Gundu, S. R., Panem, C. A., & Timmapuram, A, 2020). API and SDK, developers can create applications that capture data from sensors, enabling its transmission to the cloud. This integration allows for real-time remote monitoring, control, and data analytics, enhancing the capabilities of cloud robotics systems. Whether used in edge computing scenarios for reduced latency or integrated with popular robotics frameworks like ROS, Phidgets devices offer a versatile interface that adapts to diverse cloud robotics applications. Furthermore, their compatibility with security measures ensures the protection of data transmitted between devices and the cloud, while continuous updates and active maintenance contribute to the ongoing improvement of the Phidgets ecosystem. In summary, Phidgets play a pivotal role in augmenting cloud robotics, facilitating secure and intelligent connectivity between physical devices and cloud-based applications.

5. Standards for Communication Between Robots and Cloud Platforms

Integrating ROS (Robot Operating System) with OPC UA (Open Platform Communications Unified Architecture) requires a thoughtful technical approach to establish robust communication and interoperability between robotics systems and industrial automation platforms. A crucial aspect of this integration is the implementation of a ROS-OPC UA bridge, acting as a middleware layer that facilitates seamless data exchange between ROS nodes and OPC UA servers. This bridge is typically developed using existing ROS packages like ros_opcua_bridge or custom nodes designed to translate ROS messages into OPC UA data types and vice versa, ensuring compatibility and consistency in data transmission. The configuring an OPC UA client within the ROS environment is essential for establishing communication with OPC UA servers deployed on industrial devices or platforms. The OPC UA client node in ROS subscribes to relevant topics or services, converts ROS messages into OPC UA format, and forwards them to the OPC UA server for processing. Conversely, the OPC UA server can publish real-time data from industrial sensors or control systems to ROS nodes, enabling robots to access and utilize this information for decision-making and control tasks.

The integrating ROS with OPC UA involves defining a comprehensive data mapping and modeling strategy to align ROS data structures with OPC UA information models effectively. This includes creating a mapping layer that converts ROS-specific data elements, such as messages and topics, into OPC UA-compliant nodes, variables, and data types. Ensuring secure communication channels is also crucial, requiring the implementation of OPC UA security features like encryption, digital signatures, and user authentication within ROS nodes to establish trusted connections with OPC UA servers. Overall, by addressing these technical aspects, developers can achieve a seamless integration between ROS and OPC UA, unlocking enhanced capabilities for robotics systems in industrial automation environments.

SECTION 7: CLOUD ROBOTICS ALGORITHMS

Cloud robotics refers to a concept where robots can offload computation, storage, and communication to cloud resources. This allows robots to benefit from the vast computing power and storage capabilities of the cloud, enabling them to perform more complex tasks and share information with other robots. Various algorithms are employed in cloud robotics to make this collaboration and offloading process efficient and effective. Here are some key algorithms and techniques used in cloud robotics:

Task Offloading Algorithms:

- Dynamic Task Offloading: Algorithms dynamically decide which tasks should be offloaded to the cloud based on factors such as current robot load, available cloud resources, and task priorities.
- QoS-based Offloading: Quality of Service (QoS) metrics, such as response time and bandwidth, are considered when deciding whether to offload a task to the cloud.

Localization and Mapping:

- Simultaneous Localization and Mapping (SLAM): Algorithms that help robots build and update maps of their environment while simultaneously determining their own position within the map. Cloud assistance can be used to enhance SLAM algorithms, especially in large-scale environments.

Path Planning:

- Cloud-assisted Path Planning: Cloud resources can be used to compute optimal paths for robots by considering real-time data, such as traffic conditions, and by leveraging advanced planning algorithms.

Machine Learning and AI:

- Cloud-based Machine Learning: Training and updating machine learning models in the cloud, and then deploying these models to robots for improved perception, decision-making, and adaptation to changing environments.

Communication Protocols:

- Efficient Communication: Algorithms for optimizing data transmission between robots and the cloud to minimize latency and bandwidth usage.

At the heart of Predictive Maintenance (PdM), sensors play a pivotal role in collecting crucial data, offering vital insights into machine health, and facilitating proactive maintenance strategies(Soori, 2013).

1. The types of sensors used for monitoring machine health parameters

Various types of sensors are employed in Predictive Maintenance (PdM) for monitoring machine health, with the specific selection contingent upon the equipment and parameters under scrutiny. The commonly utilized sensors encompass:

1. Vibration Sensors: These sensors identify alterations in vibration patterns, signaling potential wear and tear, misalignment, or imbalances in rotating components.
2. Temperature Sensors: Monitoring temperature variations reveals issues like overheating, friction problems, or blockages in cooling systems.
3. Pressure Sensors: These sensors track fluid pressure levels, pinpointing leaks, blockages, or fluctuations in pump performance.
4. Acoustic Sensors: These sensors detect unusual noises from machinery, offering insights into potential internal component failures or lubrication issues.
5. Flow Sensors: These sensors monitor fluid flow rates (e.g., oil or coolant), detecting leaks, blockages, or changes in pump performance.
6. Chemical Sensors: In specific applications, these sensors monitor the presence or concentration of certain chemicals (such as lubricants or contaminants) to assess equipment health (Yan, 2017).

2. The challenges of data collection and transmission from industrial environments

The integration of data analytics and Predictive Maintenance (PdM) shows significant potential for enhancing operations in industrial settings. Nonetheless, the unique challenges associated with data collection and transmission in these environments can pose obstacles to the effectiveness of PdM systems (Chang, 2011).

Challenges in Data Collection:

- Harsh Environments: Industrial settings, characterized by dust, noise, and extreme temperatures, demand robust sensors and data acquisition equipment to ensure the accuracy and reliability of data collection.
- Limited Connectivity: Unreliable or unavailable wireless connectivity in certain industrial areas necessitates alternative solutions such as wired connections or mesh networks.
- Data Security: Industrial data often contains sensitive information, underscoring the need for robust cybersecurity measures to safeguard against breaches and data loss.

Challenges in Data Transmission:

- Bandwidth Limitations: Large volumes of sensor-generated data can strain existing network infrastructure, resulting in bottlenecks and delays in data transmission.
- Data Integrity: Transmission errors or corruption can significantly impact the accuracy of Predictive Maintenance models, requiring the implementation of error-correction mechanisms and data validation processes.

- Interference: Industrial environments are susceptible to electromagnetic interference, which can disrupt wireless data transmission and compromise data quality (Yan,2018)
  3. The importance of data quality and consistency for accurate predictive models

In the realm of predictive maintenance, where the ability to foresee equipment failures before they manifest is crucial, the pivotal role of data cannot be overstated.

Consider this scenario: You're inputting a mix of Skittles and M&Ms into a machine learning algorithm, expecting it to accurately predict the flavor of the next candy (Chang, 2011). The outcome would be, to say the least, chaotic. Similarly, incorporating disorganized data into predictive maintenance models can result in unreliable predictions, potentially leading to costly consequences such as:

- Unscheduled Downtime: Failure to detect equipment issues in advance can result in unscheduled downtime, impacting production efficiency and profitability.
- Unnecessary Maintenance: False alarms triggered by poor data quality can lead to the misallocation of resources and manpower for unnecessary repairs.
- Safety Hazards: In critical applications, inaccurate predictions rooted in subpar data quality could compromise safety and increase the risk of accidents.

Let's delve into the crucial components:

i. Data Accuracy:

Precise Measurements: Accurate and error-free sensor data from equipment is essential. Just like relying on a thermometer that constantly fluctuates between 90°F and 110°F would be unreliable for predicting fever, inaccurate sensor data disrupts the entire prediction process.

Data Integrity: The presence of missing values, outliers, and inconsistencies in the data can distort the model's understanding of normal equipment behavior (Yan,2018).

ii. Data Consistency:

Standardized Formats: Data originating from various sources, such as sensors, maintenance logs, and historical records, must adhere to consistent formats and units. Mixing Celsius and Fahrenheit for temperature readings, for instance, would compromise the reliability of equipment health assessments.

Data Cleaning and Preprocessing: Before inputting data into the model, it ensures the model learns from high-quality information, analogous to a chef using only fresh ingredients for a delectable meal (Lee,2017)

B. Advances in data analytics techniques for extracting insights from machine data:

Continuous advancements in data analytics techniques have significantly bolstered the effectiveness of Predictive Maintenance, harnessing the power of machine-generated data for valuable insights.

1. The application of statistical methods for identifying trends and patterns

Envision the task of navigating through vast data sets as akin to a detective diligently searching for clues. The statistical methods serve as an essential magnifying glass, unveiling previously unnoticed patterns and trends that hold crucial insights to deciphering the data's narrative. Let's delve into key methods and their practical applications:

Time Series Analysis: This method involves tracking data points collected over time, such as website traffic or stock prices. Picture it as scrutinizing a movie frame-by-frame to detect subtle changes in movement or expression. Utilizing techniques like ARIMA models and trend decomposition, this approach not only predicts future values but also identifies seasonality or anomalies. This enables anticipation of market fluctuations or surges in website traffic.

Clustering: This technique groups akin data points, analogous to organizing fruits by color or size, or arranging a bookshelf by genre or author. Methods such as K-means clustering and hierarchical clustering unveil concealed structures within the data, facilitating the segmentation for focused analysis. This can be applied to categorize customers based on their purchasing habits or to cluster patients with comparable medical conditions for tailored treatment approaches.

Hypothesis Testing: This approach evaluates the credibility of an assertion concerning the data. Consider examining whether a new marketing campaign genuinely boosts website traffic. Statistical tests such as t-tests and chi-square tests play a role in determining whether observed trends carry statistical significance or are merely coincidental.

2. The use of machine learning algorithms for predictive modeling and anomaly detection

Figure 4. Screenshot

Figure 5. means vs. after k-means

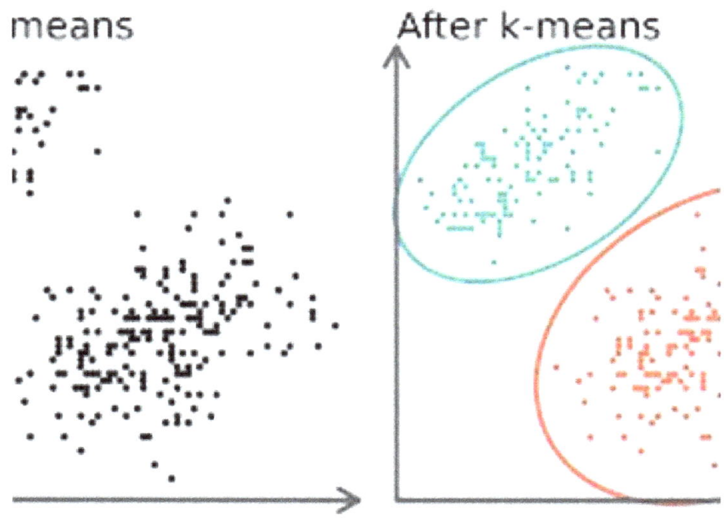

Figure 6. Distribution plot

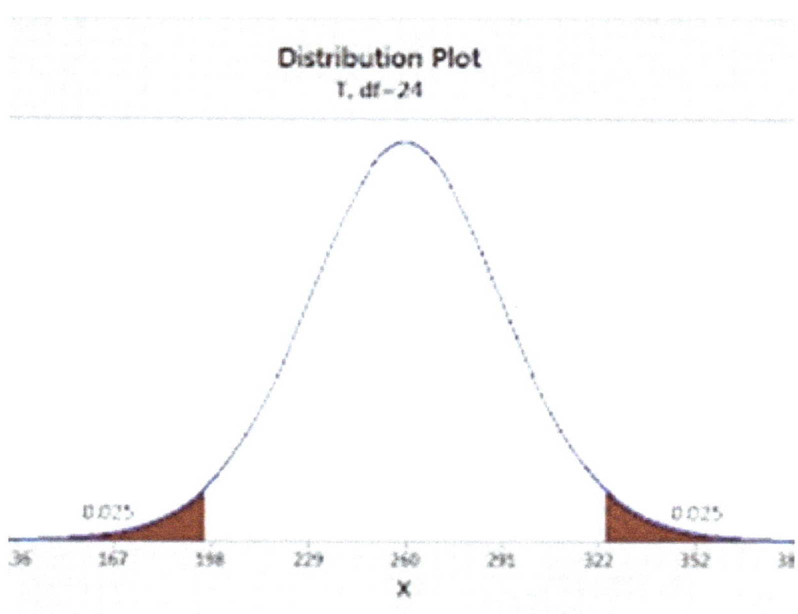

Predictive Modeling: Envision gazing into a crystal ball, not to unveil fortunes, but to anticipate future outcomes based on historical data – that encapsulates the essence of predictive modeling. Machine learning algorithms leverage past data to construct models capable of forecasting future events or values (Lin, 2005). Key algorithms for predictive modeling include:

i. Regression Algorithms: These algorithms quantify the relationships between variables, enabling predictions of future values based on existing trends.
ii. Decision Trees: Utilizing tree-like structures, these algorithms make sequential decisions based on data features, ultimately predicting specific outcomes.

Anomaly Detection: Anomalies can be likened to unexpected guests at a party – their presence disrupts the anticipated pattern. Anomaly detection algorithms excel at identifying these deviations, potentially signaling equipment malfunctions, fraudulent transactions, or cyberattacks. Algorithms tailored for anomaly detection include:

i. K-means Clustering: This algorithm groups similar data points, allowing outliers to be identified as points significantly different from any cluster.
ii. Local Outlier Factor (LOF): LOF assigns anomaly scores to data points based on their local density, spotlighting points significantly less dense than their surroundings.
iii. One-Class Support Vector Machines (OCSVM): These algorithms learn a boundary around normal data, flagging anything that falls outside as an anomaly (Gao, 2019).
    C. The use of machine learning algorithms for predictive modeling and anomaly detection:

Incorporating machine learning algorithms into predictive modeling and anomaly detection within the scope of data analytics in Predictive Maintenance amplifies the capability to derive actionable insights from machine-generated data.

1. The selection of appropriate machine learning algorithms for specific applications

Selecting the appropriate machine learning algorithm tailored to the specific application is paramount for achieving accurate and reliable results. Here are key considerations to steer your decision-making:

a. Problem Type:
    i. Predictive Modeling:
        - Linear Regression: Suitable for predicting continuous values like sales or temperature.
        - Logistic Regression: Ideal for classifying binary outcomes such as customer churn or fraud detection.
        - Decision Trees: Versatile for handling both numerical and categorical features, providing interpretable rules.
        - Support Vector Machines (SVMs): Well-suited for managing high-dimensional data and complex relationships.
    ii. Anomaly Detection:
        - K-means Clustering: Identifies outliers based on their distance from cluster centers.
        - One-Class SVM: Learns a model of normal behavior and flags deviations as anomalies.
        - Isolation Forest: Efficiently detects anomalies in high-dimensional data (Lee,2014).
b. Data Characteristics:
    - Data Volume: Some algorithms (e.g., decision trees) handle large datasets well, while scalability issues may arise with others (e.g., SVMs).

- ◦ Data Type: Algorithms exhibit different sensitivities to data types, be they numerical, categorical, text, or images.
- ◦ Missing Values: Robustness to missing data varies among algorithms; for example, decision trees are generally more tolerant than linear regression.

c. Performance Requirements:

- ◦ Accuracy: Reflects the model's proficiency in making correct predictions on new data.
- ◦ Speed: Encompasses the time required for both training and making predictions.
- ◦ Interpretability: Measures the ease of understanding the model's decision-making process.

2. The training and validation of predictive models using historical data

The primary challenge encountered in machine learning models lies in creating a meaningful representation of the input data (X, comprising time series input variables). This involves executing a transformation (denoted as the model or function f) capable of approximating the anticipated outcomes (Y, the variable of interest). The dataset provided for this purpose is commonly termed the "training set." Machine learning algorithms leverage this training set to employ a predefined set of operations, effectively constructing a model.

3. The real-time monitoring of machine data and the detection of anomalies

The simultaneous real-time monitoring of machine data, paired with the utilization of machine learning algorithms for anomaly detection, constitutes a potent synergy that elevates predictive maintenance practices.

A. Monitoring Machine Data in Real Time:

i. Essential Components:

Figure 7. Automatic validation and analysis of predictive models by means of big data and data science

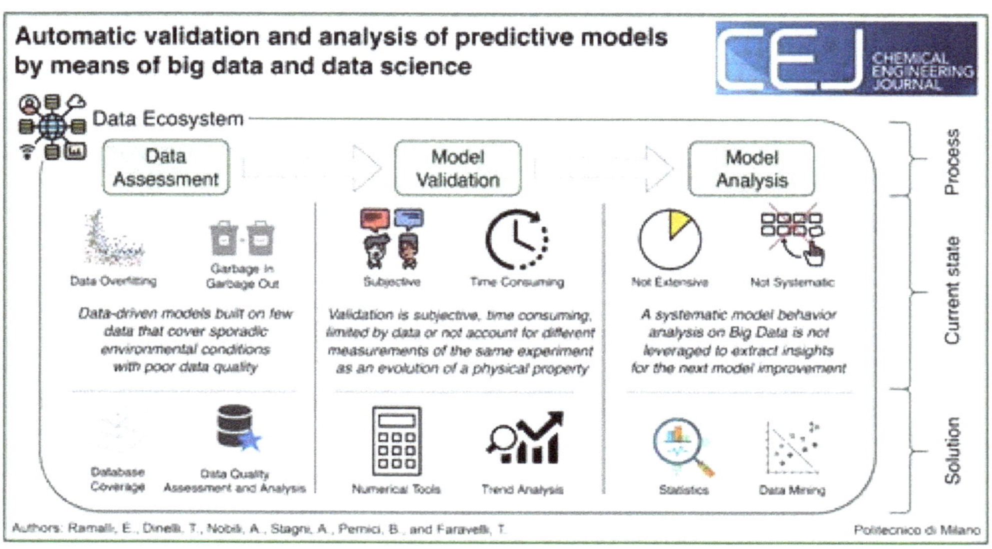

1. Sensors: Gather data from diverse machine components such as temperature, vibration, pressure, and voltage.
2. Data Transmission: Transmit collected data to a central processing unit or a cloud-based platform.
3. Data Processing and Analysis: Cleanse and prepare data, applying machine learning models for instantaneous analysis.

B. Anomaly Detection Utilizing Machine Learning:
  i. Common Algorithms:
    1. Supervised Learning: Requires labeled data (normal vs. anomalous) for training models (e.g., Support Vector Machines, Random Forests, Neural Networks).
    2. Unsupervised Learning: Identifies anomalies without prior knowledge of normal patterns (e.g., Isolation Forests, One-Class SVMs, Autoencoders).

3. IMPLEMENTING PREDICTIVE MAINTENANCE STRATEGIES

Implementing predictive maintenance (PdM) strategies involves leveraging data analytics and technology to shift from reactive maintenance (fixing things when they break) to proactive maintenance (addressing issues before they cause downtime). This proactive approach minimizes disruptions, optimizes asset performance, and increases overall efficiency.

3.1. Integrating Data Analytics and Predictive Maintenance into Existing Industrial Systems

Transforming from reactive repairs to proactive predictions, integrating data analytics and predictive maintenance into your existing industrial systems is a game-changer. But how do you navigate this journey seamlessly? Start by identifying your most critical assets, those whose failures send shivers down your production line. Assess your data landscape, understanding what flows where and how. Plug any skill gaps, be it in data analysis or model development. Then, equip those critical assets with sensors, their whispers becoming valuable insights. As data streams in, clean and organize it, ensuring quality fuels accurate predictions. Choose the right analytical tools, be it machine learning, statistics, or deep learning wizardry for complex patterns. Train, test, and refine your models, their accuracy sharpening with each data morsel. Connect your PdM system to your existing maintenance software, letting predictions seamlessly guide decision-making. Remember, it's a continuous dance, monitoring performance, retraining models, and adapting to the ever-changing industrial symphony. Don't forget your team, educating them on the power of prediction and building trust in this new conductor of efficiency. By taking these steps, you'll unlock the true potential of data analytics and predictive maintenance, ensuring your industrial systems hum with optimal performance, cost-effectiveness, and safety (Lee, 2015).

3.1.1. The challenges of integrating predictive maintenance into existing operational workflows

Predictive maintenance wants to be your new best friend. It whispers about problems with your machines before they break, saving you headaches (and repair bills!). But getting it to work with your current team and tools can be tricky. This guide will help you avoid the bumps and bruises and make the most of this helpful friend!

While predictive maintenance (PdM) promises a proactive approach to asset management, integrating it into existing workflows can present several challenges (Table 1):

A. **Data-Related Challenges** (Table 1):

Table 1. Data-related challenges

Data Quantity and Quality: Sufficient and reliable data is essential for accurate predictions. However, organizations often face	**Insufficient historical data to train models effectively. Poor data quality due to inconsistencies, errors, or missing values. Challenges in integrating data from disparate sources (sensors, maintenance records, operational logs)**
Data Integration	Connecting PdM systems with existing enterprise resource planning (ERP), maintenance management systems (CMMS), and other operational systems can be complex, requiring data synchronization and compatibility

B. **Technical and Infrastructure Challenges** (Table 2):

Table 2. Technical and infrastructure challenges

Sensor Installation and Connectivity:	**Equipping assets with sensors can be costly and time-consuming, especially in large-scale operations. Ensuring reliable data transmission and connectivity from sensors to data storage and analytics platforms can pose challenges.**
Model Development and Maintenance	Building accurate predictive models requires expertise in data science and machine learning. Models need ongoing monitoring, retraining, and updating to maintain accuracy as equipment ages and operating conditions change.
Cyber security Concerns	Protecting sensitive operational and maintenance data from cyberattacks is crucial, necessitating robust cybersecurity measures

C. **Organizational and Cultural Challenges** (Table 3):

Table 3. Technical and infrastructure challenges

Change Management:	**Transitioning from reactive to predictive maintenance often involves significant shifts in mindset, processes, and decision-making. Overcoming resistance to change and building trust in PdM systems among maintenance personnel and management can be challenging**
Skill Gaps	Organizations may lack the internal expertise to implement and manage PdM systems effectively, requiring investment in training or external support.
Integration with Existing Workflows	Aligning PdM recommendations with existing maintenance schedules, resource allocation, and procurement processes can be complex and require careful coordination

D. **Addressing These Challenges**:

i. **Comprehensive Data Strategy:** Prioritize data quality, standardization, and integration from the start.
ii. **Collaboration and Expertise:** Foster collaboration between IT, maintenance, and operations teams, and consider external expertise if needed.
iii. **Phased Implementation:** Start with pilot projects to build understanding and gradually expand PdM adoption.
iv. **Change Management:** Invest in training and communication to build trust and adoption among employees.
v. **Strong Cybersecurity Measures:** Prioritize data security and privacy throughout the implementation process.
vi. **Continuous Monitoring and Improvement:** Regularly evaluate model performance, data quality, and workflow integration to ensure effectiveness and identify areas for improvement.

By effectively addressing these challenges, organizations can realize the full benefits of predictive maintenance, leading to reduced downtime, optimized maintenance costs, improved asset reliability, and enhanced operational efficiency.

3.2. The Need for Data Integration and Standardization Across Different Systems

Imagine you have a delicious meal, but each course is served on a different plate with mismatched utensils. It sounds confusing and frustrating, right? That's exactly what happens when data isn't integrated and standardized across different systems(Li, 2014). It's messy, and inefficient, and ultimately hinders your ability to make informed decisions. Here's why data integration and standardization are crucial (Table 4):

3.2.1. Avoids Confusion and Errors

Mismatched formats, units, and naming conventions create inconsistencies and confusion. Standardizing data ensures clarity and accuracy, reducing the risk of errors and misinterpretations.

3.2.2. Enhances Collaboration and Communication

When data is consistent, different teams and departments can easily share and understand information. This fosters collaboration, improves communication, and breaks down silos within organizations.

3.2.3. Simplifies Analysis and Insights

Imagine trying to compare apples and oranges with different weights and scales. Without standardized data, analysis becomes cumbersome and unreliable. Integration allows you to combine data from different sources, leading to deeper insights and better decision-making.

3.2.4. Boosts Efficiency and Saves Time

Spending time searching for data or resolving inconsistencies wastes valuable resources. Integrated and standardized data makes it easier to access and analyze information, streamlining workflows and saving time.

3.2.5. Improves Data Governance and Compliance

Standardized data makes it easier to track, manage, and secure information. This is crucial for meeting regulatory requirements and ensuring data governance best practices.

Examples of Data Integration and Standardization

- A hospital integrating patient data from multiple systems for comprehensive care.
- A manufacturing company standardizes machine sensor data to optimize production processes.
- A retail chain integrating sales data from different stores to analyze customer behavior and improve marketing campaigns.

3.3. Challenges and Considerations for Successful Implementation: Navigating the Data-Fueled Path

Embracing data analytics and predictive maintenance (PdM) unlocks a world of efficiency, cost savings, and optimized operations. Yet, integrating these powerful tools seamlessly into your existing workflow

Table 4. Challenges and solutions for integration and standardization

Challenge	Solution
Legacy systems: Different formats and technologies in older systems create incompatibility.	Utilize data mapping tools to translate data between formats and structures. Implement data virtualization to access legacy data without direct integration.
Varied data sources: Internal and external sources may have different structures and definitions.	Master data management (MDM) creates a single source of truth for key data elements. Develop data dictionaries to document and standardize data definitions across sources.
Data quality issues: Inconsistencies, errors, and missing values make data unreliable.	Implement data cleansing processes to identify and address errors and inconsistencies. Establish data validation rules to ensure data quality at the point of entry.
Resource constraints: Technical expertise and financial investment are required for implementation.	Prioritize critical data sources and integrate them first. Consider cloud-based solutions for improved scalability and cost-effectiveness. Utilize open-source tools and frameworks where possible.
Organizational silos and resistance to change: Different departments may resist sharing data or changing workflow.	Implement strong communication and change management strategies. Highlight the benefits of data integration for all stakeholders. Provide training and support to ensure employee adoption of new processes.
Security and privacy concerns: Sharing and integrating data raises concerns about unauthorized access and misuse.	Implement data governance policies and access controls to ensure data security and privacy. Leverage data encryption and anonymization techniques when necessary.

isn't a walk in the park(Muhtaroglu,2013). Let's explore some key challenges and considerations you might encounter on this data-fueled journey:

3.3.1. Data-Centric Challenges

- **Data Quality and Quantity:** Garbage in, garbage out. Ensuring accurate, reliable, and sufficient data is crucial for accurate predictions and insightful analysis. Data cleansing, standardization, and addressing missing values are crucial steps.
- **Integration Hurdles:** Integrating data from disparate sources – sensors, maintenance records, operational logs – can be a technical puzzle. Compatibility issues, inconsistent formats, and siloed systems add complexity.
- **Model Selection and Development:** Choosing the right analytical model (machine learning, statistical, etc.) depends on your data type, desired outcomes, and available resources. Building and fine-tuning these models requires expertise and ongoing iteration.

3.3.2. Technical and Infrastructure Hurdles

- **Sensor Installation and Connectivity:** Equipping key assets with sensors often involves logistical challenges and cost considerations. Reliable data transmission and connectivity add another layer of complexity.
- **Cybersecurity Concerns:** Protecting sensitive data from cyberattacks is paramount. Robust security measures and data governance policies are essential.
- **Change Management and Training:** Transitioning from reactive maintenance to PdM requires cultural shifts and adaptation. Providing training and fostering trust in the system among personnel is crucial.

3.3.3. Organizational and Strategic Considerations

- **Initial Investment:** Setting up sensors, software, and data infrastructure requires upfront costs. Justifying the ROI and aligning PdM with your overall business strategy is key.
- **Skill Gaps:** Implementing and managing PdM systems effectively may require expertise beyond your in-house team. Hiring or seeking external support should be considered.
- **Workflow Integration:** Aligning PdM recommendations with existing maintenance schedules, resource allocation, and procurement processes demands careful planning and coordination.

3.3.4. Conquering the Challenges

- **Phased Implementation:** Start with pilot projects to build understanding and gradually expand PdM adoption. This minimizes risk and allows for course correction.
- **Collaboration and Communication:** Foster collaboration between IT, maintenance, and operations teams. Open communication and a shared vision are essential for success.

- **Continuous Monitoring and Improvement:** Regularly evaluate model performance, data quality, and workflow integration. This ensures effectiveness and identifies areas for improvement.
- **Invest in Expertise:** Don't hesitate to seek external support for model development, data analysis, and change management. The right expertise can accelerate your journey.

Remember, successfully implementing data analytics and PdM is a multifaceted endeavor. By acknowledging the challenges, planning strategically, and continuously adapting, you can unlock the immense potential of these tools, transforming your operations into a data-driven powerhouse.

3.4. The Everlasting Journey: Continuous Improvement and Model Maintenance in PdM

Imagine your predictive maintenance (PdM) model as a trusty robot assistant, diligently scanning your equipment for potential glitches. Like any good helper, though, it needs upkeep and refinement to stay sharp and effective(Sikorska,2011). That's where continuous improvement and model maintenance come in – the never-ending quest to keep your PdM engine humming at peak performance.

3.4.1. Why Continuous Improvement Matters

A. **The Data River Flows:** Your equipment operates in a dynamic world. Wear and tear, changes in operating conditions, and even seasonal variations can all shift the data landscape. If your model remains static, it's like trying to navigate a constantly changing river with an outdated map.
B. **Accuracy Drift:** Over time, even the most accurate models can experience "drift," their predictions straying from reality. This can lead to missed maintenance issues or unnecessary interventions, costing you time and resources.
C. **Evolving Challenges:** New equipment types, unforeseen failure modes, and even cyber threats can emerge, demanding your PdM system to adapt and learn.

3.4.2. The Continuous Improvement Cycle

1. **Monitor and Evaluate:** Keep a watchful eye on your model's performance. Track prediction accuracy, false positives and negatives, and overall effectiveness in preventing downtime.
2. **Analyze and Diagnose:** Dig into the data to understand why the model might be faltering. Are there changes in equipment behavior, new data patterns, or potential biases?
3. **Refine and Retrain:** Based on your analysis, update your model with new data, adjust its parameters, or even switch to a new model altogether. Remember, it's an iterative process – test, tweak, and repeat!
4. **Automate and Optimize:** Streamline the process by automating data collection, model evaluation, and retraining wherever possible. This keeps your PdM system constantly updating and ready for the next challenge.

3.4.3. Benefits of Continuous Improvement

- **Enhanced Accuracy and Reliability:** Sharper predictions mean fewer missed problems and unnecessary maintenance, saving you time, money, and frustration.
- **Improved Safety and Efficiency:** By proactively addressing potential failures, you can prevent accidents, and unplanned downtime, and optimize asset utilization.
- **Future-Proofing Your Operations:** A nimble PdM system can adapt to changing conditions and evolving threats, keeping your operations resilient and efficient in the long run.

Remember, continuous improvement and model maintenance are not one-time tasks, but an ongoing commitment. Treat your PdM system as a living, evolving entity, constantly learning and adapting to keep your industrial engine running smoothly. By embracing this never-ending journey, you'll unlock the full potential of PdM and transform your operations into a beacon of data-driven excellence.

3.5. Case Studies: Unveiling the Power of Predictive Maintenance Across Industries

Predictive maintenance (PdM) is no longer a futuristic fantasy; it's a tangible tool transforming industries across the board (Wielki, 2013). To illustrate its real-world impact, let's dive into three compelling case studies, showcasing successful PdM applications and the benefits they reaped:

3.5.1. Airlines Soaring High with PdM(Table 5):

Benefits:

- **Reduced maintenance costs:** By proactively addressing issues before they escalate, airlines can avoid costly emergency repairs and unnecessary maintenance interventions.
- **Enhanced safety:** Early detection of potential engine failures prevents catastrophic incidents, ensuring passenger safety and reducing risk.
- **Improved operational efficiency:** Predictive maintenance minimizes unplanned downtime, leading to fewer flight cancellations and delays, resulting in smoother operations and satisfied customers.
- **Quantitative Analysis:** One major airline reported a 30% reduction in engine maintenance costs after implementing PdM. Another airline saw a 50% decrease in unplanned engine removals thanks to PdM's predictive capabilities.

Table 5. Airlines soaring high with PdM

Industry	Application	Challenge	Solution
Aviation	Aircraft engine health monitoring	Aircraft engine failures can be catastrophic and costly, leading to grounded flights and hefty repair bills. Early detection of potential issues is crucial for safety and financial well-being.	Airlines utilize sensors embedded in engines to collect real-time data on vibration, temperature, and other critical parameters. PdM algorithms analyze this data, predicting potential failures before they occur.

- **Lessons Learned:** Collaboration between airlines, engine manufacturers, and data analysts is key for developing and implementing effective PdM systems. Continuous monitoring and improvement of PdM models are crucial for optimizing their accuracy and effectiveness.

3.5.2. Factories Churning Out Efficiency With PdM (Table 6):

a. Benefits:
   - **Increased uptime and production output:** PdMminimizes unplanned downtime, allowing factories to run smoothly and maximize production capacity.
   - **Improved product quality:** By ensuring consistent machine performance, PdM reduces the risk of defects and inconsistencies in manufactured products.
   - **Reduced maintenance costs:** Proactive maintenance prevents major breakdowns, saving manufacturers from costly repairs and replacements.

b. Quantitative Analysis:
   - A manufacturing company using PdM reported a 25% increase in production output due to reduced downtime.
   - Another manufacturer saw a 15% decrease in maintenance costs thanks to PdM's predictive capabilities.

c. Lessons Learned:

Figure 8. Data analysis from an aircraft engine sensor on a laptop

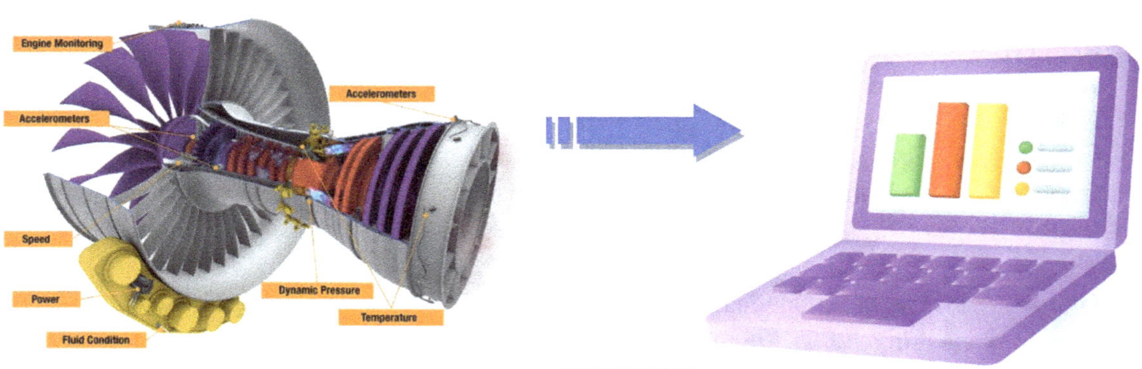

....

Table 6. Factories churning out efficiency with PdM

Industry	Application	Challenge	Solution
Manufacturing	Machine health monitoring in production lines	Unforeseen machine breakdowns disrupt production schedules, impacting output, product quality, and profitability. Predicting and preventing these disruptions is paramount.	Manufacturers install sensors on critical equipment to monitor factors like vibration, energy consumption, and temperature. PdM algorithms analyze this data to identify early signs of wear and tear, predict potential failures, and recommend preventive maintenance.

- Selecting the right sensors and data analysis tools is crucial for effectively monitoring specific equipment and processes.
- Integrating PdM with existing maintenance management systems streamlines workflow and facilitates proactive interventions.

3.5.3. Wind Farms Weathering the Elements with PdM(Table 7):

a. Benefits:
   - **Increased energy production:** PdM minimizes turbine downtime, ensuring consistent electricity generation from wind farms.
   - **Reduced maintenance costs:** Proactive maintenance prevents major breakdowns, saving wind farm operators from expensive repairs and replacements.
   - **Enhanced safety:** Early detection of potential failures reduces the risk of accidents and injuries for technicians working on wind turbines.

b. Quantitative Analysis:
   - A wind farm operator reported a 10% increase in annual energy production after implementing PdM. Another wind farm saw a 20% decrease in maintenance costs thanks to PdM's predictive capabilities.

c. Lessons Learned:

Figure 9. Robotic arm in a factory connected to sensors monitored on a screen

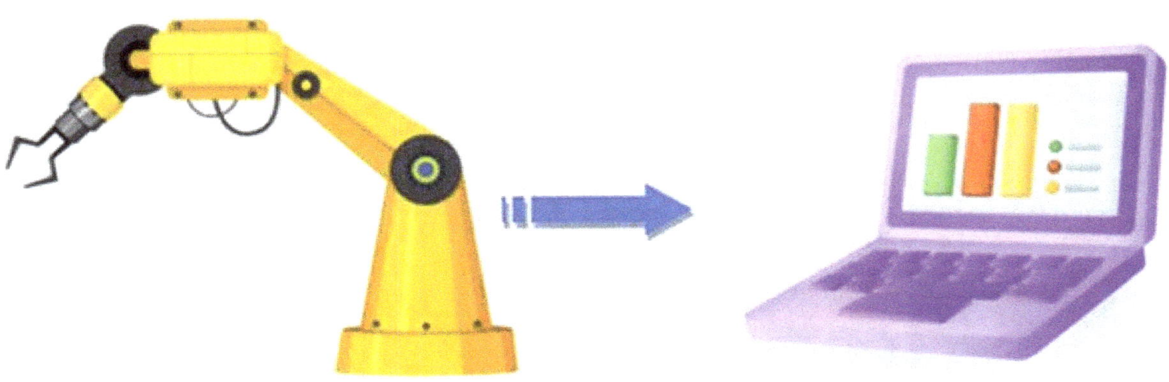

Table 7. Wind farms weathering the elements with PdM

Industry	Application	Challenge	Solution
Renewable energy	Wind turbine health monitoring	Wind turbines operate in harsh conditions, making them susceptible to wear and tear. Identifying potential issues early is crucial for preventing costly repairs and ensuring consistent energy production.	Wind turbine blades and gearboxes are equipped with sensors that collect data on vibration, temperature, and oil levels. PdM algorithms analyze this data to predict potential component failures and schedule preventive maintenance.

◦ Choosing sensors that withstand harsh weather conditions is essential for reliable data collection in wind farm environments. Collaboration with wind turbine manufacturers and data analysts is crucial for developing and implementing effective PdM systems tailored for specific turbine models.

4. THE DAWN OF MACHINE-DRIVEN MAINTENANCE: USHERING IN INDUSTRY 4.0

4.1. Predictive maintenance (PdM) isn't just a buzzword; it's the cornerstone of a transformative revolution – Industry 4.0. Imagine factories where machines seamlessly talk to each other, anticipating needs and optimizing operations through the magic of data and automation. PdM lies at the heart of this shift, paving the way for a new era of machine-driven maintenance.

4.2. The Convergence of Powerful Forces

Three key forces are fueling this revolution:

a. **Data Analytics:** PdM harnesses the power of sensor data, analyzing vast quantities of information to predict potential failures before they occur. Algorithms sift through vibration patterns, temperature changes, and other subtle cues, providing early warnings and actionable insights.
b. **Internet of Things (IoT):** A network of interconnected sensors and devices forms the nervous system of this intelligent ecosystem. They capture data in real time, transmitting it to central hubs for analysis and decision-making, blurring the lines between the physical and digital realms.
c. **Artificial Intelligence (AI):** Machine learning algorithms learn from historical data and real-time insights, constantly refining their predictive models. This self-learning capability lets PdM systems adapt to changing conditions and evolve over time, becoming increasingly accurate and efficient.

Figure 10. A wind turbine data analyzed from sensors on a laptop overlooking a wind farm

4.3. Shaping the Digital Factory of Tomorrow

The implications are profound:

a. **Enhanced Efficiency and Productivity:** Downtime becomes a relic of the past as PdM proactively prevents breakdowns, ensuring smooth production flows and maximized output.
b. **Reduced Costs and Optimized Resources:** Targeted maintenance eliminates unnecessary interventions, minimizes spare parts consumption, and optimizes resource allocation, leading to substantial cost savings.
c. **Improved Safety and Reliability:** Anticipating and addressing potential issues before they escalate minimizes the risk of accidents, environmental damage, and costly equipment failures.
d. **New Business Models and Services:** Data-driven insights unlock new ways to serve customers, offering predictive maintenance as a service or optimizing maintenance plans for maximum efficiency.

4.4. The Future of PdM

The journey is far from over. Expect to see:

a. **Integration with other Industry 4.0 technologies:** PdM will seamlessly integrate with digital twins, robotics, and process automation, leading to fully autonomous and adaptive factories.
b. **Advanced analytics and AI capabilities:** New algorithms and deep learning techniques will further refine prediction accuracy and enable even more granular insights into equipment health.
c. **Cybersecurity advancements:** Robust security measures will be crucial to protect sensitive data and ensure the integrity of this interconnected environment.

4.5. Embrace the Era of Machine-Driven Maintenance

PdM is no longer a futuristic fantasy; it's a tangible reality within reach. By embracing this transformation, you can:

a. **Boost your competitiveness:** Stay ahead of the curve in an increasingly data-driven economy.
b. **Optimize your operations:** Achieve unprecedented levels of efficiency, productivity, and cost savings.
c. **Empower your workforce:** Provide your people with the data and tools they need to make informed decisions and drive continuous improvement.

Remember, the future belongs to those who embrace change. So, step into the era of machine-driven maintenance and unlock the true potential of Industry 4.0 for your business.

5. CONCLUSION

As we near the culmination of this journey through the transformative convergence of data analytics and predictive maintenance, a clear vision emerges the dawn of a new era where machines anticipate our needs, optimizing themselves and our operations. This chapter has explored the potent cocktail of factors igniting this revolution: the ubiquitous flow of machine data, the limitations of reactive approaches, and the transformative promise of PdM. We delved into the power of data analytics, dissecting its role in collecting, extracting insights from, and ultimately predicting the whispers within the vast chorus of machine language. The challenges of integrating these technologies into existing systems and navigating cultural shifts weren't ignored, but rather, viewed as hurdles to be surmounted on the path to progress. But the true impact of PdM lies in its tangible rewards – the fruits of a proactive approach. From slashed downtime and optimized resource utilization to enhanced safety and extended equipment lifespans, the benefits ripple across operational landscapes, painting a picture of increased efficiency and robust resilience. We even glimpsed the future, where Industry 4.0 takes full flight, powered by seamless collaboration between interconnected machines, driven by the ever-evolving insights gleaned from data. Yet, the journey doesn't end here. It's an ongoing quest for deeper understanding, more refined algorithms, and ever-tightening integration with the industrial symphony. The pioneers who embrace this data-driven future, who nurture the convergence of analytics and maintenance, will stand at the forefront of a revolution that redefines the very meaning of operational excellence.

So, as you close this chapter, remember – the power lies not just in the data, nor in the algorithms, but in the transformative leap you take by harnessing both. Be the conductor of your industrial orchestra, where machines dance to the tune of data-driven insights, ushering in an era of proactive efficiency and unparalleled success.

REFERENCES

Bousdekis, A., Lepenioti, K., Ntalaperas, D., Vergeti, D., Apostolou, D., & Boursinos, V. (2019). A RAMI 4.0 View of Predictive Maintenance: Software Architecture, Platform and Case Study in Steel Industry. *International Conference on Advanced Information Systems Engineering CAiSE 2019: Advanced Information Systems Engineering Workshops*, 95-106.

Chang, C., & Lin, C. (2011). LIBSVM: A library for support vector machines. *ACM Transactions on Intelligent Systems and Technology, 2*(3).

Chen, Q., & Hsu, M. (2001). Inter-enterprise collaborative business process management. *Proceedings 17th International Conference on Data Engineering*. 10.1109/ICDE.2001.914836

Engel, Y., Etzion, O., & Feldman, Z. (2012). A basic model for proactive event- driven computing. *Proceedings of the 6th ACM International Conference on Distributed Event-based Systems,* 107-118. 10.1145/2335484.2335496

Fdhila, W., Rinderle-Ma, S., Knuplesch, D., & Reichert, M. (2015). Change and compliance in collaborative processes. IEEE SCC, 162–169. doi:10.1109/SCC.2015.31

Gao, J. (2019). *Machine Learning Applications for Data Center Optimization*. Available at: https://static.googleusercontent.com/media /research.google.com/en//pubs/ archive/42542.pdf

Guillén, A. J., Crespo, A., Gómez, J. F., & Sanz, M. D. (2016). A framework for effective management of condition based maintenance programs in the context of industrial development of E-Maintenance strategies. *Computers in Industry*, *82*, 170–185. doi:10.1016/j.compind.2016.07.003

Hribernik, K., von Stietencron, M., Bousdekis, A., Bredehorst, B., Mentzas, G., & Thoben, K. D. (2018). Towards a unified predictive maintenance system-a use case in production logistics in aeronautics. *Procedia Manufacturing*, *16*, 131–138. doi:10.1016/j.promfg.2018.10.168

Lee, J., Ardakani, H. D., Yang, S., & Bagheri, B. (2015). Industrial big data analytics and cyber-physical systems for future maintenance & service innovation. *Procedia CIRP*, *38*, 3–7. doi:10.1016/j.procir.2015.08.026

Lee, J., Kao, H.-A., & Yang, S. (2014). Service innovation and smart analytics for industry 4.0 and big data environment. *Procedia CIRP*, *16*, 3–8. doi:10.1016/j.procir.2014.02.001

Lee, K.B., Cheon, S., & Kim, C.O. (2017). A Convolutional Neural Network for Fault Classification and Diagnosis in Semiconductor Manufacturing Processes. *IEEE Transactions on Semiconductor Manufacturing, 30*(2), 135-142.

Li, L., Bagheri, S., Goote, H., Hasan, A., & Hazard, G. (2013). *Risk adjustment of patient expenditures: A big data analytics approach*. Paper presented at the Big Data, 2013 IEEE International Conference on. 10.1109/BigData.2013.6691790

Lin, C. C., & Tseng, H. Y. (2005, January). A neural network application for reliability modelling and condition-based predictive maintenance. *International Journal of Advanced Manufacturing Technology*, *25*(1–2), 174–179. doi:10.1007/s00170-003-1835-3

Makhdoom, I., Abolhasan, M., Abbas, H., & Ni, W. (2019, January). Blockchain's adoption in IoT: The challenges, and a way forward. *Journal of Network and Computer Applications*, *125*(1), 251–279. doi:10.1016/j.jnca.2018.10.019

Matsubara, A., & Ibaraki, S. (2009). Monitoring and control of cutting forces in machining processes: A review. *International Journal of Automotive Technology*, *3*(4), 445–456. doi:10.20965/ijat.2009.p0445

Muhtaroglu, F., Demir, S., Obali, M., & Girgin, C. (2013). *Business model canvas perspective on big data applications*. Paper presented at the Big Data, 2013 IEEE 10.1109/BigData.2013.6691684

Peng, Y., Dong, M., & Zuo, M. J. (2010). Current status of machine prognostics in condition-based maintenance: A review. *International Journal of Advanced Manufacturing Technology*, *50*(1–4), 297–313. doi:10.1007/s00170-009-2482-0

Sikorska, J., Hodkiewicz, M., & Ma, L. (2011). *Prognostic modelling options for remaining useful life estimation by industry*. Mechanical Systems and Signal Pro. doi:10.1016/j.ymssp.2010.11.018

Sikorski, J. J., Haughton, J., Kraft, M., Street, P., & Drive, P. F. (2016, June). Blockchain technology in the chemical industry: Machine-to-machine electricity market. *Applied Energy*, *195*(178), 234–246.

Soori, M., Arezoo, B., & Habibi, M. (2013). Dimensional and geometrical errors of three-axis CNC milling machines in a virtual machining system. *Computer-Aided Design*.

Sundararajan, A. (2014). Peer-to-peer businesses and the sharing (collaborative) economy: Overview, economic effects and regulatory issues. *The Power of Connection: Peer to Peer Businesses*.

van der Aalst, W. (2000). Process-oriented architectures for electronic commerce and inter organizational workflow. *Information Systems*, *24*(8), 639–671. doi:10.1016/S0306-4379(00)00003-X

Voisin, A., Levrat, E., Cocheteux, P., & Iung, B. (2010). Generic prognosis model for proactive maintenance decision support: Application to pre-industrial e- maintenance test bed. *Journal of Intelligent Manufacturing*, *21*(2), 177–193. doi:10.1007/s10845-008-0196-z

Wan, J., Tang, S., Shu, Z., Li, D., Wang, S., Imran, M., & Vasilakos, A. V. (2016). Software-Defined Industrial Internet of Things in the Context of Industry 4.0. *IEEE Sensors Journal, 16*(20).

Wielki, J. (2013). *Implementation of the big data concept in organizations-possi-bilities, impediments and challenges.* Paper presented at the Computer Science.

Wu, S.-j., Gebraeel, N., Lawley, M. A., & Yih, Y. (2007). A neural network integrated decision support system for condition-based optimal predictive maintenance policy. *Systems, Man and Cybernetics, Part A: Systems and Humans. IEEE Transactions on, 37*(2), 226–236.

Xin, N. Y., & Ling, L. Y. (2013). *How we could realize big data value*. Paper presented at the Instrumentation and Measurement, Sensor Network and Automation (IMSNA), 2013 2nd International Symposium on.

Yan, H., Wan, J., Zhang, C., Tang, S., Hua, Q., & Wang, Z. (2018). Industrial Big Data Analytics for Prediction of Remaining Useful Life Based on Deep Learning, Real-Time Edge Analytics for Big Data in Internet of Things. *IEEE Access, 6*, 17190 – 17197.

Yan, J., Meng, Y., Lu, L., & Li, L. (2017). Industrial Big Data in an Industry 4.0 Environment: Challenges, Schemes, and Applications for Predictive Maintenance, Complex System Health Management based on Condition Monitoring and Test Data. *IEEE Access, 5*, 23484 – 23491.

Zhang, Y., Ren, S., Liu, Y., & Si, S. (2017). A big data analytics architecture for cleaner manufacturing and maintenance processes of complex products. *Journal of Cleaner Production 142*(2), 626-641.

Chapter 4
Cloud Robotics and Automation:
A Comprehensive Study on Techniques, Taxonomy, Applications, and Research Issues

SandeepKumar Hegde
Nitte University, India

Rajalaxmi Hegde
https://orcid.org/0000-0001-5610-8748
Nitte University, India

Thangavel Murugan
United Arab Emirates University, UAE

ABSTRACT

Cloud robotics is currently regarded as one of the cutting-edge research areas within the mainstream automation and artificial intelligence fields. The new machine-to-cloud communication will increase the productivity and efficiency of industrial robots. For decades, automated devices have been developing and affecting practically every element of life. Cloud computing and associated technologies have a great deal of potential to get around hardware limitations and boost performance. This study emphasises the developments in robotic technology with a particular focus on cloud-based automation as a new development. Unexpected effects of robotic technologies on human life have been seen. This chapter discusses the fundamental ideas, the creation process, and the general architecture of cloud robotics in order to explore the potential of clouds for improving robotics for industrial systems. Finally, it examines how useful cloud robotic systems might be in various real-world scenarios.

DOI: 10.4018/979-8-3693-1914-7.ch004

1. INTRODUCTION

The improvement in the development of robots has rapidly increased in the industries, and the result causes the machines to become configured to reach a improved level of efficiency with features such as good accuracy, robustness and good compatability. Robotic automation provides a crucial role in executing the everyday tasks which are carried out on a regular basis. Robots are used to construct a network known as a cluster of robots, where each robot serves as point of data, and the robots are connected by wireless or wired communication (Herm et al., 2023). To do tasks more efficiently, the cloud robots are outfitted with making decisions and number-sensing components.

Over the past few decades, robotics has undergone significant advancements, leading to an increase in the applications of this technology to a variety of real-world issues, such as automated manufacturing, extraterrestrial operations, unsupervised operations the catastrophe devices, self-movable cars, socially aligned robots, and medical care and other specific featured robots. The main intention of upon-board hardware and computational limitations of the single robots utilized in many of these applications place restrictions on their capabilities. The area of network robotics first evolved approximately two decades ago to address this issue, connecting group of robots using secured as well wired or unsecured network (Udendhran et al., 2023). By exchanging perceived data and doing tasks collaboratively and/or in unison, networked robots solve the issue with isolated robot systems. Network enabled robots, however, would experience many issues as one hand robot systems.

Resource restrictions, communication information as well other learning constraints, and communication limits are the key issues with networked robots. The compute and storage capabilities of each robot are constrained by its technology, which has implications for collective robots connected across a network (Dittakavi, 2022). Once a robot has been conceived, constructed, and put into service, it is technically challenging to update or change available device configurations. Robots connected over a network are likewise restricted by information because they can only access the collected data using other robots with similar sensors and a similar network (Vermesan et al., 2022). This can usually result in improved performance in static settings; nevertheless, their performance rapidly deteriorates when the environment abruptly changes or if they are placed in unusual circumstances.

The combination of robotics and automation with cloud computing in recent times has surfaced as a revolutionary concept, radically altering the functionalities and capacities of conventional robotic systems. This thorough analysis explores the ever-changing topic of cloud automation and robotics, including information on methods, classification, uses, and future directions for the field. Robotics systems can overcome the constraints of onboard computing power and memory by utilising the enormous computing and storage capacities of the cloud, providing improved sensing, cognition, and decision-making abilities. This paper classifies several cloud robotics systems and methodologies, such as internet-based control, data storage, compute offloading, and collaborative learning, using a systematic taxonomy.

It also highlights the revolutionary power of cloud-enabled robots across domains by examining a wide range of industries spanning manufacturing, healthcare, the agricultural sector, disaster response, and beyond. Notwithstanding the encouraging progress, a number of research obstacles still exist, encompassing security and privacy apprehensions, latency problems, and scale constraints. Through the resolution of these issues, scientists may fully realise the endless possibilities of cloud robotics, opening the door to the creation of robotic systems that are scalable, intelligent, and adaptable to a variety of real-world situations. This study promotes creativity and advances this quickly developing subject by providing a fundamental resource for academics, professionals, and hobbyists attempting to negotiate the challenging terrain of cloud robots and automation.

Cloud robotics has the potential to revolutionise many different sectors and applications, but in order to reach its full potential, a number of issues need to be resolved. Through technological advancements, problem solving, and future direction exploration, researchers can set the stage for cloud robots to become widely used in the years to come. Investigations into collaborative robotics and human-robot interaction (HRI) to facilitate more natural and effective human-robot cooperation. To open up new possibilities in cloud robotics, interdisciplinary themes like swarm robotics, bio-inspired robotics, and cognitive robots are being investigated. The objective of a chapter on cloud robotics and automation techniques, challenges, and future issues is to provide a comprehensive understanding of the current state of the field, including its methodologies, hurdles, and potential directions for advancement. Here are the main objectives of such a chapter: Leverage cloud computing platforms' resources to enable robotic systems' smooth scalability, which will enable the deployment of massive robot fleets and the management of a variety of workloads. Optimize performance and reduce operating costs by allocating compute, storage, and bandwidth to robotic activities and processes in an efficient manner. Leverage the computational capacity of cloud servers to enable real-time processing and analysis of sensor data from robots, enabling prompt decision-making and adaptive behavior. Give operators the ability to remotely access and manage robots via cloud-based interfaces. This will allow them to keep an eye on the actions of the robots, troubleshoot any issues, and take required action from any location with an internet connection. By utilising cloud-based AI and machine learning algorithms for perception, decision-making, and learning from data, you can increase the intelligence and autonomy of robots and empower them to adapt to changing environments.

The focus of cloud robotics and automation problem description is on optimising robotic systems' capabilities, efficiency, and flexibility through the use of cloud computing resources. This entails tackling a range of possibilities and difficulties, such as but not restricted to: effectively supporting robotic tasks using cloud resources, minimising delay and cost, including computing, storage, and communication bandwidth. using cloud-based computing power and data analytics capabilities to empower robots to make wise judgements in real-time guaranteeing smooth communication and compatibility between robotic systems and cloud services, along with standardising integration interfaces and protocols. creating systems that are reliable and perform well even when they are scaled to support a huge number of users and robots.

The organization of the chapter as follows: where topics 1 serves as the introduction and objectives and problem definition, topic 2 delves into the literature review, topic 3 discusses about the features and benefits of cloud robotics, topic 4 discusses about importance of robotics and cloud computing, challenges with cloud robotics then followed by conclusion.

2. RELATED WORK

Cloud robot automation is already a well-known field that combines the two rapidly evolving ideas of robotics and computing in the cloud. Networked robots use protocols to exchange information like dynamically route or spontaneous routing to maintain device-to-device connections (Bouziane et al., 2022). Currently, the robotic service offers services for perceiving, acting, and controlling. Due to their high resilience, speed, and precision, robots can now be used for a variety of laborious and potentially hazardous activities. The highly functional digital robot performs a variety of activities, involving navigational duties, digital printing, scenario analysis, and rescue missions as well as healthcare (Siripurapu et al., 2023). Despite recent developments, it is still possible to quickly supply and release of the available system resources which includes servers, storage devices and networks. with little involvement from the service provider or administration work. On the other hand, networked robotics faces other limitations, including those related to resources, skill acquisition, and communication.

The robot can extend its capabilities and boost system performance by using the cloud. In this paper the author explained that by utilizing cloud computing technologies, networked robotics can get over these constraints. One of the most intriguing uses of cloud robotics is the concept of the smart city (Ichnowski et al., 2023). According to (Bellini, Nesi, & Pantaleo, 2022). this is a city that uses technology for communication and information to enhance public awareness, interaction, and the efficiency of the services it provides and monitoring. A smart city combines its social, business, communication technology, and physical infrastructure to tap into the city's collective knowledge (Ibrahim Khalil & Kamel, 2023). IoT and cloud innovations will enable city infrastructure to provide its residents a wide range of services.

Cloud resources can be used for cloud robotics platforms by outsourcing difficult and time-consuming compute operations for on-demand services (Mello, Ribeiro, & Frizera-Neto, 2022). Any system's performance level is maximized while task completion times are reduced as a result. Transferring a task to the cloud is not, however, as simple as it might seem. It is actually somewhat complicated since it necessitates a careful balance between computer connectivity cost and cost with regard to transmission. Therefore, it is very important for any process to identify the proper tasks to understand and to offload. Presenting a robotics analysis with framework that permits the best job offloading is one of the goals of this work.

Robotics, computing via the cloud, and artificially intelligent (AI) innovations come together in cloud robotics and automation. By utilising the processing power and data storage of cloud systems, this multidisciplinary field seeks to improve robot capabilities (Buerkle et al., 2023). The design, development, and control of physical robots to carry out particular tasks autonomously or semi-autonomously are the main emphasis of traditional robotics. This covers topics including motion planning, control algorithms, sensor integration, and robot kinematics and dynamics. Pay-as-you-go computer services (such as processing, networking, and storage) are delivered over the internet in the context of cloud computing. Scalability, adaptability, and accessibility are offered by cloud platforms, which let users access resources whenever they need them from any location with an internet connection (Militano et al., 2023). A key factor in increasing the intellect and autonomy of robots is artificial intelligence (AI) technology, such as machine learning, natural language processing, computer vision, and decision-making algorithms. Robots with artificial intelligence (AI) can see and comprehend their surroundings, learn from data, make decisions, and adjust to changing circumstances.

Robotics and cloud computing are combined in cloud robotics to enable robots to do tasks beyond their hardware constraints (Sharma, Tomar, & Hazra, 2024). Robots can take advantage of the massive computational resources that are accessible remotely by shifting labour-intensive computational operations to the cloud, such as data processing, analysis, and decision-making. Robotics and cloud computing are combined in cloud robotics to enable robots to do tasks beyond their hardware constraints. Robots can take advantage of the massive computational resources that are accessible remotely by shifting labour-intensive computational operations to the cloud, such as data processing, analysis, and decision-making.

3. FEATURES AND BENEFITS OF CLOUD ROBOTICS

Robotic features that can access the cloud have high storage capacities that they can employ to retain all of their important data for later use (Kunduru, 2023). IoT systems can be thought of as a structure that makes cloud automation possible because they are based on protocols of communication that allow physical assets to talk with other assets and transfer and share information. This can usually result in improved performance in static settings; nevertheless, their performance rapidly deteriorates when the environment abruptly changes or if they are placed in unusual circumstances. Numerous robotics applications, including continuous translation, mapping, and other detecting systems, quickly produce large amounts of sensory data that are hard to store due to the restricted storage capacity on board.

Along with these benefits, cloud robotics gives robots access to huge data, including open-source algorithms and code, worldwide maps for translation, object models for potential manipulation tasks, and global maps for localization (Garg et al., 2020). Lastly, cloud robotics enables cooperative learning across spatially dispersed robots by exchanging knowledge on how to complete challenging tasks. Cloud robotics leverages cloud crowdsourcing techniques to give robots access to human expertise. With the development of the cloud robotics, robot are no more standalone machines with pre-programmed functions.

A few of the more common uses of robotics that have evolved in recent decades are automation in producing goods, interplanetary processes, unmanned rescue operations, the catastrophic devices self-movable motor vehicles, environmentally enabled automated machinery, medically enabled treatment and health robots. The upon-board circuits and processing capabilities place restrictions on the single devices utilized in many of those applications. The area of networked robotics (Romeo et al., 2020) first evolved to address this issue roughly 20 years ago, connecting a collection of robots via a wired and portable web connection. With the help of cooperative and/or planned task completion and data communication, networked robots solve the issue with isolated robot systems. Even with recent advancements, it is still possible to offer and release computer resources (such networks, storage units, servers, and services) fast and with little administrative labor or service provider involvement. Networked robots, on the other hand, has other constraints, such as those pertaining to resources, skill gain, and communication. By utilizing the cloud, the robot can increase system performance and expand its capabilities. In fact, in paper (Fosch-Villaronga & Millard, 2019) it asserts that cloud computing technologies can help networked robotics overcome these limitations. One of the most fascinating uses of cloud robot and automation is the development of smart city paradigm.

The usage of social networking sites, the World Wide Web, and sensing in cyber physical systems are what define the World Wide Web of things. The Internet of things is not designed to identify data or gather data systems. Robots with cloud capabilities can save all the essential data in the vast storage areas that the cloud offers. In complement to these advantages, cloud robotics provides robots with access to a vast amount of information, such as open-source procedures and code, maps of the world for the purpose of practicing localization, models of objects that the computers may require in order to do jobs, and more data.

We discovered through our research in this area that the two main categories of cloud robotics research are robotics cloud applications and cloud robotics infrastructure system architectures. The first category comprises several system designs which allow cloud the field of robotics such as architectures of software for robotics and cloud services as well as usage guidelines and languages for communication between robots and the cloud. Numerous robotics and automation-related applications have also used cloud robotics. The second category explains about the robotics communication along with its features.

4. ROBOTICS AND CLOUD COMPUTING

Informally speaking, a group of processes that send online information to distributed and shared computing assets, such as those used for handling, storing, and massive datasets, can be referred to as cloud robotics and computation. Specialized frames or service models are required to enable the migration of robotic applications to the cloud (Afrin et al., 2021). The cloud infrastructure is made up of numerous high-performance servers, proxy servers, and enormous databases, all of which have the capacity to handle extremely fast processing and computing as well as vast amounts of storage.

A. Cloud Robotics Architecture

Cloud robotics is based on an artificial random cloud consisting of a centralized cloud and several networked robots. Because of this unique blend, which is appropriate for specific applications, we have a great deal of design freedom when developing computing models. Our main focus is on the following three models of elastic computing.

- **Peer-Based Model:** Every cloud virtual machine (VM) and robot is regarded as a single computing unit. Together, these VMs and robots make up a completely distributed computing mesh. It is possible to break up a task into smaller modules that can be executed over a portion of the computing mesh's nodes.
- **Proxies-Based Model:** One robot in a group of networked robots serves as the team leadership and connects with the proxy virtual machine, or VM, in the cloud architecture to enable interaction among a robotics network and the cloud. Two tiers make up the hierarchy that organizes the collection of computer units.
- **Clone-Based Model:** Each robot is the same as a system-level clone operating on the cloud. The robot can assign a task to its clone or to itself. Furthermore, a peer-to-peer network formed by the robotic clones is more interconnected than the original ad-hoc machine to machine system. Moreover, this machine to system model provides feature for recurring failures in the real machine to machine network.Each one of these flexible computing systems provides different qualities, including mobility management, interoperability, and adaptation to network connections.

Robustness is the network's link between the centrally situated cloud infrastructure and the group of connected robots. When it comes to network connectivity, the peer-based system is the most reliable, while the proxy-based technique is the least. Given that it has the greatest number of robots linked to the cloud, the clone-based approach is the most reliable.

Interoperability is the additional labor required to operate a cloud robots network in conjunction The clone-based method is the least interoperable, while the proxy-based system is the most due to its hierarchical structure. Mobility is defined as the ability to support mobile robots. The peer-based strategy provides the highest mobility flexibility because virtual machines can be established throughout the robot cloud architecture. On the other hand, because sophisticated migration of virtual machine the procedures are required to allow robot mobility, the clone-based architecture provides the least flexibility. Three main aspects influence the choice of models for elastic computing: network conditions, application requirements, and resource availability.

B. Cloud Infrastructure and Management

The set of hardware and software components that make up cloud infrastructure include the networking, processing, storage, and virtualization capabilities needed to enable cloud computing. Cloud infrastructure types often feature a user interface for controlling such virtual resources. Figure 1 shows some of the applications and the general structure of cloud robotics systems.

With cloud Infrastructure a group or business can purchase the computing infrastructure they require online, including storage, load balancing, firewalls, and processing power (on real or, more frequently, virtual servers). Instead of maintaining and provisioning their own physical infrastructure, they do this. Rather, they rent the IaaS provider's resources as needed. Although this is an established instance, secure and hybrid clouds are just a small part of the wider spectrum of platforms and settings that make up the infrastructure of clouds, or cloud architecture.

Physical robots having sensors to sense their surroundings and gather information. Proxy Server serves as a bridge that facilitates communication, data processing, and management features between the robots and the cloud. Communication management is responsible for managing the transfer of data between robots and the cloud, guaranteeing dependability, safety, and effective use of available network resources (Touhid, et al., 2023). Data retrieval coordinates communication with cloud services, preprocesses sensor data, and compiles information from several robots. Database holds and controls permanent information about cloud robotics activities, including task assignments, robot status, logs,

Figure 1. Cloud robotics system architecture

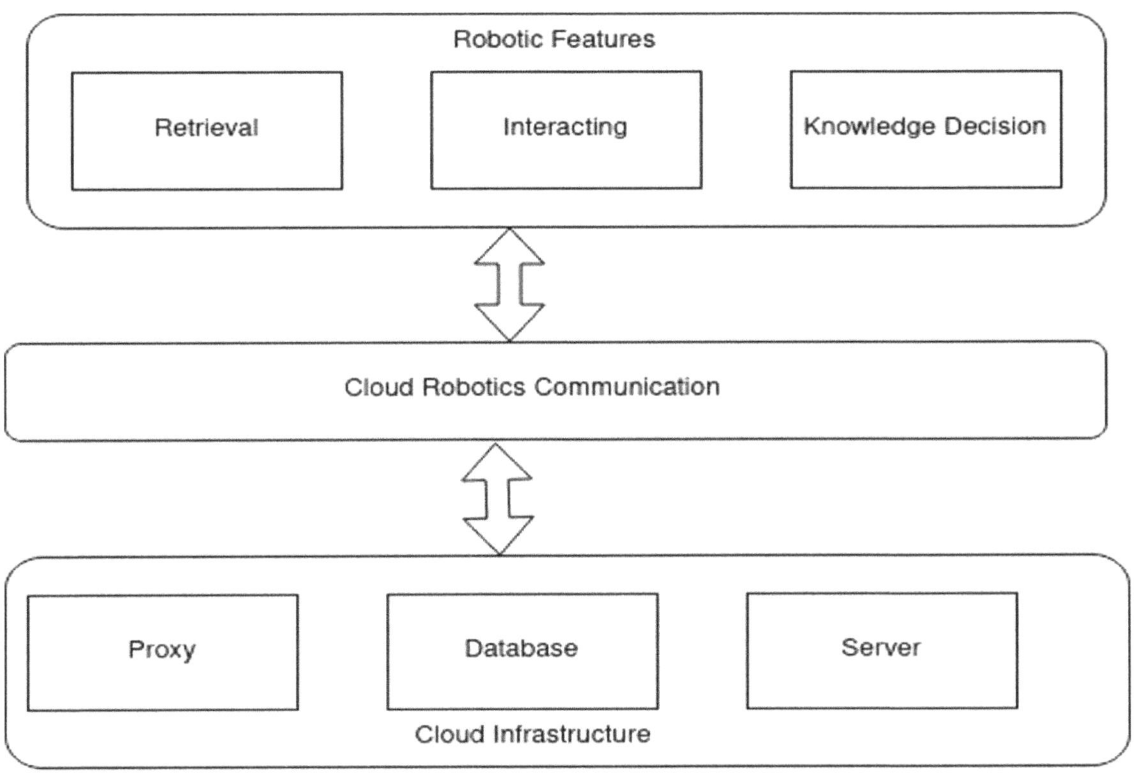

and historical information. Resources for Remote Computing offers cloud-based functions such as AI services, storage, scalable computing resources, and other services for processing sensor data, executing AI algorithms, storing data, and coordinating robotic actions.

5. CHALLENGES WITH CLOUD ROBOTICS

The current framework has intrinsic resource, information, and communication limits that impede the progress of networked robotics. In this part, we go through these restrictions in more detail.

1. Limitations on Resources and Scheduling

Even though a robot may divide its computing effort with different components in the network, the combined effectiveness of all the robots in the network is limited by their resources, which include memories, storage space, and internal processors or computer units. The dimensions, shape, source of power, moving style, and operational environment of the robots impose physical constraints on these onboard computing systems. Upgrading or changing the resource configurations of robots poses a technical challenge once they have been developed, constructed, and deployed.

2. Computing Difficulties

The capability of cloud robotics to assign computationally demanding jobs to the cloud for completion is one of its main advantages. However, the decision to assign a certain task necessitates a well-coordinated framework capable of handling a wide range of complex issues. First and foremost, the offloading strategy ought to consider several factors, including the amount of data transferred and the task's end date. Second, given the availability of cloud resources, the choice should also take into account if it makes more sense to carry out the task among the networked robots. Finally, managing live VM migrations and allocating virtual machines appropriately to carry out the activity that has been offloaded present a problem given the amount of cloud resources distributed among many data centers. To ascertain the best practices for job execution, we support the use of an optimization framework. Our goal is to reduce the robot's energy consumption as much as possible while still completing the work by the designated date.

We have taken into account the two different options of robot execution on its own and cloud execution in our preliminary inquiry. The task's delay sensitivity determines whether to execute it in the cloud or stand-alone, as was covered in the previous section. The computation request's communication time with the cloud must be considered when making the choice. We give upper boundaries in this section for the amount of time needed to send a message with a high probability. All wireless communication systems have the inherent problem of packet delivery errors and communication outages. It is possible to think of a gossip-based communication network as an architecture with more failures but much reduced overheads. Additionally, the network architecture determines the resulting increment in failure rate, and this can be efficiently controlled by using the internet as a hyper node in the network.

3. Information interpretation

A robot's computational power, storage capacity, and the kind and quantity of sensors it has placed all limit the volume of information it can access. Networked robotics uses a communication network to allow robots to exchange information and cooperate to finish a massive calculation or a global task. Nevertheless, networked robotics has limitations in its ability to learn from examples or scenarios that the network encounters, as well as by data that robots within the network observe or compute. A mechanical team studying navigation might do very well in a fixed context, where any challenges can be drawn out with improved precision over time.

4. Focuses on Communication

Proactive routing, which entails routine message exchanges to ensure that routes to every potential location in the network are maintained, and ad hoc routing, which creates dynamic routes to destination nodes only when messages need to be sent, have become standard protocols for M2M, or machine-to-machine, communications. The route creation and upkeep process of proactive routing requires a significant amount of processing power and memory. Since a route must be created prior the message can be transmitted, ad hoc routing methods have a high latency and are impractical in highly dynamic network topologies. These issues are important for mobile robotic networks and have the potential to seriously impair performance.

The capabilities, effectiveness, and application of robotic systems connected with cloud computing infrastructure are the goals of a broad range of research concerns and methodologies in cloud robotics. The research issues and challenges as follows Creating methods for effectively scaling robotic systems on the cloud so that many users and robots can be integrated while performance and dependability are maintained. Reducing the latency caused by communication between robots and the cloud, especially for apps that need to respond and make decisions quickly. Maximising the use of cloud resources, including computing, storage, and bandwidth, to enable robotic tasks with the least amount of expense and latency. Ensuring the security and privacy of robotic systems and their interactions with cloud services, including data encryption, secure communication protocols, and access control mechanisms.

Establishing common standards and protocols for communication and integration between robotic systems and cloud platforms, enabling seamless interoperability and collaboration. Developing adaptive control algorithms, machine learning techniques, and decision-making frameworks to enable robots to operate autonomously and adapt to changing environments and tasks.

6. CLOUD ROBOTICS APPLICATIONS

Today, cloud-based robotics are used in a variety of applications, including home automation, healthcare, military use, and more. By optimizing the automation process in terms of latency, processing speed, and access control, the artificial integration in cloud-based robotics will further improve the system's performance. Cloud robotics offers the following benefits over traditional networked robots, which will be useful for future robotic applications (Dawarka & Bekaroo, 2022). Cloud robotics offers multiple applications in data-heavy or computation-intensive jobs in the fields of intelligent transportation, smart homes, health care, education, entertainment, and military because of these benefits. The advantages and difficulties that cloud robotics presents for conventional robotic applications are covered in this section.

- **The capacity to transfer computationally demanding jobs to the cloud.**

All that the robots need to maintain are the essential sensors, actuators, and minimal processing power to allow for real-time operations (such real-time control). Longer battery life results in a lighter, less expensive robotic platform with easier-to-maintain gear. Software onboard the robots also requires less frequent upgrades, making maintenance easier. The operating life and functionality of the robotic network can be readily extended because the cloud hardware may be upgraded separately from the robotic network.

- **Having access to enormous data sets**. Through cloud databases, the robots may obtain the knowledge and information they need to complete tasks. They are spared from handling the collection and upkeep of such data.
- **Availability of new abilities and pooled knowledge.** The robots can exchange information and pick up new abilities and expertise from one another thanks to the cloud. A record-keeping system or library of behaviors or abilities that correspond to various work needs and environmental complexity can be hosted in the cloud.

7. ENHANCING ROBOTICS CLOUD COMPUTING

Several features help enhance the features of robotics and the cloud computing they are:

1. **Digital Transformation:** To operate your business more effectively, digital transformation includes helping a robot automate a task and provide essential data back. In order for the services to absorb the data and generate the insights it is needed to manage the business more effectively, and it will require a long-term plan.
2. **Cloud robotics innovations:** It was found that internet-based robotics would produce intelligent robots with reduced power consumption and increased processing efficiency. Since it is going to be less hardware and emissions, these features will result in cheaper production costs. The innovation in cloud robotics has gained momentum because of the efforts of major firms like IBM and Google, as well as the involvement of academic institutes in certain global ongoing projects. The main goal of research efforts has been to create robots that are both highly effective and easily accessible.
3. **Supervised Autonomy:** Even though the heads of these robots are in the clouds, a brain could still be present. Each robot is an intelligent, self-contained device. Mobile robots must safely maneuver factories, warehouses, and supply centers while avoiding people, forklifts, and other obstacles.
4. **Faster Deployment:** The cloud robotic platform is the first autonomous mobile machine (AMR) choice for carrying materials and data collection that installs in a matter of hours as opposed to months and doesn't need immediate integration to prove benefits. This type of automation is called on-demand. An automobile company's fleet of autonomous mobility robots that pull carts is efficiently managed via a cloud software platform at an external logistics warehouse.

CONCLUSION

In order to overcome the limitations that the existing networked robots have, we have put forth a cloud robotics architecture. Through cloud robotics, robots can share data, information, and computational resources as well as get access to new skills and knowledge that they would not have otherwise acquired. This sets up a new robotics paradigm that, in our opinion, will result in fascinating new advancements. Cloud infrastructure's elastic computing resources and communications network, it enables the deployment of low-cost robots with low compute and memory requirements. The utilization of distant computing resources to increase memory, processing capacity, group learning, and connection for robotics applications were important aspects of cloud robotics. It supports several features such as robust connectivity, storage, and processing capabilities of contemporary data centers etc .Cloud robotics use is increasing in several growing industries, including logistics, safety and surveillance, personal help and care, guidance and learning, pleasure, and affection. Cloud robotics has been created as a result of growing use of technologies that are cloud-based and cloud-based artificial intelligence robotics technology, which has raised demand in the worldwide cloud robotics industry. The globe can be greatly impacted by the late mechanical improvement known as cloud computing. It provides a plethora of benefits to its clients and businesses, it reduces operating expenses for organizations by allocating fewer funds for maintenance and upgrades and by directing more resources on the business itself. Nonetheless, cloud computing faces a variety of challenges. People are apprehensive about the security and privacy of their information. Overall, there are no rules or regulations that provide information via cloud computing. Furthermore, cloud robotics has been created using technologies that are cloud-based and cloud-based artificial intelligence robotics technology, which has increased need in the worldwide cloud robotics market.

REFERENCES

Afrin, M., Jin, J., Rahman, A., Rahman, A., Wan, J., & Hossain, E. (2021). Resource allocation and service provisioning in multi-agent cloud robotics: A comprehensive survey. *IEEE Communications Surveys and Tutorials*, *23*(2), 842–870. doi:10.1109/COMST.2021.3061435

Bellini, P., Nesi, P., & Pantaleo, G. (2022). IoT-enabled smart cities: A review of concepts, frameworks and key technologies. *Applied Sciences (Basel, Switzerland)*, *12*(3), 1607. doi:10.3390/app12031607

Bouziane, R., Terrissa, L. S., Ayad, S., & Brethé, J. F. (2022). Towards an architecture for cloud robotic services. *International Journal of Computers and Applications*, *44*(9), 863–874. doi:10.1080/1206212X.2021.1964790

Buerkle, A., Eaton, W., Al-Yacoub, A., Zimmer, M., Kinnell, P., Henshaw, M., Coombes, M., Chen, W.-H., & Lohse, N. (2023). Towards industrial robots as a service (IRaaS): Flexibility, usability, safety and business models. *Robotics and Computer-integrated Manufacturing*, *81*, 102484. doi:10.1016/j.rcim.2022.102484

Dawarka, V., & Bekaroo, G. (2022). Building and evaluating cloud robotic systems: A systematic review. *Robotics and Computer-integrated Manufacturing*, *73*, 102240. doi:10.1016/j.rcim.2021.102240

Dittakavi, R. S. S. (2022). Evaluating the Efficiency and Limitations of Configuration Strategies in Hybrid Cloud Environments. *International Journal of Intelligent Automation and Computing*, *5*(2), 29–45.

Fosch-Villaronga, E., & Millard, C. (2019). Cloud robotics law and regulation: Challenges in the governance of complex and dynamic cyber–physical ecosystems. *Robotics and Autonomous Systems*, *119*, 77–91. doi:10.1016/j.robot.2019.06.003

Garg, S., Sünderhauf, N., Dayoub, F., Morrison, D., Cosgun, A., Carneiro, G., ... Milford, M. (2020). Semantics for robotic mapping, perception and interaction: A survey. *Foundations and Trends® in Robotics, 8*(1–2), 1-224.

Herm, L. V., Janiesch, C., Helm, A., Imgrund, F., Hofmann, A., & Winkelmann, A. (2023). A framework for implementing robotic process automation projects. *Information Systems and e-Business Management*, *21*(1), 1–35. doi:10.1007/s10257-022-00553-8

Ibrahim Khalil, M. W., & Kamel, M. A. E. (2023). Towards smart sustainable cities vision and challenges. *International Journal of Nonlinear Analysis and Applications.*

Ichnowski, J., Chen, K., Dharmarajan, K., Adebola, S., Danielczuk, M., Mayoral-Vilches, V., ... Goldberg, K. (2023, May). Fogros 2: An adaptive and extensible platform for cloud and fog robotics using ros 2. *Proceedings IEEE International Conference on Robotics and Automation.*

Kunduru, A. R. (2023). Cloud BPM Application (Appian) Robotic Process Automation Capabilities. *Asian Journal of Research in Computer Science*, *16*(3), 267–280. doi:10.9734/ajrcos/2023/v16i3361

Mello, R. C., Ribeiro, M. R., & Frizera-Neto, A. (2022). *Implementing Cloud Robotics for Practical Applications: From Human-robot Interaction to Autonomous Navigation* (Vol. 152). Springer Nature.

Militano, L., Arteaga, A., Toffetti, G., & Mitton, N. (2023). The cloud-to-edge-to-IoT continuum as an enabler for search and rescue operations. *Future Internet*, *15*(2), 55. doi:10.3390/fi15020055

Romeo, L., Petitti, A., Marani, R., & Milella, A. (2020). Internet of robotic things in smart domains: Applications and challenges. *Sensors (Basel)*, *20*(12), 3355. doi:10.3390/s20123355 PMID:32545700

Sharma, M., Tomar, A., & Hazra, A. (2024). Edge Computing for Industry 5.0: Fundamental, Applications and Research Challenges. *IEEE Internet of Things Journal*, 1. doi:10.1109/JIOT.2024.3359297

Siripurapu, S., Darimireddy, N. K., Chehri, A., Sridhar, B., & Paramkusam, A. V. (2023). Technological Advancements and Elucidation Gadgets for Healthcare Applications: An Exhaustive Methodological Review-Part-I (AI, Big Data, Block Chain, Open-Source Technologies, and Cloud Computing). *Electronics (Basel)*, *12*(3), 750. doi:10.3390/electronics12030750

Sood, V. M., Singh, Y., Bhargava, B., & Narang, S. K. (Eds.). (2024). *Intelligent Security Solutions for Cyber-Physical Systems*. CRC Press. doi:10.1201/9781003406105

Touhid, M. T. B., Marne, M., Oskroba, T., Mirahmadi, S. A., Zhu, E., Mehrabian, A., Defersha, F., & Yang, S. (2023). Building a cloud-based digital twin for remote monitoring and control of a robotic assembly system. *International Journal of Advanced Manufacturing Technology*, *129*(9), 4045–4057. doi:10.1007/s00170-023-12611-7

Udendhran, R., Yamini, G., Badrinath, N., Jegathesh Amalraj, J., & Suresh, A. (2023). Enhancing representational learning for cloud robotic vision through explainable fuzzy convolutional autoencoder framework. *Soft Computing*, 1–11. doi:10.1007/s00500-023-08570-6

Vermesan, O., Bröring, A., Tragos, E., Serrano, M., Bacciu, D., Chessa, S., & Bahr, R. (2022). Internet of robotic things–converging sensing/actuating, hyperconnectivity, artificial intelligence and IoT platforms. In *Cognitive Hyperconnected Digital Transformation* (pp. 97–155). River Publishers. doi:10.1201/9781003337584-4

Chapter 5
Cloud Robotics and the Power of Application Offloading Strategies

Dankan Gowda V.
https://orcid.org/0000-0003-0724-0333
Department of Electronics and Communication Engineering, BMS Institute of Technology and Management, Karnataka, India

Kirti Rahul Kadam
Management Department, Bharati Vidyapeeth University (Deemed), Kolhapur, India

Nazeer Shaik
https://orcid.org/0009-0008-1361-258X
Department of CSE, Srinivasa Ramanujan Institute of Technology, India

K. D. V. Prasad
https://orcid.org/0000-0001-9921-476X
Symbiosis Institute of Business Management, Symbiosis International University (Deemed), Pune, India

Sofia Rani Shaik
https://orcid.org/0000-0002-7453-5070
Symbiosis Centre for Management Studies, Symbiosis International University (Deemed), Pune, India

ABSTRACT

In the face of the development of robotic technology, integration with cloud technologies is a big opportunity to greatly improve robots. This chapter will synthetize the spirit which robotic applications lacks to refine and add depth to it. Also, this procedure not only gives the robotic device more computing power, but also ups its storage capacities and resource management, all without rendering robotical atrophy. In this chapter, the authors examine several different forms of application running away from home. In particular, with the growing public interest in robotics, it is vital that not only participants from industry but also researchers and developers who should realize that through cloud-based offloading they really have a chance to create a revolution.

DOI: 10.4018/979-8-3693-1914-7.ch005

1. INTRODUCTION

Recently, with the advent of cloud computing, robotics and this new branch known as "cloud robots" has been creating ripples. For example, seamless combining of cloud computing technologies with robotic devices represents a completely different kind of concept that jointly integrates aptitudes, connection and collaboration in ways which have never been seen before (Kumar and R. Singhal 202 2). In this chapter, we will briefly introduce cloud robotics in general and then discuss the development of some basic concepts such as application offloading together.

Definition of Cloud Robotics: In essence, cloud robotics is synonymous with the union of robotic systems and cloud computing infrastructure. With this paradigm, robots access cloud-based computation and data storage resources as well as deep learning algorithms to improve their own capabilities (Nayyar J., A. Nagpal), 2018). Unlike robot-only robots, cloud connected robots can access and process prodigious amounts of information in real life. They are the only way to make better informed decisions about what to do and perform complicated tasks more rapidly (P. S. Kumar and A. Sharma, 2019).

The addition of cloud technologies extends the reach of robotic systems beyond the constraints of onboard computing. In exploiting the cloud, robots have access to a seemingly inexhaustible supply of computational power, can rapidly reference huge amounts of data and information and can make use of some powerful machine-learning algorithms (S. El-Gendy, 2020). In this form of symbiosis between robots and the cloud, robotic systems have reached a state where they can adapt, learn and progress over time. Their evolution is no longer limited by their onboard hardware.

Importance of Cloud Robotics: Cloud robotics allows robots to be elevated in ability, becoming intelligent entities capable of complicated operations. This is the real significance of cloud robotics. One of the main benefits is that it makes real-time data processing possible (K. N. V. Srinivas and P. K. Lakineni, 2023). Unlike traditional robots which merely relied on local computing, cloud-based robots can delegate compute-intensive tasks to remote servers, saving time and enhancing the analysis of data streams.

Figure 1. Schematic representation of a cloud robotic system, emphasizing robot connections to cloud computing through wireless network

A Cloud Robotic System (Figure. 1) is presented schematically here, with robots and cloud computing linked via a wireless network (P. S. Kumar, S. Gangadharan and S. C. N, 2023). You will find in this pictorial representation all manner of parts, including robotic objects and cloud servers, linked by lines or arrows to represent the movement of information. The cloud may be representative of the wireless network, suggesting the link between robotic systems and the cloud (Hussain, N., Samal, A., 2024). To diagram the underlying concept, this figure depicts how robots hand data and computational duties among themselves to outside resources in cloud computing via their wireless communications--the very activity that constitutes what we call cloud robotics. Furthermore the advent of cloud robotics has encouraged integration of robotic systems. Through the interconnectivity cloud networks, robots are able to exchange information as well know-hows or delegating tasks (V. S. Prasad, Y Mahajan and SuneethaS., 2023). This kind of collaborative ability becomes very useful for situations requiring robots to work together in, say, industrial automation or a search-and-rescue mission.

In addition, the cloud provides robots with a centralized information source which can be used by them to make their own choices (A. Kumar and V. Kalpana, 2019). When the cloud becomes a large-capacity collective intelligence reservoir, robots can make more intelligent choices than just based on past data (M. Kranthi and R. C Tanguturi 2023). There are also patterns, as well as current information that can be used in making decisions too. Besides broadening the scope of adaptability of robotic systems, it can also lead to more advanced applications that are context aware.

In short, the integration of cloud computing technologies with robotics represents a revolution in the capabilities of robotic systems. Besides easing real-time data processing and enabling interaction between robots, cloud robotics also provides a platform on which advanced strategies--to be explored in later sections--such as application offloading can be designed.

2. FOUNDATIONS OF CLOUD ROBOTICS

The development of robotics has been quite an adventure. Holding the most advanced technology in its grasp, it has constantly extended the scope and applications of robotic systems over these past many years (A. Ranjan, K. K. Hegge and R. Agrawal, 2023). In terms of hardware and intelligence, robots have been gradually developed from their first use in manufacturing industries to their current wide range of applications on our daily lives. Rigid programming and restricted autonomy marked the initial robotic systems (Avinash, Kumaraswamy, Anand Kumar, 2023). The first type of robots were mainly designed to perform monotonous tasks in confined spaces, like assembly lines. Along with technological development came increasingly complex and flexible robots. With the incorporation of sensors, robots could sense their surroundings and thus began to perform various flexible moves.

Figure 2 presents a detailed depiction of the traditional ROS workflow in conjunction with MATLAB-based desktop prototyping (Chinamuttevi, Veera Sivakumar, 2023). It is thus a figure representing the path of information and the processes involved in developing a robotics framework based on ROS with desktop prototyping using MATLAB. You'll see the different parts such as ROS nodes, MATLAB-based communications channels, and other things that integrate all this smoothly. In fact it probably indicates that, within the larger context of ROS-based robotic system development, MATLAB plays a role in prototyping and simulation (Shekhar, R. & Chaturvedi, Abhay, 2023). The following diagram is an attempt to provide a bird's-eye view of the workflow, with which it shall be easier to understand how ROS and MATLAB are integrated together in terms of robotics development (V. Srinivas, T. De, 2023). The

Figure 2. In-depth representation of the traditional ROS workflow integrated with MATLAB-based desktop prototyping

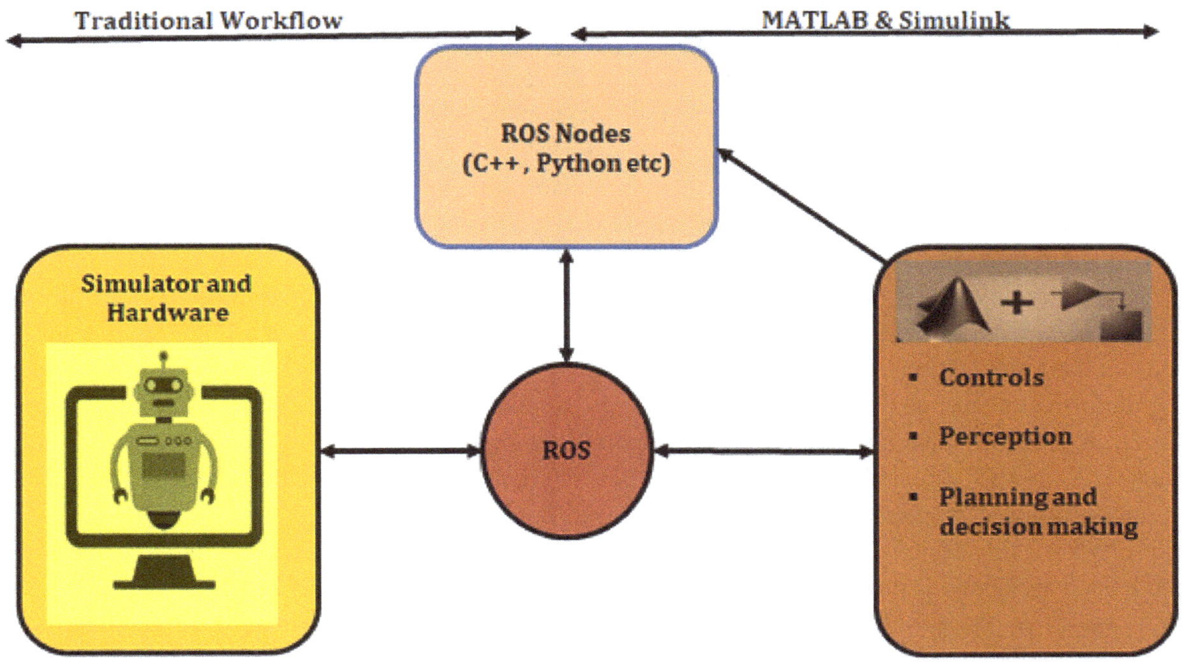

move towards cloud integration in robotics is the result of the need to increase computational power and storage capacity. With its limited computing capabilities, the traditional robot was increasingly unable to meet the needs of more complex applications (K. K. Hegge and R. Agrawal, 2023). Cloud computing offered a way out. Giving robots access to enormous computational resources, they could handle large datasets and run complex algorithms.

Thus the development of robotics is a long journey from simple, mechanical systems to autonomous artificial creatures with cloud connectivity for increased functionality (K. Prasad, and I. Gugan, 2023). Cloud integration not only overcame the calculating power limitations of robots, it also created space for new modes of cooperation, learning and adaptability.

Key Components of Cloud Robotics

To understand the basics of cloud robotics, one needs to know what goes into making up these intelligent systems (N. S. Reddy and B. Ashreetha, 2023). Working together, these parts allow robots to sense and understand their environment, to make judgements about it all by themselves and finally get access to the cloud resources they need.

Sensors: Sensors are the "sensory organs" of robotic systems, allowing them to see and making it possible for robots to gather information about their environment. Sensors are very important in cloud robotics. They gather real time data that can be sent to the cloud for processing. This covers cameras for recording visual data, LiDAR (light detection and ranging) scanners for spatial mapping, and various other sensors for environmental monitoring (S. Reddy P, P. S. Patwal, 2022).

Actuators: The actuators carry out calculations, and turn computational decisions into concrete actions. This can include motors, servos and other such devices that make movement and manipulation possible (P. Pavankumar, N. K. Darwante, 2022). With cloud-connected robots, the robots can relocate complex calculations related to motion planning and control to the cloud, so that they can run their programs more efficiently and dynamically.

Communication Protocols: Cloud robotics relies on communication to send data between the robots and cloud infrastructure (A. Singla, N. Sharma, 2022). Hobbies such as MQTT or ROS (Robot Operating System) facilitate smooth communication so that robots can send information to the cloud or receive instructions in real-time.

Figure 3 is likely an illustration detailing the three main cloud service models: For example, Infrastructure as a Service (IaaS), Platform as a Service (PaaS), and Software as a Service (SaaS). These service models should be clear from the visual representation, perhaps in the form of labeled components or layers.

Infrastructure as a Service (IaaS): This layer generally represents basic infrastructure elements such as servers, storage and networks (K. S and M. R. Arun, 2022). It is the foundation for users to install and run their applications.

Platform as a Service (PaaS): The PaaS layer can expand services constructed atop the infrastructure which are showing tools and middleware that ease development, deployment, management of applications (P. Ramesh Naidu and N. Guruprasad, 2021). These may be databases, development frameworks and application hosting environments.

Software as a Service (SaaS): SaaS is represented at the top layer. This represents fully developed and hosted applications provided to end-users (M. Shruthi and B. Naveen K, 2020). This could involve software applications such as email services, customer relationship management (CRM) tools or collaboration platforms.

The diagram is intended to serve as a visual reference for how cloud service models are arranged hierarchically, from the infrastructure elements at lower levels all the way up to fully managed applica-

Figure 3. Illustration of cloud service models, detailing infrastructure, platform, and software as a service

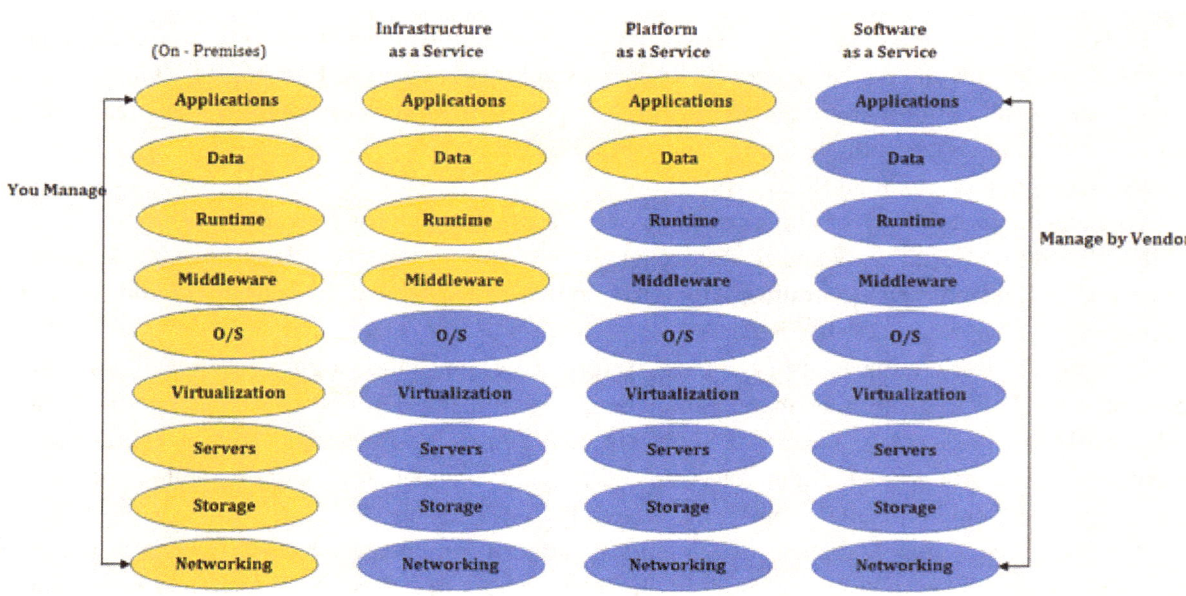

tions (M. Ramesha, and G. N. Pai, 2020). This graphic is a valuable aid in understanding the differences and relationships between these three main cloud service models.

By understanding these basic elements, you will have a better idea of how cloud robotics releases the power of these elements and boosts their abilities through cloud. In the following sections, we will explore strategies including application offloading that tap these components to really realize the potential of cloud-connected robotic systems.

3. APPLICATION OFFLOADING IN CLOUD ROBOTICS

One key maneuver in cloud robotics is called application offloading, and it redefines the capabilities of robotic systems by moving computationally intensive tasks from the onboard computing resources of a robot to external computational resources in the cloud. In this way, robots are able to take advantage of the cloud infrastructure's vast computational power and storage space (R. S. Varun, and T. Rajesh, 2018). The crux of application offloading is that while current robots have increasingly capable onboard processors, some tasks-particularly those featuring high data volume or complicated computations--are too heavy for these localized systems. Application offloading is a link to the world, providing robots with more than their immediate physical capabilities. In Figure 4, we present an infographic that gives a comprehensive and schematic presentation of Smart Cloud Architecture. In this mental picture, it is reasonable to imagine the portrayal of different important parts and their relationships within the structure of a smart cloud architecture (J. J. Jijesh, and Shivashankar, 2017). Likely the graphic contains basic cloud elements like servers, databases and networking infrastructure, which are core building blocks of architecture. Transfer elements, represented by arrows or lines, will perhaps describe the complex interaction of data and communications, from wired to wireless (A. Krishnan, S. Swarna and B. H. S, 2020). Also, don't be surprised to see symbols or pictures of smart devices like sensors or IoT things. As these interact with the cloud for data exchange.

Furthermore, the infographic needs to represent visual cues of such tools for data processing and analytics that reside within the cloud (J. Khan, F. Ali, F. Ullah, J. Bacha and S. Lee, 2019). It's clear how important these are. Security capabilities--encryption, authentication and access control Thus the security features can be visually exaggerated to reinforce just how tough the defense really is. Moreover, icons or diagrams of the user interfaces can show how end-users connect to the smart cloud network using applications or dashboards. In sum, Figure 4 provides an informative and easily understood picture of a smart cloud architecture's consequences one can quickly comprehend all its complexities and relations. Since cloud-based robots send applications off for processing, they can make full use of all the computational resources (W. Liu, 2021). As a result, tasks that would be impossible or too time consuming to accomplish solely on board are easily done. By this method, robots are made into more flexible and adaptable agents that can perform a wider range of tasks efficiently.

Benefits of Application Offloading

The adoption of application offloading in cloud robotics brings forth a myriad of benefits that contribute to the overall enhancement of robotic capabilities:

Increased Computational Power: The greatest advantage of application offloading is that it adds computational resources. By means of cloud infrastructure, robots are now able to have access to high-performance servers capable of running complex algorithms and computations in real time (H. M. Mishra, A. Qtaishat and M. Kumar, 2023). This greater computational capability means robots can

Figure 4. Infographic depicting the schematic overview of smart cloud architecture

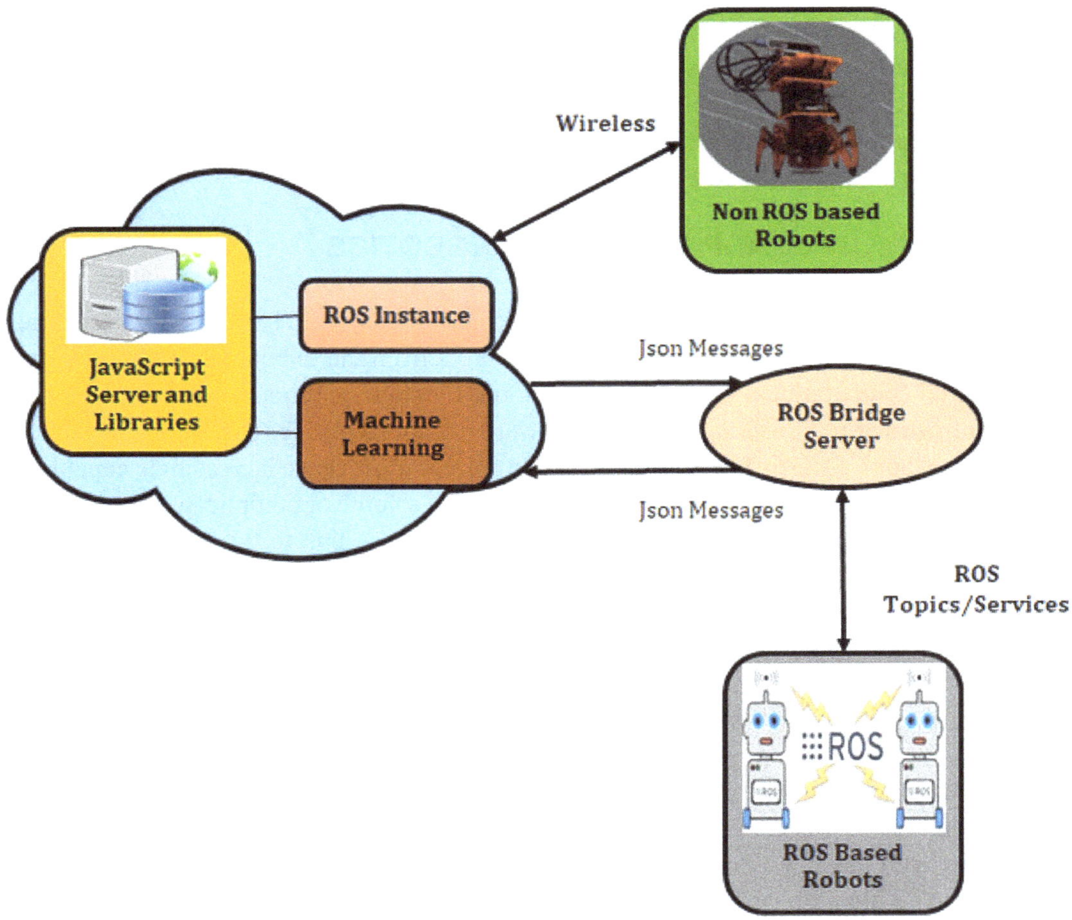

perform complex tasks, process large data sets and carry out high level calculations--and all without crippling the onboard processor.

Energy Efficiency: By sending computationally intensive tasks offshore, the onboard systems can save energy. Cloud servers, which are often specially tuned for power efficiency, can handle the load more economically than using the local power source alone for a robot (D. Pal and S. Joshi, 2023). This energy efficiency is of great importance to mobile robots, as it extends their working life and reduces the number of times they need recharging or refueled.

Scalability: Seamless scalability is achieved through application offloading. With cloud robotics, as the number of computational resources required for a task increases, additional resources can be dynamically allocated from the cloud infrastructure. Scalability is one of the important characteristics of robotic systems working in dynamic environments or performing tasks which require different levels of computation.

Real-time Responsiveness: Real-time responsiveness in robotic systems is achieved because the cloud can process data quickly (S. M. Sali and K. R. Joy, 2023). If robots are to respond quickly to changes in their environment or task requirements, human decision-making must be offloaded to the cloud. This real-time responsiveness is crucial for applications such as autonomous navigation, where decisions made in a millisecond can have an important effect on the entire robotic system.

Figure 5 serves as an illustrative depiction honing in on the middle layer of the Smart Cloud Architecture, with a specific emphasis on the integral component: the Cloud Server. In terms of this graphic picture, readers will be able to see a complete playout of the Cloud Server and its operations in the larger context of smart cloud. It probably illustrates parts naturally included in the cloud server, such as its

Figure 5. Illustration of the smart cloud architecture with a focus on the middle layer cloud server

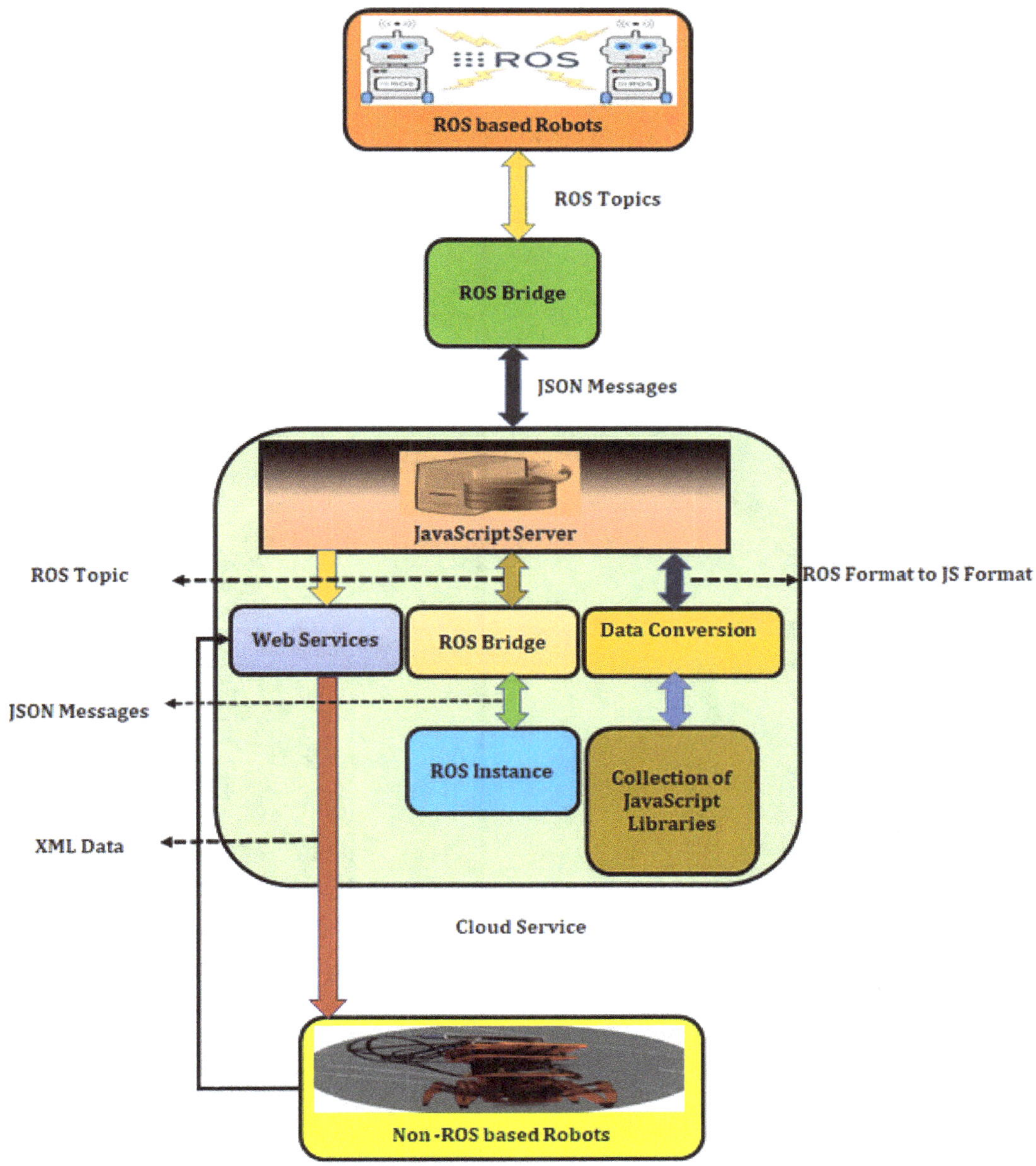

processing units, storage systems and perhaps other specialized modules for tasks like data calculations or machine learning. These connectivity functions are given more prominence, so as to highlight the flow of data between the cloud server and smart devices. This important role that the cloud plays in handling and managing data generated by smart devices is thus made clearer. Moreover, the illustration of middle layer could highlight software or services such as those which enable communication and interoperation between different layers in the smart cloud structure. More stress on data handling and storage makes clear the importance of a cloud server in dealing with all that data generated by the smart environment. Moreover, the picture may emphasize two characteristic features of the cloud server feature: its scalability and flexibility. That is to say, cloud servers can be scaled up or down depending on different workloads and demands. Cloud server security measures can also be represented visually, with encryption, access control, and secure data handling as central themes. All in all, Fig. 5 is a highly detailed and precise perspective-a microcosm, as it were-giving us an intimate appreciation of the central role played by the Cloud Server relies in this complex structure of a smart cloud architecture.

Figure 6 presents a complete picture of the information exchange path, with special emphasis given to the connection between a robot and cloud by way of RESTful Web Service. The graphic depiction probably involves pictures of the robot and cloud infrastructure, with some representation of visual elements such as the sensors, actuators and servers in the cloud. We give the most attention to portray-

Figure 6. Overview of the information exchange path - robot and cloud connectivity via RESTful web service

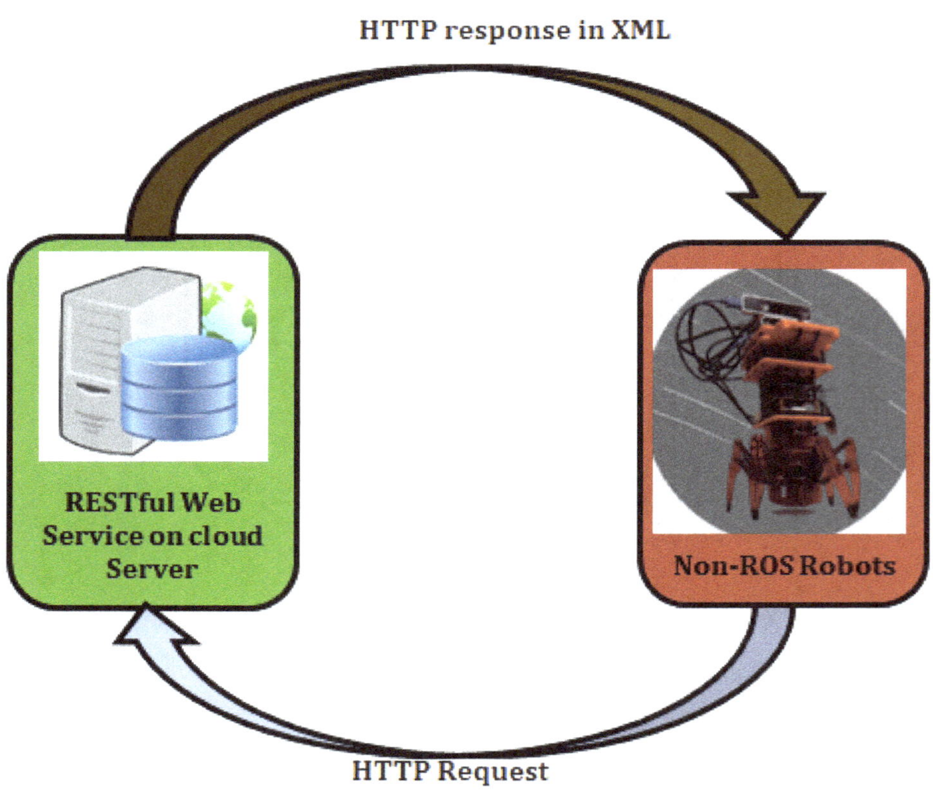

ing the function of a RESTful Web Service, which is used by the cloud and robot to communicate. It probably depicts arrows or lines to represent the transmission of information in an exclusive stateless, resource-centered manner. The flow goes back and forth, showing the bidirectionality of communication. You can see main points such as lightweight and scalable nature of RESTful protocols, which are very important in the data exchange processes. The following diagram in Figure 6 serves not just to give a clear and intuitive understanding of how information flows effortlessly back and forth between robot and cloud, but also very concisely tells us exactly how this connectivity operates.

In fact, application offloading is the initiator that releases the full power of cloud-controlled robots. All of the benefits from this strategy are not simply a way of overcoming the computational limitations inherent in onboard systems, but also lay the foundation for even more intelligent, efficient and adaptable robotic applications. In the next section we will explore a number of approaches and factors that need to be considered in practical application offloading for cloud robotics.

4. STRATEGIES FOR EFFECTIVE APPLICATION OFFLOADING

Dynamic offloading in cloud robotics Dynamic offloading refers to dynamically allocating computational tasks between the onboard system and the cloud according to real-time conditions and system requirements. In contrast to static off-loading, where tasks are predetermined before execution, smart off-loading allows for adaptive decision making while runtime.

Real-time Conditions: The environmental monitoring in dynamic offloading is continuous. In real time, various parameters like network bandwidth, computational load and latency are evaluated. Based on a dynamic analysis of these conditions, the system can decide whether it will pass off some tasks to the cloud or process them locally.

Adaptive Task Allocation: Tasks cannot be entered in a static table specifying their distribution between onboard system or cloud architecture; they must change dynamically as the environment alters. This adaptability means that the robot can make full use both of local and cloud resources. By this they mean it should be responsive as well as efficient.

Resource Availability: When offloading tasks dynamically, resources in the cloud--computational power and storage--are taken into account. The aim of the system is to balance the load and avoid resource bottlenecks.

The data flow between a robot and the cloud is probably portrayed in Figure 7 as some kind of flow chart. This flow chart should use common shapes and lines used in the formulation of a procedure as well as decision points, to illustrate steps along with choices that may occur during data exchange. In particular, dynamic offloading adds most utility when there are large fluctuations in demand for computing power or if the robot meets with an unexpected obstacle. This allows the real-time decision capabilities of such cloudically connected robotic systems, with all aspects being complementary to each other.

4.1 Load Balancing

Load balancing is one important method used in cloud robotics application offloading. It is basically a question of distributing the computational tasks among different cloud infrastructure resources as evenly and optimally possible in order to achieve both good performance at low cost. Task Distribution Algorithms: Computewhizzesitize the cloud networks A corporation possessing a great many servers

Figure 7. Visual representation of data flow between the robot and cloud in a flow chart

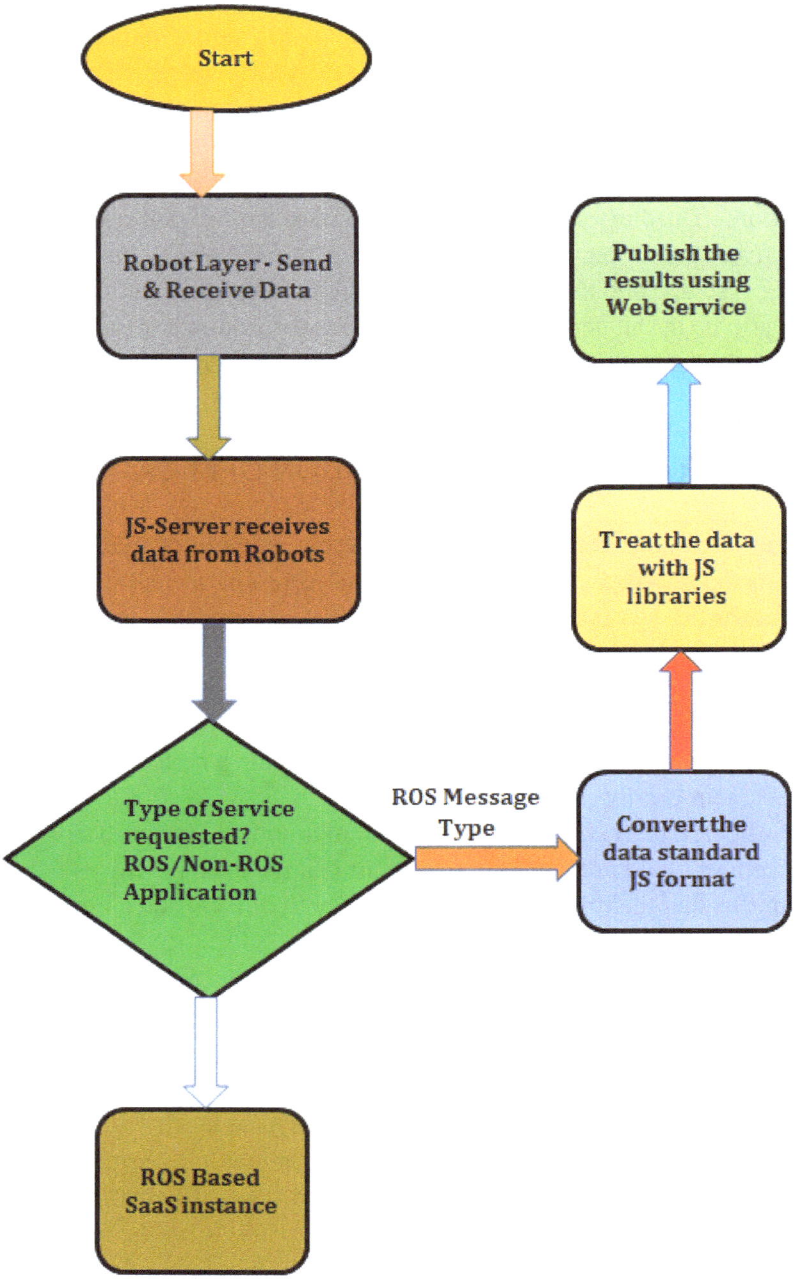

capable of accomplishing task X at its server farms and having usage data on the speed by which each individual user is able to make use of them. The power required from their computer in order for users running client applications anywhere out there over this network should be enough; otherwise, they'd go broke again fairly quickly Many algorithms are used including HRR(Round Robin), LCR (Least connections) and WRR = Weighted Round robin).

Dynamic Load Adjustment: In cloud robotics, to achieve load balancing is a continuous readjustment -- not one-time only. Operation- When the workload on cloud servers is inconsistent, the system schedules tasks to keep things in balance. In selecting this kind of load adjustment, not only do servers don't overwork through being choked with requests but cloud resources also become more balanced.

Fault Tolerance: For instance, so-called load balancing strategies include fault tolerance mechanisms designed to counteract such events as failure or breakdown of the servers. With tasks moved to a healthy server in the event of failure, redundancy and failover mechanisms are built into CFS.

Scalability and constant performance is the indicator of load balancing in cloud robotics. With the ridiculously large cloud resources, tasks can be divided and dispatched very effectively. It truly plays on every available computational power without allowing clusters to hit a bottleneck.

4.2 Latency Management

One aspect of the cloud robot is latency; which means how long it takes between doing a task and seeing results. In cloud robotics, this measurement always requires careful consideration. As one example, effective offloading should minimize latency to retain real-time responsiveness.

Proximity-aware Offloading: Now, such tasks are passed to cloud servers located only next door so that latency communication delay is reduced. Dying out: In addition to considering the distribution of cloud resources by geography, proximity-aware offloading tries as much as possible to shorten round-trip times.

Edge Computing Integration: This enables the cloud structure to take care of all aspects apart from truly low-latency tasks, along with latent decisions which can be made locally using edge devices. For instance, edge computing allows computation this closer to the robot itself. So in other words, it's huge cuts on the times spent transferring data over to a cloud-based server at an enormous distance.

Prioritization of Critical Tasks: When offloading tasks, latency-sensitive ones are given priority. By identifying the tasks which need to be done rapidly, the system can then see that important operations are completed as quickly as possible.

The multi-robot configuration shown in Fig. 8 features a variety of robots with different designs, sizes or functions, all depicted visually to give an idea of the complexities involved. The manner in which these robots are arranged and placed within the overall configuration has been thoughtfully planned out, perhaps highlighting their working together for a common objective. Arrows or lines can be used to illustrate connection and communication links between the robots, showing how they exchange information and how they work together. This illustration is intended to stress the functional diversity of each robotic entity, and its contribution to the objectives being targeted by this overall configuration. To give a feeling of the environment in which the multi-robot arrangement works, contextual elements may also be introduced, such as various real life settings or tasks. Further detail about individual robots, their functions, or the collaborative tasks they perform could be provided by labels and annotations. In sum, then, an image such as Figure 8 is a visual story. It captures well the idea of a collection of robots working together in concert with one another and with humans. Examples are teleoperation, where response speed is clearly important. How to reconcile the trade-off between offloading tasks for cloud processing and local execution is a key question in ensuring low latency on cloud-connected robotic systems. Later in this document we will examine some applications that have implemented these strategies through

Figure 8. Illustration of a diverse multi-robot configuration

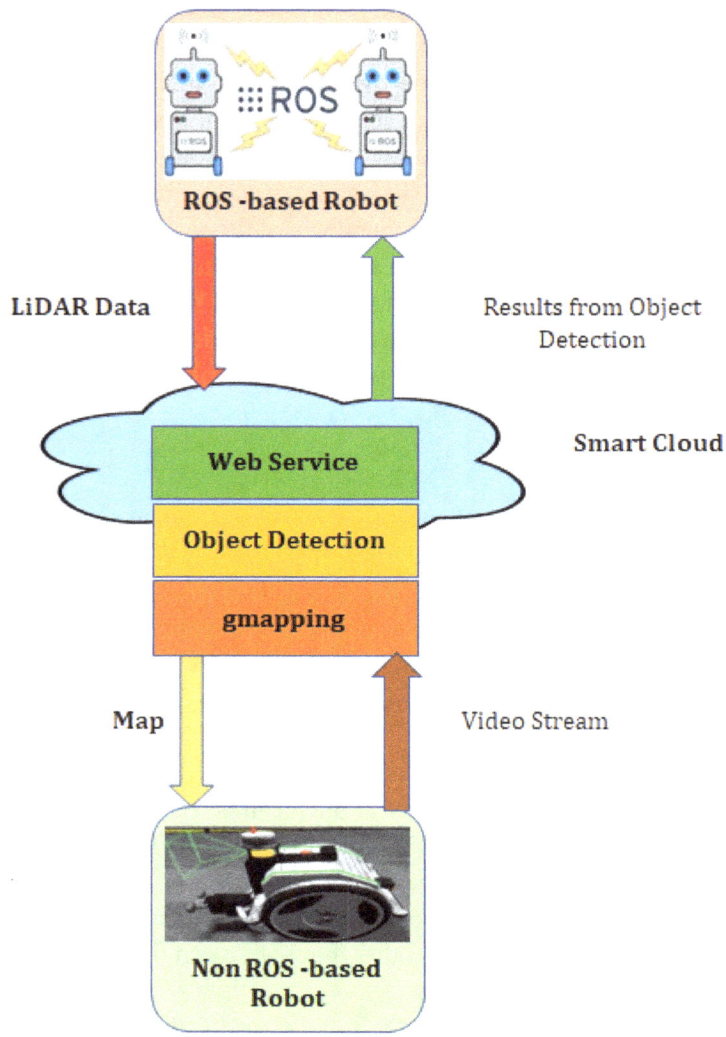

real-world case studies and examples, illustrating how they benefited cloud robotics applications' performance and adaptability.

5. CASE STUDIES AND EXAMPLES

5.1 Real-World Implementations

Amazon Robotics (Kiva Systems):

Description: Kiva Systems is an in-house division of Amazon, and uses a fleet of mobile robots that are also primarily used in fulfillment centers. Robots like these go round and around carrying shelves of products to be packed for orders.

Application Offloading Strategy: These robots delegate the tasks of navigation and coordination to the cloud. Or, high level commands are sent to the robot and general path planning is done in real-time on cloud. Avoidance of collisions between robots come from synchronization between them as well.

Benefits: Or Does it This approach does well in using cloud resources, allowing robots to deal with the rapidly changing environment of e-commerce fulfillment.

NASA's Robonaut:

Description: NASA developed Robonaut, a humanoid robot meant to work on the surface of other planets. It can do operations such as tool work and equipment handling in dangerous areas that would be beyond the reach of human astronauts.

Application Offloading Strategy: Because complex image processing and decision making are needed, therefore Robonaut uses the cloud application offloading. Through analyzing image data collected by the images its cameras take, with clouds as resources, this robot that does what it sees requires few other skills to become a space functionary capable of undertaking many missions.

Benefits: Thanks to the integration of cloud computing, the robot has a more powerful calculation ability than when it was first sent up. It is now capable of carrying out complicated work such as helping researchers to perform scientific still sea ratio into actual qinghao reserves experiments in space-which used stilts instead floorboards commonly see have its own astrology and enjoyed wide circulation throughout northern China.

5.2 Industry Use Cases

Healthcare: Remote Surgery Systems

Description: As for health care, surgeons now have remote surgery systems with which they can carry out operations on patients who are at different places around the world. For example, there is the da Vinci Surgical System.

Application Offloading Strategy: Cloud-based processes such as image analysis and instrument control make other kinds of off-loading possible. The need for: In sum, this situation means that the robo

Benefits: Graphics are projected onto the patient from a four-dimensional brain model, or detailed information about surgical sites are disclosed in highly accurate three dimensionality. These elements shift remote surgery one step further under the cloud robotics framework: experts can be far away while still at work and their handiwork is shared across more channels for specialized medical services.

Manufacturing: Collaborative Robotics (Cobots)

Description: Are called cobots or collaborative robots and work right next to human workers in factories. Assembly, material handling and quality control are just a few task they assist with.

Application Offloading Strategy: For example, path planning and optimization are handled in the cloud for more complicated tasks. This allows cobots to get used in a dynamic response fashion based on changes in production needs and for them continue peacefully working next to human beings.

Benefits: Frankly, cloud-enabled cobots are able to switch from task to task quickly and easily. While this greatly reduces downtime, it also considerably enhances the agility and efficiency of manufacturing overall.

Logistics: Autonomous Delivery Robots

Description: Last-mile delivery: For logistics and e-commerce, autonomous delivery robots make sense. Walking on sidewalks and in public, the robots carry pack one after another to customers.

Application Offloading Strategy: Cloud takes over tasks of planning the route and finding its optimal path. Real-time traffic, weather information and management data will be sent to the robots. These include things like delivery routes which are most advantageous for wind conditions or good for rainy days. This allows prompt and efficient deliveries.

Benefits: These robots will on-the fly to optimize delivery routes, and have a great affect in improving the overall stability and speed of last mile logistics. They are said by their developers cloud connected because they connect cyber space with reality telemetrically via SIM card linkage.

Equipped with this rationale, we prepare a very detailed and illuminating figure 9 which is the proposed predictive algorithm tailored to decision-making offloading. The visual representation probably illustrates the algorithmic flow, displaying in a series of steps an integral part of the decision making offloading process. Within this algorithm, decision points are probably highlighted. They are locations where the algorithm makes a judgment based on information that it is given or upon predetermined conditions. You can see how the algorithm processes information to make decisions via a representation of data input sources and the output obtained from them. It may also describe various components of the predictive model, including features, variables or mathematical operations that make up its basic structure. The flow of information within the predictive algorithm may be depicted by arrows or lines denoting connectivity and interactions between different components. If applicable, the illustration might visually take account of several performance indicators or points of assessment to assess how well the proposed predictive algorithm works. In fact, Figure 9 is a complete and detailed graphical explanation that presents in a refined way the complex workings of the algorithm designed for decision-making offloading. These case studies demonstrate the broad range of applications there is for cloud robotics

Figure 9. Illustration detailing the proposed predictive algorithm for decision-making offloading

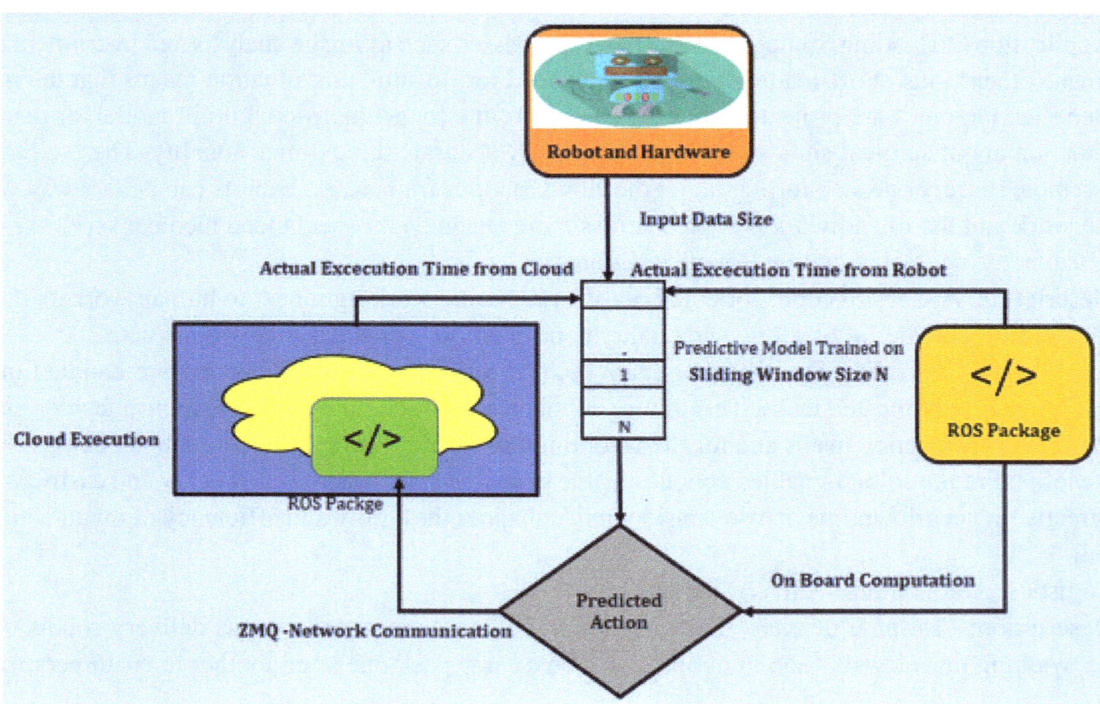

and illustrate how successful application offloading has been in so many fields. Now that technology is racing forward in several leaps, it will not be long before cloud-based robotics play a huge role in the way robots work and relate to each other on earth. Now let's see what obstacles cloud robotics faces and what potential it has in this rapidly growing branch.

6. CHALLENGES AND FUTURE TRENDS

The problems of cloud robotics and future trends paint a picture of the thin edge beyond. With security assuming such prominence, the data secrecy aspect of it needs rigid standards while its private attributes and ability to resist cyber attacks also require strict stringency. Some solutions are: strong encryption, safe communication protocols that send sensitive information only over secure connections; regular monitoring of security events on the network and requiring compliance with industry-specific regulatory structures. Looking forward, wireless applications are benefiting from the greater trend of integration and distribution called edge computing. With lower latency along with optimized bandwidth as well as decreased equipment at offices. This is all enabled by decentralized compute power located where data needs to be processed--at network edges instead of in central corporate servers However, along with the cloud robotics that follows on its heels comes a synergy between them. Tasks are allocated dynamically between local edge devices and remote cloud servers much as stars allocate tasks to guiding robots in their network of networks (Cooperative Furball Robot Network). Moreover, the battlefield is on verge of being transformed by applying AI and ML. A cloud-connected network of smart robotic systems can share continuous learning, perform predictive analysis and generate knowledge that will cooperate in this quite different encirclement campaign (to say nothing about victory prediction). After these various technologies mature, future offerings will standardize on the base structure while allowing application delegation algorithms to be customized. Robots can adjust decisions according to task complexity and environmental uncertainties in real time. As advancements in security, edge computing and AI/ML continue to drive innovation throughout the broader industries, we're all on the verge of holding cloud robotics in our hands. Whether it be smartphones or banking cards, Zhifei Jiefang is hers for the taking. What lies in store for the future could turn cloud robotics into an entirely different way of going, working with and adjusting robots than we see today.

7. CONCLUSION

The research on cloud robotics and application offloading strategies has thus opened up a revolutionary space where the combination of robotics with cloud computing tears aside all restrictions on technology. Many could use the definition of cloud robotics, based on combination with cloud technologies of robotic systems. This reflects the profound change in capabilities their applications can create. It makes possible real-time cooperation and dynamic responsiveness. However, the impact this shift has had on the adaptability and intelligence of robotic systems is profound. A look at the development of robotics from rigid programming to dynamic cloud integration shows just how much. Sensors, actuators device and communications protocols provide the foundations for all three of the dynamic strategies discussed here-application offloading. By means of case studies and industry applications, the concrete achievements of cloud-connected robotic systems demonstrate their versatility and efficiency. But there are also challenges on the horizon. As we balance security concerns and various possible directions for future developments, cloud robotics seems even more interesting. With the future likely to combine edge computing with developments in artificial intelligence and machine learning, as well as improvement of application offloading strategies, there exists good reason for optimism. Standing at the confluence of pressure and possibilities, the increasingly important next wave of cloud robotics is set to completely change certain industries, raise efficiency to a higher level, and recreate our understanding of intelligent robotic systems for a technological world in constant flux.

REFERENCES

Avinash, K., & Kumar, A. (2023). A novel approach of unsupervised feature selection using iterative shrinking and expansion algorithm. *Journal of Interdisciplinary Mathematics*, *26*(3), 519–530. doi:10.47974/JIM-1678

Chinamuttevi, V. S. (2023). A novel RF-SMOTE model to enhance the definite apprehensions for IoT security attacks. *Journal of Discrete Mathematical Sciences and Cryptography*, *26*(3), 861–873. doi:10.47974/JDMSC-1766

El-Gendy, S. (2020). IoT Based AI and its Implementations in Industries. *2020 15th International Conference on Computer Engineering and Systems (ICCES)*, 1-6. 10.1109/ICCES51560.2020.9334627

Hegge, K. K., & Agrawal, R. (2023). Industrial Automated Multipurpose Robot Using WIFI. *2023 4th International Conference for Emerging Technology (INCET),* 1-8.

Hussain, N., & Samal, A. (2024). Design and Implementation of Single to Multi-cloud Securities in Network. In N. R. Shetty, N. H. Prasad, & N. Nalini (Eds.), *Advances in Computing and Information. ERCICA 2023. Lecture Notes in Electrical Engineering* (Vol. 1104). Springer.

Jijesh & Shivashankar. (2017). Design and implementation of automatic medicine dispensing machine. *RTEICT 2017 - 2nd IEEE International Conference on Recent Trends in Electronics, Information and Communication Technology, Proceedings*, 1962–1966.

K. S & Arun. (2022). Priority Queueing Model-Based IoT Middleware for Load Balancing. *2022 6th International Conference on Intelligent Computing and Control Systems (ICICCS)*, 425-430.

Khan, J., Ali, F., Ullah, F., Bacha, J., & Lee, S. (2019). Artificial Intelligence and Internet of Things (AI-IoT) Technologies in Response to COVID-19 Pandemic: A Systematic Review. *IEEE Access : Practical Innovations, Open Solutions*, *10*, 62613–62660. doi:10.1109/ACCESS.2022.3181605

Kranthi, M., & Tanguturi, R. C. (2023). Design of Intelligent Medical Integrity Authentication and Secure Information for Public Cloud in Hospital Administration. *2023 2nd International Conference on Edge Computing and Applications (ICECAA),* 256-261.

Krishnan, A., & Swarna, S. (2020). Robotics, IoT, and AI in the Automation of Agricultural Industry: A Review. *2020 IEEE Bangalore Humanitarian Technology Conference (B-HTC)*, 1-6. 10.1109/B-HTC50970.2020.9297856

Kumar, A., Kalpana, V., & Hariram, V. (2023). Artificial Intelligence based Health Monitoring System on IoTH platform. *2023 Second International Conference on Augmented Intelligence and Sustainable Systems (ICAISS)*, 1458-1463.

Kumar, P. S., & Sharma, A. (2019). IoT Enabled Indoor Autonomous Mobile Robot using CNN and Q-Learning. *2019 IEEE International Conference on Industry 4.0, Artificial Intelligence, and Communications Technology (IAICT)*, 7-13.

Kumar & Singhal. (2020). Design and Analysis of IoT-Based Intelligent Robot for Real-Time Monitoring and Control. *2020 International Conference on Power Electronics & IoT Applications in Renewable Energy and its Control (PARC)*, 549-552.

Liu, W. (2021). Improvement of Navigation of Mobile Robotics Based on IoT System. *2021 IEEE International Conference on Robotics, Automation and Artificial Intelligence (RAAI)*, 69-72. 10.1109/RAAI52226.2021.9507826

Mishra, H. M., Qtaishat, A., & Kumar, M. (2023). IoT Based Smart Systems using Artificial Intelligence and Machine Learning: Accessible and Intelligent Solutions. *2023 6th International Conference on Information Systems and Computer Networks (ISCON)*, 1-6.

Nayyar & Nagpal. (2018). Internet of Robotic Things: Driving Intelligent Robotics of Future - Concept, Architecture, Applications and Technologies. *2018 4th International Conference on Computing Sciences (ICCS)*, 151-160.

Pal, D., & Joshi, S. (2023). AI, IoT and Robotics in Smart Farming: Current Applications and Future Potentials. *2023 International Conference on Sustainable Computing and Data Communication Systems (ICSCDS)*, 1096-1101. 10.1109/ICSCDS56580.2023.10105101

Pavankumar, P., & Darwante, N. K. (2022). Performance Monitoring and Dynamic Scaling Algorithm for Queue Based Internet of Things. *2022 International Conference on Innovative Computing, Intelligent Communication and Smart Electrical Systems (ICSES)*, 1-7.

Prasad, K., & Gugan, I. (2023). Enhancing Microcomputer Edge Computing for Autonomous IoT Motion Control. *International Journal on Recent and Innovation Trends in Computing and Communication*, *11*(3), 58–67. doi:10.17762/ijritcc.v11i3.6202

Prasad, V. S., Mahajan, Y., & Suneetha, S. (2023). A Cloud-Based UV Monitoring System for Remote Real-Time UV Exposure Tracking. *2023 4th International Conference on Smart Electronics and Communication (ICOSEC),* 1764-1770.

Ramesh Naidu, P., Guruprasad, N., & Dankan Gowda, V. (2021). A High-Availability and Integrity Layer for Cloud Storage, Cloud Computing Security: From Single to Multi-Clouds. *Journal of Physics: Conference Series*, *1921*(1), 012072. doi:10.1088/1742-6596/1921/1/012072

Ramesha, M., & Pai, G. N. (2020). Internet of things: Internet revolution, impact, technology road map and features. *Adv. Math. Sci. J.*, *9*(7), 4405–4414. doi:10.37418/amsj.9.7.11

Ranjan, A., Hegge, K. K., & Agrawal, R. (2023). Wireless and Solar-Powered Multipurpose Robot for Industrial Automation: A Sustainable Solution. *2023 8th International Conference on Communication and Electronics Systems (ICCES)*, 146-151.

Reddy, N. S., & Ashreetha, B. (2023). Technologies for Comprehensive Information Security in the IoT. *2023 International Conference for Advancement in Technology (ICONAT),* 1-5.

Reddy, S., & Patwal, P. P. S. (2022). Data Analytics and Cloud-Based Platform for Internet of Things Applications in Smart Cities. *2022 International Conference on Industry 4.0 Technology (I4Tech)*, 1-6.

Sali, S. M., & Joy, K. R. (2023). Intelligent Rover: An IoT Based Smart Surveillance Robotic Car for Military. *2023 2nd International Conference on Computational Systems and Communication (ICCSC)*, 1-6. 10.1109/ICCSC56913.2023.10143011

Shekhar, R. (2023). Securing networked image transmission using public-key cryptography and identity authentication. *Journal of Discrete Mathematical Sciences and Cryptography*, *26*(3), 779–791. doi:10.47974/JDMSC-1754

Shruthi & Naveen. (2020). Internet of Things: A Survey on Devices, Ecosystem, Components and Communication Protocols. *2020 4th International Conference on Electronics, Communication and Aerospace Technology (ICECA)*, 611-616.

Singla, A., & Sharma, N. (2022). IoT Group Key Management using Incremental Gaussian Mixture Model. *2022 3rd International Conference on Electronics and Sustainable Communication Systems (ICESC)*, 469-474.

Srinivas, K. N. V., & Lakineni, P. K. (2023). Development of a Real-time Location Monitoring App with Emergency Alert Features for Android Devices. *2023 4th IEEE Global Conference for Advancement in Technology (GCAT),* 1-8.

Srinivas, V., & De, T. (2023). Wi-Fi Router Signal Coverage Position Prediction System using Machine Learning Algorithms. *2023 International Conference on Sustainable Computing and Smart Systems (ICSCSS),* 253-258.

Varun, R. S., & Rajesh, T. (2018). Implementation of swarm intelligence in obstacle avoidance. *RTEICT 2017 - 2nd IEEE International Conference on Recent Trends in Electronics, Information and Communication Technology, Proceedings*, 525–528.

Chapter 6
Robotic Evolution Integrating IoT and Robots

Dankan Gowda V.
https://orcid.org/0000-0003-0724-0333
Department of Electronics and Communication Engineering, BMS Institute of Technology and Management, Karnataka, India

Anjali Sandeep Gaikwad
Department of Computer Application, Bharati Vidyapeeth University (Deemed), Kolhapur, India

Aparna Atul Junnarkar
Department of Information Technology, Vishwakarma Institute of Information Technology, Pune, India

K. D. V. Prasad
Symbiosis Institute of Business Management, Symbiosis International University (Deemed), Pune, India

Sofia Rani Shaik
https://orcid.org/0009-0005-2714-6442
Symbiosis Centre for Management Studies, Symbiosis International University (Deemed), Pune, India

ABSTRACT

In a landscape of technology that is changing at breakneck speed, the combination of the internet of things (IoT) with robotics represents an important step toward future automation and intelligent systems. In this chapter, the authors explain how such an integration will change things. This combination of IoT's connected devices and robotic dynamism brings unprecedented progress in many areas, such as industrial automation and healthcare. After cutting through the technical jumble and coming to terms with plausible applications, problems, and opportunities afforded by this match, readers will have a clear understanding of what this revolutionary field has today, as well as its promise for tomorrow.

DOI: 10.4018/979-8-3693-1914-7.ch006

1. INTRODUCTION

The fields of robotics and the Internet of Things (IoT) have undergone revolutionary changes in recent years, with a revolution taking place in the way we think about intelligent systems. Robotics, which had been limited to the industrial world, has also broken out of its shell. There is very little that even advanced technologies such as IoT could not naturally integrate (A. Ranjan, K. K. Hegge and R. Agrawal, 2023). Now, the contemporary image of robotics is a departure from the concept of machine limited to assembly line work. Today's robots are far more sophisticated, and have already made inroads into many other fields including medicine, logistics, entertainment or education (Avinash, Kumaraswamy, Anand Kumar, 2023). These new developments are propelled forward by advances in materials, sensors and artificial intelligence (AI), enabling robots to work with unerring precision as well as flexibility.

Meanwhile, the coming of age of the Internet of Things has altered models for connectivity and data exchange (Chinamuttevi, Veera Sivakumar, 2023). Web of Things Formally defined, IoT really is all about putting sensors and connectivity into everyday objects so data can be gathered and shared across a network. This interlinked eco-system has far reaching implications; it provides functions for smarter, more responsive systems (Shekhar, R. & Chaturvedi, Abhay, 2023). The marriage of robots and the Internet of Things represents a critical juncture in the development of intelligent systems. Robots used to only work alone, but now they fit right into the Internet of Everything as connected devices. This integration, however, is not just an incremental improvement (V. Srinivas, T. De, 2023). It represents a fundamental paradigm shift to more intelligent, plastic and integrated robotic systems.

Enabled by cloud computing. The arrival of cloud robotics allowed for centralization in terms of function and resource allocation, so that several robots could stake a claim to the same storage space or processing capacity (K. K. Hegge and R. Agrawal, 2023). But cloud robotics is lacking a crucial feature: a way for robots and other devices to have a common language when they speak with one another online (K. Prasad, and I. Gugan, 2023). One possible mechanism for such a worldwide communications system is the Internet of Things (IoT).

Connected devices, or "things," may exchange data and coordinate their actions over the Internet thanks to the Internet of Things (IoT), a worldwide network (N. S. Reddy and B. Ashreetha, 2023). This makes it possible for all parties to communicate using the same communication technology, creating a platform for conversation. As shown in figure.1, the Internet of Robotic Things (IoRT) is an integration of the Internet of Things (IoT), robotics, and cloud computing. The idea is to develop this further by bringing these groups together. The Internet of Robotic Things (IoRT) will be a global network via which high tech robots connected through current and future communications technologies can communicate with one another, exchanging data that is compatible between them (S. Reddy P, P. S. Patwal, 2022). Moreover, the IoRT architecture permits common access by robots to scalable data processing, data storage and communications resources via a common pool of cloud computing services currently provided by data centres (P. Pavankumar, N. K. Darwante, 2022).

Building an IoRT platform using cloud computing, robots, and the Internet of Things is the goal of this chapter. In order to build an infrastructure that can accommodate IoRT features, the following skills are required by the system: The dissemination of data that can be understood and used by various robots via the use of current and future communication technologies (A. Singla, N. Sharma, 2022). The following is a picture of the type of network required for this function. Dybrid media allow one to achieve 2 C, which involves centralised cloud computing services--computer processing, data warehousing and data exchange. On the other hand, the second ability requires centralised cloud computing

Figure 1. Overview of IoRT technologies: A visual representation

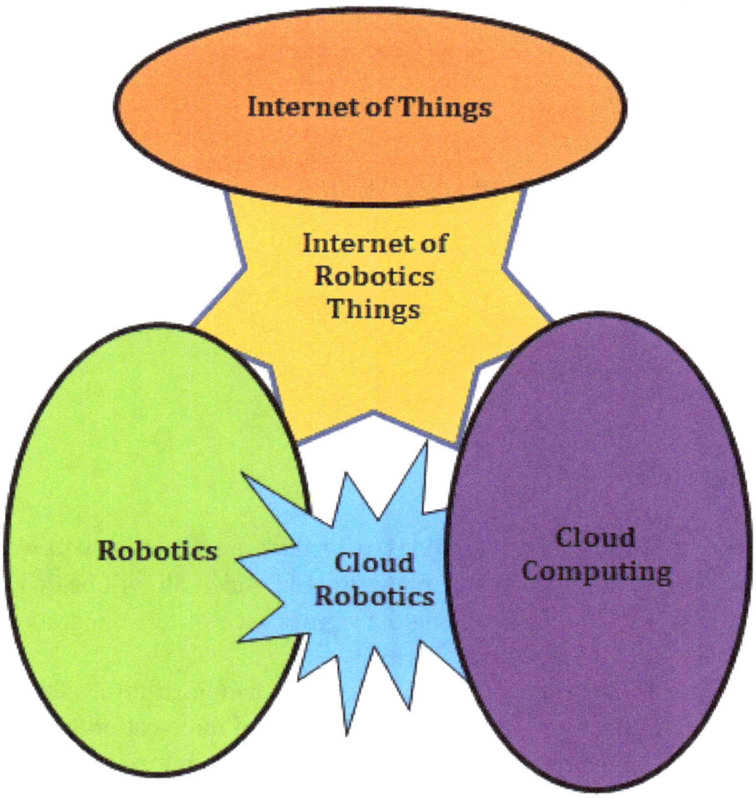

Figure 2. Overview of network structures: a) centralized, b) distributed, c) combined

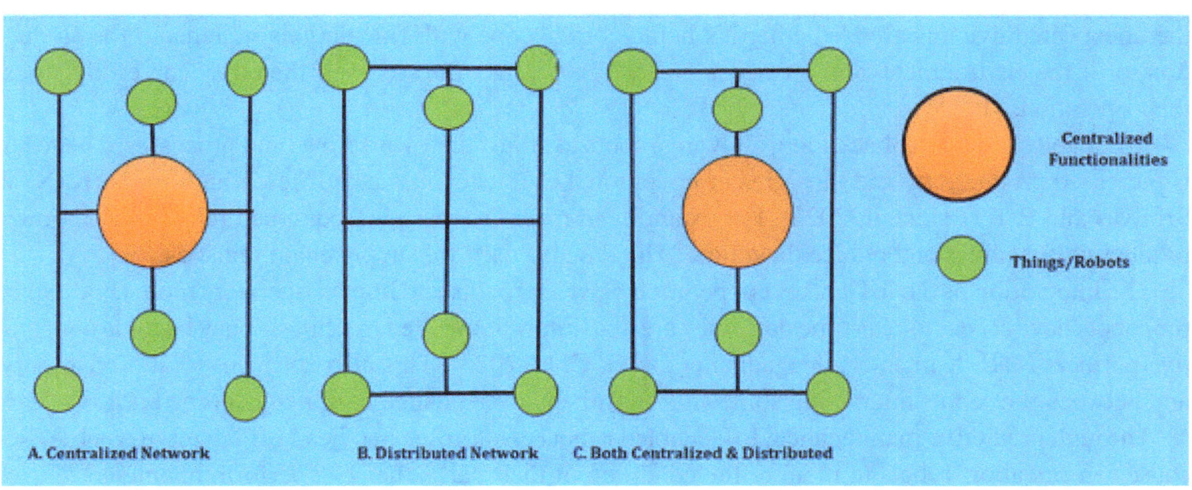

functions--see picture. Through this combination program, we could generate a network that is at once centralized and decentralized, as pictured in figure 2C. The origins of robotics can be found in industrial applications, but the scene now extends far beyond the production floor (K. S and M. R. Arun, 2022). The most advanced robots can be found in hospitals doing surgery, navigating through warehouses to organize inventory, and even inside our houses carrying out chores. The basis of this expansion lies in integrating IoT, which allows robots to feed constantly on the latest data and make cool-headed decisions in this ever changing world.

Standing at the crossroads of these technological advancements, the symbiosis between robotics and IoT is about to revolutionize our conception of intelligent systems (Kumar and R. Singhal, 2020). In this chapter, then, we shall turn our attention to the ins and outs of this integration. Origins and applications, problems it poses. But why are industries around the world looking so eagerly forward to a future with integrated computing.

Significance of Integration

The marriage of robotics with the Internet of Things (IoT) marks an mighty step forward, endowing robots with unprecedented strength. From there we enter a time in which all will be efficient and intelligent. Here it is shown just how great and far-reaching the significance of this integration will be, going into a number of aspects (Nayyar and A. Nagpal, 2018).

When robots and the IoT meet up, streams of real-time data gush forth. Sensors embedded in the environment and on robotic platforms gather a steady stream of information which is then transformed into richer, more accurate knowledge about surroundings, targets and tasks. This stream of data makes the robot more aware--it can thus choose with greater precision (P. S. Kumar and A. Sharma, 2019).

With the integrated use of IoT and robotics, adjustments can be made in real-time according to ever changing circumstances. But traditional robots are passive players who act within predetermined boundaries. With IoT added in, this changes completely (S. El-Gendy, 2020). Now they can respond to the unexpected dynamically. Whether it's being able to cope with fluctuations in demand for production, or if the environment is as changeable and flux-taking s the sea itself then they can easily adjust their operations fly.

IoT enables a data feedback loop, allowing robots to adjust their workflows and processes. The result is increased productivity and thus greater operational efficiency in many different industries (K. N. V. Srinivas and P. K. Lakineni, 2023). For example, in manufacturing IoT-equipped robots can optimize production schedules and reduce downtime. They're also used for supply chain logistics.

The integration of the IoT makes cooperation between robots and human operators more fluid. Since the robots have access to real-time data, they are not simply automated machines. They become cooperative partners (P. S. Kumar, S. Gangadharan and S. C. N, 2023). This also enhances total productivity, and between the robots interacting somewhat sensitively even change in milieu is borne back.

Through predictive maintenance, IoT-enabled robots can foresee and head off potential trouble. Because sensors allow robots to monitor the condition of their parts, they can perform maintenance even before there's a breakdown (Hussain, N., Samal, A., 2024). This initiative shortens downtime, extends the service life of robotic systems and improves use of resources.

Finally, the meaning of connecting IoT with robots transcends technical development. It opens the way for one not only of intelligent robotics, but also highly integrated in people's lives and superbly adding to efficiency and productivity so that people become better off. For this chapter, formulating a point of departure leads us to see such transformative synergy.

2. FOUNDATIONS OF IOT AND ROBOTICS

The Internet of Things (IoT), which has been around for several decades, is often called the Third Industrial Revolution. This is a reality in which taking the data generated by everyday objects, devices and equipment or gathered from around them and sending these online has become routine. To put it simply, the Internet of Things makes everything ordinary a thinking and connected thing that can chat or co-operate anywhere at any time (V. S. Prasad, Y. Mahajan and S. Suneetha, 2023).

The Internet of things is based on the complete fusion of physical and virtual worlds. Several key principles define the essence of IoT:

Connectivity: That is, IoT devices are connected and networked so that they can communicate with one another and exchange data.

Sensing: The sensors on IoT devices enable the collecting of data about their surroundings or what they come into contact with.

Data Processing: Local data can be collected, processed and interpreted on the device itself or in the cloud. Insights become actionable (A. Kumar, V. Kalpana and V. Hariram, 2023).

Actuation: The following are based on the processed data: IoT devices can adjust settings, undertake a response or communicate with other things.

Interoperability: By establishing common standards for protocols and communication formats, interoperability between different types of IoT devices is achieved. There results a integral ecology.

Security: Since what is broadcast is sensitive information, security must be a first principle. It includes data encryption, access control and device authentication (M. Kranthi and R. C. Tanguturi, 2023).

It is only after knowing several of them that one can understand the finer points of their operation. The key components include:

Sensors: These devices gather data from the world of flesh. Sensors include temperature sensors, motion sensors and cameras.

Actuators: Actions that run on data received by devices. Such as motors, servos or valves working on sensor readings.

Communication Protocols: Protocols such as MQTT, CoAP and HTTP provide a basis for data exchange between devices.

Network Infrastructure: These include wired and wireless network connected devices, such as Wi-Fi (WLAN), Bluetooth or cellular.

Edge Computing: Latency reduction and getting the bandwidth back: Local data processing on devices (P. Ramesh Naidu and N. Guruprasad, 2021).

Cloud Computing: In the cloud, massive amounts of data can be processed and stored. Scalability and accessibility are ensured.

Middleware: Middleware components are intermediaries between devices and applications (M. Shruthi and B. Naveen K, 2020). Message brokers and application programming interfaces (APIs) play an essential role in this.

User Interface: They provide a means of testing and managing IoT devices through graphical interfaces or dashboards. It can be web based, mobile applications or dedicated software.

Understanding how IoT works depends on understanding these components and principles, without which even if we add robotics to the equation it would be impossible for us to imagine intelligent systems that are able to see, hear walk and speak. In the next few chapters, we will explore some potential synergies between IoT and robotics based on this knowledge.

2.1 Robotics Basics

From the structure and control of robots through to highly sophisticated sensor systems, a rich technology lies behind robotics. Understanding Robotic Basics One must first understand the different types of robots and their varied applications (M. Ramesha, and G. N. Pai, 2020). They can roughly be divided into three categories of machines, the industrial robots which are specially designed for different tasks in various environments; medical robots and service robots. Robot capabilities today are varied, and robots often use advanced sensors such as cameras, LiDAR (light detecting and ranging), or tactile sensors to see their surroundings. Robot applications stretch from manufacturing, through healthcare and exploration to education and even entertainment (R. S. Varun, and T. Rajesh, 2018). As robotics has advanced, advancements in the fields of artificial intelligence and machine learning have made possible a shift from rigidly programmed movements to more flexible adaptive behaviours. Having already covered the robotics fundamentals, one thing becomes clear--these machines are not just implements; they are elements of intelligent systems themselves. Robots can learn, adapt, and coexist with humans in an astonishingly wide range of applications.

The messaging patterns describe the link and dialogue between two rival services. The communications patterns can be divided into two types of events: Synchronized and asynchronous, as the image indicates. 3. Synchronous: This is a situation in which A1 asks A2 for information and receives it, or goes to get information from A2 and gives it to them. Since the procedure always needs a response, A1 will remain stopped until the instance finishes processing (J. J. Jijesh, and Shivashankar, 2017). On the other hand, communication between B1 and its subscribers, B2 and B3 is asynchronous—i.e., if B1 doesn't especially care how (or whether) B2 or they respond to incoming messages; or if B1 doesn't particularly need B1 doesn't have to wait until one message has finished before sending another to the same recipient.

Cloud services often use two distinct communications patterns: push and pull. Pushes are asynchronous events, while pulls are synchronous. Within Figure 4. With these two services represented by A1 and A2, one is the publisher and one's a subscriber (A. Krishnan, S. Swarna and B. H. S, 2020). If the first scenario is correct, then whenever A2 receives a message on a certain issue or channel, it will send out a push notification to A4. With asynchronous push, A2 can independently send out many messages to A4 or other services at the same time.

A protocol is the set of rules governing how data are sent from one service to another. The most widely used communication protocols supported by IoT systems are Message Queuing Telemetry Transport (MQTT), MQTT over WebSockets, and Hypertext Transfer Protocol (HTTP). Figure 5 illustrates these protocols.

Figure 3. Illustration of pipeline instances demonstrating a) synchronous and b) asynchronous messaging models

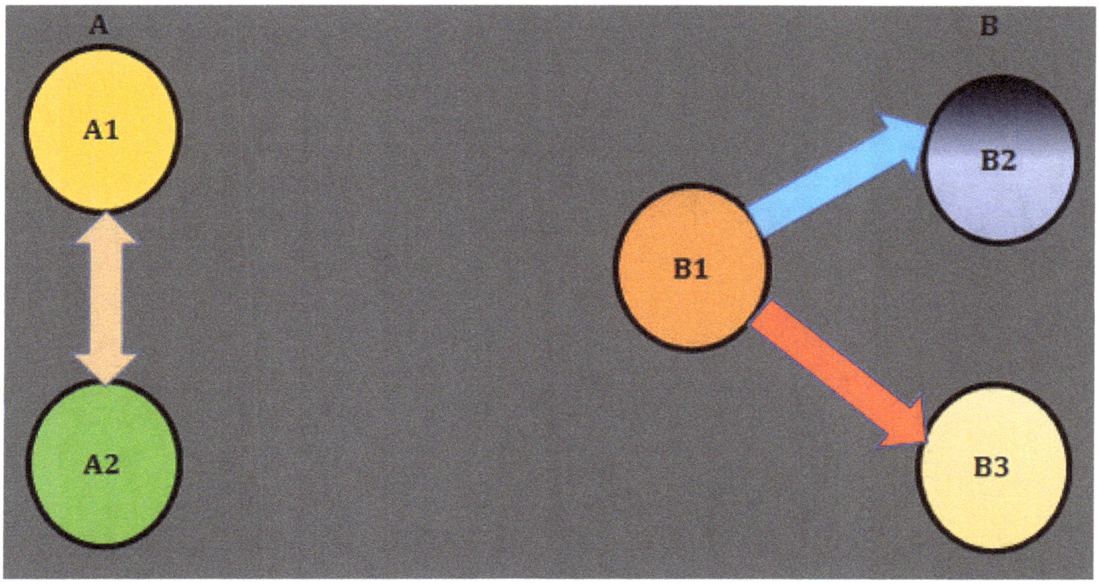

Figure 4. Push and pull messaging patterns illustrated

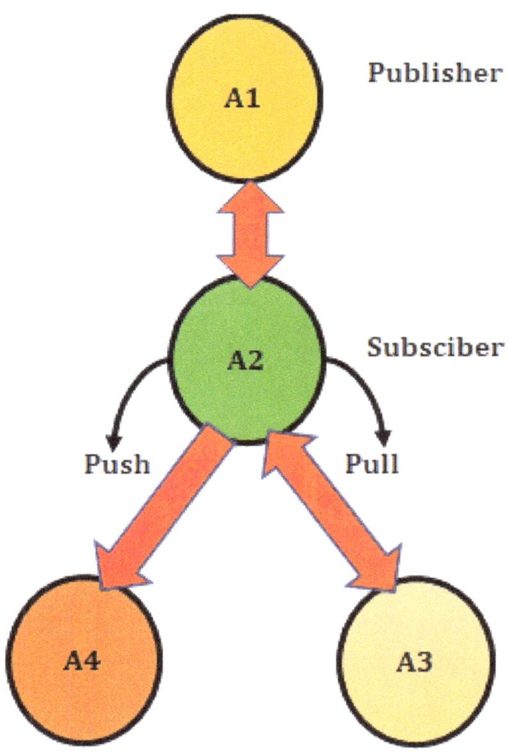

Figure 5. Illustrating various communication protocols

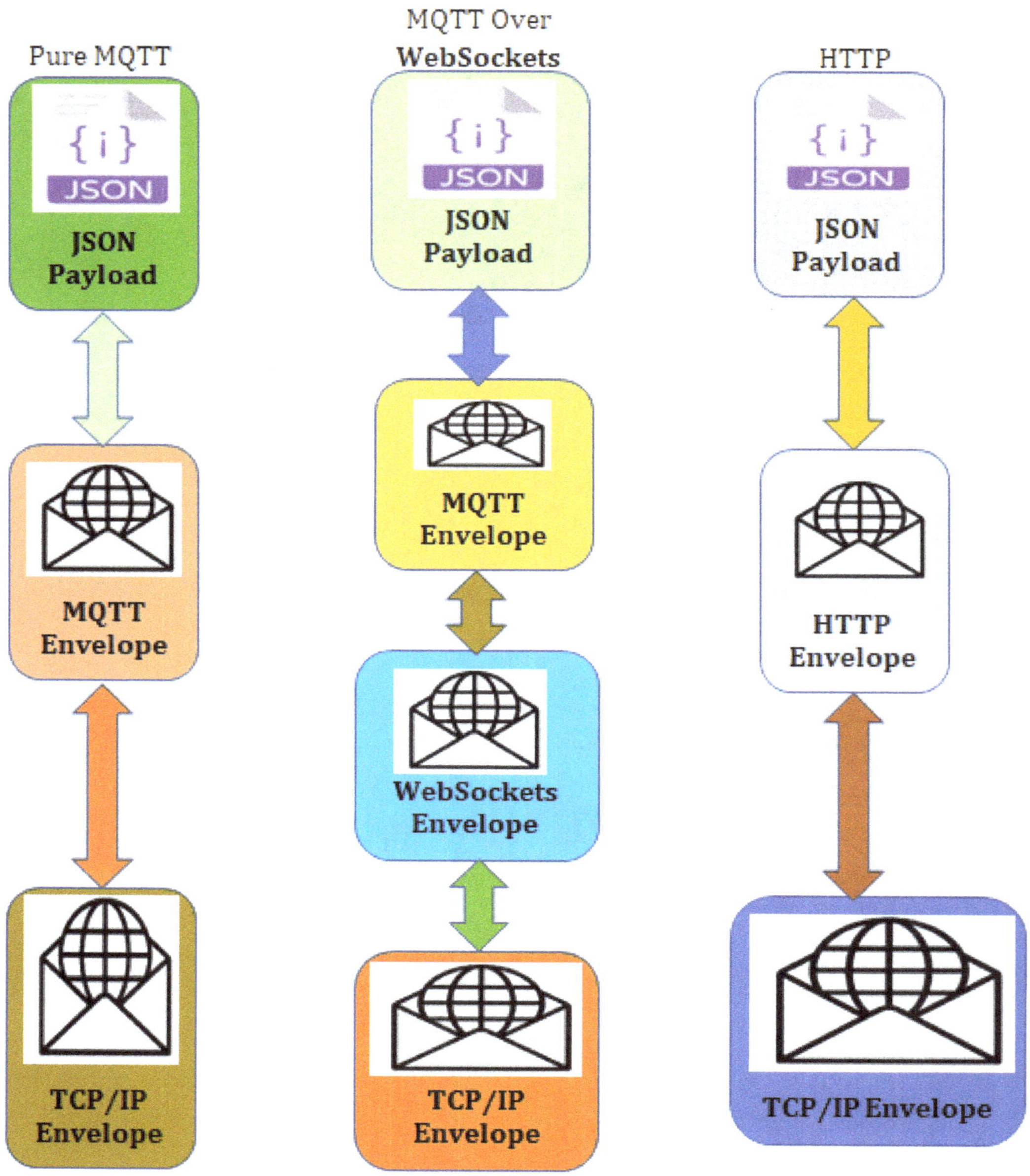

3. SYNERGIES BETWEEN IOT AND ROBOTICS

Interconnectivity is at the core of IoT and robotics integration. This section examines how IoT allows robotic devices to communicate and exchange data with ease. Some examples are provided, which demonstrate the advantages of connected robotics systems.

3.1 IoT as the Enabler of Interconnectivity

Sensor data can be shared among robotic devices through the internet of things. For example, drones carrying environmental sensors can supply real-time data to roadside robots in agricultural fields, enabling them to make accurate decisions as long as all the information is there. By combining several robotic systems together, collaborative decision making becomes possible. Robots working on different stages of production can speak to each other in manufacturing to improve workflow efficiency, adapt quickly and increase overall efficiency (J. Khan, F. Ali, F. Ullah, J. Bacha and S. Lee, 2019). IoT's interconnectivity allows for flexible task assignment among robots. For example, robots equipped with RFID technology in warehouses can communicate information about packages and optimize sorting and delivery based on real-time demand.

3.2 Examples of Interconnected Robotic Systems

IoT links robotic arms on the production line in smart manufacturing environments. They exchange data on production targets, quality control metrics, and equipment condition to ensure a smooth-flowing manufacturing process. With IoT sensors on board, agricultural robots work together in precision farming. For instance, drones are dispatched to gather data on soil health or crop conditions (W. Liu, 2021). They talk to ground-based robots which actually perform the work of planting, fertilizing and harvesting--all in order to improve agricultural methods.

IoT networked autonomous vehicles optimize traffic flow. Intelligent road vehicles collect information from each other and share it with smart infrastructure in order to avoid traffic jams, reduce collisions, and increase the overall efficiency of transportation.

Medical robots are playing an important role in operations. Medical robots endowed with IoT features send out real-time information to monitoring devices, allowing surgeons to make more objective assessments, and improving the accuracy of medical treatment (H. M. Mishra, A. Qtaishat and M. Kumar, 2023).

Warehouse cobots are netted with IoT. They exchange data on inventory levels, delivery schedules, and warehouse layouts, so that route planning can be flexibly adjusted as needed and orders can be delivered quickly and efficiently.

Starting to take shape is an original idea: The Internet of Robotic Things (IoRT), which pairs the Internet of Things (IoT) with cloud robotics, the blend of cloud computing and robotics (D. Pal and S. Joshi, 2023). For the exchange, processing and storage of interoperable data, IoRT uses preexisting communications technology and cloud computing resources provided by cloud providers. Figure 6 identifies the five levels in the architecture.

3.3 Advantages of Interconnected Robotic Systems

A series of interconnected robotic systems can adapt and change in real time, making for the best performance possible in a changing or uncertain environment (S. M. Sali and K. R. Joy, 2023). Robotic devices optimize workflows through communication, eliminating bottlenecks and improving efficiency in both tasks and processes. Robots connected to a network can exchange resources-information, energy and computing power--thereby developing more cooperative patterns. Inter-connected robots share data, which leads to more fully informed decision making and drives greater accuracy and effectiveness in their actions. With interconnected systems, even if one robot has a malfunction the others can readjust

Figure 6. Infographic illustrating every layer for a holistic system overview

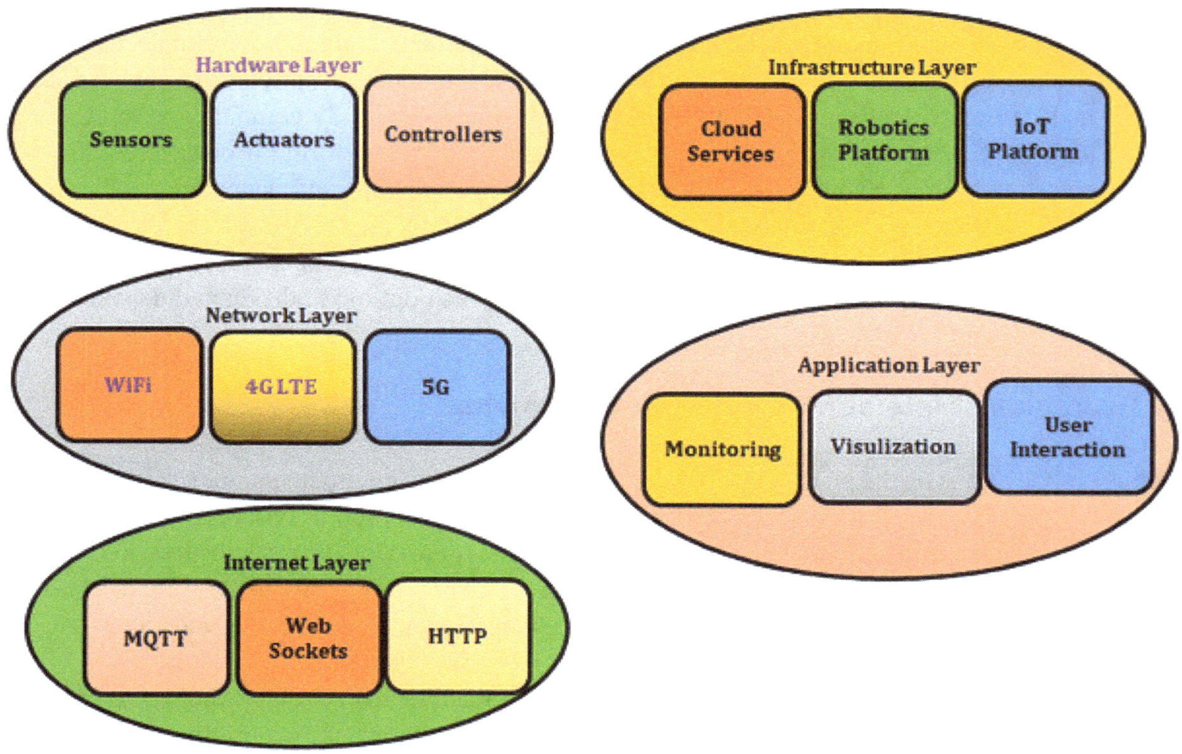

and make up to ensure continuous running with small downtime. The system is scalable, flexible and extensible due to the interconnecting capabilities of robotic devices.

The IoRT platform is comprised of three main elements, as shown in Figure 7: the Internet of Things (IoT), cloud services and robotics platforms. In the picture, ROS is represented as a network of connect robots. The key features are built on AWS cloud services like DynamoDB for storage and AWS Lambda for processing. One part is the Internet of Things (IoT) platform AWS IoT Core, which links up with the centralised function as well as acting as a bridge between all the robots.

Its publish-subscribe communications architecture means that ROS is a leading open-source robotic platform for robotics development. A master node for the network can be used as a common meeting point where nodes announce the topics and channels to which they are subscribed, and publish. Each node in the ROS network talks to every other via fixed channels and topics. The ROS master is then readied, connecting the publishing and subscribing nodes point-to-point and listening continuously for new ads. An agreed-upon transport protocol, such as TCPROS or UDPROS, is used to bring up the connection between various nodes. The setup for connection is shown in Figure 8. To put it simply, there is more to interconnection in robotic systems than just information transmission. By establishing such a synergistic network, the overall advance of the production process participating robots share data and intelligence toward common goals. These connected systems create a new age of efficiency, flexibility and precision in a wide range of fields.

Figure 7. Illustrating the components of the service oriented test platform

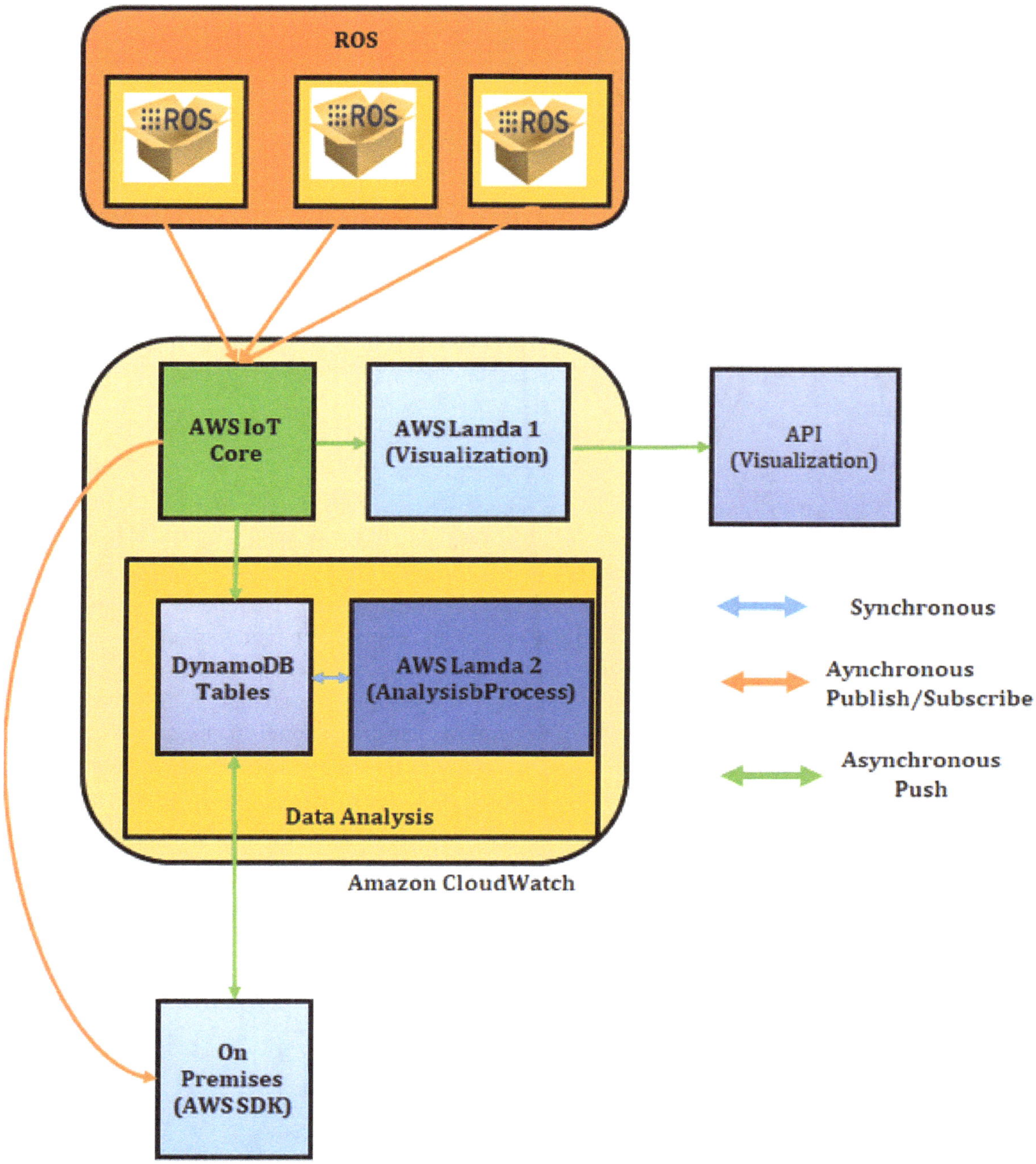

3.4 Sensor Integration

It is at the core of both IoT systems and robotics and serves as a bridge between the physical world and digital realm. But in this context, sensors are the eyes and ears of smart machines, taking data from their

Figure 8. Infographic detailing the stages of connection establishment for clarity

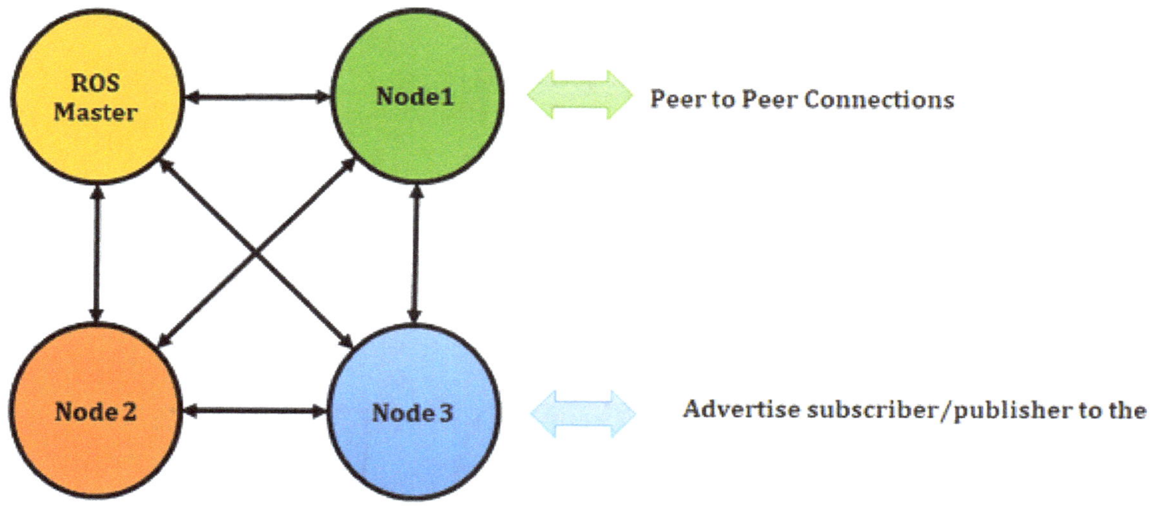

surroundings and providing the basis for well-informed choices. Sensors, including cameras, ultrasonic sensors, gyroscopes and temperature sensors are used to integrate various kinds of sensing. They enable robots and IoT devices to sense their environment. Just as cameras make visual perception possible for robots and allow them to identify objects, navigate spaces and even understand human gestures, microphones are the ears through which words can be heard. Ultrasonic sensors help with distance measuring, making obstacle avoidance and navigation easier. Gyroscopes improve spatial recognition, enabling precise maneuvering and orientation control. Temperature sensors, among others, are environmental sources of information important for all sorts of applications--from industrial processes to climate monitoring. Sensor integration in robotics and IoT has brought about the synergy of an integrated perception. A machine that understands its surroundings will adapt itself dynamically to changing conditions. The ongoing progress in sensor technology, combined with the application of machine-learning algorithms to data interpretation, enhance the sensing capabilities of these systems. Now that sensor integration is rapidly improving not only the purity, efficiency and effectiveness of robotic operations but also presenting rich possibilities for applications throughout a range of fields-a future in which intelligent systems can enhance their communication with and understanding of our flesh-and-bone world.

4. APPLICATIONS OF INTEGRATED IOT AND ROBOTS

Predictive Maintenance in Automotive Manufacturing:

Description: One major automotive manufacturer installed IoT sensors in robotic arms and machinery on their production line. These sensors monitor conditions in real time, collecting temperature, vibration and usage data.

Outcome: Predictive maintenance algorithms, based on the collected data, anticipate faults before they arise. The company reduced unexpected downtime, which meant increased productivity and cost savings.

Collaborative Robotics in Electronics Assembly:

Description: An integrated electronics assembly plant also installed collaborative robots (cobots) with IoT capabilities. In the assembly process, these robots worked with human workers to transform raw materials into finished products. IoT sensors observed and measure how humans interacted with the robots.

Outcome: Using the IoT data, workflow and worker-robot collaboration were optimized. Based on real-time data, the enhanced safety protocols reduce the risk of accidents. As a result, efficiency increased and the assembly line was more flexible and malleable.

Smart Factories with IoT-Connected Machines:

Description: A textile manufacturer equipped its manufacturing machines with IoT connectivity by attaching sensors that measured machine performance, energy consumption and yield.

Outcome: Real-time data analysis allowed preemptive decision making. Settings for best efficiency were adjusted automatically by machines, and production schedules optimized based on fluctuations in demand. This put efficiency overall up by 20%.

Supply Chain Optimization in Food Processing:

Description: A food processing facility integrated IoT sensors along their supply chain, from raw materials in storage until the finished product was out on delivery. Temperature, humidity and storage conditions were monitored by sensors.

Outcome: The company analyzed IoT data and thus better managed inventory, saved waste from spoilage, and ensured the quality of finished products. This resulted in a slicker supply chain, with lower operation costs.

Remote Monitoring and Diagnostics in Heavy Machinery:

Description: A mining company installed IoT sensors onto their heavy machinery to enable remote monitoring. These sensors monitored equipment condition, fuel levels and operational parameters.

Outcome: The company used predictive analytics based on IoT data to identify potential problems before they caused breakdowns. Remote diagnostics were used for timely maintenance, cutting unplanned downtime by 30% and increasing the lifespan of this machinery.

4.1 Smart Cities and Infrastructure

Automated City Surveys with Drones:

Description: Sensors using the Internet of Things (IoT) on drones can conduct automatic surveys of cityscapes. These sensors detect building conditions, the state of infrastructure and changes in the cityscape.

Outcome: Prepared data is used to help make decisions about urban planning. Drone technology is a cheaper and more efficient way of seeing whats happening in the city, identifying areas needing attention or development.

Environmental Monitoring for Sustainable Development:

Description: As for the environmental factors of air quality, noise levels and temperatures in urban areas, integrated robot systems consisting of ground sensors and drones are adopted.

Outcome: Real-time environmental data has informed urban planners of how development threatens the ecology. This approach provides a foundation for sustainable development, and allows data-driven decision making to minimize environmental impact.

Smart Traffic Lights and Adaptive Signage:

Description: These traffic lights are also equipped with IoT sensors and cameras that create a smart system for the control of traffic. The lights can time it to real-time traffic, and electronic billboards show drivers changing information.

Outcome: As a result of the adaptive traffic management system, roads are less congested and travel time is reduced. This integration increases citywide transportation efficiency.

Autonomous Vehicles for Public Transportation:

Description: Equipped with IoT sensors, autonomous buses and shuttles plough out preprogrammed routes at the owner's whim. Passengers can interact with the system through mobile applications or smart kiosks.

Outcome: Making public transit more accessible and efficient. Routes are adjusted dynamically according to the patterns of demand, reducing wait times and improving transportation lines.

IoT-Enabled Predictive Maintenance for Utilities:

Description: Like water pipes and electrical grids, utility networks are equipped with sensors. Some of these sensors monitor such elements as pressure, flow rates and electric current.

Outcome: The raw data from the sensors are used by predictive maintenance algorithms to find potential problems before anything shuts down. Taking this approach is better for downtime, lower repair costs and more reliable utilities.

Robotic Inspection of Bridges and Highways:

Description: Equipped with cameras and sensors, drones and robotic devices inspect the condition of bridges, highways or other important structures.

Outcome: By performing frequent and automatic checks, damage caused by wear and tear, corrosion or structural problems can be detected early. With this method, maintenance is done quickly. Disasters are averted, and the life of assets is prolonged.

4.2 Benefits of Integrated Robotics and IoT in Smart Cities

Efficient Resource Management: Cities with integrated systems can efficiently manage resources such as energy, water and transportation. They save money while promoting sustainability.

Improved Quality of Life: Smart city technologies allow people to live better in terms of mobility and congestion, as well as in a more sustainable, eco-friendly urban environment.

Enhanced Safety and Security: Through IoT sensors and integrated robotics, safety is enhanced by measuring environmental conditions, automating emergency responses and performring crime prevention in cities.

Sustainable Development: A data-centered approach to urban planning, traffic organization and infrastructure upkeep promotes sustainable development while keeping a low impact on the environment.

Informed Decision-Making: When combined, robotics and IoT enable city planners and administrators to have real-time information, allowing for informed decisions as they begin building their cities.

To have ROS and AWS IoT work together, in both systems you must specify the topics and channels. This is because they both subscribe to a publish/subscribe communications model. A publisher selects a subject or channel to publish to, while a subscriber chooses a channel or topic to follow. Graphic 9 shows the ROS communication pipeline. The graphic comes with a detailed explanation.

These strategies are used in data analysis in various ways and may be represented very differently depending on the approach and application. A popular method of data analysis, data mining is both synonymous with and an integral component of the Knowledge Discovery in Databases (KDD) process, as explained in the book Data Mining: Concepts and Techniques. The figure.10. shows the procedure. As such, the actions will not be comprehensively evaluated because they exceed its scope. Nevertheless, the case study adopts this approach as a test to measure the capabilities of IoRT platform. Although the

Figure 9. Visualizing the ROS message workflow: interaction between local hardware (orange) and the cloud (green)

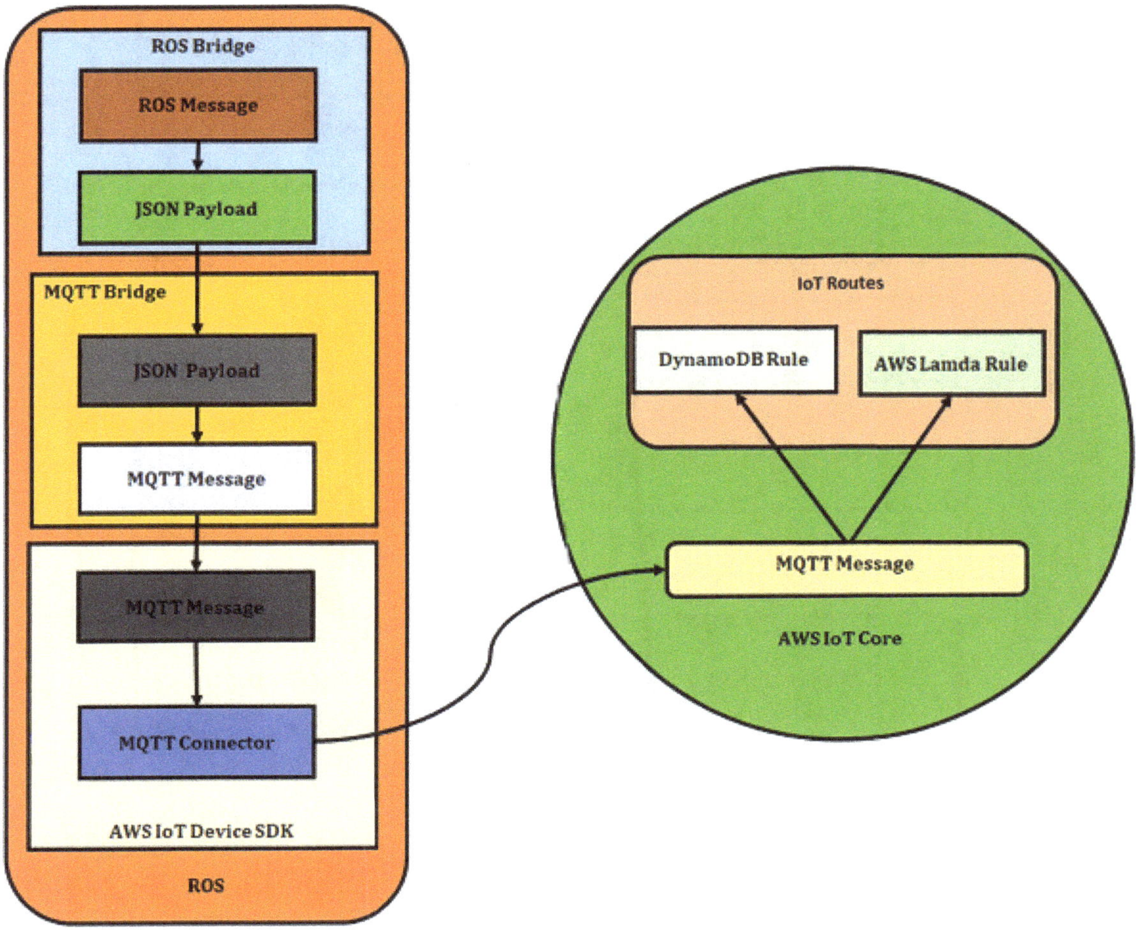

combination of robotics and IoT in smart cities certainly represents a technological breakthrough, it is just as much a strategic attempt by smart planners to counteract the multifaceted challenges of urbanization. Leveraging these technologies, cities can become more efficient, environmentally friendly and responsible to their citizens.

5. CHALLENGES AND SOLUTIONS

The combination of robotics with the Internet of Things (IoT) can lead to many advantages, but also creates security and privacy concerns about data exchanged in these systems.

Figure 10. Data analysis process

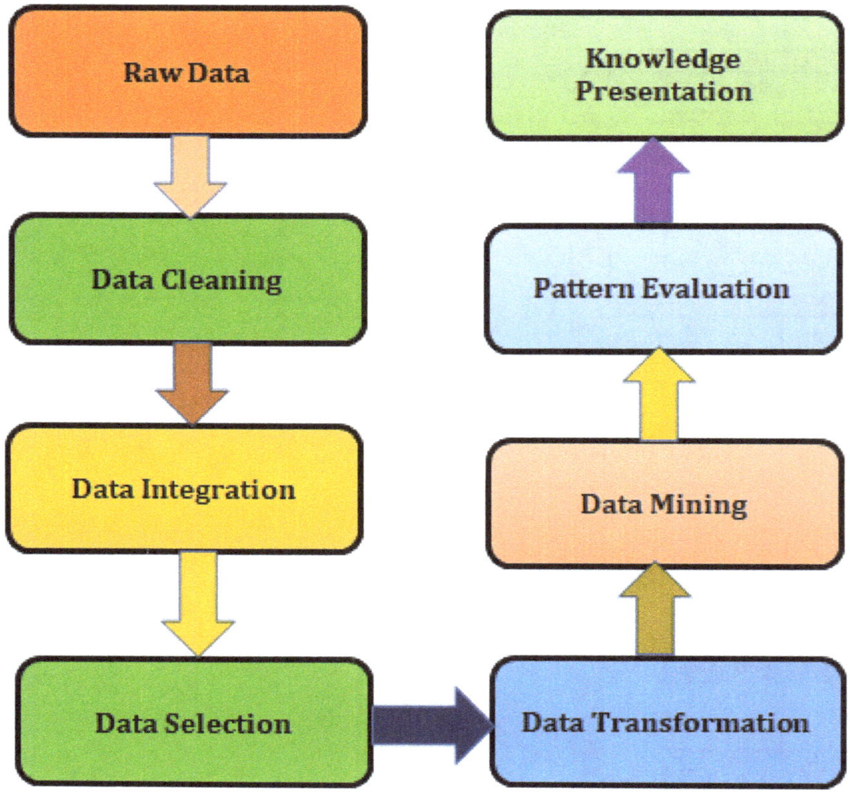

5.1 Technical Challenges

Following are a few problems created by the combination of the Internet of Things (IoT) and robotics that must be resolved before these technologies can reach revolutionary heights. Below are some key challenges and potential solutions:

Interoperability and Standardization:

Challenge: Many different kinds of IoT devices and robotic systems use multiple communication protocols and standards, raising interoperability problems.

Solution: Standardized communication protocols (i.e., MQTT, OPC UA) have been widely developed and adopted, making it easy for devices to talk with each other. This is all the more important as it requires industry-wide cooperation to formulate common standards.

Connectivity and Latency

Challenge: There are many applications of robots in which real-time communication is necessary. The performance of integrated systems can be hampered by latency.

Solution: An edge computation approach, by doing more of the data processing closer to the source, eliminates latency dependent on centralized cloud services. In addition, 5G technology provides higher network speeds and reduced latency make connected devices more responsive.

Security and Privacy

Challenge: Maintaining the security and privacy of data for integrated systems. Furthermore, massive amounts of data are generated by IoT devices and robots.

Solution: Enforcing strong security includes encryption, authentication, and access controls to protect data. Technologies such as secure hardware and decentralized identity can help to strengthen security and privacy.

Power Consumption and Energy Efficiency

Challenge: Most IoT devices and robotic systems run on battery so energy efficiency is very important to long-life operation.

Solution: Improvements in energy efficiency can be made through innovations in low-power design, the use of energy-efficient components, and the choice of advanced batteries or alternative sources of power such as solar or kinetic harvesting.

Cost of Implementation

Challenge: One potential obstacle involves the initial cost of implementing integrated systems.

Solution: With technologies maturing and adoption rising, costs go down. Furthermore, open-source platforms and joint efforts can lower implementation costs by offering shared resources and information.

Ethical Considerations and Regulation

Challenge: Dealing with ethical considerations when deploying IoT and robotics, and keeping up to date regulations.

Solution: By developing and implementing ethical guidelines for the development and use of integrated systems, and actively participating in regulations concerning their use, we will be able to use these technologies responsibly and acceptably as a society.

5.2 Future Innovations and Emerging Technologies

AI and Machine Learning Advancements: Advanced AI and machine learning technology is bringing improved decision making, predictive analytics and pattern recognition, all in integrated systems.

Quantum Computing: In particular, the potential of quantum computing can solve many complex computational problems. This will improve optimization, simulation and data processing by leaps and bounds.

Biologically Inspired Robotics: This draws on biological systems with its designs and control mechanisms that are as efficient and flexible as those in nature. These all belong under the banner of biomimetic robotics.

Swarm Robotics: Swarm robotics is the art of building large teams of relatively uncomplicated robots to accomplish tasks collectively, following the lead set by social insects. This strategy boosts scalability and replicability in robotic systems.

Neuromorphic Computing: Aiming to manufacture processors that recap the structure and function of neural networks, neuromorphic computing is based on human brain. This could result in robots that are more productive and flexible.

Blockchain for Security: In integrated systems, blockchain technology is a decentralized and safe way to do data transactions. It can further increase the security and transparency of this process.

Biohybrid and Soft Robotics: A: There have been innovations in soft robotics, as well as biohybrid systems that combine biological and synthetic components. Robots can now handle more demanding environments than ever before, and interact with the natural world more smoothly.

Advanced Materials: Improvements in lightweight, durable and flexible materials increase the design potential of robotic components. As a result, robotic systems become more versatile, capable and adaptable to different applications.

Explainable AI: Where decision making involves AI, explainable AI models provide transparency and grity in terms of the rationale behind decisions; they speak to questions of accountability and trust.

Edge AI and Processing: Local processing of data, or edge AI as it is known, improves the speed of decision making and decreases reliance on Internet connection. Dealing with these technical barriers will help bring IoT and the robotics industry to a whole new realm.

6. FUTURE TRENDS

From the standpoint of new technique invention, this is a mighty time for artificial intelligence (AI) and Web-based applications; 3D printing has strip away an important hindrance to production. The Internet of Things (IoT), robotics will undoubtedly shake up traditional human societies It has far-reaching effects across many industries, combining intelligent algorithms, data analytics and robotic systems into one. Looking ahead, then, at the world of integrated IoT and robotics where machine learning and AI create ever expanding flexibility as well higher levels of dynamic movement toward increasing fulfillment.

Sensor Data Fusion for Enhanced Perception: Sensor data fusion becomes a critical element as integrated Internet of Things and robotics coalesce in the evolution of intelligent systems, improving their sensors. In this paradigm, various sensors work together synergistically to yield a nuanced view of its environment. However, now that data from different sensors such as cameras, LiDAR and tactile sensors is integrated together, a more complete perception is achieved with capabilities even greater than what individual sensors can offer.

Cognitive Computing for Human-Robot Interaction: Among integrated IoT and robotics systems, Cognitive computing is on the front lines of changing human-robot interaction. This Scenario is based on cutting-edge AI techniques that can equip robots with higher level cognitive capabilities and includes NLP, gesture recognition and more which copy some human thought processes. With NLP, human commands can be understood by robots and intuitive, natural communication becomes possible. Cognitive computing- enabled robots can decipher and understand complex language forms. This allows the robots to not only approach people for dialogue, but even carry out tasks according to verbal commands. Moreover, gesture recognition will make machines able to read and react appropriately to human gestures increasing the smoothness of communication between man and machine. With the development of these reasoning abilities, robots are not only more responsive but also better adapted to human speech.

Future Innovations in AI and Robotics Integration: In the future, combining AI and robotics integration will bring about revolutionary developments that will revolutionize industries and raise the limits of intelligence. Moreover, with technology gradually catching up, the focal points are Explainable AI (XAI), AI-assisted sensor routing, edge-based learning for real time response and the definition of a cognitive robot. With explainable AI, the decision processes of AI systems become more transparent and explicable for humanity. Critical applications such as healthcare and autonomous vehicles can rely on this to build trust. Under the impetus of AI, sensor fusion is becoming much more advanced. It helps robots overcome limitations in their perceptual powers and become successful participants in complex environments. To reduce the latency involved in robotics applications such as drones and autonomous vehicles, Edge AI moves AI processing closer to the data source, enabling real-time decision making.

Cognitive robotics, focused on endowing robots with human-style thinking ability as well, makes possible applications in the medical field of areas where robots understand people's needs and can adapt to changing circumstances. These future innovations both enhance the adaptability and learning abilities of robotic systems and resolve some major explainability, real-time processing, and human-robot collaboration problems. In the future, intelligent machines will run across industry and society in increasingly complex ways. As AI and robotics integration continues to grow by leaps and bounds, these innovations are poised to play a crucial role defining this new era of intelligent machines.

6.1 Human-Robot Collaboration

In a world of integrated systems, the changing relationship between humans and robots is remaking industries, upgrading productivity, and transforming society. The human and robot demonstrated their synergistic potential in collaborative applications. In this section, we explore both examples of human-robot collaboration and the social impact they have.

Logistics and Warehousing: The combination of IoT and robotics has completely overturned how logistics and warehousing works, ushering in a new age of efficiency and automation. Where intelligent robots, furnished with IoT sensors, relax in a comfortable chair and put their feet up. Their main functions are to help reduce redundant work preparing materials for use and to keep track of inventory management, order fulfillment, and material transportation. Seamlessly negotiating the warehouses, autonomous mobile robots optimize the flow of goods and require less manual intervention. The power of the Internet of Things (IoT) Overcoming the limitations of geographically distributed delivery centers, such as slower service and lower response times, suggests that in the future drone deliveries will be provided collaboratively. With IoT devices able to collect and share real-time data, visibility across the supply chain is improved and future demand can be forecasted more accurately as well as inventory optimized. This, in turn, means lower costs of businesses, higher levels of reliability and faster reaction speeds to customer needs. The combination of IoT and robots in logistics not only increases operational efficiency, it also serves as a precursor to the smart warehouses of tomorrow, with their array of connected machines humming away together in response to one another's movements.

Education and Training: The combination of IoT and robotics is transforming the face of learning and teaching, providing strategic approaches to improve learning opportunities and equip people with the skills they need in a changing world. Robots fitted with IoT functions act as interactive teachers of the first order, guiding pupils in a variety of tailored learning activities. These robotic teaching assistants are able to adjust their approaches depending on students 'learning styles, making for a more personalized and flexible educational experience. Also, use of IoT sensors enables robots to collect information on student performance. Teachers can then analyse the data and determine areas for improvement as well as design plans specifically for each individual. Regarding workforce training, for example, in a variety of situations collaborative robots can quite realistically perform as furniture. The trainees themselves are then able to come up and make actual contact with this synthetic reality, obtaining hands-on experience yet still staying safely within the previously determined boundaries of the environment. Therefore, the application of IoT and robotics to education was able not only to enhance the existing methods of instruction, but also brought forth a more stimulating and flexible teaching format for students as well as prospects for workers in terms of industrial development.

Entertainment and Social Interaction: The combination of IoT and robotics is changing how we entertain ourselves, interact with one another, and experience new dimensions. Iot-equipped social robots are soon coming out into public spaces. They can be seen there as companions and entertainers, chatting people up in shopping malls, hotels or large festivals etc.. These robots can sense human emotions and reactions, gestures, and speech. The robot uses this information to adapt the way that it behaves, creating a more enjoyable interaction for everyone involved. In homes, AI assistants powered by IoT offer entertainment and company. This enhances the quality of life in general. Moreover, the introduction of robotics into entertainment places including theme parks makes for exhibit attractions like no other. Such robots are used to provide fascinating and lively shows. With IoT and robotics becoming ever more advanced, the line between technology and fun is blurring. Interpersonal interaction gets back to its roots in novel ways for both leisure and a friend.

Societal Implications and Considerations: The confluence of IoT and robotics opens up a whole range of societal implications and considerations, including but not limited to the following. With the increasing occurrence of applications that require collaborative work by humans and robots, in addition to considering it as an outcome of the development of science and technology, we also have to carefully analyze how it's changing our lives. For example, the changing working environment poses one challenge. As robots are increasingly used in a cooperative role, employment dynamics will change and job roles will likely be upgraded, which means that the workforce must be retrained and restructured. When robots enter daily life, there are necessary considerations of ethics that must be considered at a broad level: transparency, accountability and privacy. Hence, if the two are to work effectively together, trust between humans and robots becomes of vital importance.

7. CONCLUSION

In conclusion: The combination of robotics and the Internet of everything is a watershed moment in mankind's technological advance. It brings us into an era where intelligent systems are linked indissolubly to people themselves! In this chapter we have travelled over the transformative land of robotic evolution. We knew its historical setting and understood the meaning of integrating IoT. Afterwards we looked at the basic notions which make up this paradigm shift. From industrial automation to smart cities, health care to entertainment. Chapters introduced the key concepts. Sensors, artificial intelligence (AI) and machine learning were all woven into a lengthy dotted line of action so indeed perception can be sharpened. It allowed decisions to become flexible; adaptability soared. Taking us from theory to real-world applications in the manufacturing, health care, logistics, education and entertainment industries. It is there that we saw the practical results of collaborative robots with our own eyes. Yet, as is the case with all technological transformations, social anxieties are everywhere. The chapter also emphasized the need to clear up moral doubts, untangle relationships and establish trust between man and robot. From employment structure to accessibility, this paper reviewed how development should be put into effect from a balanced and responsible angle. The potential for future integrated IoT and robotics is endless. The trajectory shows continuous development, from Explainable AI innovation to the enhancement of security and anomaly detection. Such a fusion of robots, Internet-of-Things devices and human beings living in interconnected smart ecosystems has at its heart something like the best kind of future combining efficiency and sustainability while maintaining quality life.

REFERENCES

Avinash, K., & Kumar, A. (2023). A novel approach of unsupervised feature selection using iterative shrinking and expansion algorithm. *Journal of Interdisciplinary Mathematics*, *26*(3), 519–530. doi:10.47974/JIM-1678

Chinamuttevi, V. S. (2023). A novel RF-SMOTE model to enhance the definite apprehensions for IoT security attacks. *Journal of Discrete Mathematical Sciences and Cryptography*, *26*(3), 861–873. doi:10.47974/JDMSC-1766

El-Gendy, S. (2020). IoT Based AI and its Implementations in Industries. *2020 15th International Conference on Computer Engineering and Systems (ICCES)*, 1-6. 10.1109/ICCES51560.2020.9334627

Hegge, K. K., & Agrawal, R. (2023). Industrial Automated Multipurpose Robot Using WIFI. *2023 4th International Conference for Emerging Technology (INCET),* 1-8.

Hussain, N., & Samal, A. (2024). Design and Implementation of Single to Multi-cloud Securities in Network. In N. R. Shetty, N. H. Prasad, & N. Nalini (Eds.), *Advances in Computing and Information. ERCICA 2023. Lecture Notes in Electrical Engineering* (Vol. 1104). Springer.

Jijesh & Shivashankar. (2017). Design and implementation of automatic medicine dispensing machine. *RTEICT 2017 - 2nd IEEE International Conference on Recent Trends in Electronics, Information and Communication Technology, Proceedings*, 1962–1966.

K. S & Arun. (2022). Priority Queueing Model-Based IoT Middleware for Load Balancing. *2022 6th International Conference on Intelligent Computing and Control Systems (ICICCS)*, 425-430.

Khan, J., Ali, F., Ullah, F., Bacha, J., & Lee, S. (2019). Artificial Intelligence and Internet of Things (AI-IoT) Technologies in Response to COVID-19 Pandemic: A Systematic Review. *IEEE Access : Practical Innovations, Open Solutions*, *10*, 62613–62660. doi:10.1109/ACCESS.2022.3181605

Kranthi, M., & Tanguturi, R. C. (2023). Design of Intelligent Medical Integrity Authentication and Secure Information for Public Cloud in Hospital Administration. *2023 2nd International Conference on Edge Computing and Applications (ICECAA),* 256-261.

Krishnan, A., & Swarna, S. (2020). Robotics, IoT, and AI in the Automation of Agricultural Industry: A Review. *2020 IEEE Bangalore Humanitarian Technology Conference (B-HTC)*, 1-6. 10.1109/B-HTC50970.2020.9297856

Kumar, A., Kalpana, V., & Hariram, V. (2023). Artificial Intelligence based Health Monitoring System on IoTH platform. *2023 Second International Conference on Augmented Intelligence and Sustainable Systems (ICAISS)*, 1458-1463.

Kumar, P. S., & Sharma, A. (2019). IoT Enabled Indoor Autonomous Mobile Robot using CNN and Q-Learning. *2019 IEEE International Conference on Industry 4.0, Artificial Intelligence, and Communications Technology (IAICT)*, 7-13.

Kumar & Singhal. (2020). Design and Analysis of IoT-Based Intelligent Robot for Real-Time Monitoring and Control. *2020 International Conference on Power Electronics & IoT Applications in Renewable Energy and its Control (PARC)*, 549-552.

Liu, W. (2021). Improvement of Navigation of Mobile Robotics Based on IoT System. *2021 IEEE International Conference on Robotics, Automation and Artificial Intelligence (RAAI)*, 69-72. 10.1109/RAAI52226.2021.9507826

Mishra, H. M., Qtaishat, A., & Kumar, M. (2023). IoT Based Smart Systems using Artificial Intelligence and Machine Learning: Accessible and Intelligent Solutions. *2023 6th International Conference on Information Systems and Computer Networks (ISCON)*, 1-6.

Nayyar & Nagpal. (2018). Internet of Robotic Things: Driving Intelligent Robotics of Future - Concept, Architecture, Applications and Technologies. *2018 4th International Conference on Computing Sciences (ICCS)*, 151-160.

Pal, D., & Joshi, S. (2023). AI, IoT and Robotics in Smart Farming: Current Applications and Future Potentials. *2023 International Conference on Sustainable Computing and Data Communication Systems (ICSCDS)*, 1096-1101. 10.1109/ICSCDS56580.2023.10105101

Pavankumar, P., & Darwante, N. K. (2022). Performance Monitoring and Dynamic Scaling Algorithm for Queue Based Internet of Things. *2022 International Conference on Innovative Computing, Intelligent Communication and Smart Electrical Systems (ICSES)*, 1-7.

Prasad, K., & Gugan, I. (2023). Enhancing Microcomputer Edge Computing for Autonomous IoT Motion Control. *International Journal on Recent and Innovation Trends in Computing and Communication*, *11*(3), 58–67. doi:10.17762/ijritcc.v11i3.6202

Prasad, V. S., Mahajan, Y., & Suneetha, S. (2023). A Cloud-Based UV Monitoring System for Remote Real-Time UV Exposure Tracking. *2023 4th International Conference on Smart Electronics and Communication (ICOSEC),* 1764-1770.

Ramesh Naidu, P., Guruprasad, N., & Dankan Gowda, V. (2021). A High-Availability and Integrity Layer for Cloud Storage, Cloud Computing Security: From Single to Multi-Clouds. *Journal of Physics: Conference Series*, *1921*(1), 012072. doi:10.1088/1742-6596/1921/1/012072

Ramesha, M., & Pai, G. N. (2020). Internet of things: Internet revolution, impact, technology road map and features. *Adv. Math. Sci. J.*, *9*(7), 4405–4414. doi:10.37418/amsj.9.7.11

Ranjan, A., Hegge, K. K., & Agrawal, R. (2023). Wireless and Solar-Powered Multipurpose Robot for Industrial Automation: A Sustainable Solution. *2023 8th International Conference on Communication and Electronics Systems (ICCES)*, 146-151.

Reddy, N. S., & Ashreetha, B. (2023). Technologies for Comprehensive Information Security in the IoT. *2023 International Conference for Advancement in Technology (ICONAT)*, 1-5.

Reddy, S., & Patwal, P. P. S. (2022). Data Analytics and Cloud-Based Platform for Internet of Things Applications in Smart Cities. *2022 International Conference on Industry 4.0 Technology (I4Tech)*, 1-6.

Sali, S. M., & Joy, K. R. (2023). Intelligent Rover: An IoT Based Smart Surveillance Robotic Car for Military. *2023 2nd International Conference on Computational Systems and Communication (ICCSC)*, 1-6. 10.1109/ICCSC56913.2023.10143011

Shekhar, R. (2023). Securing networked image transmission using public-key cryptography and identity authentication. *Journal of Discrete Mathematical Sciences and Cryptography*, *26*(3), 779–791. doi:10.47974/JDMSC-1754

Shruthi & Naveen. (2020). Internet of Things: A Survey on Devices, Ecosystem, Components and Communication Protocols. *2020 4th International Conference on Electronics, Communication and Aerospace Technology (ICECA)*, 611-616.

Singla, A., & Sharma, N. (2022). IoT Group Key Management using Incremental Gaussian Mixture Model. *2022 3rd International Conference on Electronics and Sustainable Communication Systems (ICESC)*, 469-474.

Srinivas, V., & De, T. (2023). Wi-Fi Router Signal Coverage Position Prediction System using Machine Learning Algorithms. *2023 International Conference on Sustainable Computing and Smart Systems (ICSCSS)*, 253-258.

Varun, R. S., & Rajesh, T. (2018). Implementation of swarm intelligence in obstacle avoidance. *RTEICT 2017 - 2nd IEEE International Conference on Recent Trends in Electronics, Information and Communication Technology, Proceedings*, 525–528.

Chapter 7
Scaling Robot Intelligence With Cloud-Based Cloning Platforms

Dankan Gowda V.
https://orcid.org/0000-0003-0724-0333
Department of Electronics and Communication Engineering, BMS Institute of Technology and Management, Karnataka, India

Kirti Rahul Kadam
Management Department, Bharati Vidyapeeth University, Kolhapur, India

Suma S. G.
Department of Computer Science and Engineering, Mahaguru Institute of Technology, India

Shrikant D. Bhopale
https://orcid.org/0009-0004-4130-0605
Dr. D.Y. Patil Pratishthan's College of Engineering, Kolhapur, India

R. Nithya
Department of Computer Science and Engineering, Vivekananda College of Engineering for Women, India

ABSTRACT

Robotics needs to be applied in many different areas which requires the existence of better-quality platforms whose performance and function can be refined. The cloud and robotic cloning in this chapter offer an innovative idea. As a result, utilizing the immense capabilities and scalability of cloud computing can only make robotic systems better and more capable in multi-robot cooperative scenarios. The chapter offers an outline of the theoretical foundations, advantages, possible concurrent applications, and future direction for cloud-based cloning platforms. By exploring the subject in depth, readers will discover how cloud-based cloning platforms can play the role of a major enabling factor making scaling up intelligence and capability possible as multi-robot systems open new frontiers for robotics.

DOI: 10.4018/979-8-3693-1914-7.ch007

1. INTRODUCTION

Robotics and artificial intelligence (AI) have experienced a complete world over the past decade. It is an age where machines are more than smart tools they must learn and adapt just like we humans. These technologies coming together thus offer great potential for revolutionizing many industries, from manufacturing to healthcare. But as we explore the frontier of AI embodied robotics, there is an immediate problem-scalability. In today's blanket of robotic systems, the bottleneck to widespread application and efficiency lies in the limitations of individual devices (Kumar and R. Singhal, 2020). These limitations, from basic computational limitations to the low capabilities of memory itself, affect robots' ability to cope with more and more complex tasks. Enter the paradigm of cloud-based cloning platforms, a visionary concept that addresses the scalability dilemma and opens new horizons for the deployment of intelligent robotic systems (Nayyar and A. Nagpal, 2018). The idea of cloud robotics is a new paradigm that has opened up a lot of doors for robotics research. Without a doubt, the availability of storage, ease of access, and remote computing assistance provide an appealing domain of study that aims to improve the on-board robot's capacity to do complicated jobs. The first operational cloud robotics project, RoboEarth, was announced in the middle of 2009, however networked robotics and online assistance in robotic systems may be dated to the end of the 1990s. Cloud computing is expanding at a rapid pace, which is attracting more and more academic attention to the idea of cloud robots. One such definition of "cloud computing" is a "cloud service" that facilitates a distributed virtualized computing environment with the purpose of improving a robot's ability to do complicated tasks. Some of the current issues for on-board robots include, for instance, creating a 3D representation of an item or determining the journey direction in unstructured environments (P. S. Kumar and A. Sharma, 2019). The shift away from on-board or local network-based services and towards cloud-based ones is being driven by the effects of the cloud on various computer systems and services (S. El-Gendy, 2020). With the cloud's scalability, big storage capacity, cheap maintenance cost, and high processing power, the robotics field might potentially benefit from these features. In most cases, the on-board robot still has storage and processing limitations (K. N. V. Srinivas and P. K. Lakineni, 2023). So, it's more important than ever to load the robot's sensor data into an external computer system that can handle the processing, store, and share of knowledge, and then send the results back to the robot. Figure.1 shows that distributed computation and storage resources are the backbone of cloud robotics infrastructure (P. S. Kumar, S. Gangadharan and S. C. N, 2023). These resources are made available to different kinds of robots so that they may carry out activities in the cloud or communicate with one another.

Because it enables cooperation in a more scalable form, where the robot is not limited by its physical location, the notion of cloud robotics is particularly advantageous in environments with several robots (Hussain, N., Samal, A., 2024). One possible use case for cloud services is providing robots with access to a database where they may store, retrieve, and share information (V. S. Prasad, Y. Mahajan and S. Suneetha, 2023). This paves the way for robots to gain knowledge from one another's experiences, such as when it comes to gathering data from their environs, exchanging relevant control rules and course plans, and so on. Cloud computing also allows them to exchange data on results, which might lead to better automation systems and robots all around the globe (A. Kumar, V. Kalpana and V. Hariram, 2023). Over time, the idea of cloud robots is starting to take off. Figure.2 shows that the RoboEarth project, in relation to modern literature, created a "world wide web" that served robots from a massive distributed database and repository (M. Kranthi and R. C. Tanguturi, 2023). An integrated database for storing and transforming human or robotic-created information into a machine-readable format is part

Figure 1. A comprehensive overview of the architectural framework and diverse applications in cloud robotics

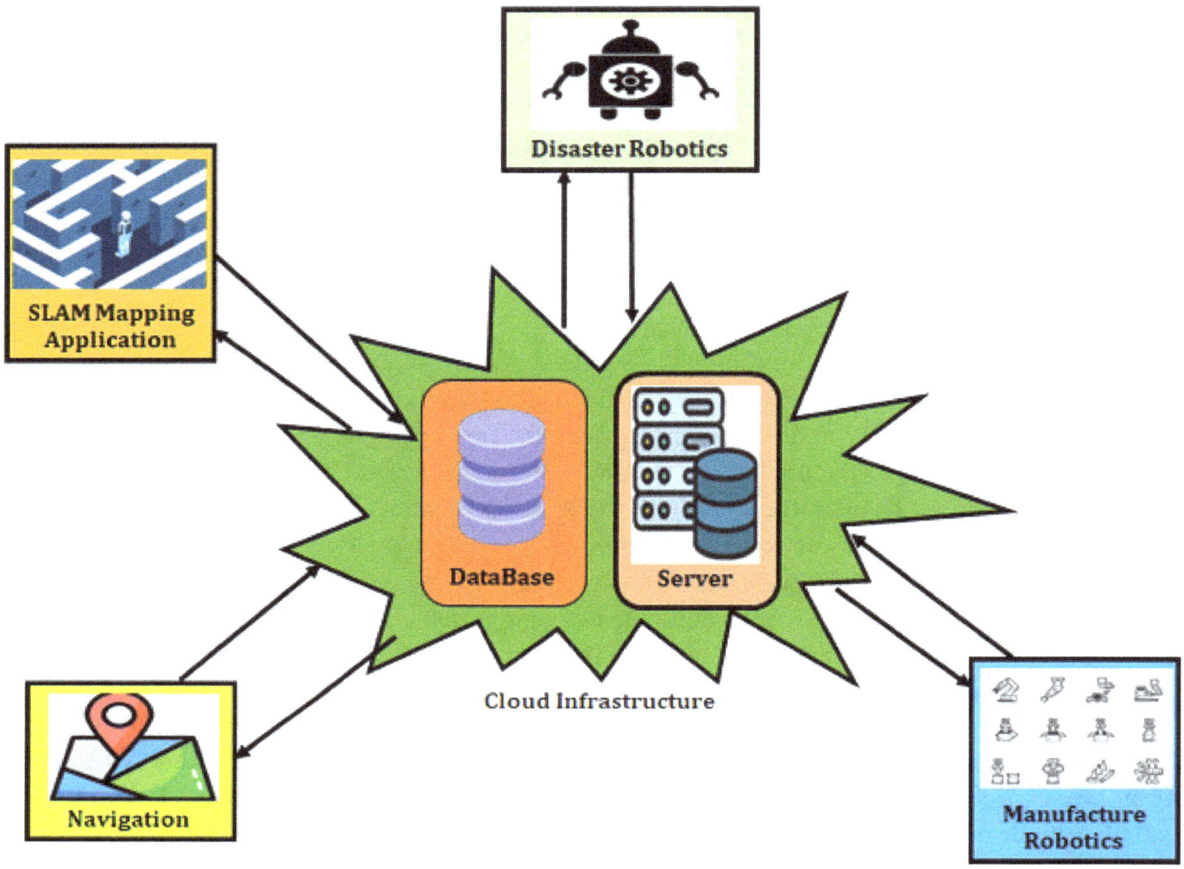

of the Cloud Robotics architecture that this offers. The RoboEarth knowledge base stores a variety of data, including maps, object recognition models, places, and task information(A. Ranjan, K. K. Hegge and R. Agrawal, 2023).

Cloud-based cloning platforms represent a breakthrough approach to amplify the intelligence of robots by harnessing the power of cloud computing. Instead of relying solely on onboard computing resources, these platforms enable robots to leverage the vast computational capabilities offered by the cloud. In fact, the robots are just avatars and clones (Avinash, Kumaraswamy, Anand Kumar, 2023). As long as you can get information into them, they themselves have no boundaries imposed by their various different hardware.

The importance of scalability in the development and implementation of intelligent robotic systems cannot be overemphasized. A scalable system is robust; it can easily expand and adapt to meet the rising demands of increasingly complex tasks, larger datasets or evolving learning algorithms (Chinamuttevi, Veera Sivakumar, 2023). Dynamic scalability is essential for the development of AI-based robots. It opens up a world of continuous improvement, high performance, and vast applications across industries (Shekhar, R. & Chaturvedi, Abhay, 2023). In what follows, as we learn more about the concepts behind

Figure 2. Exposition of the system architecture underlying RoboEarth

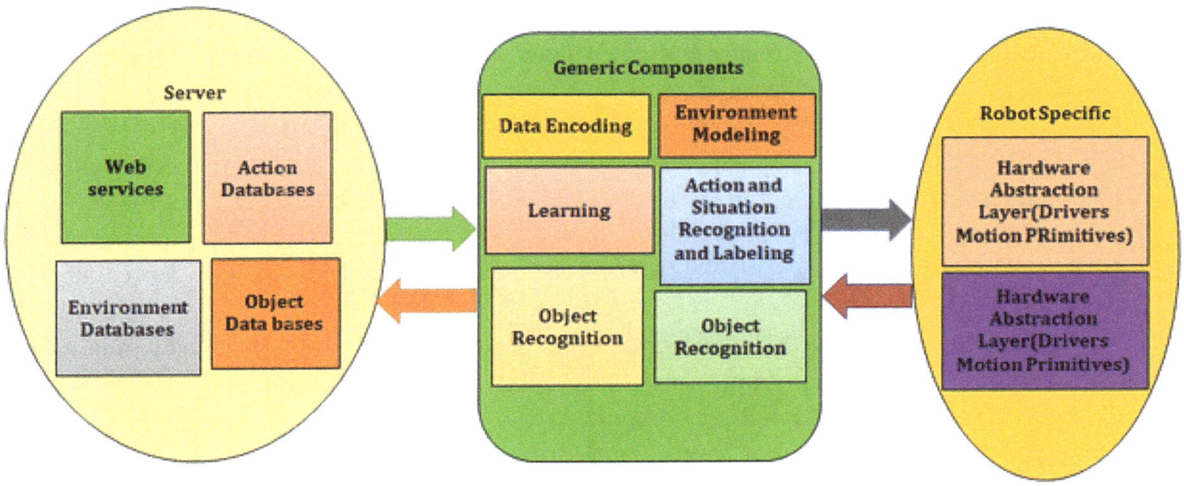

cloud-based cloning, explore its many applications for robotics and take a look at case studies of this approach in practice, we shall also examine some of the obstacles that still stand in its way to complete fulfillment. ahead lie not cnly a series of technical breakthroughs, but also the ethical issues involved and a systematic grasp of how cloud-based cloning platforms will affect intelligent robotic systems.

2. FOUNDATIONS OF CLOUD-BASED CLONING

Cloud Computing in Modern Technological Advancements: Today, cloud computing underlies a great many technological innovations. In principle, it is about delivering computing services over the internet. For example: storage, processing power and software applications(V. Srinivas, T. De, 2023). In terms of robotics, cloud computing provides a scalable and elastic platform that compensates for the natural finite limits onboard robots 'on-chip capability. The cloud is a massive supply that can be flexibly reallocated according to the computational needs for robotic tasks(K. K. Hegge and R. Agrawal, 2023). In addition to providing robots with greater processing power, it also permits real-time data analysis, making decision-making more sophisticated.

Overview of Cloning Platforms and their Relevance to Robotic Intelligence: In robotics, platforms are systems that create virtual copies, or clones, of robotic entities in the cloud. These virtual counterparts--or avatars--mimic the behavior and capabilities of their physical selves and have the added advantage of residing in a virtualized, cloud-based environment(K. . Prasad, and I. . Gugan, 2023). This approach means that intelligence is no longer tied to the limitations of an individual robotic hardware. With the need for robots to cooperate and share in the pooled intelligence, it is clear why platforms for cloning are necessary. They serve as a mechanism for robots to send their experiences, learning and insights up into the cloud, building a common information store. Such shared intelligence enables the robotic fleet to have higher-level capabilities, because each robot can profit from the experiences of others.

Seamless Scalability Enabled by Cloud-Based Solutions: One of the most important advantages of cloud-based cloning platforms is that they allow robotic systems to scale effortlessly(N. S. Reddy and B. Ashreetha, 2023). One of the major problems with traditional robotic systems is that they are unable to accommodate increasing computational requirements or changing tasks. This limitation is overcome by cloud-based solutions, which enable robots to offload computationally intensive processes to the cloud. This scalable architecture thus allows robotic systems to dynamically change with varying workloads. As required tasks become more complex, or new challenges arise, robots will match the nearly unlimited computational capabilities of the cloud (S. Reddy P, P. S. Patwal, 2022). In addition to improving the performance for individual robots, it also allows simultaneous improvement of the entire robotic work force (P. Pavankumar, N. K. Darwante, 2022). The system develops a collective intelligence that grows with every new experience.

In the subsequent sections, we will delve deeper into the advantages of cloud-based cloning, exploring how this innovative approach transforms collaborative learning, real-time updates, and resource optimization within the realm of intelligent robotics.

the existing machine-to-cloud We have identified models for M2C cloud robotic connections. A machine-to-cloud (M2C) connection mechanism may be either peer-based, proxy-based, or clone-based(A. Singla, N. Sharma, 2022). The mobility, interoperability, and robustness of a model are all defined by its own unique characteristic.

The three models of elasticity Were defined according to the information shown in Figure.3. The Peer-Based Model entails a single system establishing a connection to a virtual machine (VM) in the cloud, which is seen as an extra computing unit. Either the robot or the cloud might carry out the task(K. S and M. R. Arun, 2022). The proxy-based architecture uses a single virtual machine (VM) with a single connection to each robot, making it function as an edge device (proxy) for many robots. approach Based on Cloneing(P. Ramesh Naidu and N. Guruprasad, 2021). In this approach, every robot connects to a virtual machine (VM) in the cloud that serves as a copy of itself. The M2C peer-to-peer network connects the robot to the cloud. On top of that, a cluster of identical copies creates an ad hoc M2M network inside the cloud.

The most reliable approach is the clone-based one as it achieves the maximum number of connections between the cloud infrastructure and the robots(M. Shruthi and B. Naveen K, 2020). Conversely, when it comes to network availability, the proxy-based strategy is the least resilient. Quality of Service determines how quickly an ooted job may be completed or how well the cloud service is maintained. Due to two reasons: network and resource availability, the clone-based and peer-based models provide the greatest quality of service(M. Ramesha, and G. N. Pai, 2020). While several robots share the same network and resources, the proxy-and gateway-based models provide the lowest quality of service.

Figure.4 depicts the DAvinCi framework, which includes features like mapping and computer vision (for example, face or object recognition). In order to complete a job that calls for the use of several robots, the data collected by their sensors is sent to a controller (R. S. Varun, and T. Rajesh, 2018). This controller, in turn, activates the robots in response to individual requests. DAvinCi is a platform as a service (PaaS) that may facilitate the creation of a world map via the execution of actions in a cloud setting. A server links the robot to a Hadoop cluster, which is the core of the system.

A ROS master node is operating on the server. In order for the robot and the server to communicate via the Hypertext Transfer Protocol (HTTP), ROS messages are included into the requests and answers (J. J. Jijesh, and Shivashankar, 2017). With Hadoop Distributed File System (HDFS), data is stored and processed in parallel using MapReduce. Keep in mind that processing speed relies heavily on parallel

Figure 3. Examining various cloud robotics computing paradigms

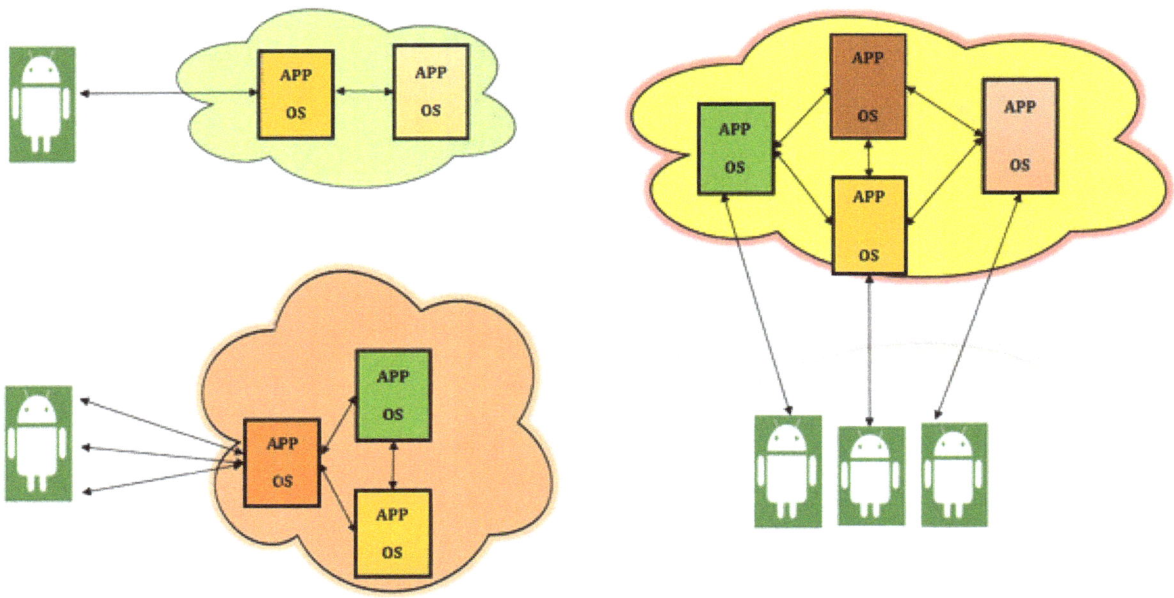

Figure 4. Architecture overview of DAvinCi: A detailed insight

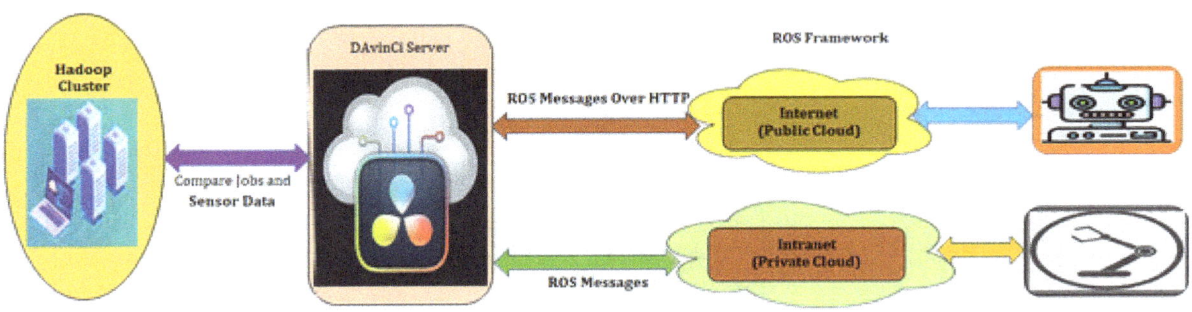

techniques. Data is partitioned into blocks for the sake of both storage management and simplifying parallel processing(A. Krishnan, S. Swarna and B. H. S, 2020).The ROS messaging system is used to link the DAvinCi server to robots, as previously stated. Nodes that are ROS subscribers on the server side act as HDFS clients, collecting these messages and pushing data to the parallel system so that Map Reduce processes may be executed. The platform is built around four primary cloud-based modules. An HTTP server, capable of operating on mobile robots, is part of the integrated package(J. Khan, F. Ali, F. Ullah, J. Bacha and S. Lee, 2019). Given that these robots are only capable of performing basic operations, the cloud handles the whole treatment procedure(W. Liu, 2021). The Protocol Handler suite takes care of all the proxy stuff, network address translation, and security controls. The APIs make the programming interfaces accessible to users. Lastly, the management package is used to perform system and resource access. For direct connections to the virtualized REALcloud environment, each client is restricted to using their own virtual machine (VM), as seen in Figure 5. Because sharing resources

Figure 5. A visual representation of the REALcloud framework: Schematic overview

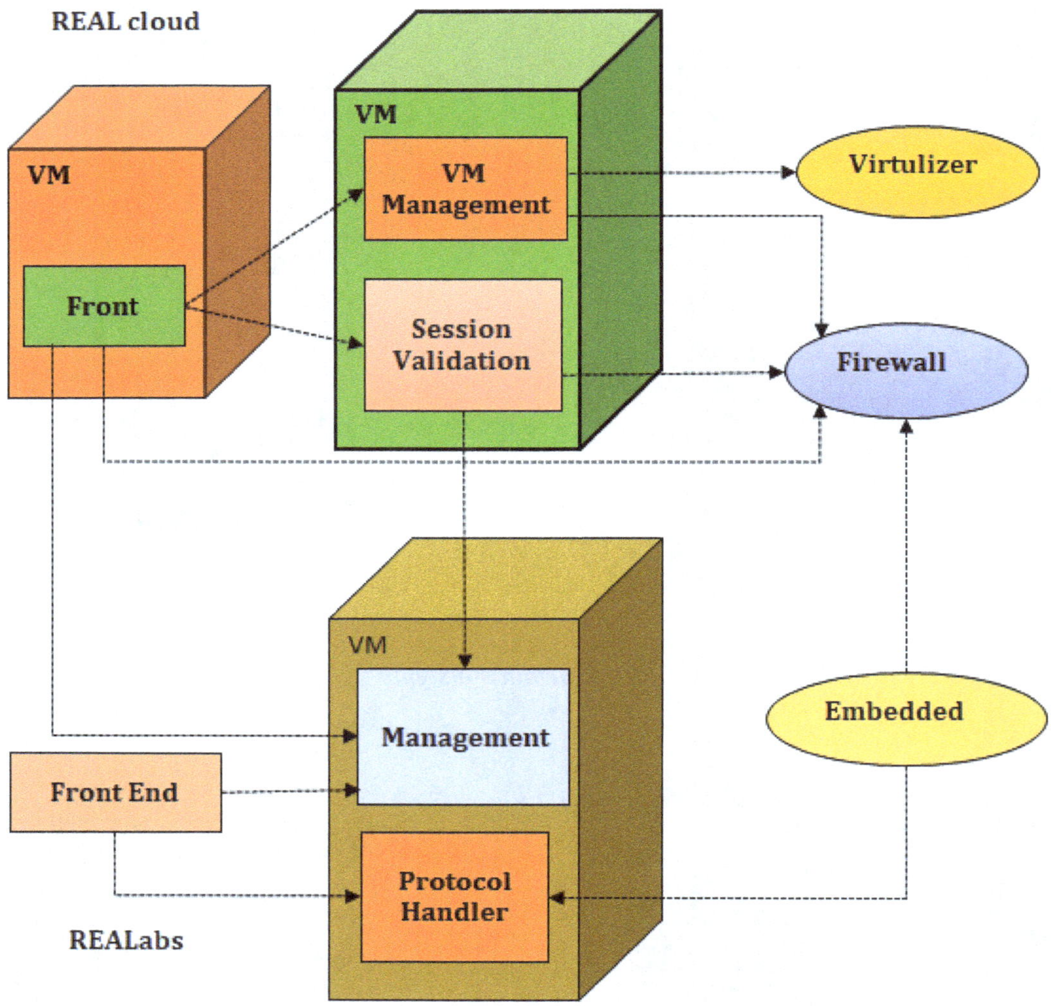

across virtual machines becomes more difficult. It is easier to see how the aforementioned packages work in this context: client packages run on the client side, controlled packages and Protocol Handler on the server side.

3. ADVANTAGES OF CLOUD-BASED CLONING FOR ROBOTICS

Scalability: Easy Scaling of Robotic Intelligence

The scalability of robotic intelligence is a pivotal factor in addressing the evolving demands of various applications. Cloud-based cloning platforms offer an elegant solution to the scalability challenge by providing an infrastructure that adapts dynamically to the computational needs of robotic systems.

Dynamic Resource Allocation: With cloud-based cloning, robots can increase their intelligence at will by borrowing the vast computational resources of the cloud(H. M. Mishra, A. Qtaishat and M. Kumar, 2023). Whether more complicated tasks, larger data volumes or higher-standard algorithms are involved, robots can alter their resource distribution as needs arise.

Parallel Processing: It is parallel processing, using many robotic instances working in concert on different tasks. Only cloud environments make this possible. These parallelisms increase the total computational throughput, helping robotic systems to execute complex operations and learn from experience at a faster pace.

Cost Efficiency: Cloud scalability means cost efficiency for robot development and deployment. To avoid the need to invest in endless amounts of expensive onboard hardware for each robot, organizations can take advantage of the pay-as-you-go model so characteristic of cloud computing, only paying for what they use(D. Pal and S. Joshi, 2023). Not only does this cut initial costs, it also allows the infrastructure to grow in proportion.

Resource Optimization: Efficient Utilization of Resources

How well robotic systems function depends greatly on the economical utilization of resources it achieved. Cloud-based cloning platforms play a significant role in resource optimization with laborious computations being done by the cloud.

On-Demand Computing: The cloud gives robots access to whatever additional computing power they need on-demand, eliminating the need for an overloaded onboard computer(S. M. Sali and K. R. Joy, 2023) . On-demand access allows resources to be applied just when they are needed and reduces idle time, improving the overall efficiency with which robots can work.

Centralized Data Storage: Through cloud-based solutions the robots can store data centrally, thus offload and retrieve information easily. Through this centralized approach, you reduce the need for local storage on each individual robot, which thus simplifies their design and lowers your hardware costs.

Energy Efficiency: Taking advantage of the cloud for computation-intensive tasks, robots need less energyfeficient hardware. Besides prolonging the useful lifetime of robotic systems, this also helps intelligent robotics to become greener and more environmentally friendly.

Real-time Updates: Updating and Upgrading Robot Intelligence in Real-Time

Never before has there been such a wonderful development: the ability to update and upgrade robot intelligence in real-time. Now, robotic systems can remain adaptive and responsive to new challenges. Through seamless updates, robots equipped with cloud-based solutions are provided the flexible ability to evolve.

Agile Development: Cloud-based cloning platforms also suit agile development methodologies. Developers can make changes and improvements more rapidly. Algorithm, learning model or even whole software stack updates can be rolled out in real-time across the robotic fleet. All robots use cutting-edge technology through this process.

Remote Patching: The system can send critical security patches, bug fixes or performance enhancements to robots remotely, without any physical intervention. This not only saves time, but also improves the security and stability of robotic systems in general.

Adaptive Learning Models: Cloud-based solutions enable robots to leverage the latest machine learning models and algorithms. As advancements occur in AI research, robots can seamlessly integrate these improvements, staying at the forefront of intelligent capabilities without the need for extensive hardware upgrades.

In order to enable low-cost robots to do computation-intensive jobs using cloud resources, RobotCloud was developed. The typical workflow of the RobotCloud architecture is shown in Figure 6. with the ROS Master Node functioning as the system's nerve centre (Cloud Controller). After receiving authorization, the unit forwards all service requests to the Service Management Point. The next step in the registration or removal process is to determine if the service is available. If it isn't, the request will be added to the queue. When the service begins to respond to the robot's request after receiving authorization, the Cloud Controller is notified. Services provided by the RobotCloud system, such as route planning and map construction, as well as a Map Reduce processing cluster.

In order to enable low-cost robots to do computation-intensive jobs using cloud resources, RobotCloud was developed. The typical workflow of the RobotCloud architecture is shown in Figure 6. with the ROS Master Node functioning as the system's nerve centre (Cloud Controller).

After receiving authorization, the unit forwards all service requests to the Service Management Point. The next step in the registration or removal process is to determine if the service is available. If it isn't, the request will be added to the queue.

When the service begins to respond to the robot's request after receiving authorization, the Cloud Controller is notified. Services provided by the RobotCloud system, such as route planning and map construction, as well as a Map Reduce processing cluster. A client gateway to the RoboEarth based on knowledge repository and increased processing capacity for the connected robots—that's the goal of this open-source framework. Because RoboEarth can manage the processing power needed to communicate with it, the Rapyuta platform is called Cloud Engine for RoboEarth.

This is because extra processing power is provided via the cloud. Moreover, it helps the connected robots out with loading duties in the same manner that other cloud robotic frameworks do, and it disguises itself as a PaaS, or cloud computing platform, tailored to use with high-bandwidth, multi-process robotic applications.You may link ROS robots and other browser machines to Rapyuta, an environment that is compatible with ROS and uses web socket-based communications protocols. When compared to other cloud computing frameworks, the system's performance may be hindered by the direct network connection. Linux containers provide a virtualization environment that allows the safe and scalable separation of processes from different robots. The computer systems are integrated with these containers. As if the processes were not segregated, containers enable the distribution of memory and processing time while keeping application performance constant. Smart solutions to the issues of scalability and security in cloud computing are provided by this approach. Since the Rapyuta is using ROS, all processes inside a single container are able to interact with one other simply. As seen in Figure.7, the communication protocols are managed by the endpoint processes. This Rapyuta was put in place to ensure that the system's communication delays and complexity were balanced. The Rapyuta internal and external processes interact using the VM-specific interfaces that use the WebSockets protocol. A number of functionalists, together known as systems management procedures, make up the essential job sets. In a typical scenario, a PaaS is created by each of the four main responsibilities. One of the most important parts of the Rapyuta ecosystem for managing links amongst robots is the ROS Master node. The Rapyuta function system is structured centrally. Each component makes use of the component network's layer of abstraction. A user is a person who links the robotic devices that he owns to the network. On a computer, the LoadBalancer

Figure 6. A comprehensive overview: Schematic diagram for the architecture of the RobotCloud framework

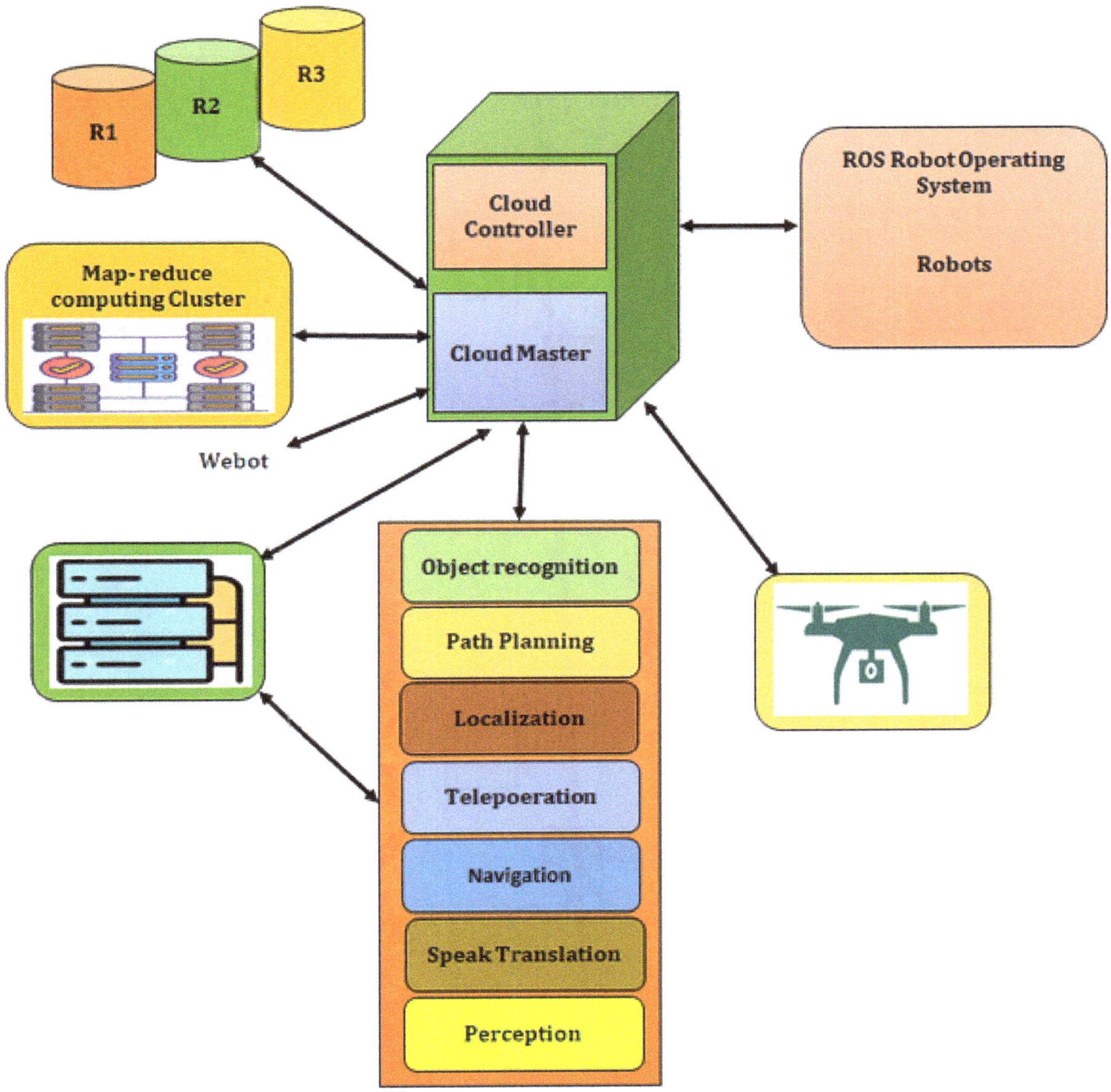

controls computers that are actively running. After that, the distributor gives these available endpoint processes to the incoming connections.

In the following chapters, we will look at concrete applications in a number of real-world settings that highlight the advantages and versatility of cloud-based cloning platforms.

Figure 7. Overview of the core infrastructures powering the rapyuta platform

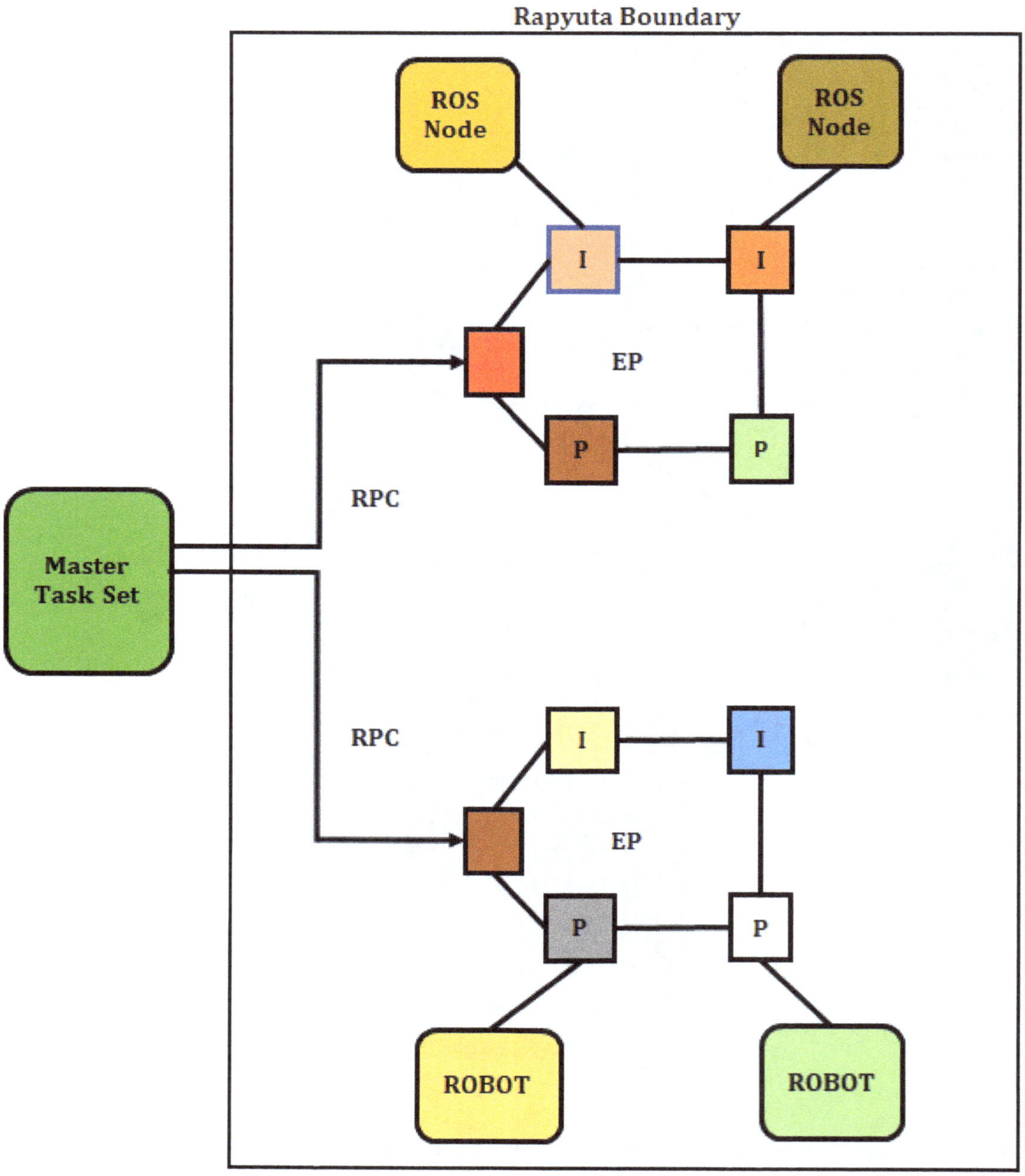

4. COLLABORATIVE LEARNING AND KNOWLEDGE SHARING

Cloud-based cloning, whose collective intelligence is produced not from one robot in isolation but rather through the continual exchange of knowledge and experience between robots, has an especially powerful collaborative potential. Here, we explore how these platforms enable robotic systems to collaborate in learning and introduce case studies illustrating the breakthrough effect of knowledge sharing on system-wide performance.

Shared Knowledge Repositories: Cloud-based cloning platforms are thus places in which robots can deposit and access large quantities of information about their experiences, performance on given tasks, and the environments in which they operate. This pool of shared knowledge serves as a framework for collective learning.

Decentralized Learning: Not only from their own experiences but also the experience of their opposite number can robots learn. Through accessing the cloud they can feed on the knowledge already accumulated by other units, avoiding expensive trial and mishap when performing new tasks or in unfamiliar surroundings.

Distributed Training: Training of machine learning models can be done jointly between various robotic instances. With this method, robots can participate in training shared models, thereby improving the learning efficiency and allowing the application of more robust and generalized algorithms.

Case Studies and Examples

Industrial Automation: In the industrial environment, robots whose cloning platforms are cloud-based can exchange ideas about management of production lines, identify areas which can be improved, and collectively raise overall operational efficiency. If one robot finds a better way to assemble parts, for example, it can inform the other robots on the same production line.

Autonomous Vehicles: Cloud-based cloning makes collective learning possible between self-driving vehicles. When a vehicle experiences an exceptional and difficult situation, it can pass this experience on to other vehicles. The whole fleet is prepared better for situations that may arise in the future. This shared pool of knowledge boosts the safety and confidence of autonomous driving systems.

Healthcare Robotics: In a health care setting, robots can provide information about patient care routines, treatment results, and procedural techniques. This joint learning model causes robots performing assistance duties or carrying out medical procedures to perfect their capabilities as a result of the experience acquired from all kinds of styles and locations for providing healthcare.

Search and Rescue Operations: They can all benefit from things in common about environmental conditions, potential dangers and how to find them. Through this joint learning robots are able to adapt to uncertain and changeable situations, giving them a more effective role in crucial operations.

Advantages of Collaborative Learning

Adaptability: Collaborative learning, by increasing the adaptability of robot systems, allows those robots to react appropriately to novel problems and scenarios.

Efficiency: Through learning from others 'experiences, robots eliminate redundancy in task-learning and can fine-tune their performance faster.

Collective Intelligence: That which outstrips even the sum of individual robots' abilities is group intelligence borne through mutual learning. The entire network becomes a bunch of interconnected elements, each with access to the same body of knowledge.

5. SECURITY AND PRIVACY CONCERNS

As cloud-based cloning platforms penetrate robotic systems, problems of security and privacy deserve close attention. In this section we will analyze a number of possible defects in cloud-based intelligent robots.

Data Security

Data Transmission: It is easy to intercept data transmission between robots and the cloud. To prevent unauthorized access or tampering, data must be guaranteed in terms of its confidentiality and integrity while being transmitted.

Data Storage: If that sensitive information includes proprietary algorithms or mission-critical data, then we have every right to worry about it ending up in the wrong hands. Stored data requires strong encryption and access controls.

Cybersecurity Risks

Cyber Attacks: Cloud-based systems are subject to cyber threats including DDoS attacks and attempts at intrusions. Such attacks could impair communications between robots and the cloud, so that robot systems may lose their functionality or even become dangerous.

Identity and Authentication: Robots connecting to the cloud must be identified, and security authentication mechanisms are needed. This means unauthorized control over robotic entities, compromised credentials or intrusions.

Privacy Concerns

Data Ownership: Who owns and controls the data generated by robots is a complicated question. In other words, we need policies and agreements that can clearly define who owns the data and how it is to be used. Otherwise there are bound to be violations of people's privacy.

Sensitive

Information Exposure: Because these cloud-based systems deal with and store sensitive data, people fear that they might unintentionally have their private information leaked. Such risks can be reduced through the adoption of privacy-protecting means.

Tackling with concerns of security and privacy about the cloud-based cloning platforms for robotics requires a multipronged approach. A first line of defense to protect data integrity and privacy in transmission from robots to the cloud can be robust encryption mechanisms. Security is further strengthened by the use of different communication protocols. They also prevent sensitive information from being intercepted, or tampered with along the way. Robotic system security must include access controls and authentication mechanisms. Rigorous, strict verification procedures are required to prevent unauthorized entry or control over robots in the field. Through the use of privacy-enhancing techniques, privacy is protected. There are also various anonymizing techniques and the technique of homomorphic encryption, which allows data to be processed securely without sacrificing individual or proprietary information. The overall security posture is also affected by legal and ethical frameworks, as well as technological factors and battlefield practice. They determine the rules governing data ownership, sharing and privacy. Since real-time security lapses need to be checked in a preemptive fashion, such that both on-the spot monitoring systems and incident response plans are required. But as technologies change, so should models of explainable AI and an ethical framework for artificial intelligence be adopted. These can indeed open up decision-making processes and keep development practices responsible. These different strategies and technologies form flexible lines of defense against all kinds of security threats, as one example is guarding privacy. What makes both trust and dependability possible is represented in this combination of cloud-based intelligent systems especially when the robot system involved has been developed with security considerations.

6. REAL-WORLD APPLICATIONS OF CLOUD-BASED CLONING IN ROBOTICS

Surprisingly enough, cloud-based cloning platforms have already been declared the winners in virtually every industry that you can imagine. Thus, let's take a look at some examples of their application in manufacturing and think tanks as well as healthcare.

Manufacturing: Collaborative Robotic Assembly Lines

Scenario: Cloud-based cloning platforms work together with robots on the assembly lines.

Implementation: When products are being assembled, robots can pass real-time information on to everyone and the entire process becomes more efficient. Everyone learns at once.

Benefits: It makes the system more adaptable to variations in product, reduces operating time necessary for different tasks and increases overall production efficiency.

Quality Control and Defect Detection

Scenario: Robots with internet connectivity perform quality inspections on production lines.

Implementation: For example, robots exchange information about defects in products found by them. Other robots can benefit from these examples and improve their ability to recognize defects.

Benefits: Improved defect detection accuracy, fewer false positives, and continuous improvement to the quality control process.

Healthcare: Patient Assistance and Monitoring

Scenario: In many healthcare settings, robotic assistants connected to the cloud provide support to patients and monitor vital signs.

Implementation: Robots communicate information about encounters with patients, thus enhancing the care provided to individuals.

Benefits: adaptability to patient needs is increased, reactions to medical emergencies are improved, and personalized healthcare assistance is provided.

Telepresence Robots in Remote Consultations

Scenario: Cloud-cloned telepresence robots provide remote consultations between doctors and patients.

Implementation: Robots exchange information from medical sessions for a joint diagnosis and plan of treatment.

Benefits: Improved access to medical resources, better distribution of medical resources and the elimination of geographical obstacles in patient care.

Logistics: Warehouse Automation and Inventory Management

Scenario: Cloud-linked robots process inventory and perform order fulfillment for warehouses.

Implementation: Robots exchange data on inventory levels optimizing stock management and order processing.

Benefits: More accurate inventory-tracking, speedier order processing and the ability to adjust easily to changes in demand.

Collaborative Sorting and Packaging

Scenario: In logistics centers, robots that can clone themselves work in groups performing sorting and packaging tasks.

Implementation: Robots provide themselves with knowledge of the most efficient sorting strategies, optimizing thus in concert the entire sorting and packaging procedure.

Benefits: Efficiency in order fulfillment, reduced processing times and adaptability to changed product assortments.

Environmental Monitoring and Agriculture: Precision Agriculture

Scenario: Precision agriculture uses cloud-connected robotic systems to monitor and manage crops.

Implementation: Soil conditions, crop health and environmental factors are all shared among robots for joint decision making in the management of crops.

Benefits: Better crop-yield forecasts, smarter resource use and eco-friendly farming.

Environmental Surveillance

Scenario: Pollution levels in urban areas are monitored and analyzed by environmental sensors attached to robots that work together.

Implementation: Robots gather real-time data on air and water quality for complete environmental monitoring.

Benefits: Detecting environmental hazards promptly, better planning for cities and the ability to confront pollution sources ahead of time.

As for robots 'operating systems, ROS is a highly regarded licence framework for reuse of open source code which enhances productivity. The distribution and maintenance of the system is one part of ROS. There are two parts of ROS that deal with the code development process:at the Willow Garage lab, first and foremost, the core component of ROS. It includes the whole ROS distribution and essential software components, in addition to most of the basic tools for remote computing. The second component is the "Universe," which encompasses all of the code that has been created and is being maintained by the ROS community across the globe. Libraries, software dependencies, and hardware definitions are often part of the Universe. These define the physical architecture and provide a basic degree of control for managing the connected devices, whether it's a robot or a single sensor connected to a smartphone or computer. Linux distributions like Ubuntu and Debian are ROS's mainstays, but ROS hydro is compatible with a wide range of operating systems, including Android and Windows. A large variety of hardware, sensors, and actuators are supported by ROS. It specifies the hardware design and a minimal degree of management to manage the connected devices, whether it a robot or a single sensor connected to a smartphone or PC.

Figure.8 depicts the basic ROS connection mechanism that enables the ROS node to transmit and receive messages from other nodes; hence, at least one ROS master must be active in the network for the ROS packages to execute. Direct message transmission between nodes is possible after registration on the master. An operational framework is necessary to govern the packages in ROS because of its various parts. Ruby on Rails code is organised into packages. The many nodes, dependencies, configurations, libraries, etc. that make up a package. A package's purpose is to provide a user-friendly framework for software reuse. ROS is adaptable enough to operate with pre-existing algorithms, which may be used to enhance robots that carry out complicated jobs. One example of a method used for robot navigation and mapping is SLAM. Nodes, messages, services, tools, and libraries in the ROS Framework are all easily managed by code. Packages, stacks, and repositories are three fundamental ideas in the ROS le system. One of the packages is the ROS framework. Nodes, libraries of dependencies, setup files, third-party software, and other components are all part of a pack. A package's goal is to facilitate software distribution by providing a user-friendly framework. An suitable stack is a collection of packages that have all the characteristics needed.

Figure.9 shows the ROS file system, which is the foundational level and an important building piece for running a function. The executable files and folder structures are included in this function. The six distinct kinds of folders each have specific purposes: Packages One of the building blocks of ROS is a package. Everything a ROS programme needs to execute, including configuration files, nodes, and libraries, is in this package. Essential package data, like as licences, repositories, and dependencies, are presented in manifests. Stacks A stack is a collection of interconnected and combined packages. Essential stack data, including dependencies and licences, is provided via the stack manifest. Messages may be transmitted in a variety of data forms. In order to process the request and reply, services decide on the data structure. A single ROS master is responsible for maintaining all the connections among the nodes, according to the overall network architecture for the ROS framework. The ROS master and nodes may coexist on a single machine or across several machines in a network, as seen in Figure 9. To function in a ROS environment, every computer and robot must have an Ethernet or wireless connection to the network.

As previously stated, in the ROS community, each ROS environment is typically governed by a single ROS master. Therefore, the notion of exchanging information across different ROS environments presents a new issue. Nevertheless, a method for developing cooperative multi-ROS settings was suggested in 2015 by a project known as Multi ROS master multimaster fkie. In order for each ROS master to communicate at least part of their ROS topics, actions, or services across different ROS settings, the project makes use of an extra network as a common network, as seen in Figure 10. Since multimaster fkie doesn't get in the way of ROS node launches in other ROS environments, master nodes don't have

Figure 8. Overview of the core elements in the ROS basic system

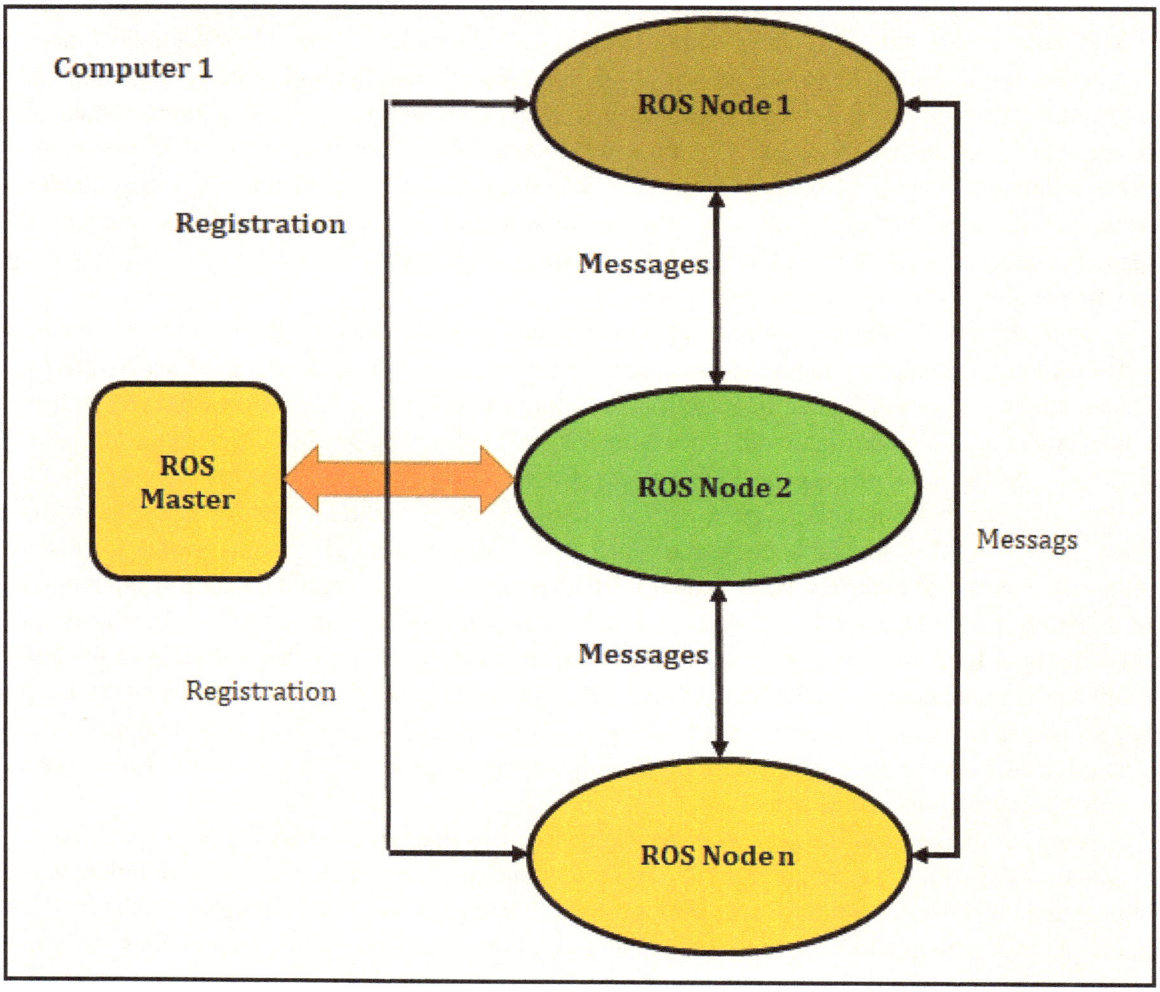

to exchange public keys; instead, users may get password-free access to published topics that are shared amongst ROS masters.

By using the ROS master in each ROS environment which is connected to the ROS multi-master system over the shared network all of the computers, cell phones, and robots in that environment will be able to communicate the subjects they have chosen. When the multi-master system communicates over wireless interfaces, the framework's main drawback becomes obvious owing to interference and power problems. Although it is unrelated to the framework, users should exercise extreme caution when using wireless connections. But when connected by cable, it has shown to be dependable. When it comes to safety, scalability, installation, and use, the multimaster fkie are tops. By allowing the operator of each ROS master to choose which topics may be shared and updated, the local ROS environment can be kept in sync with the other ROS environments. All the way up to the most recent release, "Melodic," this multi-master solution has been supported by ROS since the "Indigo" framework. Consequently, this project was chosen to be a component of the suggested solution in this thesis to build a collaborative

Figure 9. Comprehensive insight into the hardware arrangement of a single master ROS system

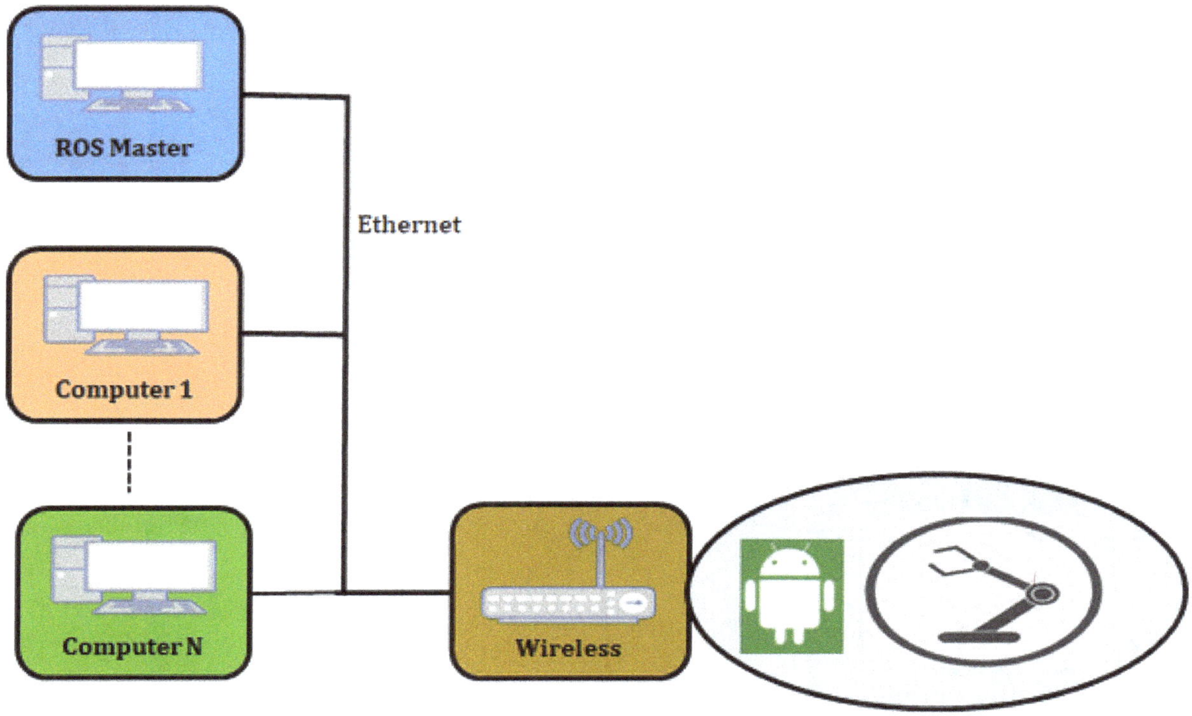

environment between the robot clone cloud photos after significant study and testing about its potential to be transported to the cloud network. The identified system constraint does not apply since the connections between ROS-masters will only take place in the cloud, which has a robust network connectivity architecture.

These examples demonstrate the widespread applicability of cloud-based cloning to robotics. It can be adapted to suit the requirements of a variety of industries, including manufacturing, healthcare, logistics and environmental monitoring. Cloud-connected robots are flexible and collaborative, creating new possibilities for innovation and efficiency in every field.

7. CHALLENGES AND FUTURE DIRECTIONS

One great misfortune about cloning robot intelligence was that to scale up, we had no choice but to go through the maze of cloud-based platforms. Walking into this minefield offended my sensibilities all too often and uncovered a string of problems which forced me carefully reconsider whether or not what I still thought could indeed be future reality. The first is the need for reduced latency. Applications must respond just as quickly--a severe problem in cases where lives or fortunes can change within seconds of a key being pressed. Also at the same time, data security and privacy are unavoidable. They require advanced encryption techniques; secure communication protocols and an absolute minimum of unauthorized accessibility in order to prevent intrusions or loss. The network reliability problem persists,

Figure 10. Overview of the hardware arrangement in multi-ROS master sub-systems collaboration

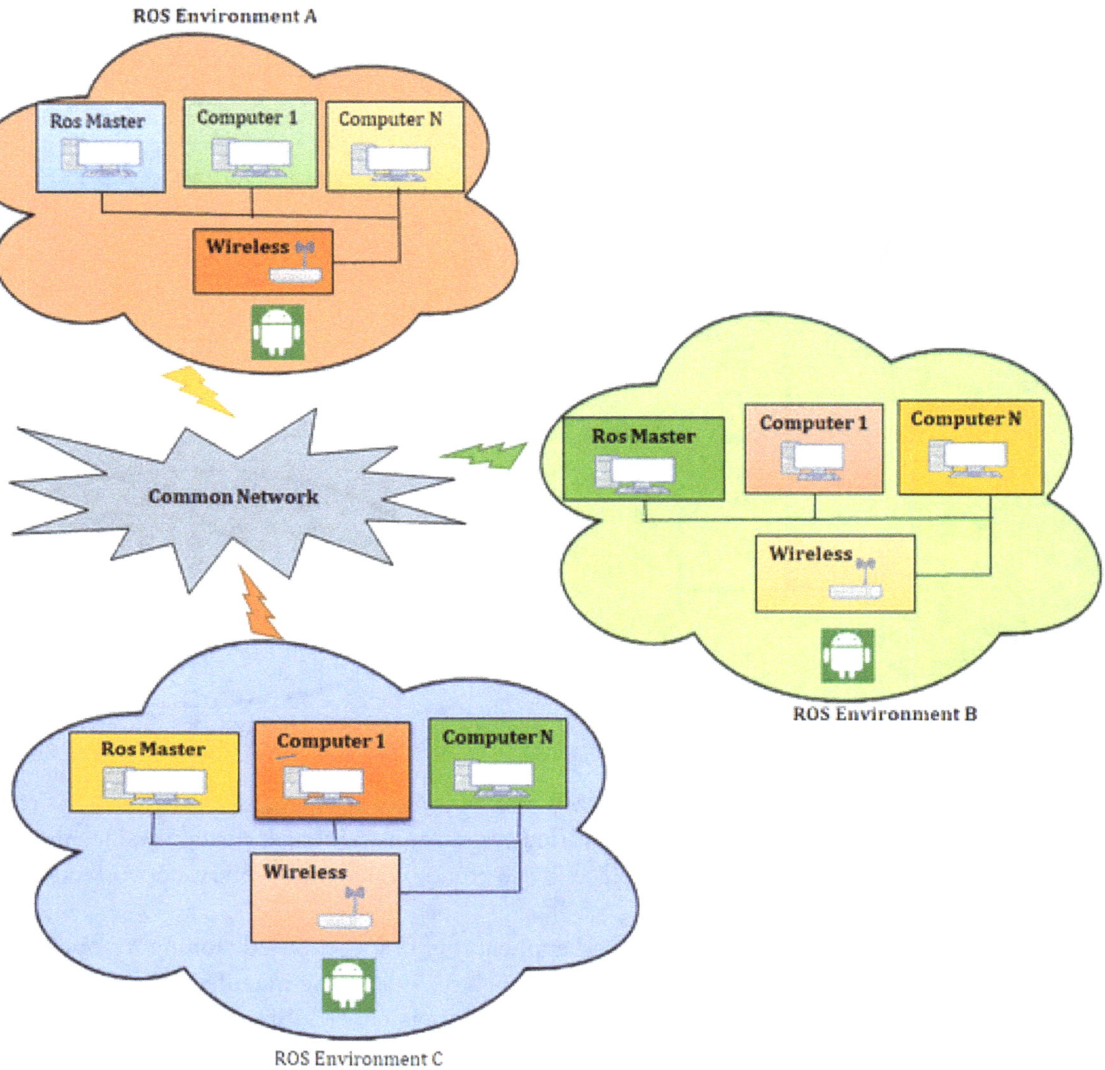

however. Applications for robust connectivity need to be developed so that robots can keep connected with the cloud. There are also scalability complexities; resource allocation intricacies and the need for standardized interfaces. All of these call for delicate answers before it can be everywhere.

The cloud-based cloning platforms of the future go on to show all manner of promise and further development. The most important fundamental change might be the integration of 5G networks and edge computing. This means that data can be processed near its source in real time, complemented by cloud-based intelligence. Such federated learning combined with edge AI will help robots learn together while avoiding the need to send data for centralized processing, and free deep-thinking power on devices themselves. Through this they can act more independently in decision making. It also promises to dramatically increase privacy. One way to solve the problem of regional outages is decentralized cloud architectures which improve reliability. Who Knows? With unparalleled computational capabilities, quantum computers could transform resource allocation and task scheduling algorithms overnight. Also, the need to develop explainable AI models and ethical frames for IA increases transparency requirements in responsible development. Branching off in such directions will require hard efforts from researchers, developers and policy makers to overcome obstacles. But it would mark the beginning of an era in which cloud-based replicating platforms transform intelligent robotic systems, making them less brittle, more flexible and better meeting ethical norms.

8. CONCLUSION

In conclusion, an investigation of cloning platforms in the cloud for scaling robot intelligence marks a revolutionary new frontier for robots. This chapter has been about scalability and the basic needs of cloud computing. The advantages it provides in resource optimization, real-time updates and collaborative learning have certainly not been ignored. The development process of intelligent robotics through cloud-based cloning brings to mind the difficulties we encountered with data security, privacy and network reliability. This isn't going be all smooth sailing ahead for us either. Yet the potential of this paradigm is enormous. New opportunities for innovation have come cross-robot learning experiences, across the board industrial applications and platform features that are always on their toes. The call to action is clear: In this way, researchers, developers and policymakers can work together to create an ethical future for intelligent robotics. The Distributed Cloud represents an innovative technology that has been a big part of history since the beginning of agriculture. Moving forward, we cannot but adopt new technologies; break through our own short-sighted habit of not budging--or else it doesn't count as advancement. In this way, the continued era of intelligent robots will still be beneficial, multi-functional and useful for all people.

REFERENCES

Avinash, K., & Kumar, A. (2023). A novel approach of unsupervised feature selection using iterative shrinking and expansion algorithm. *Journal of Interdisciplinary Mathematics*, *26*(3), 519–530. doi:10.47974/JIM-1678

Chinamuttevi, V. S. (2023). A novel RF-SMOTE model to enhance the definite apprehensions for IoT security attacks. *Journal of Discrete Mathematical Sciences and Cryptography*, *26*(3), 861–873. doi:10.47974/JDMSC-1766

El-Gendy, S. (2020). IoT Based AI and its Implementations in Industries. *2020 15th International Conference on Computer Engineering and Systems (ICCES)*, 1-6. 10.1109/ICCES51560.2020.9334627

Hegge, K. K., & Agrawal, R. (2023). Industrial Automated Multipurpose Robot Using WIFI. *2023 4th International Conference for Emerging Technology (INCET),* 1-8.

Hussain, N., & Samal, A. (2024). Design and Implementation of Single to Multi-cloud Securities in Network. In N. R. Shetty, N. H. Prasad, & N. Nalini (Eds.), *Advances in Computing and Information. ERCICA 2023. Lecture Notes in Electrical Engineering* (Vol. 1104). Springer.

Jijesh & Shivashankar. (2017). Design and implementation of automatic medicine dispensing machine. *RTEICT 2017 - 2nd IEEE International Conference on Recent Trends in Electronics, Information and Communication Technology, Proceedings*, 1962–1966.

K. S. & Arun. (2022). Priority Queueing Model-Based IoT Middleware for Load Balancing. *2022 6th International Conference on Intelligent Computing and Control Systems (ICICCS)*, 425-430.

Khan, J., Ali, F., Ullah, F., Bacha, J., & Lee, S. (2019). Artificial Intelligence and Internet of Things (AI-IoT) Technologies in Response to COVID-19 Pandemic: A Systematic Review. *IEEE Access : Practical Innovations, Open Solutions*, *10*, 62613–62660. doi:10.1109/ACCESS.2022.3181605

Kranthi, M., & Tanguturi, R. C. (2023). Design of Intelligent Medical Integrity Authentication and Secure Information for Public Cloud in Hospital Administration. *2023 2nd International Conference on Edge Computing and Applications (ICECAA),* 256-261.

Krishnan, A., & Swarna, S. (2020). Robotics, IoT, and AI in the Automation of Agricultural Industry: A Review. *2020 IEEE Bangalore Humanitarian Technology Conference (B-HTC)*, 1-6. 10.1109/B-HTC50970.2020.9297856

Kumar, A., Kalpana, V., & Hariram, V. (2023). Artificial Intelligence based Health Monitoring System on IoTH platform. *2023 Second International Conference on Augmented Intelligence and Sustainable Systems (ICAISS)*, 1458-1463.

Kumar, P. S., & Sharma, A. (2019). IoT Enabled Indoor Autonomous Mobile Robot using CNN and Q-Learning. *2019 IEEE International Conference on Industry 4.0, Artificial Intelligence, and Communications Technology (IAICT)*, 7-13.

Kumar & Singhal. (2020). Design and Analysis of IoT-Based Intelligent Robot for Real-Time Monitoring and Control. *2020 International Conference on Power Electronics & IoT Applications in Renewable Energy and its Control (PARC)*, 549-552.

Liu, W. (2021). Improvement of Navigation of Mobile Robotics Based on IoT System. *2021 IEEE International Conference on Robotics, Automation and Artificial Intelligence (RAAI)*, 69-72. 10.1109/RAAI52226.2021.9507826

Mishra, H. M., Qtaishat, A., & Kumar, M. (2023). IoT Based Smart Systems using Artificial Intelligence and Machine Learning: Accessible and Intelligent Solutions. *2023 6th International Conference on Information Systems and Computer Networks (ISCON)*, 1-6.

Nayyar & Nagpal. (2018). Internet of Robotic Things: Driving Intelligent Robotics of Future - Concept, Architecture, Applications and Technologies. *2018 4th International Conference on Computing Sciences (ICCS)*, 151-160.

Pal, D., & Joshi, S. (2023). AI, IoT and Robotics in Smart Farming: Current Applications and Future Potentials. *2023 International Conference on Sustainable Computing and Data Communication Systems (ICSCDS)*, 1096-1101. 10.1109/ICSCDS56580.2023.10105101

Pavankumar, P., & Darwante, N. K. (2022). Performance Monitoring and Dynamic Scaling Algorithm for Queue Based Internet of Things. *2022 International Conference on Innovative Computing, Intelligent Communication and Smart Electrical Systems (ICSES)*, 1-7.

Prasad, K., & Gugan, I. (2023). Enhancing Microcomputer Edge Computing for Autonomous IoT Motion Control. *International Journal on Recent and Innovation Trends in Computing and Communication*, *11*(3), 58–67. doi:10.17762/ijritcc.v11i3.6202

Prasad, V. S., Mahajan, Y., & Suneetha, S. (2023). A Cloud-Based UV Monitoring System for Remote Real-Time UV Exposure Tracking. *2023 4th International Conference on Smart Electronics and Communication (ICOSEC)*, 1764-1770.

Ramesh Naidu, P., Guruprasad, N., & Dankan Gowda, V. (2021). A High-Availability and Integrity Layer for Cloud Storage, Cloud Computing Security: From Single to Multi-Clouds. *Journal of Physics: Conference Series*, *1921*(1), 012072. doi:10.1088/1742-6596/1921/1/012072

Ramesha, M., & Pai, G. N. (2020). Internet of things: Internet revolution, impact, technology road map and features. *Adv. Math. Sci. J.*, *9*(7), 4405–4414. doi:10.37418/amsj.9.7.11

Ranjan, A., Hegge, K. K., & Agrawal, R. (2023). Wireless and Solar-Powered Multipurpose Robot for Industrial Automation: A Sustainable Solution. *2023 8th International Conference on Communication and Electronics Systems (ICCES)*, 146-151.

Reddy, N. S., & Ashreetha, B. (2023). Technologies for Comprehensive Information Security in the IoT. *2023 International Conference for Advancement in Technology (ICONAT),* 1-5.

Reddy, S., & Patwal, P. P. S. (2022). Data Analytics and Cloud-Based Platform for Internet of Things Applications in Smart Cities. *2022 International Conference on Industry 4.0 Technology (I4Tech)*, 1-6.

Sali, S. M., & Joy, K. R. (2023). Intelligent Rover: An IoT Based Smart Surveillance Robotic Car for Military. *2023 2nd International Conference on Computational Systems and Communication (ICCSC)*, 1-6. 10.1109/ICCSC56913.2023.10143011

Shekhar, R. (2023). Securing networked image transmission using public-key cryptography and identity authentication. *Journal of Discrete Mathematical Sciences and Cryptography*, *26*(3), 779–791. doi:10.47974/JDMSC-1754

Shruthi & Naveen. (2020). Internet of Things: A Survey on Devices, Ecosystem, Components and Communication Protocols. *2020 4th International Conference on Electronics, Communication and Aerospace Technology (ICECA)*, 611-616.

Singla, A., & Sharma, N. (2022). IoT Group Key Management using Incremental Gaussian Mixture Model. *2022 3rd International Conference on Electronics and Sustainable Communication Systems (ICESC)*, 469-474.

Srinivas, K. N. V., & Lakineni, P. K. (2023). Development of a Real-time Location Monitoring App with Emergency Alert Features for Android Devices. *2023 4th IEEE Global Conference for Advancement in Technology (GCAT),* 1-8.

Srinivas, V., & De, T. (2023). Wi-Fi Router Signal Coverage Position Prediction System using Machine Learning Algorithms. *2023 International Conference on Sustainable Computing and Smart Systems (ICSCSS),* 253-258.

Varun, R. S., & Rajesh, T. (2018). Implementation of swarm intelligence in obstacle avoidance. *RTEICT 2017 - 2nd IEEE International Conference on Recent Trends in Electronics, Information and Communication Technology, Proceedings*, 525–528.

Chapter 8
Communication Protocols in Cloud Robotics

C. S. Savidhan Shetty
https://orcid.org/0009-0007-3032-1736
National Institute of Technology, Karnataka, India

Manjunatha Badiger
Sahyadri College of Engineering and Management, India

ABSTRACT

The advent of cloud robotics has transformed traditional robotic systems by integrating them with the vast computational power of cloud infrastructure. This chapter explores the pivotal role of communication protocols in facilitating seamless interactions between physical robots and the cloud. The foundation of cloud robotics lies in addressing specific challenges such as low-latency, reliability, scalability, and security. Various communication protocols, including MQTT, RESTful APIs, and WebSocket, are examined for their suitability in different applications, ranging from data offloading to real-time monitoring. Security considerations are paramount in the transmission of sensitive data between robots and cloud servers, necessitating encryption, authentication, and authorization mechanisms. The chapter also delves into the importance of standardization and interoperability to foster collaboration among diverse robotic systems.

INTRODUCTION

In the ever-evolving landscape of robotics, the convergence of cloud computing and robotic systems has given rise to a transformative paradigm known as cloud robotics. This marriage of technologies has redefined the capabilities of robots, unlocking unprecedented levels of computational power, data access, and collaborative potential. At the heart of this revolution lie communication protocols, serving as the vital conduits that enable seamless interactions between physical robots and the expansive capabilities of cloud infrastructure.

DOI: 10.4018/979-8-3693-1914-7.ch008

The Rise of Cloud Robotics

As robotics transcends traditional boundaries, the symbiotic relationship between robots and cloud computing has propelled the field into uncharted territories. Cloud robotics harnesses the collective intelligence of the cloud to amplify the capabilities of individual robots, enabling them to perform complex tasks, access vast datasets, and benefit from real-time processing. This collaborative fusion has not only revolutionized existing applications but has also paved the way for novel use cases in sectors ranging from manufacturing and healthcare to space exploration (Hu, Tay, & Wen, 2012).

The Significance of Communication Protocols

At the epicentre of this synergy lies the significance of communication protocols. Beyond serving as mere conduits for connectivity, these protocols dictate the efficiency, security, and scalability of cloud robotics applications. They represent the language through which robots converse with cloud infrastructure, facilitating the exchange of critical information essential for real-time decision-making, task execution, and collaborative endeavours (Gouda & Multari, 1991). As the technological tapestry of cloud robotics continues to unfold, the role of communication protocols stands out as a pivotal determinant of success.

Navigating Challenges in Cloud Robotics Communication

However, the integration of cloud computing into robotics presents a unique set of challenges for communication protocols. The dynamic and distributed nature of cloud robotics amplifies concerns related to latency, bandwidth, security, and reliability (Zereik et al., 2018). As robots become increasingly reliant on cloud services, addressing these challenges becomes imperative to ensure the seamless and efficient operation of cloud-connected robotic systems.

Enabling Collaborative Potential Through Communication Protocols

Beyond their technical functionalities, communication protocols play a foundational role in unlocking the collaborative potential inherent in cloud robotics. These protocols serve as the silent architects of cooperation, allowing robots to seamlessly share information, coordinate actions, and operate in harmony within a cloud-connected ecosystem. Whether engaged in collaborative manufacturing tasks, coordinating movements in a swarm, or sharing insights across a network of autonomous vehicles, communication protocols are the conduits of collaboration (Tardioli, 2014).

The ability of robots to communicate effectively and in real-time fosters a collaborative intelligence that transcends individual capabilities. It empowers a collective robotic force to tackle complex challenges, distribute workloads efficiently, and adapt dynamically to changing conditions. This collaborative potential is not just a technical achievement; it marks a shift in the paradigm of how robots operate, emphasizing cooperation and synergy as essential elements in the realm of cloud robotics (Amaran et al., 2015).

As we delve into the intricate world of communication protocols in cloud robotics, we will explore not only the technical aspects of data exchange but also the profound impact these protocols have on enabling robots to work together seamlessly. The journey ahead encompasses the challenges, innovations, and future trends that will further enhance the collaborative potential of cloud-connected robotic systems, making them not just advanced machines but integral components of intelligent, cooperative ecosystems.

Evolution of Communication Protocols

The dynamic nature of technology propels communication protocols in cloud robotics toward a continual process of evolution. Emerging trends, technological advancements, and the growing demands of diverse applications shape the trajectory of these protocols. From the integration of edge computing to advancements in 5G technology and the infusion of machine learning, the evolution of communication protocols is a dynamic landscape responding to the evolving needs of cloud-connected robotic systems (Chung et al., 2010).

As we navigate through the evolution of communication protocols, we will explore how these advancements impact not only the technical aspects of data exchange but also the broader capabilities of cloud robotics. The chapter will uncover the role of protocols in enabling adaptability, intelligence, and resilience in the face of technological and environmental challenges.

Security Considerations in Communication Protocols

The collaborative potential of cloud robotics also brings forth security considerations. Communication protocols must not only facilitate seamless collaboration but also ensure the confidentiality, integrity, and authenticity of transmitted data. The chapter will delve into the security measures embedded within communication protocols, addressing challenges related to data privacy, secure authentication, and the robustness of encryption mechanisms (Dragomir et al., 2016).

Some important aspects of communication protocols in cloud robotics

1. Foundations of Cloud Robotics:

The integration of cloud computing with traditional robotic systems, forming what is known as cloud robotics, represents a groundbreaking synergy that significantly enhances the capabilities of robotic applications. At the core of this transformative paradigm is the fusion of physical robots with the expansive computational and storage resources provided by cloud infrastructure. The foundation of cloud robotics rests upon the concept of seamless communication between the physical robot and the cloud. This communication serves as the linchpin that enables the real-time exchange of data, commands, and information between the robot and the cloud servers. The significance of this seamless communication cannot be overstated, as it forms the backbone for various aspects that drive the effectiveness and versatility of cloud robotics (Kehoe et al., 2015).

A. Enhanced Computation:

Cloud robotics liberates robots from the constraints of onboard computing power. By offloading computationally intensive tasks to the cloud, robots can harness the immense processing capabilities available. This results in a substantial augmentation of computational resources, empowering robots to execute complex algorithms, simulations, and analyses.

B. Extended Storage Capacity:

Traditional robots often grapple with limited onboard storage. The integration with cloud computing introduces virtually limitless storage, allowing robots to store and retrieve vast amounts of data. This expanded storage capacity is particularly advantageous for applications that involve extensive data logging, mapping, and long-term operation.

C. Real-time Communication:

The communication protocols employed in cloud robotics establish real-time connections between robots and cloud servers. This instantaneous exchange of information is critical for applications that demand swift decision-making, such as autonomous navigation, response to dynamic environments, and collaborative tasks. Robust communication protocols ensure the reliability and efficiency of this real-time interaction.

D. Scalability and Adaptability:

Cloud robotics enables scalability by providing the flexibility to scale computational resources based on the requirements of specific tasks. As the complexity of a task increases, robots can seamlessly tap into additional cloud resources. This scalability ensures that robotic systems remain adaptable to varying workloads and dynamic environmental conditions.

E. Optimized Collaboration:

The seamless communication facilitated by robust protocols enhances collaboration between robots and the cloud. Multiple robots can share information, insights, and updates through the cloud, fostering a collaborative environment. Human operators can also engage in real-time collaboration, providing guidance, supervision, and intervention when necessary.

In essence, the foundation of cloud robotics lies in creating a symbiotic relationship between physical robots and the virtually limitless capabilities offered by cloud computing resources. The seamless communication facilitated by robust protocols is the conduit through which enhanced computation, extended storage, and real-time capabilities are realized. This foundation not only expands the horizons of what robots can achieve but also sets the stage for a new era of intelligent, interconnected, and highly adaptable robotic systems.

2. Requirements and Challenges:

The implementation of cloud robotics introduces a set of requirements and challenges that shape the design, deployment, and performance of robotic systems leveraging cloud computing resources. Addressing these requirements and overcoming challenges are crucial for the success of cloud robotics applications. Let's delve into the intricacies of the requirements and challenges associated with cloud robotics (Chen et al., 2018):

Requirements

1. **Low-Latency Communication:** Real-time applications, such as teleoperation or tasks requiring quick responses, necessitate low-latency communication between the robot and the cloud. Minimizing delays in data transmission is vital for maintaining the responsiveness and effectiveness of the robotic system.
2. **Reliability and Robustness:** Reliable communication is paramount for the seamless operation of cloud robotics. The system must be resilient to network fluctuations, disruptions, and potential downtimes to ensure continuous functionality. Robust communication protocols and error-handling mechanisms are essential components.
3. **Scalability:** Cloud robotics applications should be designed to scale seamlessly with the increasing demands of computational resources. Scalability ensures that the system can handle a growing number of robots, diverse tasks, or fluctuations in data processing requirements without compromising performance.
4. **Security and Privacy:** Given the transmission of sensitive data between robots and cloud servers, robust security measures are imperative. Encryption, authentication, and authorization protocols must be in place to safeguard data against unauthorized access, ensuring the privacy and integrity of information.
5. **Interoperability:** Cloud robotics often involves the collaboration of heterogeneous robotic systems and devices. Therefore, interoperability becomes crucial. Standardized communication interfaces and protocols facilitate seamless integration, allowing robots from different manufacturers to work together effectively.
6. **Resource Efficiency:** Efficient utilization of computational resources is essential to optimize costs and energy consumption. Cloud robotics applications should be designed to minimize unnecessary data transfer and leverage cloud resources judiciously, ensuring efficiency in resource usage.

Challenges

1. **Network Latency and Bandwidth Constraints:** The reliance on cloud infrastructure introduces potential challenges related to network latency and bandwidth limitations. High-latency connections can impact the responsiveness of robots, particularly in time-sensitive applications.
2. **Data Security Concerns:** Transmitting sensitive data over networks raises concerns about data security. Protecting data integrity, confidentiality, and preventing potential cyber threats become critical challenges that need to be addressed effectively.
3. **Dependency on Internet Connectivity:** Cloud robotics systems are inherently dependent on internet connectivity. Any disruptions in connectivity can hinder real-time communication and pose challenges, especially in scenarios where consistent and reliable connectivity is crucial.
4. **Cost Management:** While cloud resources offer scalability, the associated costs can accumulate, especially in applications with high computational demands. Managing and optimizing costs without compromising performance is a persistent challenge for cloud robotics developers.
5. **Adaptation to Dynamic Environments:** Cloud robotics often operate in dynamic and unpredictable environments. Adapting to changes in the environment, such as variations in network conditions or the introduction of new tasks, poses a challenge that requires robust system design and flexibility.

6. **Data Governance and Compliance:** Cloud robotics applications may involve handling data subject to regulatory requirements and compliance standards. Ensuring that the storage and processing of data adhere to legal and ethical standards poses an ongoing challenge in cloud robotics development.

Effectively addressing these requirements and overcoming challenges is crucial for unlocking the full potential of cloud robotics, ensuring the reliability, security, and scalability of these innovative systems across diverse applications.

Communication Protocol Types:

a. Message Queuing Telemetry Transport (MQTT):

MQTT, a lightweight and efficient messaging protocol, has emerged as a key player in the realm of communication protocols for cloud robotics. Originally designed for low-bandwidth, high-latency, or unreliable networks, MQTT's simplicity and flexibility make it well-suited for the diverse communication needs of robotic systems interfacing with cloud infrastructure.

Key Characteristics

1. **Publish-Subscribe Model:** MQTT operates on a publish-subscribe messaging pattern. In the context of cloud robotics, this means that robots (publishers) can send messages to specific topics, and cloud servers or other robots (subscribers) interested in those topics receive the messages. This decoupling of senders and receivers promotes scalability and flexibility in communication.
2. **Quality of Service (QoS) Levels:** MQTT supports different QoS levels for message delivery. This ensures that messages can be sent with varying degrees of reliability, addressing the dynamic requirements of cloud robotics applications. QoS levels range from "At Most Once" (fire and forget) to "Exactly Once" (guaranteed delivery).
3. **Low Bandwidth and Overhead:** One of MQTT's strengths is its minimal protocol overhead. The lightweight nature of the protocol makes it suitable for scenarios where bandwidth is limited, making it a practical choice for communication between resource-constrained robots and cloud servers.
4. **Last Will and Testament (LWT):** MQTT includes a feature called Last Will and Testament, allowing a device (robot) to specify a message to be sent in case of an unexpected disconnection. This ensures that the cloud infrastructure or other connected devices are aware of the robot's status, contributing to the reliability of the system.
5. **Retained Messages:** MQTT supports the concept of retained messages, where the last message sent on a topic is retained by the messaging broker. This is particularly useful in cloud robotics scenarios, ensuring that new subscribers receive the most recent state or information about a specific topic upon connecting.

Applications in Cloud Robotics

1. **Real-time Monitoring:** MQTT's ability to facilitate real-time communication is invaluable in cloud robotics applications where monitoring and receiving immediate feedback are crucial. This is particularly relevant in scenarios such as teleoperation or surveillance.

2. **Sensor Data Transmission:** For robotic systems that generate and transmit sensor data, MQTT provides an efficient means of sending this information to the cloud for processing and analysis. This is vital for applications like environmental monitoring or data-driven decision-making.
3. **Command and Control:** MQTT's publish-subscribe model allows for effective command and control mechanisms. Cloud servers can publish commands to specific topics, and robots subscribe to those topics, enabling seamless control and coordination in distributed robotic systems.
4. **Energy-Efficient Communication:** Due to its lightweight nature, MQTT is well-suited for scenarios where energy efficiency is a priority. In resource-constrained robotic devices, such as those operating on battery power, MQTT helps minimize communication overhead, extending the device's operational life.

Challenges and Considerations

1. **Security:** While MQTT itself does not enforce security features, it can be implemented over secure connections (MQTT over TLS/SSL) to address security concerns. Implementing proper authentication and encryption is crucial to safeguarding communication in cloud robotics.
2. **Scalability:** While MQTT is inherently scalable, effective scalability in cloud robotics requires careful consideration of the messaging broker's capabilities and deployment architecture to handle a growing number of connected devices.

In conclusion, MQTT's simplicity, efficiency, and adaptability make it a powerful communication protocol in cloud robotics. Its features align well with the requirements of robotic systems interfacing with cloud infrastructure, facilitating real-time communication, efficient data exchange, and seamless coordination in a wide range of applications.

b. Representational State Transfer (REST):

Representational State Transfer (REST) is an architectural style commonly used in web services, and it has found extensive application in the domain of cloud robotics. REST is characterized by its simplicity, scalability, and statelessness, making it an effective communication protocol for systems with distributed and resource-constrained components (Mohanarajah, et al., 2015).

Key Characteristics

1. **Statelessness:** One of the fundamental principles of REST is statelessness. Each request from a client to a server contains all the information needed to understand and fulfill that request. This property simplifies communication, enhances scalability, and supports efficient handling of requests in cloud robotics scenarios.
2. **Resource-Based:** REST models the interaction between components as the manipulation of resources. In the context of cloud robotics, these resources can represent various entities such as robots, sensors, or tasks. Resources are uniquely identified by URIs (Uniform Resource Identifiers), providing a standard way to access and manipulate them.

3. **CRUD Operations:** RESTful services typically support CRUD (Create, Read, Update, Delete) operations. This aligns well with the needs of cloud robotics applications, where operations like retrieving sensor data, updating robot states, or triggering specific actions are common requirements.
4. **Uniform Interface:** REST emphasizes a uniform and consistent interface, which simplifies interactions between clients and servers. The uniformity is achieved through standard conventions, including the use of HTTP methods (GET, POST, PUT, DELETE) and a set of status codes indicating the outcome of a request.
5. **Stateless Communication:** RESTful communication is stateless, meaning each request from a client to a server is independent. The server does not store any information about the client's state between requests. This property enhances scalability and allows for efficient load balancing in cloud environments.

Applications in Cloud Robotics

1. **Resource Management:** RESTful APIs in cloud robotics facilitate resource management by providing standardized endpoints to access and manipulate robotic entities. Robots, sensors, and other components can be represented as resources, and their states can be queried or modified using standard HTTP methods.
2. **Command and Control:** RESTful services enable command and control functionalities in cloud robotics. Cloud servers can expose RESTful endpoints that allow external entities to send commands to robots or retrieve information about their current state, contributing to effective coordination and collaboration.
3. **Interoperability:** REST's uniform interface and reliance on standard protocols, such as HTTP, contribute to interoperability. This enables different robotic systems, regardless of their underlying technologies, to communicate seamlessly through shared RESTful APIs.
4. **Scalability:** The stateless nature of RESTful communication supports scalability in cloud robotics applications. With each request containing all necessary information, load balancing and distributed processing can be efficiently managed to handle a growing number of connected devices.
5. **Data Exchange:** RESTful services are effective for data exchange between robots and cloud servers. Sensors on robots can publish data to RESTful endpoints, and cloud applications can retrieve this data in a standardized format (often JSON or XML), enabling easy integration and analysis.

Challenges and Considerations

1. **Limited Real-Time Capabilities:** REST may not be optimal for applications requiring extremely low-latency communication or real-time responsiveness. Other protocols like MQTT or WebSockets might be more suitable for such scenarios.
2. **Overhead in Small Payloads:** REST may introduce overhead in scenarios where communication involves small payloads, as the protocol carries additional metadata. This consideration is particularly relevant for resource-constrained robotic devices.

In conclusion, REST's simplicity, uniformity, and support for resource-based interactions make it a valuable communication protocol in cloud robotics. Its wide adoption, standardized interface, and compatibility with HTTP contribute to its effectiveness in enabling interoperability and facilitating various applications within the dynamic landscape of cloud robotics.

c. WebSocket Protocol:

WebSocket is a communication protocol that provides full-duplex communication channels over a single, long-lived connection. Unlike traditional HTTP, which follows a request-response paradigm, WebSocket allows for bidirectional, real-time communication between clients (such as robots) and servers (cloud infrastructure). This characteristic makes WebSocket particularly suitable for cloud robotics applications where continuous, low-latency data exchange is crucial.

Key Characteristics

1. **Full-Duplex Communication:** WebSocket enables simultaneous communication in both directions—clients can send messages to servers, and servers can push messages to clients. This bidirectional communication is maintained over a single, long-lived connection, avoiding the overhead of establishing a new connection for each interaction.
2. **Low Latency:** WebSocket's real-time nature makes it well-suited for scenarios where low-latency communication is essential. In cloud robotics, this capability is crucial for applications such as teleoperation, where quick responses are necessary for effective robot control.
3. **Efficient Use of Resources:** WebSocket minimizes resource usage compared to protocols that require opening and closing connections for each communication. The persistent connection reduces the need for repetitive handshakes, conserving bandwidth and improving overall efficiency.
4. **Event-Driven Model:** WebSocket operates on an event-driven model, allowing clients and servers to react to messages and events in real-time. This makes it a versatile choice for applications requiring rapid response to changing conditions or events in the robotic system.
5. **Compatibility with Web Technologies:** WebSocket is designed to work seamlessly with web technologies, making it easy to integrate with existing web-based interfaces and applications. This compatibility is advantageous for cloud robotics systems that leverage web-based control interfaces or monitoring dashboards.

Applications in Cloud Robotics

1. **Teleoperation:** WebSocket's low-latency bidirectional communication is ideal for teleoperation scenarios where operators control robots remotely. The real-time nature of WebSocket ensures that control commands from the operator are quickly transmitted to the robot, enabling responsive and precise control.
2. **Real-Time Monitoring:** Cloud robotics applications often require real-time monitoring of sensor data and robot states. WebSocket facilitates the continuous streaming of data from robots to cloud servers, allowing operators or monitoring systems to receive immediate updates and respond to changing conditions.
3. **Collaborative Robotics:** WebSocket supports collaborative robotic scenarios where multiple robots need to coordinate their actions. The bidirectional communication allows robots to share information in real-time, enhancing their ability to work together on complex tasks.

4. **Live Updates and Notifications:** WebSocket is well-suited for delivering live updates and notifications to robots or operators. Cloud servers can push real-time alerts or information to robots based on events or changes in the environment, enabling timely responses.

Challenges and Considerations

1. **Connection Management:** While WebSocket reduces the need for connection establishment overhead, proper connection management is essential. Handling disconnections, reconnecting after network failures, and managing a potentially large number of concurrent connections are considerations for robust WebSocket implementations.
2. **Security:** WebSocket connections can be secured using protocols such as Secure WebSocket (WSS) to encrypt data during transmission. Security considerations, including authentication and encryption, are vital to protect the integrity and confidentiality of data in cloud robotics applications.

WebSocket's real-time, bidirectional communication capabilities make it a powerful choice for cloud robotics applications requiring low-latency interactions, continuous data streaming, and rapid responsiveness. When properly implemented and secured, WebSocket enhances the efficiency and effectiveness of communication in dynamic and collaborative robotic systems interfacing with cloud infrastructure.

Security Considerations:

Security considerations in cloud robotics are paramount to safeguard sensitive data, ensure the integrity of communications, and protect against potential cyber threats. The distributed nature of cloud robotics, involving the exchange of data between physical robots and cloud servers, poses unique challenges that must be carefully addressed. Here are key security considerations in the context of cloud robotics:

1. **Data Encryption:** Employing strong encryption mechanisms is fundamental to protecting data transmitted between robots and cloud servers. The use of protocols like Transport Layer Security (TLS) or its predecessor, Secure Sockets Layer (SSL), helps ensure that communication channels are encrypted, preventing unauthorized access or tampering of data in transit.
2. **Authentication Mechanisms:** Implementing robust authentication mechanisms is essential to verify the identities of both the robots and the cloud servers. This helps prevent unauthorized access to the cloud infrastructure and ensures that only authenticated and authorized entities can interact with the robotic systems.
3. **Authorization Controls:** Authorization mechanisms play a critical role in controlling access to resources and functionalities within the cloud robotics ecosystem. Fine-grained access controls should be implemented to restrict actions based on roles and permissions, preventing unauthorized manipulation of robot behaviors or access to sensitive information.
4. **Secure Communication Protocols:** Carefully selecting and implementing secure communication protocols is crucial. For example, using protocols like HTTPS for web-based interfaces and secure versions of messaging protocols (e.g., MQTT over TLS) helps protect against man-in-the-middle attacks and eavesdropping.
5. **Secure Software Development Practices:** Adhering to secure software development practices is essential in preventing vulnerabilities. This includes regularly updating software, patching known vulnerabilities, and conducting security audits to identify and address potential weaknesses in both robot firmware and cloud-based applications.

6. **Physical Security of Robots:** Considering the physical security of robotic devices is often overlooked but critical. Robots should be physically protected to prevent tampering or unauthorized access. Physical security measures, such as robust authentication for physical access or tamper-evident designs, can mitigate risks.
7. **Secure Boot and Firmware Integrity:** Ensuring the integrity of the robot's firmware is essential. Implementing secure boot processes and mechanisms to verify the integrity of firmware before execution helps prevent unauthorized modifications and safeguards against malicious attacks attempting to compromise the robot's software.
8. **Incident Response and Monitoring:** Establishing robust incident response plans and continuous monitoring mechanisms is crucial for detecting and responding to security incidents promptly. Intrusion detection systems, log monitoring, and anomaly detection can aid in identifying and mitigating security threats in real-time.
9. **Regular Security Audits and Penetration Testing:** Conducting regular security audits and penetration testing is essential to identify and address vulnerabilities. These assessments should cover both the robot's firmware and the cloud-based infrastructure, helping ensure the overall security resilience of the cloud robotics system.
10. **Compliance with Standards and Regulations:** Adhering to relevant cybersecurity standards and regulations is essential. Depending on the application domain, industry-specific standards (such as ISO 27001) or regional regulations may prescribe specific security measures that must be followed to achieve compliance.
11. **Data Privacy and Compliance:** Given the potential for handling personal or sensitive data in cloud robotics, compliance with data protection regulations (e.g., GDPR, HIPAA) is essential. This involves implementing measures to ensure the privacy and lawful processing of data throughout the robotic system.

In summary, a comprehensive security strategy for cloud robotics should encompass encryption, authentication, access controls, secure software development, physical security, incident response, and compliance with relevant standards. By addressing these considerations, developers and operators can create a resilient and secure cloud robotics ecosystem, mitigating risks and ensuring the trustworthiness of the robotic systems.

Standardization and Interoperability:

Standardization and interoperability are critical factors in the development and deployment of cloud robotics systems. Standardization involves the establishment of common frameworks, protocols, and specifications, while interoperability ensures that diverse robotic systems and components can seamlessly work together. These elements contribute to the scalability, flexibility, and collaborative potential of cloud robotics (Hunkeler, Truong, & Stanford-Clark, 2008).

Key Aspects

1. **Communication Protocols:** Standardizing communication protocols is essential for ensuring that robots, sensors, and cloud services can interact efficiently. Common protocols, such as MQTT, RESTful APIs, or WebSocket, promote interoperability by providing a shared language for devices and systems to exchange information .

2. **Data Formats and Representations:** Standardizing data formats and representations is crucial for achieving interoperability in cloud robotics. Consistent data structures, such as JSON or XML, allow different components to understand and process information uniformly. This standardization facilitates seamless data exchange between robots and cloud services.
3. **Semantic Interoperability:** Semantic interoperability involves standardizing the meaning of data and ensuring a common understanding of information across different robotic systems. Establishing ontologies or defining standardized vocabularies helps achieve semantic consistency, enabling robots and cloud services to interpret data uniformly.
4. **Device Profiles and APIs:** Standardizing device profiles and APIs (Application Programming Interfaces) enables the integration of various robotic devices into a cloud robotics ecosystem. This allows developers to create applications that can interact with diverse robots, sensors, and actuators using a common set of interfaces.
5. **Security Standards:** Standardizing security measures is crucial for maintaining a secure cloud robotics environment. Common security standards and best practices, such as those defined by ISO 27001 or NIST, help ensure that security mechanisms, encryption methods, and access controls are consistent across different robotic systems.
6. **Robot Operating System (ROS):** ROS, an open-source middleware framework, serves as a standard platform for developing robotic software. Its widespread adoption fosters interoperability by providing a common framework for robotic applications. ROS allows different robotic components to communicate and share data seamlessly.

Benefits

1. **Enhanced Collaboration:** Standardization and interoperability facilitate collaboration between different robotic systems and components. A standardized communication framework allows robots from different manufacturers to share information, work together on tasks, and contribute to a collaborative robotic ecosystem.
2. **Scalability:** By adhering to standardized protocols and interfaces, cloud robotics systems become more scalable. New robots or devices can be easily integrated into existing systems without extensive modifications, promoting scalability in terms of both the number of devices and the complexity of tasks.
3. **Flexibility and Adaptability:** Interoperability allows for flexibility and adaptability in cloud robotics. Robots with different capabilities can be seamlessly added or replaced, and cloud services can be extended or updated without disrupting the entire system. This flexibility ensures that cloud robotics solutions can evolve over time.
4. **Reduced Development Time and Costs:** Standardized interfaces and protocols streamline the development process. Developers can leverage existing standards, APIs, and frameworks, reducing the time and cost associated with creating custom solutions for each robotic component. This accelerates the overall development cycle.
5. **Market Growth and Innovation:** Standardization fosters a more open and competitive market. When robots and cloud services adhere to common standards, it encourages innovation as developers can focus on creating value-added features rather than reinventing basic communication or integration mechanisms. This, in turn, accelerates the growth of the cloud robotics market.

Challenges

1. **Diverse Use Cases and Requirements:** Cloud robotics spans various industries and applications with diverse use cases and requirements. Developing standards that cater to this diversity while maintaining interoperability can be challenging.
2. **Rapid Technological Evolution:** The field of robotics is rapidly evolving, with new technologies and capabilities emerging regularly. Standardization efforts must keep pace with these changes to remain relevant and effective.
3. **Global Adoption and Consensus:** Achieving global adoption and consensus on standards can be challenging due to varying regulatory environments, cultural differences, and divergent industry practices.
4. **Balancing Flexibility and Standardization:** Striking the right balance between standardization and the need for flexibility can be a delicate task. Overly rigid standards may stifle innovation, while too much flexibility may hinder interoperability.

In conclusion, standardization and interoperability are foundational elements in the success of cloud robotics. Efforts to establish and adhere to common standards contribute to the development of cohesive, collaborative, and scalable cloud robotics ecosystems, fostering innovation and growth in this dynamic field.

Future Trends and Emerging Protocols

The field of cloud robotics is dynamic, and as technology continues to advance, several future trends and emerging protocols are poised to shape the landscape. These trends and protocols are likely to influence the capabilities, efficiency, and scalability of cloud robotics applications. Let's explore some of these anticipated developments:

1. **Edge Computing Integration:** Future cloud robotics systems are expected to integrate more closely with edge computing resources. Edge computing involves processing data closer to the source (robots) rather than relying solely on distant cloud servers. This trend reduces latency, enhances real-time processing, and enables more efficient utilization of network bandwidth.
2. **5G Technology:** The deployment of 5G networks is anticipated to have a significant impact on cloud robotics. With faster and more reliable connectivity, 5G facilitates low-latency communication, high data transfer rates, and improved network reliability. This is particularly crucial for applications requiring real-time control and responsiveness (Raunholt et al., 2021).
3. **Advanced Networking Protocols:** Emerging networking protocols specifically tailored for the requirements of cloud robotics are likely to be developed. These protocols may address challenges such as ultra-low latency, high reliability, and efficient communication in dynamic and resource-constrained environments.
4. **Federated Learning:** Federated learning is an emerging approach where machine learning models are trained collaboratively across decentralized devices. In the context of cloud robotics, federated learning allows robots to learn and adapt to their environments without transmitting sensitive data to centralized servers. This privacy-preserving approach is expected to gain prominence.

5. **Decentralized Communication Models:** As cloud robotics ecosystems grow, decentralized communication models may become more prevalent. Instead of relying on a single centralized server, robots may communicate directly with each other in a peer-to-peer fashion. This approach enhances fault tolerance, reduces dependencies, and contributes to a more resilient system.
6. **Reinforcement Learning for Autonomy:** Future trends in cloud robotics may see an increased reliance on reinforcement learning for autonomous decision-making. Robots could leverage cloud-based reinforcement learning models to adapt and improve their behavior based on experiences and environmental changes (Kormushev, Calinon, & Caldwell, 2013).
7. **Domain-Specific Protocols:** As cloud robotics applications become more specialized across various industries, domain-specific communication protocols may emerge. These protocols would be tailored to the unique requirements and challenges of specific applications, optimizing performance and efficiency.
8. **Explainable AI for Transparency:** With the increasing adoption of artificial intelligence in cloud robotics, there is a growing emphasis on explainable AI. Future trends may involve the development of protocols that provide transparency and insights into the decision-making processes of AI algorithms, promoting trust and accountability.
9. **Quantum Communication for Security:** Quantum communication, leveraging the principles of quantum mechanics, holds promise for enhancing the security of communication in cloud robotics. Quantum key distribution protocols may be explored to provide secure and unbreakable encryption for data transmitted between robots and cloud servers (Gonçalves, 2018).
10. **Blockchain for Trust and Accountability:** Blockchain technology may find applications in cloud robotics to establish trust, transparency, and accountability. Decentralized ledgers could be used to record and verify transactions, ensuring the integrity of data and fostering trust in the interactions between robots and the cloud.

In conclusion, the future of cloud robotics is likely to witness a convergence of cutting-edge technologies, advanced communication protocols, and innovative approaches to address evolving challenges. As these trends unfold, the field is expected to continue pushing the boundaries of what is possible, enabling more intelligent, efficient, and collaborative robotic systems interfacing with cloud infrastructure.

Case Studies

Certainly, let's explore a couple of case studies that highlight the implementation and impact of cloud robotics in real-world scenarios:

1. **Amazon Robotics (formerly Kiva Systems):** Amazon Robotics, formerly known as Kiva Systems before its acquisition by Amazon, is a pioneering example of cloud robotics in e-commerce fulfilment centers. The company developed a fleet of mobile robots designed to automate the movement of goods within warehouses (Integrating Different Levels of Automation: Lessons From Winning the Amazon Robotics Challenge, 2016)

Implementation

- Thousands of mobile robots navigate autonomously in the warehouse, transporting shelves of products to human workers for order fulfilment.
- The robots communicate with a central control system that manages the entire fleet and optimizes the movement of goods, reducing the time it takes to fulfil customer orders.

Impact

- Significantly increased operational efficiency: The use of cloud-connected robots has drastically reduced the time it takes to pick, pack, and ship orders.
- Scalability: Amazon can easily scale its robotic workforce by deploying additional robots and updating control algorithms through cloud-based systems.

2. **Rethink Robotics - Sawyer:** Rethink Robotics developed collaborative robots (cobots) with a focus on ease of use and adaptability. One notable example is the Sawyer robot, designed to work alongside human workers in various manufacturing environments.

Implementation

- Sawyer robots are equipped with sensors and cameras, enabling them to perceive their surroundings and collaborate safely with humans.
- The robots can be programmed and trained by non-expert users, making them accessible for a wide range of tasks without the need for specialized programming skills.

Impact

- Increased flexibility: Sawyer's adaptability allows it to be quickly redeployed for different tasks, making it suitable for small-batch manufacturing and agile production environments.
- Enhanced safety: The robot's ability to work safely alongside humans is a testament to the collaborative nature of cloud-connected robots.

3. **NASA's Robonaut:** NASA's Robonaut is an example of cloud-connected robots used in space exploration. Robonaut is designed to work in environments too hazardous for humans and has been deployed on the International Space Station (ISS) (Diftler et al., 2011).

Implementation

- Robonaut is equipped with a variety of sensors and cameras for perception and features dexterous robotic hands for intricate tasks.
- The robot is remotely operated by human astronauts on the ISS from a ground-based control center, showcasing the use of cloud-based communication for space exploration.

Impact

- Remote operation: Cloud connectivity enables real-time remote control of Robonaut, allowing astronauts to perform tasks in space without physically being present.

- Improved efficiency: The ability to teleoperate Robonaut for specific tasks enhances the overall efficiency of activities on the ISS.

These case studies highlight the diverse applications of cloud robotics, ranging from e-commerce and manufacturing to space exploration. The implementation of cloud-connected robots in these scenarios has demonstrated improvements in efficiency, scalability, safety, and adaptability, showcasing the transformative potential of cloud robotics across various industries.

Machine Learning and Adaptive Protocols in Cloud Robotics

The integration of machine learning (ML) into communication protocols is a transformative development in the realm of cloud robotics. This synergy enhances the adaptability, efficiency, and responsiveness of communication systems, introducing a new paradigm where protocols can dynamically evolve based on learned patterns and real-time conditions (Velásquez Villagrán, Pesado, & Estevez, 2020).

1. Adaptive Learning Mechanisms

Traditional communication protocols often operate with predefined parameters, which may not be optimal for the diverse and dynamic conditions encountered in cloud robotics. Machine learning brings adaptability to these protocols by introducing learning mechanisms that analyze historical data and adapt communication strategies based on observed patterns. For example, the protocol may learn to adjust data transfer rates during specific operational scenarios or optimize communication paths based on the network topology.

2. Predictive Analytics for Communication Needs

Machine learning algorithms integrated into communication protocols can perform predictive analytics to anticipate communication needs. By analyzing patterns in data exchange, traffic, and task requirements, these algorithms can make informed predictions about future communication demands. This predictive capability allows protocols to proactively allocate resources, optimize data transmission, and prepare for upcoming changes in the robotic system's communication patterns.

3. Dynamic Optimization of Parameters

Adaptive protocols leverage machine learning to dynamically optimize communication parameters. These parameters include data transfer rates, packet prioritization, and network paths. By continuously monitoring the performance of the communication network and learning from real-time feedback, adaptive protocols can autonomously adjust these parameters to ensure optimal performance. This dynamic optimization enhances the efficiency of data exchange while responding to variations in network conditions.

4. Context-Aware Communication

Machine learning enables communication protocols to become context-aware, taking into consideration the specific context in which robotic systems operate. Contextual factors, such as the type of task being performed, the environment, or the urgency of communication, can influence the protocol's behavior. Adaptive protocols, through machine learning, learn to recognize and adapt to different contextual cues, ensuring that communication strategies align with the specific requirements of each scenario.

5. Fault Tolerance and Anomaly Detection

Machine learning algorithms contribute to enhanced fault tolerance in communication protocols. By learning normal patterns of communication and system behavior, adaptive protocols can detect anomalies indicative of potential faults or cyber threats. This capability enables the protocol to respond to abnormal situations, mitigate risks, and maintain the integrity and reliability of communication in the face of unexpected challenges.

6. Energy-Efficient Communication

In resource-constrained environments, energy-efficient communication is paramount. Adaptive protocols, powered by machine learning, can optimize communication patterns to minimize energy consumption. This is particularly crucial for robotic systems operating on battery power. By learning from historical data and real-time energy consumption patterns, the protocol can intelligently manage communication processes to conserve energy without compromising performance.

7. Continuous Improvement Through Feedback

Machine learning facilitates continuous improvement of communication protocols through feedback loops. As the robotic system operates and interacts with the environment, the protocol receives feedback on its performance. This feedback is used to update the machine learning models, allowing the protocol to iteratively improve its adaptability and decision-making capabilities over time.

Challenges and Considerations

While the integration of machine learning into communication protocols brings significant benefits, it also poses challenges. The design and implementation of adaptive protocols require careful consideration of factors such as model interpretability, training data quality, and the potential for biases in learning algorithms. Additionally, the real-time nature of cloud robotics demands efficient learning algorithms that can adapt quickly to changing conditions without introducing unnecessary latency (Sorrentino, Cavallo, & Fiorini, 2020; Wan et al., 2016; Qureshi & Koubâa, 2014).

In summary, the fusion of machine learning and communication protocols in cloud robotics represents a paradigm shift toward more intelligent, adaptive, and efficient communication systems. These adaptive protocols, driven by machine learning algorithms, have the capacity to learn, evolve, and optimize their behavior in response to the dynamic and complex nature of cloud-connected robotic environments. As machine learning continues to advance, the synergy with communication protocols is poised to play a pivotal role in shaping the future of cloud robotics, unlocking new levels of adaptability, performance, and intelligence in robotic systems.

CONCLUSION

In conclusion, communication protocols serve as the lifeline that intricately weaves together the fabric of cloud robotics, orchestrating a symphony of seamless interactions between physical robots and the vast capabilities of cloud infrastructure. These protocols represent the language through which data is exchanged, commands are transmitted, and information flows in real-time, forming the backbone of a dynamic and interconnected robotic ecosystem.

As technology marches forward, the landscape of communication in cloud robotics stands on the precipice of continuous evolution. The advent of groundbreaking advancements, such as 5G technology and the integration of edge computing, heralds a new era of possibilities. These developments promise to address challenges related to latency, bandwidth, and scalability, further enhancing the responsiveness and efficiency of cloud-connected robotic systems.

Yet, with every stride forward, novel challenges and opportunities emerge. Security considerations become paramount as the volume of transmitted data grows, necessitating robust encryption, authentication, and authorization mechanisms to safeguard the integrity and confidentiality of information traversing the cloud robotics network. Striking a delicate balance between standardization and flexibility becomes crucial, ensuring that communication protocols can adapt to the diverse needs of applications while fostering interoperability and collaboration.

In this ever-evolving landscape, the careful selection and implementation of communication protocols stand as pivotal determinants of success. The right protocols not only facilitate efficient data exchange but also pave the way for innovations in teleoperation, real-time monitoring, and collaborative robotics. As industries embrace the transformative potential of cloud robotics, the journey forward hinges on the continual refinement and adaptation of communication protocols—a journey that holds the promise of unlocking the full potential of intelligent, interconnected, and adaptable robotic systems across a myriad of domains. Through strategic protocol choices, the future of cloud robotics is poised to transcend current limitations, ushering in an era where human-robot collaboration and innovation know no bounds.

REFERENCES

Amaran, M. H., Noh, N. A. M., Rohmad, M. S., & Hashim, H. (2015). A comparison of lightweight communication protocols in robotic applications. *Procedia Computer Science*, *76*, 400–405. doi:10.1016/j.procs.2015.12.318

Chen, W., Yaguchi, Y., Naruse, K., Watanobe, Y., Nakamura, K., & Ogawa, J. (2018). A study of robotic cooperation in cloud robotics: Architecture and challenges. *IEEE Access : Practical Innovations, Open Solutions*, *6*, 36662–36682. doi:10.1109/ACCESS.2018.2852295

Chung, J.-M., Park, K., Won, T., Oh, W., & Choi, S. (2010). New protocols for future wireless systems. *2010 53rd IEEE International Midwest Symposium on Circuits and Systems*.

Diftler, M. A., Permenter, F. A., Hargrave, B. K., Platt, R., Savely, R. T., Ambrose, R. O., Mehling, J. S., Abdallah, M. E., Radford, N. A., Bridgwater, L. B., Sanders, A. M., Askew, R. S., Linn, D. M., & Yamokoski, J. D. (2011). Robonaut 2 - The first humanoid robot in space. *2011 IEEE International Conference on Robotics and Automation*. 10.1109/ICRA.2011.5979830

Dragomir, D., Gheorghe, L., Costea, S., & Radovici, A. (2016). A survey on secure communication protocols for IoT systems. *2016 International Workshop on Secure Internet of Things (SIoT).* 10.1109/SIoT.2016.012

Gonçalves, C. P. dos S. (2018). Quantum robotics, neural networks and the quantum force interpretation. SSRN *Electronic Journal*. doi:10.2139/ssrn.3244327

Gouda, M. G., & Multari, N. J. (1991). Stabilizing communication protocols. IEEE Transactions on Computers. *IEEE Transactions on Computers*, *40*(4), 448–458. doi:10.1109/12.88464

Hu, G., Tay, W., & Wen, Y. (2012). Cloud robotics: Architecture, challenges and applications. *IEEE Network*, *26*(3), 21–28. doi:10.1109/MNET.2012.6201212

Hunkeler, U., Truong, H. L., & Stanford-Clark, A. (2008). MQTT-S — A publish/subscribe protocol for Wireless Sensor Networks. *2008 3rd International Conference on Communication Systems Software and Middleware and Workshops (COMSWARE '08).*

Kehoe, B., Patil, S., Abbeel, P., & Goldberg, K. (2015). A survey of research on cloud robotics and automation. *IEEE Transactions on Automation Science and Engineering, 12*(2), 398–409. doi:10.1109/TASE.2014.2376492

Kormushev, P., Calinon, S., & Caldwell, D. (2013). Reinforcement learning in robotics: Applications and real-world challenges. *Robotics (Basel, Switzerland)*, *2*(3), 122–148. doi:10.3390/robotics2030122

Mohanarajah, G., Hunziker, D., D'Andrea, R., & Waibel, M. (2015). Rapyuta: A cloud robotics platform. *IEEE Transactions on Automation Science and Engineering, 12*(2), 481–493. doi:10.1109/TASE.2014.2329556

Qureshi, B., & Koubâa, A. (2014). Five traits of performance enhancement using cloud robotics: A survey. *Procedia Computer Science*, *37*, 220–227. doi:10.1016/j.procs.2014.08.033

Raunholt, T., Rodriguez, I., Mogensen, P., & Larsen, M. (2021). Towards a 5G mobile edge cloud planner for autonomous mobile robots. *2021 IEEE 94th Vehicular Technology Conference (VTC2021-Fall).*

Sorrentino, A., Cavallo, F., & Fiorini, L. (2020). A plug and play transparent communication layer for cloud robotics architectures. *Robotics (Basel, Switzerland)*, *9*(1), 17. doi:10.3390/robotics9010017

Tardioli, D. (2014). A wireless communication protocol for distributed robotics applications. *2014 IEEE International Conference on Autonomous Robot Systems and Competitions (ICARSC).* 10.1109/ICARSC.2014.6849795

Velásquez Villagrán, N., Pesado, P., & Estevez, E. (2020). Cloud robotics for industry 4.0 - A literature review. In *Communications in Computer and Information Science* (pp. 3–15). Springer International Publishing.

Wan, J., Tang, S., Yan, H., Li, D., Wang, S., & Vasilakos, A. V. (2016). Cloud robotics: Current status and open issues. *IEEE Access : Practical Innovations, Open Solutions*. doi:10.1109/ACCESS.2016.2574979

Zereik, E., Bibuli, M., Mišković, N., Ridao, P., & Pascoal, A. (2018). Challenges and future trends in marine robotics. *Annual Reviews in Control*, *46*, 350–368. doi:10.1016/j.arcontrol.2018.10.002

Chapter 9
Security and Privacy in Cloud Robotics

G. Suchetha
Sahyadri College of Engineering and Management, India

A. Masooda
Sahyadri College of Engineering and Management, India

C. Harinakshi
https://orcid.org/0009-0003-0321-766X
Sahyadri College of Engineering and Management, India

ABSTRACT

Cloud robotics, the fusion of cloud computing and robotics, defines a transformative era. Empowering robots with unprecedented cloud access enables intricate tasks but poses security challenges. Transmitting sensitive data demands robust security. Machine learning aids real-time threat detection; authentication, access control, data encryption, and storage best practices are explored. In conclusion, the chapter guides stakeholders in cloud robotics security by addressing critical issues, promoting best practices, and anticipating trends. Offering insights into data security and privacy, it equips professionals to harness the potential of cloud robotics while mitigating risks. This concise exploration, spanning the digital and physical realms, serves as a beacon in the dynamic field of cloud robotics security and privacy.

1. INTRODUCTION

The fusion of robotics with cloud computing has ushered in a new era of innovation, allowing robots to tap into the immense computational resources available in the cloud. Cloud robotics promises unprecedented capabilities, from autonomous navigation to complex problem-solving, transforming various sectors ranging from manufacturing and healthcare to service industries. Cloud computing, data storage in the cloud, and other internet-based technologies can be used to create new-generation robots that are more cost-effective, power-efficient, and have improved communication, storage, and mobility than existing

DOI: 10.4018/979-8-3693-1914-7.ch009

robots and industrialization frameworks. Modern industries like business, medicine, space exploration, and many more use robots. Numerous examples demonstrate how dangerous it is to deploy technologies like the Internet of Things (IoT), cloud computing, big data, etc. in robotics, which also have negative effects on people and organizations as mentioned in (Jain, S., & Doriya, R. (2019)).

A networked robotic system is defined by a collection of robotic devices connected through both wired and wireless communication networks. These systems are typically categorized into teleoperated robots and multi-robot systems. In teleoperation scenarios, a human operator remotely controls or manipulates a robot by sending commands and receiving real-time measurements via the communication network. Examples include scenarios like operating planetary rovers and conducting remote medical surgeries. On the other hand, in multi-robot systems, a coordinated team of networked robots collaboratively completes tasks by exchanging sensory data and information through the communication network. Applications of multi-robot systems include cooperative robot manipulators, robot teams engaged in search and rescue operations, and clusters of microsatellites cooperating to form particular configurations. This interconnected and collaborative nature of networked robotic systems opens up diverse possibilities, from precision surgery to intricate search and rescue operations.

However, as the potential of cloud robotics is embraced, the profound implications it carries for security and privacy cannot be ignored. With robots becoming increasingly autonomous and interconnected, they generate and transmit a wealth of sensitive data, encompassing everything from environmental sensor readings to audio and video feeds. This data, while invaluable for decision-making and problem-solving, is also a potential goldmine for malicious actors seeking unauthorized access or exploitation. Therefore, addressing the multifaceted security and privacy challenges associated with cloud robotics comprehensively and proactively is paramount.

Deploying cloud robotics presents a significant challenge in terms of ensuring security and privacy. The collaboration and interaction of cloud robots with physically distant counterparts raise sensitive issues related to the protection of robotic data and activities. In scenarios where co-located robots communicate through wireless message broadcasts, it is crucial to safeguard these messages from eavesdropping or tampering by external robots or unauthorized devices.

Unlike the simplified model where all robots in a group share the same security privileges, real-world cloud robot groups comprise robots with diverse capabilities, functions, roles, and attributes, each requiring different levels of access privileges. Managing security in such a heterogeneous environment becomes a complex problem, particularly when considering the mobility of robots. Policies governing role assignment, attributes, authorization, and access privileges must accommodate dynamic changes in group membership.

The utmost priority is to prevent security breaches in cloud robots, as such breaches could result in irreversible disasters in the physical world. Regrettably, there is a scarcity of relevant research addressing the dynamic establishment of a secure cloud robot network. Key aspects, such as determining eligible robots for collaboration, specifying required resources and services, and defining access to these robotic services, remain unexplored. (Ko, H., Keoh, S. L., & Jin, J. (2017)) mentioned resolving various security issues, including authentication, membership management, privacy protection, and access control within the cloud robot network, is imperative to ensure a secure and robust deployment.

This chapter embarks on a journey through the intricate landscape of security and privacy within the domain of robotics, examining the fundamental principles, challenges, best practices, and emerging trends that are paramount to understanding and addressing these critical issues. As this domain is explored, it becomes evident that the implications extend far beyond the confines of the robotics industry; they permeate our daily lives, affecting healthcare, transportation, industry, and even our personal privacy.

In this chapter, the foundations of cloud robotics are briefly introduced, followed by an exploration of security challenges and privacy concerns in cloud robotics, case studies, and practical implementations. This is followed by a discussion on legal frameworks and regulations.

2. FOUNDATIONS OF CLOUD ROBOTICS: BRIDGING INTELLIGENCE AND CONNECTIVITY

In the realm of technological evolution, cloud robotics stands as a transformative paradigm, seamlessly intertwining the capabilities of cloud computing with the agility and functionality of robotic systems. This section delves into the foundational aspects of cloud robotics, unraveling its definition, characteristics, evolutionary trajectory, key components, and the intricate integration of cloud computing with robotic frameworks.

2.1 Definition and Characteristics of Cloud Robotics

At its core, cloud robotics is a paradigm where robotic systems collaborate with cloud-based services to augment their computational capabilities and functionalities. This collaboration extends beyond the confines of individual robots, allowing them to access shared resources, information, and collective intelligence hosted in the cloud. A break from conventional, isolated robotic systems, cloud robotics is defined by real-time data exchange, shared learning, and adaptive decision-making.

The characteristics of cloud robotics are rooted in their interconnected nature. These systems exhibit dynamic responsiveness, leveraging cloud-based resources to enhance their adaptability to diverse environments. The ability to process vast datasets in real-time, access powerful computational resources, and engage in collaborative problem-solving propels cloud robotics into the forefront of intelligent automation.

2.2 Evolution of Cloud Robotics and Its Impact on Various Industries

The evolution of cloud robotics has been marked by a trajectory of continuous innovation with a profound impact on diverse industries. Initially emerging as a concept in research laboratories, cloud robotics has transitioned into practical applications across sectors such as manufacturing, healthcare, agriculture, and logistics.

In manufacturing, cloud robotics enables seamless collaboration between robotic devices, facilitating adaptive manufacturing processes and efficient resource allocation. Healthcare witnesses the integration of cloud robotics in telemedicine, remote surgeries, and patient care, extending the reach of medical expertise. Agriculture benefits from precision farming techniques, where robotic systems equipped with sensors and actuators collaborate with cloud-based analytics for optimized crop management. Logistics experiences streamlined warehouse operations and intelligent transportation systems, where cloud-connected robots enhance efficiency and decision-making.

The impact of cloud robotics reverberates across industries, reshaping conventional practices and introducing a new paradigm of intelligent collaboration between machines and cloud-based services.

2.3 Key Components of Cloud Robotics Systems

Cloud robotics systems encompass a set of key components that synergize to create a cohesive and intelligent framework. These components are designed to enhance the capabilities of robotic systems, providing them with access to cloud-based resources and facilitating seamless collaboration.

1. Robotic Devices

- The physical robots form the foundational layer, equipped with sensors, actuators, and processing units. These devices interact with their surroundings and generate data that is transmitted to the cloud for analysis and decision-making.

2. Cloud Computing Infrastructure

- The cloud computing infrastructure serves as the backbone, providing computational power, storage, and services for processing data and executing complex algorithms. Cloud servers play a pivotal role in augmenting the processing capabilities of individual robotic devices.

3. Communication Networks

- The communication networks act as conduits for data exchange between robotic devices and the cloud. Whether through wired or wireless connections, these networks facilitate real-time communication, enabling robots to access cloud-based services seamlessly.

4. Data Storage

- Data storage components in the cloud allow for the accumulation and retrieval of information generated by robotic devices. This centralized data repository becomes a valuable resource for analytics, machine learning, and collaborative decision-making.

5. Software Frameworks

- Software frameworks govern the interactions between robotic devices and cloud-based services. These frameworks include communication protocols, application programming interfaces (APIs), and middleware that enable seamless integration and collaboration.

2.4 Integration of Cloud Computing, Data Storage, and Communication Networks With Robotic Systems

The integration of cloud computing, data storage, and communication networks with robotic systems is the linchpin of cloud robotics' functionality. This integration occurs at multiple levels, orchestrating a symphony of interactions that enhance the capabilities of individual robots and contribute to the collective intelligence hosted in the cloud.

1. Real-Time Data Exchange

- Robotic devices generate a continuous stream of data through their sensors and actuators. This data is transmitted in real-time to the cloud, where it undergoes analysis, contributing to informed decision-making and adaptive behaviors.

2. Computational Offloading

- Resource-intensive tasks that exceed the processing capacity of individual robots are offloaded to the cloud. This computational offloading empowers robots with the ability to execute complex algorithms, machine learning models, and simulations, leveraging the expansive computational resources in the cloud.

3. Collaborative Learning

- Cloud robotics facilitates collaborative learning, where insights gained by one robotic device can be shared with others through the cloud. This collective learning process enhances the adaptability and intelligence of the entire robotic ecosystem.

4. Remote Control and Monitoring

- The integration allows for remote control and monitoring of robotic devices. Operators can leverage cloud-based interfaces to monitor, control, and reconfigure robotic systems from anywhere with an internet connection.

5. Adaptability to Dynamic Environments

- The dynamic nature of cloud robotics enables adaptability to changing environments. As robotic devices encounter new scenarios, the cloud provides a reservoir of knowledge and computational power to facilitate real-time adjustments and decision-making.

The integration of cloud computing, data storage, and communication networks with robotic systems creates a symbiotic relationship where the strengths of each component amplify the capabilities of the others. This integration is fundamental to the evolution of cloud robotics, enabling a paradigm where intelligence and connectivity converge to redefine the landscape of automation.

In conclusion, the foundations of cloud robotics rest on the principles of intelligent collaboration and seamless connectivity. The evolution of this paradigm has rippled across industries, reshaping conventional practices and introducing a new era of intelligent automation. The key components of cloud robotics systems work in tandem, creating a cohesive framework where robotic devices, cloud computing infrastructure, communication networks, data storage, and software frameworks synergize to unlock unprecedented capabilities. The integration of these components facilitates real-time data exchange, computational offloading, collaborative learning, remote control, and adaptability to dynamic environments. Navigating the intricate foundations of cloud robotics unveils a future where the convergence of intelligence and connectivity holds the promise of transforming industries and redefining the possibilities of automation.

3. SECURITY CHALLENGES IN CLOUD ROBOTICS

3.1 Security Issues in Cloud Computing

Cloud computing has become a prominent focus in the realm of research, characterized by the division of its services into three main categories: Software as a Service (SaaS), Platform as a Service (PaaS), and Infrastructure as a Service (IaaS). These services, managed by Cloud Service Providers (CSPs), allow users to access and utilize them, paying only for the specific services they consume. The cloud offers various advantages to both end-users and organizations, including elasticity, on-demand services, Security features, scalability, pay-as-you-go pricing, and cost-effectiveness.

Despite the numerous benefits, the cloud environment is not immune to vulnerabilities. Notably, (Al Awadhi, E., Salah, K., & Martin, T. (2013, November)) conducted research deploying honeypots into cloud networks and concluded that clouds are susceptible to various attacks and malicious activities. The cloud computing landscape faces several security challenges and vulnerabilities, among which distributed denial of service (DDoS) attacks stand out as one of the most perilous threats.

This discussion delves into the security issues prevalent in cloud computing, categorized into four levels: network, data storage, virtualization, and operating system. The identified threats pose potential risks to the integrity and reliability of cloud services, emphasizing the need for robust security measures in the evolving landscape of cloud computing.

Network security is paramount in defending cloud-based applications. The National Institute of Standards and Technology (NIST) identifies distributed denial-of-service (DDoS) attacks as a major threat to cloud networks. Malicious activities like spoofing, malware injection, and flooding attacks jeopardize cloud integrity. (Jensen, M., Schwenk, J., Gruschka, N., & Iacono, L. L. (2009, September)) stressed the need for robust security in web browsers and servers. Addressing these concerns is vital for fortifying cloud networks and ensuring the security of applications and sensitive data.

Data Storage Security: Due to the storage of data from internet devices in a dispersed computing environment, ensuring the security of data storage in cloud computing is a critical challenge. Furthermore, (Jain, S., & Doriya, R. (2019)) have noted that, despite their widespread use, platforms like Dropbox are vulnerable to security flaws such as hash value manipulation, stolen host IDs, and direct download attacks. Data breach incidents, such as the ones involving Epsilon, highlight the dangers of cloud-based services by exposing private data, including email addresses and bank account information. Vulnerabilities on sites like Quora and Reddit were observed by even well-known services like Amazon EC2, underscoring the widespread nature of security risks in cloud data storage.

Data Virtualization Security: Cloud computing primarily depends on virtual environments, especially virtual machines under hypervisor management; hence, virtualization security is essential. The hypervisor, which sits between actual hardware and virtual computers, has vast privileges, making it vulnerable to a variety of flaws. Exploiting these flaws could provide attackers access to the real host, jeopardizing other virtual machines that coexist. Additionally, attackers may use placement techniques, such as cross-VM side-channel attacks, to steal data from a targeted VM on the same machine. These flaws highlight the significance of having strong security measures in virtualized environments to prevent future attacks.

Application Security: Application security is critical to preventing potential harm from unauthorized access by attackers. The extensive lines of programming code within applications pose a fundamental security challenge. (Ouedraogo, M., Mignon, S., Cholez, H., Furnell, S., & Dubois, E, (2015)) addressed the fact that application security requires the use of both hardware and software resources. Unauthorized users may utilize a variety of methods, such as XSS attacks, DoS attacks, SQL injection attacks, and backdoor attacks, to make unauthorized changes to the application. Implementing strong anti-attack procedures is critical for protecting apps from potential security breaches.

The integration of cloud computing with robotics has ushered in a new era of capabilities, yet it is not without its challenges, especially in the realm of security. This section provides a comprehensive overview of the unique security challenges specific to cloud robotics, delving into potential threats, vulnerabilities, and risks. Furthermore, it explores shared vulnerabilities in networked robotic systems and examines the critical role of insider threats in compromising the security of cloud-connected robots.

3.2 Security Challenges in Cloud Robotics

The landscape of cloud robotics introduces a dynamic set of security challenges that require careful consideration to ensure the integrity, confidentiality, and availability of data and operations. Understanding these challenges is crucial for deploying secure and resilient cloud robotics systems.

1. **Data Security:** Cloud robotics systems involve the transmission and storage of sensitive data in the cloud. Ensuring the confidentiality and integrity of this data is a paramount concern. Threats such as unauthorized access, data breaches, and eavesdropping during data transmission pose significant risks.
2. **Communication Security:** Communication between robotic devices and the cloud occurs over networks, which are susceptible to interception and manipulation. Securing the communication channels is essential to prevent data tampering, man-in-the-middle attacks, and unauthorized access to control commands.
3. **Unauthorized Access:** The interconnected nature of cloud robotics introduces the risk of unauthorized access to robotic devices and cloud resources. Malicious actors gaining control over robots can not only compromise data but also manipulate physical actions, leading to safety hazards and operational disruptions.

4. **Cloud Service Security:** Leveraging cloud services introduces dependencies on third-party providers. Ensuring the security of these services, including authentication mechanisms, access controls, and data encryption, is vital. Vulnerabilities in cloud platforms can have cascading effects on the security of robotic systems.
5. **Insider Threats:** The human element introduces insider threats, where individuals with access to the system may intentionally or unintentionally compromise security. This could involve malicious actions, negligent behavior, or inadvertent sharing of sensitive information.
6. **Networked Robotic Systems:** In a networked environment, the security of one robotic device can impact the entire system. Shared vulnerabilities among networked robots can be exploited to compromise the entire network, emphasizing the need for a collective security posture.
7. **Resource Limitations:** Robotic devices often operate with limited computational resources. Implementing robust security measures without impacting performance is a delicate balance. Resource-efficient encryption, secure protocols, and lightweight security mechanisms become essential.

3.3 Discussion on Potential Threats, Vulnerabilities, and Risks

Understanding the potential threats, vulnerabilities, and risks is pivotal in devising effective security strategies for cloud robotics.

1. Threats

- Malware and Ransomware: Malicious software can be injected into robotic systems, affecting their operation or holding them hostage for ransom.
- Denial of Service (DoS) Attacks: Overloading robotic devices or cloud services with requests can disrupt normal operations.
- Physical Tampering: Physical access to robotic devices can lead to unauthorized modifications, posing risks to safety and security.
- Social engineering: Manipulating human operators or users to gain unauthorized access or divulge sensitive information.

2. Vulnerabilities

- Insecure Communication Protocols: Weak or insecure communication protocols can be exploited for eavesdropping or manipulation.
- Lack of Authentication: Insufficient authentication mechanisms can lead to unauthorized access to robotic devices or cloud resources.
- Inadequate Update Mechanisms: Failure to regularly update software and firmware exposes systems to known vulnerabilities.
- Unsecured Data Storage: Storing sensitive data without proper encryption can lead to data breaches and privacy violations.

3. Risks

- Operational Disruptions: Security breaches can disrupt the normal operation of robotic systems, leading to downtime and financial losses.
- Safety Concerns: Compromised robotic devices may pose safety risks, especially in critical applications such as healthcare or manufacturing.
- Data Breaches: Unauthorized access to sensitive data can result in privacy violations, financial losses, and reputational damage.
- Regulatory Non-Compliance: Security incidents may lead to non-compliance with data protection and industry regulations, attracting legal consequences.

3.4 Shared Vulnerabilities in Networked Robotic Systems

In networked robotic systems, interconnectedness introduces a unique set of challenges, where the security of one device can impact the entire network.

1. **Inter-Device Communication:** Communication between robotic devices within a network can be exploited if it is not adequately secured. Shared communication channels may expose vulnerabilities to eavesdropping or unauthorized access.
2. **Centralized Control Points:** Networked robotic systems often have centralized control points or cloud-based controllers. Compromising these control points can give adversaries significant influence over the entire network.
3. **Collaborative Learning:** Shared learning mechanisms among robotic devices can inadvertently propagate vulnerabilities. If one device learns from a compromised source, the entire network may be affected.
4. **Dependency on Shared Resources:** Networked robotic systems often depend on shared resources in the cloud. If these resources are compromised, it can impact the functionality and security of multiple devices.

3.5 The Role of Insider Threats

Insider threats pose a nuanced challenge in cloud robotics, where human operators or individuals with access to the system can intentionally or unintentionally compromise security.

1. **Malicious Insiders:** Individuals with access to the system may exploit their privileges for malicious purposes, such as unauthorized data access, manipulation of robotic operations, or introducing malware.
2. **Negligent Behavior:** Insider threats may arise from negligence, where individuals inadvertently compromise security through actions such as misconfigurations, failure to apply updates, or accidental data exposure.
3. **Inadvertent Information Sharing:** Even well-intentioned sharing of information by insiders can inadvertently expose sensitive data, introducing risks of unauthorized access or data breaches.

4. **Lack of Awareness:** Insiders who are unaware of security best practices may unintentionally engage in activities that compromise the integrity of the system, such as falling victim to social engineering attacks.

3.5 Mitigation Strategies

Addressing the multifaceted security challenges in cloud robotics requires a proactive and multifaceted approach.

1. **Encryption and Secure Communication:** Implementing end-to-end encryption and secure communication protocols helps protect data during transmission. Utilizing technologies like Transport Layer Security (TLS) ensures the confidentiality and integrity of communication, as addressed byOuedraogo, M., Mignon, S., Cholez, H., Furnell, S., & Dubois, E. (2015).
2. **Authentication Mechanisms:** Robust authentication mechanisms, including multi factor authentication, help prevent unauthorized access to robotic devices and cloud resources.
3. **Regular Software Updates:** Establishing a systematic process for regular software and firmware updates is critical to patching vulnerabilities and ensuring the security of robotic systems.
4. **Secure Data Storage:** Implementing secure data storage practices, including encryption of stored data and access controls, protects against unauthorized data access and breaches.
5. **Network Segmentation:** Segregating networks and employing network segmentation strategies help contain potential security breaches and limit the impact of compromised devices.
6. **Insider Threat Awareness Training:** Providing training and awareness programs for individuals with access to the system helps mitigate insider threats by promoting responsible behavior and awareness of security risks.
7. **Collaborative Security Practices:** Promoting collaborative security practices among networked robotic devices involves establishing secure communication channels, sharing threat intelligence, and collectively addressing vulnerabilities.

In navigating the complex landscape of security challenges in cloud robotics, these mitigation strategies provide a foundational framework for developing robust and resilient security postures. By integrating these strategies, stakeholders in the field can bolster the security of cloud-connected robotic systems, ensuring their potential benefits are realized without compromising safety, privacy, or operational integrity.

4. PRIVACY CONCERNS IN CLOUD ROBOTICS

The intersection of cloud computing and robotics has paved the way for unprecedented technological advancements, propelling us into an era of innovation and efficiency. However, this convergence raises profound concerns about privacy, particularly in the realm of cloud robotics. In this exploration, the nuanced landscape of privacy implications is delved into, addressing the collection, storage, and processing of sensitive data by robotic devices. The privacy risks inherent in the integration of cloud technologies in robotics are also dissected, and crucially, the legal and ethical considerations that underscore data privacy in the intricate domain of cloud robotics are scrutinized.

4.1 Examination of Privacy Implications in Cloud Robotics

Privacy, a fundamental human right, takes center stage as cloud robotics ushers in a new wave of interconnected devices and intelligent systems. The collection of data by robotic devices equipped with a plethora of sensors introduces a myriad of privacy implications. From environmental sensor readings to audio and video feeds, robotic devices generate a continuous stream of data that, while invaluable for decision-making, raises concerns about the inadvertent intrusion into individuals' private spaces.

According to (Floridi, L., & Taddeo, M. (2016)) the centralization of data storage in the cloud, a hallmark of cloud robotics, offers benefits in accessibility and collaboration but concentrates vast amounts of potentially sensitive information. This prompts a crucial question about the security of data repositories and the implications of having a centralized reservoir of information about individuals. Additionally, the reliance on cloud-based services for data processing and analysis introduces the risk of extracting patterns, trends, or insights from sensitive data, potentially leading to privacy infringements.

4.2 Privacy Risks Associated With Cloud Technologies in Robotics

The integration of cloud technologies in robotics amplifies privacy risks, necessitating a careful balance between innovation and the protection of individual rights. Unauthorized access to cloud resources becomes a tangible threat, as the interconnected nature of cloud robotics exposes systems to potential breaches. Inadequate data encryption during transmission and storage opens avenues for unauthorized interception and access, further magnifying privacy risks.

The involvement of third-party cloud service providers introduces an additional layer of complexity. The privacy policies and security practices of these providers become pivotal factors influencing the overall privacy posture of robotic systems. Prolonged data retention in the cloud raises concerns about the duration for which sensitive information is exposed, potentially conflicting with data protection regulations.

4.3 Legal and Ethical Considerations Related to Data Privacy in Cloud Robotics

The ever-evolving landscape of data privacy is shaped not only by technological advancements but also by legal and ethical considerations. Ensuring compliance with data protection laws is imperative in cloud robotics deployments. Adhering to frameworks such as the General Data Protection Regulation (GDPR) in the European Union is not only a legal requirement but also a cornerstone of safeguarding individual privacy rights.

Informed consent emerges as an ethical imperative in the context of cloud robotics. Transparent communication about the purposes of data collection and the rights of individuals empowers them to make informed choices about their data. (Mittelstadt, B. D., Allo, P., Taddeo, M., Wachter, S., & Floridi, L. (2016)) addressed the concept of "privacy by design" and gained prominence, advocating for the integration of privacy considerations into the very fabric of the design and development of cloud robotics systems.

Ethical use of data becomes a guiding principle, emphasizing that data collected by robotic systems should be used for legitimate and intended purposes. This requires organizations to establish stringent controls to prevent the misuse or exploitation of data for unethical practices, ensuring the trust and confidence of individuals and stakeholders.

Conclusion: Navigating the Ethical Horizon of Cloud Robotics Privacy Concerns

In conclusion, as cloud robotics continues to redefine automation across industries, the ethical considerations surrounding privacy become increasingly paramount. The examination of privacy implications, risks, and legal and ethical considerations serves as a compass for navigating the complex terrain of cloud robotics. Striking a delicate balance between technological innovation and privacy protection is imperative to ensure that the benefits of this convergence are realized without compromising the fundamental rights and trust of individuals.

Embarking on this transformative journey requires stakeholders in the realm of cloud robotics to collectively engage in responsible practices, transparency, and a commitment to upholding the highest standards of privacy. By doing so, the power of cloud robotics can be harnessed while safeguarding the sensitive data that fuels its advancements, thus forging a path forward that is both innovative and ethically grounded.

5. CASE STUDIES AND PRACTICAL IMPLEMENTATIONS

The exploration of case studies and practical implementations in the realm of cloud robotics provides invaluable insights into real-world scenarios, shedding light on both successful security and privacy measures as well as lessons learned from security incidents. By delving into specific instances, this section enriches the understanding of how theoretical concepts manifest in practical applications, offering a nuanced perspective on the dynamic landscape of cloud robotics.

5.1 Case Studies Highlighting Successful Security and Privacy Measures

In this subsection, let's delve into documented case studies where cloud robotics have been successfully implemented with robust security and privacy measures. These cases serve as exemplars for best practices, illustrating how organizations can navigate the intricate challenges of securing robotic systems connected to the cloud. Examining the specific security protocols, encryption methods, and access controls implemented in these cases provides a practical guide for stakeholders seeking to enhance the security posture of their own cloud-connected robotic systems.

Case Study 1

Manufacturing Sector: Security Breach in a Cloud-Connected Robotic System

In 2023, a prominent manufacturing company, XYZ Corp, implemented a cloud-connected robotic system in its production facility to automate assembly line processes. The robotic system utilized cloud-based services for data storage, processing, and remote monitoring. However, a vulnerability in the authentication system allowed unauthorized access to the cloud platform, leading to a breach of sensitive production data. The challenges faced are:

1. **Inadequate Authentication Mechanisms:** The lack of robust authentication mechanisms exposed the cloud-connected robotic system to unauthorized access, compromising the integrity and confidentiality of production data.
2. **Dependency on Cloud Services:** The reliance on cloud-based services for data storage and processing increased the attack surface and vulnerability to external threats.
3. **Impact on Operations:** The security breach resulted in disruptions to manufacturing operations, including downtime, loss of productivity, and potential financial losses.

The impact of the security breach was significant, causing disruptions to manufacturing operations. XYZ Corp incurred financial losses due to halted production, as well as damage to its reputation and trust among clients and partners. Investigations revealed that the security breach could have been prevented with stronger authentication measures, such as multi-factor authentication and robust access controls. XYZ Corp subsequently implemented enhanced security protocols and conducted comprehensive employee training on cybersecurity best practices to prevent future incidents.

This example highlights the critical importance of robust security measures in cloud-connected robotic systems deployed in the manufacturing sector. It underscores the need for proactive cybersecurity strategies to safeguard sensitive production data and mitigate the risk of security breaches.

Case Study 2

Privacy Breach in a Cloud-Connected Robotic System: Telemedicine Application

In 2022, a leading healthcare provider, MedTech Solutions, implemented a cloud-connected robotic system to enhance its telemedicine services. The robotic system facilitated remote patient monitoring, diagnostic imaging, and consultation services through cloud-based platforms. However, a privacy breach occurred due to inadequate data encryption during transmission, compromising patient confidentiality.

MedTech Solutions' cloud-connected robotic system experienced a privacy breach during a telemedicine consultation session. While transmitting patient health data between the robotic device and the cloud server, the data was intercepted by a malicious actor due to weak encryption protocols. As a result, sensitive patient information, including medical records and diagnostic images, was exposed. The privacy breach raised concerns about patient confidentiality and compliance with healthcare data protection regulations, such as the Health Insurance Portability and Accountability Act (HIPAA). MedTech Solutions faced regulatory scrutiny and potential legal consequences for failing to adequately protect patient data. The Challenges are:

1. **Inadequate Data Encryption:** The primary challenge faced by MedTech Solutions was the inadequacy of data encryption during the transmission of patient health data between the robotic device and the cloud server. Weak encryption protocols left patient information vulnerable to interception by malicious actors.
2. **Compliance with Healthcare Regulations:** The privacy breach raised concerns about compliance with healthcare data protection regulations, particularly the Health Insurance Portability and Accountability Act (HIPAA). Failure to ensure patient confidentiality and data security could result in regulatory penalties and legal consequences for MedTech Solutions.

Solutions

1. **Enhanced Encryption Protocols:** To address the challenge of inadequate data encryption, MedTech Solutions implemented enhanced encryption protocols for transmitting patient health data. This included adopting industry-standard encryption algorithms and implementing end-to-end encryption to protect sensitive information from unauthorized access.
2. **Compliance Measures:** MedTech Solutions took proactive steps to ensure compliance with healthcare regulations, particularly HIPAA. This involved conducting comprehensive audits of its cloud infrastructure and telemedicine platforms to identify and address any gaps in data security and privacy practices. The company also established protocols for regular compliance assessments and updates to ensure ongoing adherence to regulatory standards.

By implementing these solutions, MedTech Solutions strengthened the security and privacy of its cloud-connected robotic system, mitigating the risk of privacy breaches and regulatory non-compliance. These measures not only safeguarded patient confidentiality but also upheld trust in the telemedicine services provided by the company.

5.2 Current Technological Solutions and Best Practices

1. **Blockchain Technology:** Blockchain technology offers decentralized and tamper-resistant storage of data, making it an attractive solution for enhancing security and privacy in cloud robotics. Implementing blockchain in cloud robotics can provide immutable records of transactions and data exchanges between robotic devices and cloud servers, ensuring data integrity and traceability. Smart contracts deployed on blockchain networks can automate and enforce security policies, such as access control and data sharing permissions, enhancing the security posture of cloud-connected robotic systems.
2. **Advanced Encryption Methods:** Advanced encryption methods, such as homomorphic encryption and post-quantum cryptography, offer enhanced data security by enabling computations on encrypted data without decryption. Homomorphic encryption allows cloud servers to perform computations on encrypted data received from robotic devices without decrypting it, preserving data confidentiality during processing. Post-quantum cryptography provides protection against quantum computing threats by utilizing cryptographic algorithms resistant to quantum attacks, ensuring long-term data security in cloud robotics deployments.
3. **Secure Communication Protocols:** Deploying secure communication protocols, such as Transport Layer Security (TLS) and Datagram Transport Layer Security (DTLS), helps encrypt data transmission between robotic devices and cloud servers, preventing eavesdropping and data tampering. Implementing mutual authentication ensures that both robotic devices and cloud servers verify each other's identities before establishing communication, mitigating the risk of man-in-the-middle attacks.
4. **Privacy-Preserving Techniques:** Privacy-preserving techniques, such as differential privacy and secure multiparty computation (SMC), safeguard sensitive data in cloud robotics environments. Differential privacy adds noise to aggregated data to prevent the identification of individual data points, preserving privacy while enabling data analysis and sharing. SMC allows multiple parties

to jointly compute a function over their private inputs without revealing sensitive information to each other, facilitating collaborative data analysis without compromising privacy.

Concrete Guidance for Implementation

1. **Assess Security and Privacy Requirements:** Conduct a comprehensive risk assessment to identify security and privacy requirements specific to the cloud robotics deployment, considering factors such as data sensitivity, regulatory compliance, and threat landscape.
2. **Select Appropriate Technologies:** Evaluate the suitability of blockchain technology, advanced encryption methods, and secure communication protocols based on the identified security and privacy requirements, considering factors such as scalability, interoperability, and computational overhead.
3. **Integrate Security into Design:** Incorporate security and privacy considerations into the design phase of cloud robotics systems, ensuring that security measures are integrated at the architectural level and throughout the development lifecycle.
4. **Implement Robust Access Controls:** Establish granular access controls and authentication mechanisms to restrict unauthorized access to robotic devices, cloud resources, and sensitive data, minimizing the attack surface and enforcing the principle of least privilege.
5. **Monitor and Update Security Measures:** Implement continuous monitoring and proactive security measures, such as intrusion detection systems and security patch management, to detect and mitigate security threats in real time. Regularly update security protocols and encryption keys to address emerging vulnerabilities and evolving threat landscapes, ensuring the resilience of cloud-connected robotic systems against security breaches.

By implementing these technological solutions and best practices, stakeholders can effectively address security and privacy challenges in cloud robotics deployments, enhancing the resilience and trustworthiness of their systems in an increasingly interconnected and data-driven environment.

5.3 Innovative Approaches and Best Practices for Secure Deployments

The dynamic nature of technology necessitates constant innovation in security practices. This subsection explores innovative approaches and emerging best practices in the deployment of secure cloud robotics systems. From advancements in encryption algorithms to novel methods of authentication and access control, this section provides a forward-looking perspective on how security in cloud robotics can evolve to stay ahead of emerging threats. By highlighting pioneering strategies, this section empowers stakeholders to anticipate and counteract future challenges in the ever-evolving landscape of cloud robotics security.

6. REGULATORY COMPLIANCE IN CLOUD ROBOTICS

The integration of cloud robotics introduces not only technical challenges but also a complex regulatory landscape. This section navigates the legal frameworks and regulations applicable to cloud robotics, emphasizing the importance of compliance with data protection laws and industry-specific regulations. By exploring the role of international standards in ensuring security and privacy, this section provides a roadmap for organizations to navigate the intricate web of legal requirements governing cloud-connected robotic systems.

6.1 Discussion on Legal Frameworks and Regulations

This subsection delves into the legal frameworks and regulations that directly impact cloud robotics deployments. It explores regional and global regulations, emphasizing the need for organizations to be cognizant of legal requirements that govern data protection, privacy, and cybersecurity. (Butun, I., Österberg, P., & Song, H. (2019)) addressed the General Data Protection Regulation (GDPR) to industry-specific standards, understanding the legal landscape is crucial for mitigating legal risks and ensuring ethical conduct in cloud robotics.

6.1.1 Regulatory Framework

Various regulations and standards govern the development and deployment of cloud robotics systems, ensuring compliance with legal requirements and industry best practices. The General Data Protection Regulation (GDPR) in the European Union sets stringent requirements for the collection, processing, and storage of personal data, impacting cloud robotics deployments involving data subjects in EU member states.

In the United States, regulations such as the Health Insurance Portability and Accountability Act (HIPAA) and the Federal Trade Commission (FTC) Act impose obligations on healthcare providers and businesses to protect patient health information and consumer data privacy. Industry-specific standards, such as ISO/TS 15066 for collaborative robots and ISO/IEC 27001 for information security management, provide guidelines for ensuring safety, security, and compliance in cloud robotics applications.

6.1.2 Transparency and Accountability

- Transparency about data collection practices, processing algorithms, and decision-making mechanisms is essential for building trust and accountability in cloud robotics systems.
- Providing clear and accessible privacy notices, consent mechanisms, and opt-out options empowers users to make informed decisions about their data and privacy preferences, promoting transparency and user autonomy.
- Implementing robust data governance frameworks, including data minimization, purpose limitation, and data retention policies, enhances accountability and regulatory compliance while safeguarding privacy and ethical principles.

6.1.3 Continuous Monitoring and Improvement

- Regular audits, assessments, and evaluations of cloud robotics systems are essential for monitoring compliance with regulatory requirements and ethical standards.
- Establishing mechanisms for reporting and addressing privacy incidents, security breaches, and ethical dilemmas promotes a culture of transparency, accountability, and continuous improvement in cloud robotics deployments.

- Engaging stakeholders, including users, data subjects, and regulatory authorities, in ongoing dialogue and collaboration fosters trust, transparency, and mutual understanding of the ethical and regulatory challenges associated with cloud robotics.

By integrating compliance with regulatory requirements and adherence to ethical principles into the design and implementation of cloud robotics systems, stakeholders can mitigate risks, build trust, and ensure responsible innovation in this rapidly evolving field. This approach not only enhances the security, privacy, and ethical integrity of cloud robotics deployments but also contributes to the long-term sustainability and societal acceptance of these transformative technologies.

6.2 Compliance With Data Protection Laws

Building on the discussion of legal frameworks, this subsection specifically focuses on the compliance aspects of data protection laws. It outlines the steps organizations must take to align their cloud robotics systems with stringent data protection requirements. Through illustrative examples and best practices, stakeholders gain insights into creating a robust framework that not only adheres to legal standards but also establishes a foundation for responsible and ethical data handling. Integration of Compliance and Ethics requires:

- Compliance with regulatory requirements and adherence to ethical principles should be integrated into the design and implementation of cloud robotics systems from the outset.
- Privacy by Design (PbD) and Security by Design (SbD) principles advocate for embedding privacy and security considerations into the development lifecycle of cloud robotics systems, promoting proactive risk mitigation and data protection.
- Conducting privacy impact assessments (PIAs) and ethical risk assessments (ERAs) helps identify and address potential privacy and ethical concerns early in the design process, enabling informed decision-making and risk management.
- Establishing governance structures and appointing data protection officers (DPOs) or ethics committees within organizations fosters accountability and oversight, ensuring that compliance and ethical considerations are prioritized throughout the lifecycle of cloud robotics projects.

6.3 The Role of International Standards

International standards play a pivotal role in harmonizing practices and setting benchmarks for security and privacy in cloud robotics. This subsection explores the significance of adhering to established standards, such as ISO/IEC 27001 for information security management and ISO/TS 15066 for collaborative robots. By aligning with internationally recognized standards, organizations enhance their credibility, ensure interoperability, and contribute to the establishment of a global framework for secure and compliant cloud robotics deployments.

7. CONCLUSION

In conclusion, this comprehensive exploration of security and privacy in cloud robotics has traversed the foundational aspects, security challenges, privacy concerns, and regulatory dimensions of this transformative paradigm. The integration of cloud computing with robotics opens new frontiers of innovation but demands a meticulous approach to addressing the multifaceted challenges.

Navigating this complex landscape requires stakeholders in the field of cloud robotics to embrace a holistic perspective, incorporating best practices, learning from both successes and failures, and remaining attuned to evolving legal and ethical considerations. By doing so, an era where cloud robotics not only revolutionizes industries but does so with a steadfast commitment to security, privacy, and ethical conduct can be collectively ushered in.

REFERENCES

Butun, I., Österberg, P., & Song, H. (2019). Security of the Internet of Things: Vulnerabilities, attacks, and countermeasures. *IEEE Communications Surveys and Tutorials*, *22*(1), 616–644. doi:10.1109/COMST.2019.2953364

Floridi, L., & Taddeo, M. (2016). What is data ethics? *Philosophical Transactions. Series A, Mathematical, Physical, and Engineering Sciences*, *374*(2083), 20160360. doi:10.1098/rsta.2016.0360 PMID:28336805

Jain, S., & Doriya, R. (2019). Security issues and solutions in cloud robotics: a survey. In *Next Generation Computing Technologies on Computational Intelligence: 4th International Conference, NGCT 2018, Dehradun, India, November 21–22, 2018, Revised Selected Papers 4* (pp. 64-76). Springer Singapore. 10.1007/978-981-15-1718-1_6

Jensen, M., Schwenk, J., Gruschka, N., & Iacono, L. L. (2009, September). On technical security issues in cloud computing. *Proceedings. IEEE International Conference on Cloud Computing*, 109–116.

Ko, H., Kech, S. L., & Jin, J. (2017). Privacy Protection for Cloud-Based Robotic Networks. *Security and Privacy in Cyber-Physical Systems: Foundations, Principles and Applications,* 195-222. 10.1002/9781119226079.ch10

Mittelstadt, B. D., Allo, P., Taddeo, M., Wachter, S., & Floridi, L. (2016). The ethics of algorithms: Mapping the debate. *Big Data & Society*, *3*(2). doi:10.1177/2053951716679679

Ouedraogo, M., Mignon, S., Cholez, H., Furnell, S., & Dubois, E. (2015). Security transparency: The next frontier for security research in the cloud. *Journal of Cloud Computing (Heidelberg, Germany)*, *4*(1), 1–14. doi:10.1186/s13677-015-0037-5

Section 2
Applications of Cloud Robotics

Chapter 10
Applications of Cloud Computing in Industrial Robotics

S. Sethuraman
Vellore Institute of Technology, India

S. Vikram
Vellore Institute of Technology, India

Dharma Surya R.
Vellore Institute of Technology, India

Brijendra Singh
https://orcid.org/0000-0003-2608-3388
Vellore Institute of Technology, India

ABSTRACT

In the past few decades, robots have played a significant role in both industrial and technological advancements. Cloud-integrated industrial robotics provide a vast network of machines to industries. The integration of robotics with cloud computing offers increased computational power for various applications of industrial, communication, healthcare, disaster management, defence systems, and space exploration missions. It also provides reliable and extensive accessibility across a network of devices. In this chapter, the authors delve into the fundamentals of robotics and cloud computing. They introduce the core concepts of robotics and cloud computing, followed by their integration. Subsequently, they explore communication in cloud robotics, encompassing machine-to-machine (M2M) and machine-to-cloud (M2C) communication. They then examine the major applications in industrial robotics and types of robots that leverage cloud robotics technology. They conclude by addressing security issues and challenges in cloud robotics.

DOI: 10.4018/979-8-3693-1914-7.ch010

1. INTRODUCTION

In the past few decades, robots have played a significant role in both industrial and technological advancements. Cloud-integrated robotics provides a vast network of machines to industries. The integration of robotics with cloud computing offers us increased computational power for various applications, including communication, healthcare, disaster management, defence systems, space exploration and many more applications. It is also provides reliable and extensive accessibility across a network of devices. Robotics is a technology utilized to mitigate human risk in real-world scenarios. Robots are frequently employed for the transportation of materials, components, or tools from one location to another (Alkhafaji & F. S. M., 2021). They excel in executing tasks with greater speed and precision compared to the average human. Cloud robotics represents the concept of integrating Internet of Things (IoT), Cloud Computing, and industrial wireless networks.

Robots are devices used for the transportation of materials, components, or tools from one place to another. Robotics is an engineering field dedicated to the study and development of robots. The integration of robotics and cloud computing involves a network of robots or machines interconnected via wired or wireless mediums for the exchange of information and data to facilitate coordinated actions (Hu, G. et al., 2012; Edwards, M.,1984).

Robots are programmable multifunctional devices employed for the movement of materials, components, and tools, as well as a wide range of other tasks. However, like any field, robotics is grounded in fundamental principles, with rotation type, workspace, and accuracy serving as some of its essential foundations. There are various types of robots was built for different types of works such as Industrial robots, Domestic robots, Medical and Service robots, Security robots and more.

1.1. Fundamentals of Robotics

Fundamentals in Robotics, also referred as the foundational principles, encompass key factors essential to the field. These fundamentals include rotation type, workspace, and accuracy.Robotics encompasses a wide range of applications, including industrial automation, exploration, security, healthcare, and surveillance.

1.1.1. Rotation Type

Robots are capable of performing various movements, both in terms of their own mobility and in manipulating materials (Yalcin, M., & Patoglu, V., 2012). The movements of robots can be categorized into three distinct types: forward and backward motion, rotation, and upward and downward motion, which are commonly referred to as yaw, pitch, and roll, respectively. Yaw pertains to the upward and downward movement of the robot, pitch involves side-to-side motion, and roll concerns the robot's front-to-back movement.

1.1.2. Workspace

The workspace refers to the area where robots can operate and is primarily defined by the degrees of freedom of the joints in the robot arm (Yalcin, M., & Patoglu, V., 2012). Degrees of freedom can be classified into several types, including the Cylindrical joint (C), Universal joint (U), and the Spherical joint (S).The Cylindrical joint provides a single degree of freedom, allowing the robot to move along the x and y axes. The Universal joint offers more than one degree of freedom, enabling the robot to move along the x and y axes as well as upward and downward. The Spherical joint, on the other hand, boasts three degrees of freedom, resulting in Universal joint motion with the added capability of spinning movements (Hernández, E. et al., 2017) .For instance, consider a robot with two revolute joints, which permits it to move within a range of 180 to 150 degrees.

1.1.3. Accuracy

Accuracy is measured based on the alignment between the programmed task and the task executed by the robot. Let's consider a scenario in which a robot is programmed to connect dots displayed on a screen. If the robot successfully connects nearly all of the dots, it is considered to have achieved a high level of accuracy. However, if the robot only manages to connect half of the dots, its accuracy is deemed to be at an average level. In cases where the robot misses all of the dots on the screen, there is effectively no accuracy at all.

1.2. Fundamentals of Cloud Computing

Cloud computing is the delivery of cloud services, such as computing power, storage, networking, databases, software, and more. Cloud computing enables the use of computing power and storage without requiring changes to users' physical computers (Choudhary, A. et al., 2022). It provides worldwide connectivity, allowing you to access or store your data anywhere in the world and at any time over the internet.There are some components that define cloud computing, including computing power, storage, databases, networking, and security. Let's understand these components more elaborately.

1.2.1. Compute

In cloud computing, the compute provides computing power to users. Users can purchase as much computing power as they need from cloud service providers and utilize it for handling complex tasks (Choudhary, A. et al.,2022 ; Reitze, D. D.,2014). For example, if a user wants to render high-definition videos or work on 3D rendering, they can acquire computational power from cloud service providers. Computing power enables high-performance computing without changing the physical components of the user's computer.

1.2.2. Storage

Cloud storage is a platform where users can store their documents, photos, and videos. It provides limitless storage space to users, who can purchase the required amount of storage from cloud service providers (Choudhary, A. et al.,2022 ; Al-Anzi et al., 2014; Reitze, D. D.,2014). Several advantages attract users to cloud storage services, including the ability to access and store as much as they need. Users can access their data anytime and anywhere in the world over the internet. Storing data in the cloud is highly reliable because cloud storage utilizes data replication techniques, where users' data is stored in multiple locations for enhanced reliability. If one database goes down, another database will connect with the users to provide reliable access.

1.2.3. Database

A cloud database is a repository where users' data is stored. It is utilized to organize and store all three types of data: structured, semi-structured, and unstructured formats (Choudhary, A. et al., 2022 ; Reitze, D. D.,2014). There are two types of databases used in cloud database management: Relational and Non-relational cloud databases. A relational database is employed to store structured data in tables with rows and columns. Structured Query Language (SQL) is used to access and manage these types of databases. Oracle, SQL, MySQL are some of the popular examples of relational databases. On the other hand, a non-relational database is used to store unstructured data such as images, text files, and more. MongoDB and HBase are some of the popular example of non-relational databases.

1.2.4. Networking and Security

Cloud networking is a backend service where organizations maintain their software or websites for seamless operation. It involves hosting the networking procedures of individuals or organizations on public or private cloud services (Choudhary, A. et al.,2022 ; Reitze, D. D.,2014). Cloud computing offers encryption and decryption security levels to users. It also includes security features such as access control, ensuring that only authorized users can access the data (Al-Anzi et al.,2014). Identity and access management guarantee that selected permissions are provided to specific users. However, storing sensitive data in the cloud is not advisable.

We have presented introduction in Section 1. Rest of the book chapter is organized as follows. In Section 2 we have depicted literature review on cloud robotics. Section 3 presents various types of robots. Section 4 enlightens the application of cloud robotics in industry. Communication technologies in cloud robotics are explained in Section 5. Section 6 depicted security of industrial robotics in cloud computing. Challenges of industrial robotics in cloud computing applications are presented in Section 7. Finally conclusion is presented in Section 8.

2. LITERATURE REVIEW

(Alkhafaji, F. S. M., 2021), proposed a model to achieve high speed and efficiency in a robotic arm for industrial robotics. The proposed work introduces a robotic arm with 5 Degrees of Freedom (DOF). To improve speed, the model proposes a high-speed controller, which is enhanced by Genetic Algorithm. The proposed model was created using applications of SolidWorks. The model involves the following steps: mechanical configuration, estimating TF, controller design, and modelling the arm. In testing, the model provides the most accurate and fastest robotic arm with 5 Degrees of Freedom (DOF) for industrial applications.

(Hernández, E., et al., 2017), presented a research paper detailing the creation of a humanoid robot named KUBO, equipped with 20 Degrees of Freedom. The robot is entirely constructed from aluminium and is equipped with sensors. Capable of mimicking human actions such as walking, balancing, and turning, among others, KUBO is described in this paper along with its kinematic analysis, mechanical design, and workspace. The proposed model serves to anticipate the motion of the humanoid robot during specific tasks and to prevent collisions. The robot incorporates various sensors including a webcam, compass, servo motors, as well as controllers with DC/DC converters, batteries, and force sensors. Kinematics is employed to calculate the robot's motion predictions.

(Choudhary, A., et al., 2022), discussed a comparative study of various cloud service providers in their research paper. The paper highlights multiple advantages of utilizing cloud computing, which offers scalable solutions ranging from small-scale to enterprise-level applications for non-cloud-based issues. It presents a comparative analysis of prominent cloud service providers, including Amazon Web Services, Oracle Compute, Microsoft Azure, Google Cloud Platform, and IBM Cloud. Furthermore, the paper includes a detailed case study covering pricing, compute services, storage services, database services, and networking services offered by these providers. Ultimately, the paper aids in understanding cloud service providers comprehensively, enabling users to choose services according to their requirements.

(Alseddiqi, M., et al., 2022), documented the importance of medical robots as an application of industrial robotics. The paper emphasizes the significance of Information and Communication Technology (ICT) in this context. It elaborates on both the current and future uses of robotics in healthcare. Over the past couple of decades, robotics in the medical field has played a significant role. There is a growing need for highly efficient robotics in healthcare to save lives on a large scale and with great efficiency. This research paper highlights the importance of quality treatment to minimize risks and enhance efficiency. Additionally, it provides insights into reducing time for specific tasks to save lives.

This paper introduced (Arunajyothi, G., 2016) a internet-based robotic framework that expands the computing and sharing of data capabilities of networked robots. The ad-hoc cloud created by machine-to-machine (M2M) communications between participating robots and the infrastructure cloud made possible by machine-to-cloud (M2C) communications are the two components that make up the cloud robotic architecture. Cloud robotics use an elastic computing approach to facilitate job dumping along with knowledge collaboration in robotic applications. Resources are dynamically assigned from a collective asset pool in the universal cloud. To address various applications, we suggest and assess a number of elastic computing models as well as communication protocols. We go over the technological difficulties in processing, connectivity, and security and show how cloud robotics may be useful in many contexts.

For bettering linked robots' estimation and sharing of data abilities they suggest a novel cloud robotic architecture (Hu, G., et al., 2012). This design blends a central cloud made possible by machine-to-cloud (M2C) communication with an ad hoc cloud created by robot-to-machine (M2M) communication. Using an adaptive computing design, cloud automation supports delegating duties and information interchange in robotic activities by randomly assigning assets from a common repository on the public cloud. We present and assess several models of adaptive computing and communication strategies for a range of applications. We also discuss the technological issues related to computing, interaction, and privacy, demonstrating the potential merits of cloud robotics in an array of circumstances.

Robots are able to offload complex data processing and storage obligated to the cloud via an evolving field called cloud automation (Aindrila Mukherjee, 2017). The variety of memory, storage, and cognitive power that robots have is often minimal. There are almost boundless intellectual memory, storage, and interaction prospects when employing the cloud. Standalone and linked robots are the two main classes of robots with cloud-based features. Cloud automated systems are reviewed, and solitary and interconnected robotic applications are examined. One such application is simultaneous localization and mapping (SLAM), in which a robot records an unnamed environment while simultaneous tracking its own location beneath it.

The emergence of robotics and cloud-based computing tools (Ma, Y., et al., 2015) has led to an increasing fascination in human-focused healthcare providers as a means of overcoming the serious deficiencies in conventional healthcare, particularly with the issue of scarce medical resources. The healthcare system described in this paper is built on a cloud computing and robotics platform that incorporates mobile cutaneous networks, robots, software, and virtualization. With the aid of a robot equipped with an array of sensors, this system is going to reliably monitor the vital signs of the user for interpretation and guidance. The study presents an original approach to distribute real-time filmed content using a better UDP-based norms, which could boost the viability of multifaceted broadcasting in the context of healthcare. Finally, a pilot study using actual testbed is used in order to evaluate the offered strategy.

The study presented cloud robotics architecture to overcome the limitations of current networked robots (Hu, G., et al., 2012). Cloud robotics allows robots to communicate computing resources, information, and data with one another via M2M (machine-to-machine) and M2C (machine-to-cloud), as well as gain new knowledge and abilities that they would not have learned on their own. This is a new paradigm in robotics, which I feel will lead to interesting future advances. It enables the deployment of low-cost robots with minimal compute power and memory needs by using the communications network and elastic computing resources provided by the cloud architecture. The cloud robotics technique may be used for a variety of applications, including SLAM, grasping, navigation, and many more that we have not addressed, such as weather monitoring, intrusion detection, surveillance, and formation control.

This research paper looks into the usage of cloud technology for robotics and automation, outlining both the possible benefits and downsides of this strategy (Zhang, H., & Zhang, L., 2019, April). The study examines the architecture design procedure for cloud robotic systems and introduces a domain-specific architecture description language for architecture-centric cloud robotics called CRALA. Additionally, it looks at how cloud robotics architecture is implemented in the cloud and how future efforts to extend CRALA by developing tools to support self-generating cloud robotic systems that are architecture-centric will be carried out. covers a variety of cloud robotics topics, such as cloud robotic system architecture, the usage of software architectures, and the deployment of cloud robotics architecture in the cloud. The significance of architecture centred development for cloud robotics is also covered, as is the requirement for a language that encompasses all essential elements of cloud robotics, such as cloud descriptions, robotic architecture, robots, and components/web services.

In this paper, the study on cloud security focuses on problems, dangers, and assaults specific to cloud computing (Patel, A., et al., 2020) is done. It offers a thorough examination of all the many facets of cloud security, including the dangers, vulnerabilities, and possible assaults related to cloud computing. Cloud security principles, data storage methods, virtualization, trust management, security policies, and particular security risks and attacks, including session hijacking, domain name system assaults, and structured query language injection attacks, are just a few of the subjects covered in the study. The survey covers cloud computing's benefits, rapid adoption, and security threats related to unauthorised access and data misuse. It highlights how crucial it is to comprehend and deal with the complexities of cloud security, reflecting the increasing importance of safeguarding cloud-based systems in the current digital environment.

This study, in particular, focuses on the state of the field and unresolved difficulties (Wan, J., et al., 2016). It goes into the architecture of these systems, how cloud robotics is developed, and the main motivations for its development. The main problems and obstacles facing modern cloud robotic systems are also covered in this study, including resource allocation, data exchange between robots and cloud platforms, cloud security, and techniques and effects analysis for ensuring service quality. It also looks at how cloud robotics may be used in real-world situations for tasks like grasping, navigation, and SLAM. In conclusion, the paper offers a thorough analysis of the advancement, difficulties, and prospective uses of cloud robotics, emphasising the crucial role that cloud computing plays in resolving the drawbacks of conventional robotic systems and facilitating high-performance, high-complexity multirobot operations.

3. TYPES OF ROBOTS

There are numerous advantages to using robots instead of human employees. Robots are designed to perform various tasks in the real world, providing us with enhanced safety, efficiency, accuracy, and speed. In this discussion, we will explore different types of robots utilized in real world applications.

3.1. Industrial Robots

Industrial robots are programmed for specific tasks, including manufacturing, testing, and packing (Chung, J., & Lee, Y. S., 2023). The robots used in industrial applications have the ability to move in three dimensions, known as Spherical (S) movements. These robots can also be employed in welding, painting, printing, material handling, and various other tasks.

3.2. Medical Robots

Robots used in medical applications are often two-way robots. In this setup, a surgeon operates one robot on the receiver side, while the patient-side robot performs the surgery under the surgeon's guidance. This setup necessitates high-speed data transfer and maximum accuracy of the medical robot. However, due to the latency in data transfer, such robots are not currently functional in real-world scenarios (Alseddiqi, M. et al., 2022).

3.3. Space Exploration Robots

Space robots are used to reduce human activities in outer space, such as conducting research on non-terrestrial planets and transmitting environmental information about other planets and moons. There are two types of robots used in outer space exploration: spacecraft-type robots, such as satellites, and planetary robots, such as rovers (Yalcin, M., & Patoglu, V. 2012; Harbin, J., et al., 2023). Spacecraft-type robots are used to collect information about outer space, while planetary robots collect data about planets and moons.

3.4. Military Robots

Military robots are remotely accessible machines designed for defense systems, search and rescue operations, and the reduction of human casualties through attack systems (Madhiarasan, M. 2021). These robots are controlled remotely to direct the operations forward.There are some types of robots that have been used in military missions, including transportation robots, defense robots, armed robots, and more. The transportation robots are used to supply guns and ammunition; they can also transport human soldiers from one place to another. Defense robots are used to reduce casualties by clearing mines.

4. APPLICATIONS OF CLOUD ROBOTICS

Now we will discuss various applications of cloud computing in robotics. Figure 1 demonstrates various applications of cloud robotics.

4.1 Grasping

Robot grasping research has endeavoured to comprehend and replicate the other two tasks. The work of handling items with fingers (as opposed to manipulating them with a robot arm) will be distinguished from the duty of restricting objects, which is also referred to as fixturing (Bicchi, A., & Kumar, V. 2000). Grasping in cloud robotics refers to a robot's ability to manipulate and handle objects. Cloud robotics can help with everything from planning and decision-making to execution and learning. Here are some instances of cloud robotics grasping techniques.

4.1.1 Cloud-Based Object Recognition

To identify objects in the environment, cloud-based image and object recognition services can be used. This information is then used to plan and execute grasping tasks. Cloud resources can provide access

Figure 1. Applications of cloud robotics

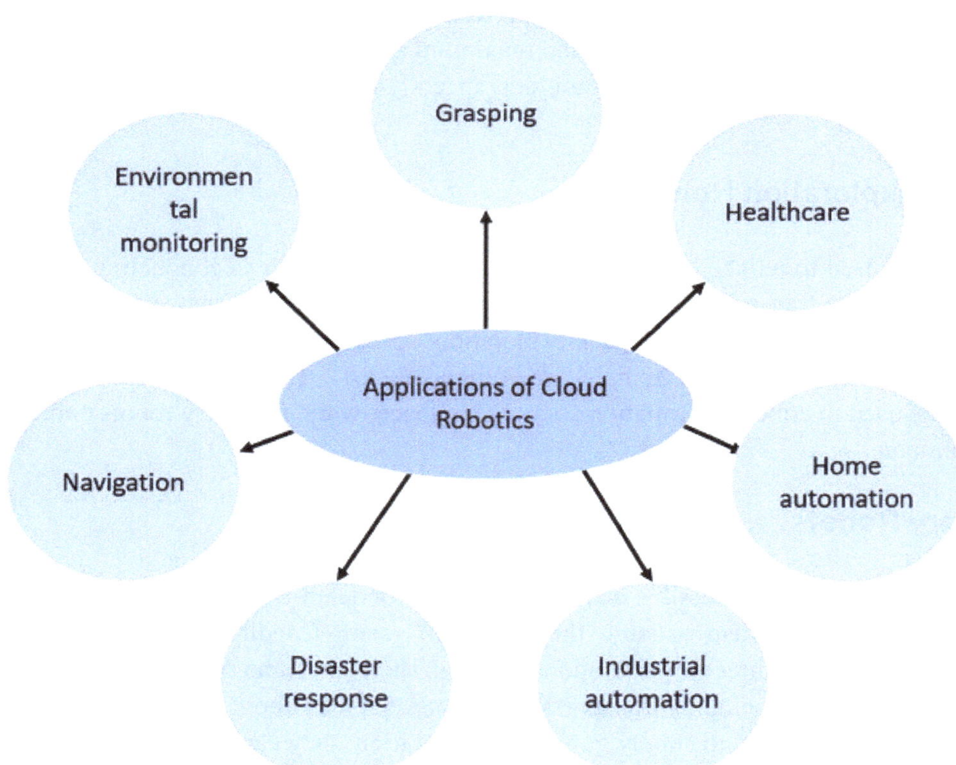

to massive databases of object images and recognition models, allowing robots to recognize a wide variety of objects.

For instance, agricultural machines with object recognition capabilities can detect and pick out weeds, fruits, or vegetables for harvesting or tailored spraying, improving farming operations' efficiency and efficacy. Subsequently, better yields and more efficient farm operations can result from farmers using this data to make well-informed decisions regarding methods of crop management including conception, sprinkler timing, and control of pests.

4.1.2 Cloud-Based Grasping Planning

Grasping planning entails determining the best way for a robot to grasp an object. Cloud resources can be used to gain access to advanced algorithms and databases of 3D object models. Cloud-based grasping planning can take into account factors such as object shape, size, weight, and the robot's capabilities to plan efficient and reliable grasps.

Cloud resources may be utilized to perform simulations and train reinforcement learning models for gripping. Simulated grasping attempts in the cloud can aid in the improvement of the robot's gripping tactics. Reinforcement learning models may be developed in the cloud and then deployed to robots for real-world gripping tasks.

4.2 Healthcare

The costs of healthcare services have been steadily rising globally in recent years, imposing a significant financial strain on both the government and society. To effectively raise the level of healthcare service digitization and balance the unequal distribution of medical resources, it is imperative to develop affordable, practical, and scalable home medical systems (Chen, M., 2013). Healthcare Cloud robotics has the ability to alter the medical business. Robots connected to the cloud may access patient data, medical records, and advanced algorithms for illness diagnosis. The creation of centred around humans' healthcare organizations is gaining pace due to the expanding influence of robots and cloud computing innovations (Ma, Y., 2015). This is because these services seek to tackle significant issues seen in traditional healthcare systems, among which is the scarcity of medical resources. This healthcare system, which blends wireless body area networks, robots, software systems, and a cloud platform, incorporates use of robotics and cloud computing. This system's main objective is to thoroughly gather and examine the physiological data of its users, with the help of robots that are fitted with a variety of sensors.

This research further suggests a novel way to offer real-time video material utilizing an upgraded UDP-based technique, which aims to increase the feasibility of multimedia data transfer inside the healthcare system(Ma, Y., 2015). In the event of equipment failures, severe weather, or online assaults, cloud-based EHR systems provide automated backup as well as recovery skills, guaranteeing accessibility and authenticity. They also promote interoperability by facilitating the easy exchange of patient data between different providers of healthcare, electronic drug prescribing systems, pharmacies, and other healthcare providers, thereby enhancing integration of care and results for patients.

4.2.1 Telemedicine and Remote Surgery

Cloud robotics presents telemedicine, allowing doctors to remotely diagnose and treat patients, and surgeons may utilize cloud-connected robotic equipment for remote surgery, across large distances.

Through the telemedicine platform, patients can interact with healthcare providers via desktops, tablets, or cell phones. Patients can access medication, follow-up care, and medical advice virtually thanks to the internet, which securely sends videos, audio recordings, and medical data through the internet. Cloud-based robotic surgical systems enable doctors to perform procedures that are minimally invasive on patients who are located in distant places in scenarios involving remote surgery.

4.2.2 Data Management and Storage

Cloud solutions allow secure and scalable data storage for healthcare information which includes medical imaging, and patient histories, electronic health records (EHRs). This ensures data accessibility and continuity of care.

4.3 Home Automation

Home automation refers to operating electronic gadgets within the home via the internet. With home automation, we can control electronic devices from anywhere and at any time. This technology facilitates the connection of electronic gadgets through the internet, eliminating the need for manual control. Home automation offers several features, including enhanced security, functionality, and efficiency. Cameras can be utilized to monitor wirelessly, accessible even from mobile phones, regardless of location or time. Home automation simplifies tasks that would otherwise require hours of waiting by enabling programming of robots to perform these tasks efficiently. These robots are utilized for various purposes such as cleaning, washing, monitoring, and more. A significant advantage is the ability to control these robots wirelessly from any location (Gunge, V. S., & Yalagi, P. S., 2016).

4.4 Industrial Automation

When it comes to industry, robots are driving incredible changes for companies' growth. Industrial automation involves accessing and controlling industrial robots wirelessly over the internet. This technology is particularly aimed at reducing human labour while increasing productivity. Automated machines are commonly employed to expedite tasks and enhance efficiency. Industrial robots vary in their functions and the environments they operate in. These robots are faster, more efficient, programmable, and contribute to higher productivity compared to human workers. Industry automation in cloud robotics entails integrating industrial robots with cloud computing to optimize efficiency. Cloud robotics enables global connectivity, allowing for programming and control of industrial robots from anywhere at any time (Bangemann, T., et al,2014).

4.5. Disaster Response

Cloud robotics disaster response provides a potent blend of innovative technology, instantaneous analysis, distant functioning capabilities, and collective decision-making, making it a critical tool for diminishing the detrimental impacts of catastrophes and saving lives. In cloud automation, disaster response pertains to utilizing robotic systems that are linked to cloud computing assets in order to efficiently adapt to natural or human-caused catastrophes. Drones, ground-based robots, self-driving cars, and other tools having an assortment of sensors and actuators are all suitable for robotic systems. Cloud robotics disaster response uses cloud computing to handle enormous volumes of data, such as sensory assessments, shots, and clips, obtained by robots systems in catastrophe sites. This allows for the acquisition of pertinent data for their choices. This method makes it feasible to control and manage robots electronically from any location with internet access, which is crucial for risky or remote areas. Although cloud automated machines are able to integrate modern innovations like cloud computing, it can process data from multiple sensors, including cameras and atmospheric analyses, in a timely manner. This is exactly cloud robotics has applications in disaster response. This technology permits instantaneous analysis of key information, such survivor authentication and harm evaluations, facilitating informed choice-making. As an illustration, humanitarian Cloud robotics crises can help aid organizations in places devastated by malnutrition, epidemics of illness, war zones, and refugee camps. Robots that are outfitted with food, water purification systems, and healthcare amenities can save lives of those who have been uprooted (Cheikhrouhou, O.,et al., 2020).

4.6 Navigation

Robotic navigation is the capacity of a robot to determine its position with respect to a reference point and plan its journey to a desired destination. It entails a variety of activities such as adaptation, planning of paths, and modelling. There are essentially two approaches: map-less and map-based.

Map-less navigation depends on data from perception sensors for direction, however it frequently suffers from dependability concerns because of limited onboard resources. Map-based navigation, on the other hand, is more reliable when an accurate map is provided. This map might be pre-existing or created throughout the navigation process. However, map-building requires a significant amount of storage space and computer power, whereas map-based navigation requires access to vast amounts of data, which can be difficult in broad navigation regions (Hu, G.,2012).

Cloud robotics promises an achievable solution to these difficulties for future cloud-enabled navigation. The cloud is capable of offering both substantial storage space and the computing power essential for effective map generation and navigation(Hu, G.,2012). Furthermore, cloud-connected systems can use commercially accessible maps such as Google Maps to create reliable. open to change, and distant autonomous navigation solutions.

Maps can be downloaded for offline usage by cloud-based navigation systems, which allows users to find their way even in places with patchy or unavailable connection to the internet. Through Dropbox, users can also consolidate their navigation data—including saved locations and preferences—across multiple devices, assuring an identical journey across systems.

4.7 Environmental Monitoring

Rapid economic growth in emerging countries has resulted in severe environmental damage in recent years. These nations frequently lack rigorous environmental rules as well as the financial resources needed to adopt pollution control measures comparable to those used in industrialized countries. As a result, the significance of clean air, clean water, and healthy soil as basic components of a high quality of life has been diminished in these areas. Pollution has become a serious danger to world health. Given the fact that reliable pollution treatments are well-established, always known, and highly affordable, pollution remains an acute problem.

Pollution is especially harmful to children and expecting women, causing chronic diseases and other health problems that impair people's well-being and lower their economic output. It can also lower life expectancy dramatically. Pollution control measures and interventions, on the other hand, have led to increases in life expectancy.

While pollution is frequently linked with chemical exposure, it is critical to acknowledge that the energy associated pollution, such as noise, also leads to serious negative effects on health, including cardiovascular disease. Open-source Arduino microcontrollers play a key role in environmental data collecting and distributed control of mobile robots linked to a cloud server in this research project. Arduino, noted for its low power consumption and price, provides simplicity in robotics applications. It may be used to control motors and RC servos via PWM or analog outputs and can accept numerous environmental sensors that can be attached to its analog inputs. The platform's adaptability is further boosted by the availability of low-cost shields for a variety of applications such as GPS technology, GSM/GPRS, Xbee, Bluetooth as well, and RFID. On the other hand, the benefits of pollution management and intervention have increased life expectancy and it can considerably reduce longevity (Correia, A. W., et al., 2013). Although chemical contamination is commonly linked to pollution, energy-related pollution, including noise, may also lead to acute health problems like heart ailments (Sørensen, M.,et al., 2012).

5. COMMUNICATIONS IN CLOUD ROBOTICS

Once robots in a network are close to one other, they are capable of communicating and they are able to communicate with cloud servers if they are close to the cloud infrastructure's access points. This creates a wireless machine-to-machine (M2M) communication network in which robots collaborate to transport and transmit data. In this way, a node inside the M2M/Machine-to-Cloud (M2C) network could serve as either a robot or an access point. To allow wireless communication over short distances among robots, many protocols such as Zigbee, Bluetooth, and Wi-Fi Direct have been created. Technologies based on electromagnetic waves and radiation may be used for distant communication. It is vital to remember that robot networks are frequently created flexibly and on the fly. There is no centralized administrator that controls the network's communication flow (Arunajyothi, G., 2016).

When a robot must transmit an inquiry to an expected destination node, that's another robot or a cloud access point, it uses a protocol that selects one of its adjacent nodes at random. The message is transmitted along with a header that includes the destination and robot identifiers, as well as a time value indicating how long the message is valid. The message is sent to all nodes next to it in a different instance of the protocol. However, depending on the application, this method may result in overbearing web page traffic. This essay will focus primarily on the form that chooses a random neighbour. At every step, each node chooses one of its neighbours at random to retransmit messages that were not intended for it but are still valid. After a sufficient number of time steps, it will be fairly likely that all messages will be passed to each destination.

In this regard, we recommend applying gossip protocols for machine-to-machine (M2M) and machine-to-cloud (M2C) connections in cloud robotics. Gossip protocols evacuate the need for route figuring and maintenance, making them ideal for dynamic mobile robotic networks. These protocols are straightforward to build, wanting just a modest percentage of additional processing and storage resources. There is a trade-off, though: gossiping may cause a rise in message delay in networks with poor conductance. This problem is much reduced with cloud robotics since the cloud's backend architecture acts as the M2M/M2C communication network's central super node. As we'll see, having a super node significantly cuts down on the amount of time needed for messages to go throughout the network. In a proxy-based cloud computing paradigm, for example, routes to frequently visited nodes, such the group leader, can be kept up to date by using a hybrid gossip algorithm. Every node selects one of its neighbours at random at each time step to resend messages that have no significance for it but are still legitimate. It's quite likely that all messages will be carried to their intended addresses after a sufficient number of time steps. Here, we recommend using gossip protocols for cloud robotics machine-to-machine (M2M) and machine-to-cloud (M2C) communication (Arunajyothi, G., 2016). Gossip protocols are very useful for flexible mobile robotic networks since they require no route discovery or maintenance. These methods are easy to put into execution, needing no extra processing and storage capacity.

6. SECURITY OF INDUSTRIAL ROBOTICS IN CLOUD COMPUTING

Figure 2 demonstrates security of robotics in cloud computing architecture.

6.1. Introduction to Security Concerns

A significant aspect of today's world is robotics, where robots perform a variety of tasks that humans are unable to do, while also exhibiting human-like behaviour and communication. Cloud computing plays a fundamental role in integrating robotics. In the realm of robotics, cloud computing is crucial as the functioning of robots relies on stored data. Presently, robotics is extensively used in industrial manufacturing, where ensuring security for these industrial robots within cloud computing systems is

Figure 2. Security of industrial robotics in cloud computing architecture

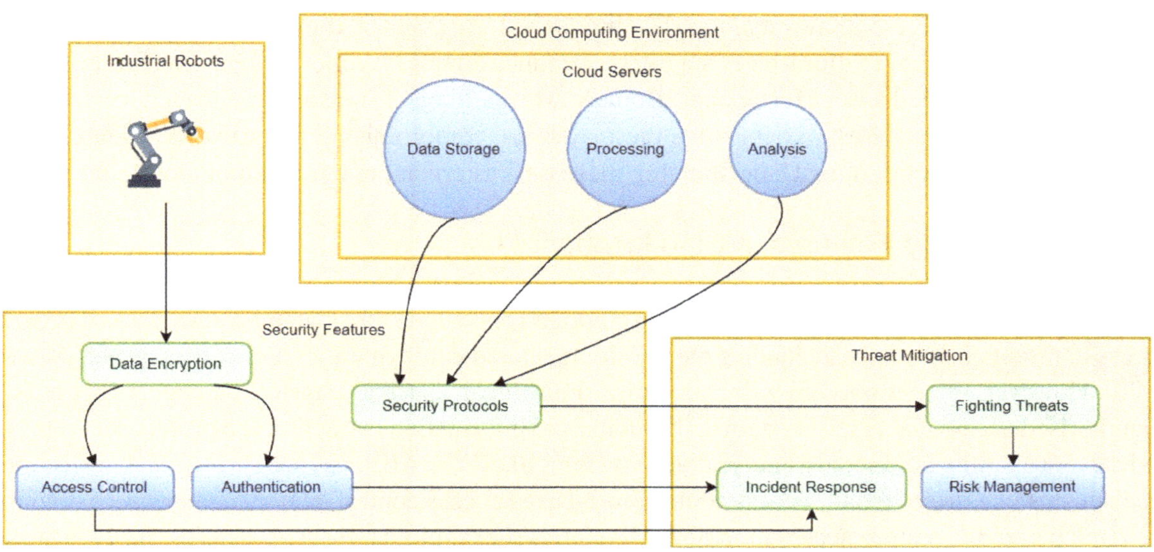

imperative (Zhang, H., & Zhang, L. 2019). The ongoing progression of robotics research will not cease in a generation; rather, it will continue to advance. However, considering that the cloud serves as its foundation, addressing security issues must take centre stage.

6.1.1. The Transformation of Robotics in the Cloud

The history of robot's dates back to the ancient world. During the industrial revolution, humans developed machines that could control electricity through small motors. Subsequently, in the early 20th century, the development of humanoid robotics commenced, which was recognized as 'Cloud-enabled Robots' during the 2010 IEEE international conference. Cloud robotics was introduced for wireless communication, leading to the creation of new experimental remote robots. Utilizing small onboard computing resources for safety and robot control, new types of experimental 'remote brain' robots were formed. According to the International Federation of Robotics (IFR), an estimated 3,903,633 industrial robots were in use globally in 2022. These resources were wirelessly connected to a more powerful remote computer for heavy processing. The integration of robotics into the cloud has the potential to revolutionize human-robot interaction and undoubtedly shape the future of the field. The robotics industry is increasingly adopting cloud technologies (Yang, P., et al. 2020). Cloud computing supports the development and deployment of new robots by robotic firms, providing access to remote resources. Smaller businesses without the capacity to establish their own data centres will find this particularly advantageous.

6.1.2. Understanding the Risks

Robotics is not limited to specific vulnerabilities; they can be prone to various risks such as cyber-attacks and unauthorized access, representing significant hazards in the field of robotics. Inadequate security within networks exposes human-robot communication to different types of assaults. Network flaws and insufficient protection pose a threat to the normal functioning and operation of industrial robots, potentially leading to production and business complications, resulting in financial losses. Specifically, these risks may result in physical harm, data theft, system blockages, and information extraction. Robotic devices are vulnerable to physical loss, hijacking, and control (Yang, P., et al. 2020). For instance, the de-authentication process allows malicious individuals to disconnect and reclaim legal ownership of these devices. Furthermore, third-party suppliers introduce numerous robot applications, some of which may be scams disguised as legitimate ones. Botnets, worms, Trojans, ransomware, backdoors, spyware, and other types of malwares may infiltrate these systems, compromising the privacy, availability, and authentication of robotic users. Understanding these risks is crucial in the realm of cloud robotics.

6.1.3. Key Security Features in Cloud Robotics

The integration of industrial robotics and cloud computing is crucial for smaller businesses, as it provides access to remote resources and enables the development and deployment of new robots (Patel, A., et al, 2020). However, these robots are vulnerable to various threats, such as physical damage, cyber attacks, unauthorised access, data theft, system disruptions, and information extraction. To protect data integrity and prevent unauthorised access, encryption algorithms like DES, AES, and RSA are used. Access control mechanisms like discretionary access control, role-based access control, and mandatory access control are implemented to restrict access to industrial robotics (Ferraiolo, D. F., et al, 1999)Security protocols like TLS, SSL, and IPsec are used to prevent unwanted access and changes. Standards like ISO 27001 and the NIST cybersecurity framework are also followed to improve data security. Proactive security measures in the cloud computing industry include continuous risk management, risk assessment, monitoring, and mitigation strategies. An incident response plan helps identify and mitigate vulnerabilities before they escalate into security incidents. For example, in a manufacturing plant using industrial robotics, role-based access control could be used to limit access to certain machines to authorized personnel only. Mandatory access control could restrict changes or updates to the robotics systems to approved individuals with proper clearance. Additionally, implementing security protocols like TLS and SSL could ensure that communication between different devices in the manufacturing process is encrypted and secure, preventing unauthorized manipulation of data or commands.

6.2. Data Encryption and Privacy

The security of industrial robotics relies on data transmission between cloud computing and industrial systems. Encrypting this data is crucial for safeguarding its integrity. Data encryption plays a significant role in ensuring the security of industrial robotics, particularly as the data is input by manufacturing industries. If someone were to hack this data, they could easily manipulate activities such as halting production or causing errors in manufacturing processes. As a result, encrypting the data used in industrial robotics is imperative to ensure data privacy and overall system security.

6.2.1. Importance of Data Encryption

Data encryption plays a crucial role in robot communication. Data encryption is a security mechanism intended to safeguard the information stored in robots linked to the cloud from theft and interception. Data encryption during communication is possible with the use of encryption algorithms like DES (Data Encryption Standard), AES(Advanced Encryption Standard), and RSA(Rivest-Shamir) (Mehmood, M. S., et al, 2019). This method primarily focuses on the key verification procedure, wherein the key verifies the decryption and encryption processes before granting permission to view the data. The mathematical formulae constitute the foundation of these algorithms. The problem here is that without the key, we are unable to decrypt the data. The control of keys is essential to the data encryption process. Secure encryption of information is achieved by using a different key for each piece of data. In the encryption process, keys are digital locks that safeguard secure communication between robotics. Certificates are used in cloud-connected settings to safeguard transactions with data. Secure connections are made using the SSL/TLS protocols. Understanding the data encryption between robots and cloud computing is also aided by this certificate. Table 1 demonstrates various security algorithms with their differences.

6.2.2. Data Privacy

When it comes to robotic data transactions, privacy of information is crucial. When it comes to safeguarding information, humanoid robots are not well defined. If a third party manages to crack the key, the data will be stolen since the third party will be able to use the key to decode the information and commit theft. The primary concern in terms of privacy of information is that robots manage and safeguard data (Patel, A., et al., 2020)We may utilise the protocol as a way to safeguard our communications and improve our interactions with the robotics. Encryption from end to end is used by the transport layer security (TLS)

Table 1. Comparison of security algorithms

Details	DES	AES	RSA
Developed Date	1997	2000	1978
Size of Key(In bits)	56	128,192 and 256	1024 (Minimum)
Size of Blocks(In bits)	64	128	512
Techniques	Symmetric	Symmetric	Asymmetric
Key for Encryption and Decryption	One Key	One key	Multiple Keys
Energy Consumption	Low	Low	High
Built-in Vulnerabilities	Brute Forced Attack, Cryptanalysis attacks	Brute Forced Attack	Brute Forced Attack

to safeguard data during transmission. Future topics including authentication, access control, and data security will be further discussed.

6.3. Access Control and Authentication

Access control and authentication are crucial in industrial robots to safeguard manufacturing processes against hackers looking to tamper with or alter output. Ensuring appropriate personnel have authority over industrial robotics is crucial. We will examine how access control and authentication are implemented in industrial robots in this section.

6.3.1. Access Control Mechanisms

Access control plays the major part of the industrial robotics in the cloud computing. Resources (data, information, sensors, actors, etc.) must have restricted or regulated access. Not everyone needs to be able to perform all tasks. There are clear guidelines. Depending on rights, and time, integrity protection is the aim of industrial production. Access control mechanisms are Discretionary Access Control (DAC), Role Based Access Control (RBAC),Mandatory Access Control (MAC). Discretionary Access Control (DAC) is also called as identity based access control. This DAC has many advantages in the robotics to safe the access but in this some are disadvantages like The administrator's file contains the hash values of the login passwords. and Users are not allowed to access that file (otherwise, someone may change the password of another user). Role Based Access Control (RBAC) was first proposed in 1992 by D.F. Ferriolo and D.R. Kuhn (Ferraiolo, D. F., et al. 1999). This RBAC is important in the cloud since it manages and assigns appropriate access while also assigning roles. Mandatory Access Control (MAC) it is also called rule- based access control (Babaeizadeh, M., et al., 2015). Decisions are made in accordance with a mandated central rule foundation. It may be approximately compared to an improved version of DAC and RBAC combined. Rights and permissions are granted in DAC, whereas roles are defined in RBAC before rights and permissions are granted. The combination of these two elements is known as a MAC. These are the three mechanisms used in the industrial robotics field.

6.3.2. User Authentication Methods

The safest method for conducting production without hacking is user Authentication in industrial robots. Ensuring that the correct person is accessing industrial robots and safeguarding security is the purpose of user authentication. Multi-Factor Authentication (MFA) is a technique of authentication. Users that work with industrial robotics are required to employ multi-factor authentication, which includes RSA tokens, biometrics, face recognition, and one-time passwords (Mehmood, M. S., et al., 2019). The user's biometric can be used in the manufacturing process when the user inputs production values, provides a password, the system verifies the information using the cloud, asks for confirmation again, and requests the authorised user's fingerprint. Therefore, in order to protect production, the user managing industrial robots must employ this kind of multi-factor authentication. To begin and end manufacturing, the user can provide their biometric. The cloud-connected industrial robots can provide the finest defence and safeguard the production by implementing these kinds of authentication techniques.

6.4. Security Protocols and Standards

In order to ensure the safety of industrial robots in cloud computing, security protocols and standards are more crucial. The security procedures for industrial robots in cloud computing will be covered in this part.

6.4.1. Protocols and Standards Overview

Security in cloud computing with industrial robotics proved difficult. Having a thorough grasp of security protocols is essential for overcoming security difficulties. In this part, we'll examine the security protocol that would be used to get around the security problems. The protocols known as TLS (Transport Sockets Layer) and SSL (Secure Sockets Layer) were employed to prevent unauthorised access and modifications to industrial robots. Industrial robots and the cloud used IPsec (Internet Protocol Security) to communicate data (Alosaimi, R., & Alnuem, M., 2016). The IPsec layer is designed to secure data transfers between cloud computing and industrial robots are the data security measures likewise based on the NIST cybersecurity framework. To ensure safe data transport, several processes are required.

6.4.2. Fighting Threats and Problems

In security protocols and standards are not just for idea; they are must for preventing the frequent problem in the industrial robotics which connected to the cloud. Utilising effective solutions is necessary to safeguard industrial robotics against unintended manufacture (Hu, G., et al., 2012). Three potential Threats exist that might compromise industrial robots during data transfers to the cloud: Man-in-the-Middle (MITM) Attacks: An unauthorised user intercepted and altered data during an MITM attack, allowing them to halt the manufacturing of industrial robots. We can defend against this attack by utilising SSL/TSL, which was primarily designed to thwart it. Attacks known as distributed denial-of-service (DDoS): these attacks aim to steal data by generating excessive traffic. With the use of load balancing and traffic filtering, we can protect Data breaches occurred when an unauthorised individual gained access to the data and modified it. With the aid of NIST framework and ISO 27001, we can safeguard data and stop unauthorised users from accessing it.

6.5. Incident Response and Risk Management

Even with robust security, errors can occur sometimes. It's critical to be ready for these circumstances and know how to respond to them.

6.5.1. Incident Response Plan

Having a plan to solve a problem is a good move. It's not a just a move it is an efficient to propose a plan to solve the security problems (vulnerabilities) before it happens. This part explains that how the plan work before a security problem arises and reduce the time of wasting the security problem after arise in the industrial robots (Alouffi, B., et al., 2021). To identify the vulnerabilities such as unauthorized access attempts and deception efforts, a complete review of the industrial robot and cloud system is required at the start of the process called "Identifying Potential Incidents." Creating a cohesive and accountable team requires the establishment of a Coordinated Response. This needs strong communication channels, clear duties for each member, and effective leadership. The main goal of this response plan is to give a solution of the security plan before it arises.

6.5.2. Best Practices for Risk Management

Risk management is the best thing to maintain the secure for industrial robotics in cloud computing. In this section we explore the Risk Management and how we can efficiently follow the risks in industrial robotics. Risk assessment has to be done and the industrial robotics can be saved while in the risk. If the risks are founded, we must do strategies to solve the risk which arises such as we must keep the software up to date and follow the protocols which we discussed in above sections. An essential component of on-going risk management is continuous risk monitoring, which includes vulnerability assessments, security audits, and threat information. Empirical examples of companies adjusting to novel hazards are a potent reminder of the need for proactive risk management (Sendi, A. S., & Cheriet, M., 2014). By implementing this risk management things this will lead us to secure our industrial robotics in cloud computing.

7. CHALLENGES OF INDUSTRIAL ROBOTICS IN CLOUD COMPUTING APPLICATIONS

In current industrial robotics, we encounter numerous challenges related to the integration of cloud computing. In this section, we will delve into the various challenges faced by industrial robotics in the context of cloud computing applications. Some of these challenges include latency and connectivity, with the connectivity of cloud computing being crucial for industrial robotics. If connectivity issues arise, such as network disruptions, production in industries may come to a halt due to latency and connectivity issues. Data security is another critical aspect; it is imperative to safeguard data during transmission between industrial robots and cloud computing applications. Scalability is also a key consideration: as the manufacturing capacity of industrial robots increases, there is a corresponding need for additional support in cloud computing applications to facilitate manufacturing processes.

7.1. Technical Challenges

The technical challenges faced by industrial cloud robotics systems include maintaining dependable communication and minimizing latency between robots and cloud computing applications. Data must be transmitted smoothly and promptly to enable real-time control, communication, and seamless coordination and collaboration among robotics systems. Achieving low latency is essential to meet manufacturing targets. Ensuring privacy and robust data security are critical challenges in industrial robotics connected to cloud computing applications (Zhang, H., & Zhang, L., 2019). There is a considerable risk of data theft during the transmission of information between robots and the cloud for manufacturing products. Various types of data, such as sensor readings, control commands, and system configurations, are collected for robots. Scalability poses a fundamental technical difficulty in cloud robotics, especially in managing the dynamic and diversified computing and storage requirements of robotic systems. Keeping the cloud infrastructure scalable to meet the increasing demands for computing and storage becomes a major technical challenge as the complexity of robots and jobs intensifies (Choudhary, S., & Singh, N., 2022). This involves addressing issues like load balancing, resource allocation, and optimizing the utilization of cloud resources to accommodate fluctuating workloads and system needs.

7.2. Operational Challenges

Industrial robotics systems face various operational challenges in managing and maintaining cloud computing application infrastructure, encompassing virtual machines (VMs) and servers. Service breakdown is another operational challenge that can lead to reduced manufacturing output. Scalability is also a significant operational challenge in industrial robotics, where the storage requirements of the industrial robots become a major concern as the number of robots and the complexity of activities increase (Zhang, H., & Zhang, L., 2019). The need for additional storage in industrial robots presents another operational challenge. The utilization of multiple industrial robots also necessitates sufficient cloud applications, adding to the operational challenges. Another operational difficulty in cloud robotics is adaptability, particularly when dealing with the dynamic and changing nature of robotic systems and their requirements. Cloud robotics systems must be capable of adjusting to shifting environments, tasks, and situations to function effectively (Yang, P., et al., 2020). We explore the need for intelligent development environments to construct adaptable cloud robotic systems, underscoring the necessity of flexibility in addressing the diverse design decisions and requirements of industrial cloud robotics systems (Wan, J., et al., 2016). The use of multiple industrial robots also requires adequate cloud applications, contributing to the operational challenges.

7.3. Strategies and Solutions

The strategies and solutions for industrial robotics in cloud computing applications involve employing hybrid solutions, which utilize a combination of cloud computing and local computing to manage various environmental circumstances and achieve real-time sensitive applications. This approach represents a continuously evolving field of study that aims to optimize the trade-off strategy between cloud computing and local computing based on individual circumstances. Emphasizing big data and open-source resources is considered a crucial response to the challenge of limited onboard resources in cloud robotics (Wan, J., et al., 2016). The integration of cloud technology with multi-robot systems enhances energy efficiency, real-time performance, and cost-effectiveness. Industrial best practices recommend taking precautions against potential issues in industrial robotics, such as latency, connectivity, data security, privacy, scalability, maintenance and management, and adaptability. For instance, a case study can illustrate how to enhance the overall efficiency of cloud-based industrial robotics.

7.4. Future Outlook

With the advancement of information technology and its applications in various sectors like healthcare, academics and agriculture (Singh, B., et al., 2023; Magesh, G., et al., 2015; Singh, B., &Acharjya, D. P. 2020; Jain, I., & Singh, B., 2023; Singh, B., & Senthil, J., 2015), it is essential to realize the potential benefits of cloud robotics in these sectors. Internet of Things (IoT) and deep learning plays an important role in the development of smart agriculture and healthcare applications (Sethuraman, S., & Singh, B., 2023; Mishra, A. et al., 2023; Singh, B., et al., 2023). Autonomous cloud robotics system could be implemented at larger scale in smart agriculture and healthcare systems with the integration of internet of things (IoT) and deep learning techniques. Electronic health records (Singh B., et al., 2013; Singh B., et al., 2016) and utilization of information technology by nurses play an important role in healthcare information management systems. For example intelligent medical system can be implemented by majority of healthcare providers to maintain patient electronic health records for efficient service delivery. The future outlook for industrial robots involves addressing challenges such as overcoming on-board processing and storage limitations, enabling remote brain functionality, and fostering collective learning behaviour. In the current global landscape, AI plays a significant role, with some industries already merging AI with industrial cloud robotics (Zhang, H., & Zhang, L., 2019). This integration may bring about new challenges that need careful consideration. For instance, the integration of advanced technologies, like the upcoming 5G technology in industries, raises the potential for stronger and faster connectivity. However, this may also give rise to various new challenges. Addressing these challenges requires potential solutions, as discussed in the sections above. As challenges evolve, innovative solutions and strategies are likely to merge, and advancements in machine learning algorithms can play a crucial role in enhancing the security and efficiency of industrial cloud-robot connections. Consider, for example, adaptive security measures in the future. Imagine machine learning techniques embedded into the security architecture of cloud-connected industrial robots. These algorithms would continuously learn and adapt to new cybersecurity threats, modifying security protocols in real-time (Wan, J., et al., 2016). Such an adaptive security framework could significantly reduce the risk of cyber-attacks while enhancing the overall resilience of the system.

8. CONCLUSION

We have discussed the fundamentals of robotics, its associated technologies and challenges involved in the development of cloud robotics applications. To understand the problems of Industrial Robotics in cloud computing applications, we identified different types of challenges, including technical challenges such as latency, connectivity, data security, and privacy. Operational challenges encompass scalability, maintenance and management, and adaptability. Additionally, we delve into strategies and solutions, emphasizing the need for precautionary measures, surveys, and case studies to better understand challenges arising in industrial cloud robotics. Looking ahead, we examine future challenges like those presented by AI, and utilizing case studies remains a pivotal process for improving these challenges.

REFERENCES

Alkhafaji, F. S. M. (2021, December). Modeling and control high speed robotic arm for industrial applications. In *2021 Global Congress on Electrical Engineering (GC-ElecEng)* (pp. 121-130). IEEE. 10.1109/GC-ElecEng52322.2021.9788417

Alosaimi, R., & Alnuem, M. (2016). Risk management frameworks for cloud computing: A critical review. *AIRCC's International Journal of Computer Science and Information Technologies*, *8*(4), 1–11. doi:10.5121/ijcsit.2016.8401

Alouffi, B., Hasnain, M., Alharbi, A., Alosaimi, W., Alyami, H., & Ayaz, M. (2021). A systematic literature review on cloud computing security: Threats and mitigation strategies. *IEEE Access : Practical Innovations, Open Solutions*, *9*, 57792–57807. doi:10.1109/ACCESS.2021.3073203

Alseddiqi, M., AlMannaei, B., Najam, O., & Al-Mofleh, A. (2022, December). The Importance of Medical Robots in Improving Healthcare Services. In *2022 International Conference on Automation, Computing and Renewable Systems (ICACRS)* (pp. 1397-1401). IEEE. 10.1109/ICACRS55517.2022.10029105

Arunajyothi, G. (2016). A study on cloud robotics architecture, challenges and applications. *International Journal of Computer Science and Network Security*, *16*(9), 110. doi:10.18535/ijecs/v5i9.27

Babaeizadeh, M., Bakhtiari, M., & Mohammed, A. M. (2015). Authentication methods in cloud computing: A survey. *Research Journal of Applied Sciences, Engineering and Technology*, *9*(8), 655–664. doi:10.19026/rjaset.9.1451

Bangemann, T., Karnouskos, S., Camp, R., Carlsson, O., Riedl, M., McLeod, S., ... Stluka, P. (2014). State of the art in industrial automation. *Industrial Cloud-Based Cyber-Physical Systems: The IMC-AESOP Approach*, 23-47.

Bicchi, A., & Kumar, V. (2000, April). Robotic grasping and contact: A review. In *Proceedings 2000 ICRA. Millennium conference. IEEE international conference on robotics and automation. Symposia proceedings (Cat. No. 00CH37065)* (Vol. 1, pp. 348-353). IEEE.

Cheikhrouhou, O., Koubâa, A., & Zarrad, A. (2020). A cloud based disaster management system. *Journal of Sensor and Actuator Networks*, *9*(1), 6. doi:10.3390/jsan9010006

Chen, M., Ma, Y., Ullah, S., Cai, W., & Song, E. (2013, September). ROCHAS: Robotics and cloud-assisted healthcare system for empty nester. In *Proceedings of the 8th international conference on body area networks* (pp. 217-220). 10.4108/icst.bodynets.2013.253922

Choudhary, S., & Singh, N. (2022). Analysis of security-based access control models for cloud computing. *International Journal of Cloud Applications and Computing*, *12*(1), 1–19. doi:10.4018/IJCAC.2022010104

Chung, J., & Lee, Y. S. (2023). The evolving impact of robots on jobs. *Industrial & Labor Relations Review*, *76*(2), 290–319. doi:10.1177/00197939221137822

Correia, A. W., Pope, C. A. III, Dockery, D. W., Wang, Y., Ezzati, M., & Dominici, F. (2013). Effect of air pollution control on life expectancy in the United States: An analysis of 545 US counties for the period from 2000 to 2007. *Epidemiology (Cambridge, Mass.)*, *24*(1), 23–31. doi:10.1097/EDE.0b013e3182770237 PMID:23211349

Edwards, M. (1984). Robots in industry: An overview. *Applied Ergonomics*, *15*(1), 45–53. doi:10.1016/S0003-6870(84)90121-2 PMID:15676500

Ferraiolo, D. F., Barkley, J. F., & Kuhn, D. R. (1999). A role-based access control model and reference implementation within a corporate intranet. *ACM Transactions on Information and System Security*, *2*(1), 34–64. doi:10.1145/300830.300834

Gunge, V. S., & Yalagi, P. S. (2016). Smart home automation: A literature review. *International Journal of Computer Applications*, *975*(8887-8891).

Harbin, J., Gerasimou, S., Matragkas, N., Zolotas, T., Calinescu, R., & Alpizar Santana, M. (2023). Model-driven design space exploration for multi-robot systems in simulation. *Software & Systems Modeling*, *22*(5), 1665–1688. doi:10.1007/s10270-022-01041-w

Hernández, E., Velázquez, R., Giannoccaro, N. I., & Gutíerrez, C. A. (2017, November). Kinematic analysis and workspace simulation of humanoid robot KUBO. In *2017 IEEE 37th Central America and Panama Convention (CONCAPAN XXXVII)* (pp. 1-6). IEEE.

Hu, G., Tay, W. P., & Wen, Y. (2012). Cloud robotics: Architecture, challenges and applications. *IEEE Network*, *26*(3), 21–28. doi:10.1109/MNET.2012.6201212

Jain, I., & Singh, B. (2023). Analysing the Applications of Cloud Computing in Smart Agriculture. In Convergence of Cloud Computing, AI, and Agricultural Science (pp. 100-119). IGI Global. doi:10.4018/979-8-3693-0200-2.ch005

Ma, Y., Zhang, Y., Wan, J., Zhang, D., & Pan, N. (2015). Robot and cloud-assisted multi-modal healthcare system. *Cluster Computing*, *18*(3), 1295–1306. doi:10.1007/s10586-015-0453-9

Madhiarasan, M. (2021). Design and development of IoT based solar powered versatile moving robot for military application. *International Journal of System Assurance Engineering and Management*, *12*(3), 437–450. doi:10.1007/s13198-021-01089-9

Magesh, G., Muthuswamy, P., & Singh, B. (2015). Use of Information Technology among school students in the State of Tamil Nadu, India. *International Journal of Applied Engineering Research: IJAER*, *10*(1), 2201–2209.

Mehmood, M. S., Shahid, M. R., Jamil, A., Ashraf, R., Mahmood, T., & Mehmood, A. (2019, November). A comprehensive literature review of data encryption techniques in cloud computing and IoT environment. In *2019 8th International Conference on Information and Communication Technologies (ICICT)* (pp. 54-59). IEEE. 10.1109/ICICT47744.2019.9001945

Mishra, A., Banerjee, S., & Singh, B. (2023). Deep Learning Techniques for Smart Agriculture Applications. In Machine Learning and Deep Learning for Smart Agriculture and Applications (pp. 1-23). IGI Global.

Mukherjee, A. (2017). Cloud Robotics Architectures. *Challenges and Applications,TTIC*, *1*, 27–33.

Patel, A., Shah, N., Ramoliya, D., & Nayak, A. (2020, November). A detailed review of cloud security: issues, threats & attacks. In *2020 4th International conference on electronics, communication and aerospace technology (ICECA)* (pp. 758-764). IEEE. 10.1109/ICECA49313.2020.9297572

Reitze, D. D. (2014, September). *Using commercial web services to build Automated Test Equipment cloud based applications. In 2014 IEEE AUTOTEST*. IEEE.

Sendi, A. S., & Cheriet, M. (2014, March). Cloud computing: A risk assessment model. In *2014 IEEE International Conference on Cloud Engineering* (pp. 147-152). IEEE. 10.1109/IC2E.2014.17

Singh, B., & Acharjya, D. P. (2020). Computational intelligence techniques for efficient delivery of healthcare. *Health and Technology*, *10*(1), 167–185. doi:10.1007/s12553-018-00280-6

Singh, B., Lopez, D., & Ramadan, R. (2023). Internet of things in Healthcare: A conventional literature review. *Health and Technology*, *13*(5), 699–719. doi:10.1007/s12553-023-00771-1

Singh, B., & Senthil, J. (2015). Use of information technology by nurses in private hospitals in the state of Tamil Nadu in India. *Mediterranean Journal of Social Sciences*, *6*(4), 658–666. doi:10.5901/mjss.2015.v6n4s2p658

Sethuraman, S., & Singh, B. (2023). Smart Agriculture Applications Using Internet of Things. In Sustainable Science and Intelligent Technologies for Societal Development (pp. 380-392). IGI Global.

Singh, B., Lopez, D., & Ramadan, R. (2023). Internet of things in Healthcare: A conventional literature review. *Health and Technology*, *13*(5), 699–719.

Singh, B., & Muthuswamy, P. (2013). Factors affecting the adoption of electronic health records by nurses. *World Applied Sciences Journal*, *28*(11), 1531–1535.

Singh, B. (2016). Nurse's attitude towards computerization in private hospitals of Tamil Nadu, India. Research journal of pharmacy and technology, 9(12), 2205-2210.

Sørensen, M., Andersen, Z. J., Nordsborg, R. B., Jensen, S. S., Lillelund, K. G., Beelen, R., Schmidt, E. B., Tjønneland, A., Overvad, K., & Raaschou-Nielsen, O. (2012). Road traffic noise and incident myocardial infarction: A prospective cohort study. *PLoS One*, *7*(6), e39283. doi:10.1371/journal.pone.0039283 PMID:22745727

Wan, J., Tang, S., Yan, H., Li, D., Wang, S., & Vasilakos, A. V. (2016). Cloud robotics: Current status and open issues. *IEEE Access : Practical Innovations, Open Solutions*, *4*, 2797–2807. doi:10.1109/ACCESS.2016.2574979

Yalcin, M., & Patoglu, V. (2012, June). Kinematics and design of AssistOn-SE: A self-adjusting shoulder-elbow exoskeleton. In *2012 4th IEEE RAS & EMBS International Conference on Biomedical Robotics and Biomechatronics (BioRob)* (pp. 1579-1585). IEEE. 10.1109/BioRob.2012.6290928

Yang, P., Xiong, N., & Ren, J. (2020). Data security and privacy protection for cloud storage: A survey. *IEEE Access : Practical Innovations, Open Solutions*, *8*, 131723–131740. doi:10.1109/ACCESS.2020.3009876

Zhang, H., & Zhang, L. (2019, April). Cloud robotics architecture: trends and challenges. In *2019 IEEE International Conference on Service-Oriented System Engineering (SOSE)* (pp. 362-3625). IEEE. 10.1109/SOSE.2019.00061

Chapter 11
Application of Swarm Robotics in Supply Chain and Logistics

Rakshith K.
Sahyadri College of Engineering and Management, India

Shreeraj N. K.
Sahyadri College of Engineering and Management, India

Shifan Mohammed
Sahyadri College of Engineering and Management, India

Rathishchandra Ramachandra Gatti
https://orcid.org/0000-0002-4086-2778
Sahyadri College of Engineering and Management, India

ABSTRACT

The advent of swarm robotics in supply chain and logistics marks a transformative shift in how these critical industries operate. This chapter explores the integration of swarm robotics into supply chain management and logistics, highlighting the paradigm shift from conventional methods to more efficient, automated systems. The chapter begins by defining swarm robotics, emphasising its characteristics such as decentralisation of control, scalability, and robustness. It then delves into the specific applications of these robotic systems in various aspects of supply chain and logistics. The chapter illustrates how swarm robotics revolutionises inventory management and warehousing procedures by enabling automated storage, retrieval, and sorting processes. The chapter also discusses the role of artificial intelligence (AI), machine learning (ML), and IoT in augmenting the capabilities of swarm robotic systems. The chapter addresses the challenges and limitations of implementing swarm robotics in the supply chain and logistics sectors. By optimising routes and reducing redundant processes, these systems significantly lower energy consumption and carbon emissions, contributing to environmentally sustainable operations.

DOI: 10.4018/979-8-3693-1914-7.ch011

1. INTRODUCTION

Inspired by social insects' collective behaviours, Swarm robotics has emerged as a fascinating field aiming to replicate nature's efficiency in robotic systems. Drawing inspiration from ants, bees, and termites, researchers in swarm robotics explore the potential of decentralised systems comprising numerous simple robots collaborating to achieve complex tasks. Unlike traditional robotics reliant on centralised control, swarm robotics emphasises decentralised decision-making, enabling scalability, adaptability, and robustness. This approach leverages collective intelligence, where individual robots follow basic rules and interact locally to exhibit intricate behaviours such as aggregation, dispersion, or foraging. The field's allure lies in its diverse applications across industries like logistics, search and rescue, and agriculture(Emam et al., 2020). However, challenges persist, including coordination and scalability in dynamic environments. Despite these obstacles, ongoing research advancements in algorithms, communication protocols, and hardware promise to propel swarm robotics toward revolutionising various sectors with the potential for cost-effective, flexible, and adaptive solutions.

Swarm robotics represents a transformative shift in supply chain and logistics management, providing innovative solutions to boost efficiency and streamline operations, as depicted in Figure 1. This technology offers several advantages: Firstly, it optimises warehouse management by enabling autonomous inventory control, efficient sorting, and swift goods retrieval. The decentralised nature of swarm robotics facilitates agile movement within confined spaces, enhancing precision in handling items and reducing operational costs while improving inventory turnover rates(Latsou et al., 2023). In transportation and distribution, swarm robotics has the potential to revolutionise last-mile delivery services. Collaborative robot fleets can navigate traffic, optimise delivery routes, and ensure timely deliveries, addressing congestion challenges and enhancing overall logistics efficiency(Deiva et al., 2022). Crucially, swarm robotics offers scalability and adaptability, allowing systems to easily scale up or down by adding or removing individual robots. This flexibility enables quick responses to changing demands and shifts in warehouse layouts without significant disruptions. Integration with data analytics and AI enhances predictive capabilities. Real-time data analysis provides insights into inventory trends, demand forecasting, and operational optimisations, empowering logistics managers to make informed decisions, leading to efficient resource allocation and reduced operational risks. Applying swarm robotics in supply chain and logistics promises heightened efficiency, adaptability, and responsiveness, paving the way for a more agile and optimised logistics infrastructure capable of meeting modern commerce's evolving demands.

Furthermore, research concerning the seamless integration of human-robot collaboration within swarm systems for logistics is relatively scarce(Oliff et al., 2020). Understanding how humans interact and collaborate with swarms of robots in complex logistics environments remains an essential area requiring further exploration. Moreover, while swarm robotics shows promise in optimising specific logistical operations, more comprehensive studies need to explore the holistic integration of swarm systems across the entire supply chain, encompassing multiple nodes and processes. Bridging these research gaps will be pivotal in advancing the practical implementation and scalability of swarm robotics within the broader logistics and supply chain management spectrum.

Figure 1. Role of swarm robotics in supply chain management

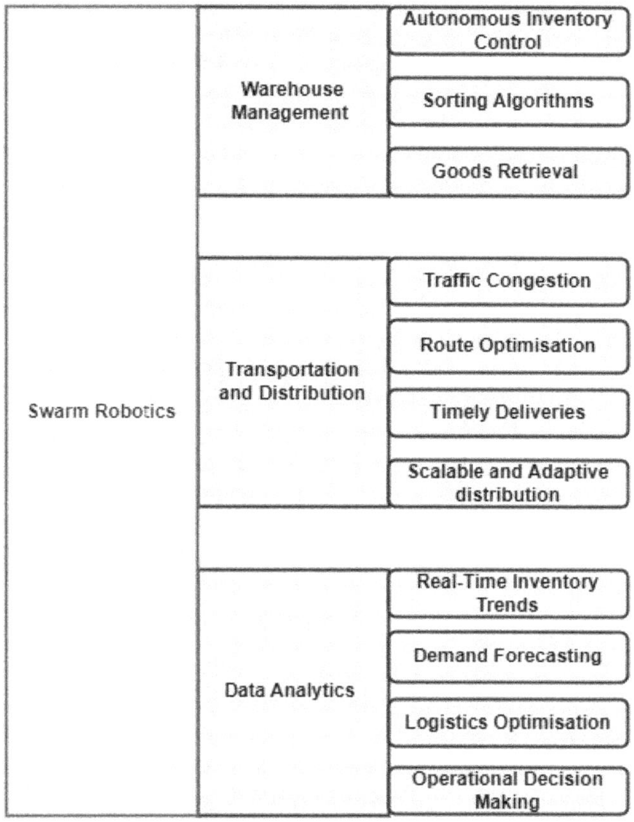

2. THEORETICAL FRAMEWORK

Fundamentally, swarm robotics operates on key principles underpinning its logistics functionality. Emergent behaviour stands out, highlighting how individual robots, guided by simple rules, collectively exhibit sophisticated behaviours within a group, fostering emergent properties that contribute to efficiency in logistics operations. Decentralised control mechanisms enable autonomous decision-making based on local interactions and limited information, resulting in self-organisation and cohesive group behaviour among the robots. Local communication and interaction play a crucial role, allowing robots to exchange information, cooperate, and coordinate actions, ultimately contributing to the emergence of collective behaviour essential for navigating complex logistical environments. The theoretical framework for the requirements of the swarm robotics can be concisely represented as in Figure 2. Adaptability and robustness are intrinsic qualities of swarm robotics, enabling these systems to adapt to dynamic logistics environments while maintaining resilience against failures. The concept of stigmergy, utilising indirect communication through the environment, further enhances collective intelligence, fostering problem-solving abilities among the swarm. Additionally, swarm systems showcase scalability and dynamic adaptability, allowing them to adjust robot numbers and behaviours to suit various logistical tasks, thus contributing significantly to their effectiveness within logistics applications.

Figure 2. Requirements of the swarm robotics

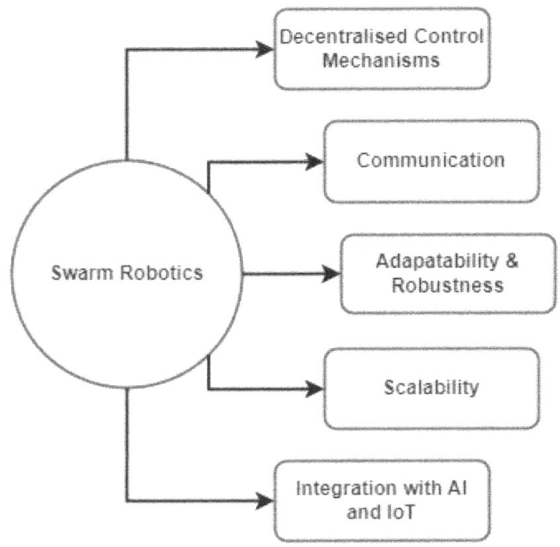

Swarm robotics represents a disruptive force in supply chain management and logistics, offering transformative solutions to tackle intricate challenges within the operational landscape. These systems redefine efficiency in warehouse management by orchestrating autonomous inventory control, proficient item sorting, and swift goods retrieval, optimising storage spaces and operational expenses. The transportation and last-mile delivery realm witnesses a revolution through swarm robotics, fostering adaptable and agile delivery networks. These networks streamline route planning, navigate through traffic, and ensure punctual deliveries, elevating the efficacy of logistics operations(Liu et al., 2023). Swarm systems' inherent scalability and adaptability present a dynamic edge, empowering seamless adjustments to meet evolving demands and adapt to diverse warehouse configurations without operational disruptions. Moreover, the potential synergy between human collaboration opens new avenues for enhanced logistics efficiency, emphasising a harmonious coexistence benefiting from both human expertise and robotic precision. Swarm robotics' relevance to supply chain management and logistics thus manifests as a catalyst for heightened efficiency, agility, and responsiveness, reshaping the logistical landscape to meet the ever-evolving demands of modern commerce.

The integration of swarm robotics with AI (Artificial Intelligence) and machine learning signifies a critical advancement in enhancing the capabilities and efficiency of logistical operations (Lara & Wassick, 2023). AI and machine learning algorithms complement swarm robotics by empowering these systems with enhanced decision-making abilities, predictive analytics, and adaptive learning. The amalgamation enables swarm robotic systems to process and analyse vast amounts of data in real time, facilitating informed decision-making in dynamic logistics environments. AI algorithms enable robots to learn from experiences, refine their behaviours, and adapt to varying operational scenarios, improving overall performance and responsiveness. Machine learning models assist in predictive analytics, forecasting demand patterns, optimising routes, and predicting maintenance needs, optimising logistics processes. Additionally, the integration fosters cognitive abilities within swarm systems, allowing for intelligent coordination, better task allocation, and improved resource utilisation. The fusion of swarm robotics with AI and machine learning holds promise in revolutionising logistics by enhancing adaptability, optimising operations, and enabling autonomous decision-making capabilities that are crucial in the ever-evolving supply chain management landscape.

3. ALGORITHMS USED IN SWARM ROBOTICS

The algorithms used in swarm robotics encompass a variety of optimization and coordination techniques that enable groups of robots to work collectively. Some key algorithms are discussed briefly here.

3.1 Particle Swarm Optimization (PSO): Particle Swarm Optimisation (PSO) is a computational technique employed for solving optimisation problems. Particle Swarm Optimisation (PSO) involves a collective movement of particles inside a search space. Each particle modifies its position by considering its individual experience and the most optimal position discovered by the entire group. The algorithm's objective is to repeatedly enhance candidate solutions in order to discover the most optimal solution. Every particle in the swarm possesses a position and velocity within the search space. The best-known positions for each particle (pbest) and the entire swarm (gbest) are updated according to the evaluation of the objective function. The procedure entails modifying particle velocities and positions using cognitive and social coefficients, inertia weight, and random numbers. PSO is derived from the collective behaviour observed in flocks of birds or swarms of insects, where individual particles are drawn towards the most optimal location discovered by themselves and the group as a whole. The algorithm has the ability to converge on one or more specific locations or to continue exploring the entire search field. The Particle Swarm Optimisation (PSO) algorithm is renowned for its simplicity, as it only requires a small number of hyperparameters. Additionally, it is highly versatile and can be applied to a diverse set of optimisation issues.

PSO can be utilised to optimise the control and coordination of robotic swarms in swarm robotics. Swarm robotics utilises PSO to improve the behaviour and performance of groups of robots by optimising their movements and interactions. Particle Swarm Optimisation (PSO) in swarm robotics entails adapting the algorithm's parameters to align with the dynamics and limitations of physical robots. This adaptation incorporates factors such as regulating velocity and acceleration, asynchronously updating robot states, and guaranteeing compliance with stability standards. By incorporating Particle Swarm Optimisation (PSO) into swarm robotics, robots may collaboratively traverse intricate landscapes, optimise the allocation of tasks, and enhance the overall efficiency of the swarm. The iterative improvement of solutions based on individual and group experiences in PSO makes it a significant tool for boosting the autonomy and efficacy of robotic swarms in diverse applications.

3.2 Ant Colony Optimization (ACO): Ant Colony Optimisation (ACO) is a metaheuristic algorithm that draws inspiration from the foraging behaviour exhibited by ant colonies. ACO, introduced by Marco Dorigo in the 1990s, is a method that utilises a group of artificial ants to collaboratively discover effective solutions for intricate optimisation issues. ACO, or Ant Colony Optimisation, involves the use of artificial ants to search for the most optimal solutions. These ants construct paths on a graph with assigned weights, with the goal of finding the path that yields the best outcome for the given optimisation issue. The algorithm is founded upon the notion of pheromone transmission, wherein ants deposit pheromone trails on roads in accordance with the quality and amount of food discovered. Subsequent ants then track these trails, their choice of path being determined by the concentration of pheromones and the rate of evaporation. This technique enables the colony to collaboratively determine the most direct route to a food supply, imitating the effective foraging behaviour of actual ant colonies. ACO has developed into a full algorithmic framework with diverse components and applications, rendering it a potent tool for resolving optimisation challenges.

The application of Ant Colony Optimisation (ACO) in swarm robotics has the potential to significantly improve the coordination and efficacy of a collection of robots. ACO algorithms can enhance the performance of swarms, assist robotics in navigating complex environments, and optimise path planning by studying the foraging behaviour of ants. ACO algorithms have the capability to aid robotics in locating the most efficient routes through dynamic or unfamiliar surroundings by emulating the way in which ants locate food sources. This facilitates effective obstacle avoidance and navigation within the swarm. By facilitating the exchange of information regarding effective routes or resolutions, ACO promotes collaboration among robotic agents operating in a swarm, thereby enhancing their capacity for collective decision-making and problem-solving. ACO techniques have the potential to improve fault tolerance in swarm robotics systems through enabling robots to adjust and discover alternative routes or resolutions in the presence of challenges or malfunctions. Decentralised control is a key principle in swarm robotics, wherein decisions are made by individual robots on the basis of local information and interactions. By implementing ACO algorithms, the swarm's overall efficiency and adaptability are enhanced. By utilising heuristic algorithms that draw inspiration from ant behaviour models, ACO can assist in the localisation of targets in perilous environments. This enables the swarm to carry out target tracking and localization tasks more efficiently.

3.3 Artificial Bee Colony Optimisation: Artificial Bee Colony Optimisation (ABC) is an optimisation method that draws inspiration from the foraging behaviour of honey bees. The ABC algorithm, developed by Karaboga in 2005, models the collaborative behaviour of honey bees in locating food sources. The algorithm utilises three categories of bees: worker bees, observer bees, and explorer bees. Foraging bees actively seek out food from familiar locations and communicate this knowledge to other bees observing their behaviour. Observer bees subsequently choose the most optimal food sources by utilising the information provided by worker bees. Scout bees are industrious bees that forsake their food sources in order to seek out new ones. The method functions by creating an initial set of solutions (food sources) and progressively enhancing these solutions through the search procedures carried out by worker, observer, and scout bees. Worker bees adapt their locations in response to the quantity of nectar in newly discovered sources, and the most optimal solutions are sent to observer bees. The method iterates until specific criteria are satisfied, with the goal of identifying optimal solutions for optimisation issues. ABC is renowned for its simplicity, efficacy, and versatility in addressing a diverse set of practical challenges, rendering it a useful asset in swarm intelligence systems.

Artificial Bee Colony Optimisation (ABC) improves robot cooperation and optimisation in swarm robotics. ABC helps many robots explore difficult situations by optimising path planning. ABC can guide robots around obstacles by emulating bee foraging. ABC can help swarms of robots assign work based on quality. ABC distributes robot work to maximise performance by replicating spectator bee choosing. ABC optimises swarm robot energy, time, and sensor distribution. ABC can assist robots distribute resources efficiently to achieve communal goals by modifying nectar amount evaluation. ABC optimises swarm robot communication. ABC can help robots communicate and coordinate by replicating bee dance communication. ABC helps swarm robots cooperate. ABC can improve swarm intelligence and task performance by using honey bees' social cooperation.

The other optimization algorithms that can be used for the control of Swarm robotics are Fish School Search Optimization, Cuckoo Search Optimization, Firefly Algorithm, Bat Algorithm, Flower Pollination

Algorithm, Gray Wolf Optimization, Elephant Swarming Optimization, Crow Search Optimization, and Raven Roasting Optimization Algorithm.

4. APPLICATIONS IN SUPPLY CHAIN AND LOGISTICS

Swarm Robotics is widely used across different supply chain and logistics areas such as warehousing and inventory management, automation in storage and retrieval, inventory handling, transportation and distribution, route optimisation, autonomous delivery systems, fleet management and tracking, real-time data analytics and fleet management and tracking.

4.1 Warehousing and Inventory Management

Applying swarm robotics in warehousing and inventory management in supply chain and logistics presents a transformative approach to streamlining and optimising operations. Swarm robotics introduces innovative solutions in warehouse management by facilitating autonomous and efficient inventory control(Latsou et al., 2023). These systems enable precise item sorting, rapid goods retrieval, and optimised storage, enhancing the overall efficiency of warehouse operations. The decentralised nature of swarm robotics allows robots to navigate and operate within confined spaces(Vásárhelyi et al., 2018), optimising floor space utilisation and ensuring efficient inventory handling. Moreover, swarm systems excel in adapting to dynamic inventory needs, ensuring accurate stock management, and reducing errors in inventory tracking, utilising technologies such as RFID tracking(Khazetdinov et al., 2020). They contribute significantly to inventory turnover rates, minimising delays in product retrieval and shipment processes. Additionally, the autonomous nature of swarm robotics in inventory management mitigates the need for constant human intervention, reducing labour costs and enhancing the speed and accuracy of inventory-related tasks. The integration of swarm robotics in warehousing and inventory management signifies a paradigm shift, offering cost-effective, efficient, and agile solutions that optimise space utilisation, improve inventory accuracy, and expedite logistics processes within supply chains.

4.2 Automation in Storage and Retrieval

Applying swarm robotics introduces groundbreaking automation capabilities in storage and retrieval processes within logistics operations. Swarm robotics revolutionises the automation of storage and retrieval tasks by enabling autonomous and efficient handling of goods in warehouses(Milner et al., 2022). These systems excel in automating the movement and organisation of inventory, facilitating swift storage and precise retrieval of goods. Through decentralised control mechanisms, robots collaborate seamlessly to navigate storage shelves, accurately place items, and retrieve goods with agility and precision. The adaptability of swarm robotics ensures dynamic optimisation of storage layouts, enabling efficient space utilisation and minimising human intervention in repetitive tasks. Moreover, swarm systems optimise retrieval processes by efficiently locating and retrieving items, reducing retrieval times and errors while improving overall operational efficiency. By automating storage and retrieval tasks, swarm robotics minimises manual labour requirements, mitigates operational bottlenecks, and enhances the speed and accuracy of warehouse operations(Latsou et al., 2023). The integration of swarm robotics in automation for storage and retrieval signifies a transformative leap towards streamlined, efficient, and agile logistics operations within supply chains.

4.3 Efficiency in Inventory Handling

Within the supply chain and logistics realm, the integration of swarm robotics significantly enhances efficiency in inventory-handling processes. Swarm robotics introduces unprecedented efficiency by optimising various aspects of inventory handling within warehouses. These systems autonomously streamline inventory handling, ensuring precise sorting, organised storage, and rapid retrieval of goods. Their decentralised nature allows agile movement in confined spaces, enabling robots to handle items with exceptional accuracy and speed(Wior et al., 2012). This agility minimises delays in inventory processing, significantly reducing idle time and enhancing overall operational efficiency.

Moreover, swarm robotics systems excel in adapting to fluctuating inventory demands, efficiently managing inventory turnover and ensuring accurate stock management. Their ability to swiftly locate and retrieve items reduces errors and delays, thus optimising inventory handling efficiency. Additionally, the integration of swarm robotics minimises reliance on manual labour, mitigating potential human errors and significantly improving the speed and accuracy of inventory-related tasks. By automating inventory handling processes, these systems enhance the overall efficiency of warehouse operations, contributing to reduced operational costs and increased throughput. Applying swarm robotics in inventory handling embodies a transformative shift towards more efficient, precise, and agile logistics operations, revolutionising inventory management within supply chains.

4.4 Transportation and Distribution

Swarm robotics brings innovative solutions to transportation and distribution in logistics by establishing adaptable and dynamic delivery networks. These networks utilise collaborative robot fleets adept at navigating diverse terrains and optimising delivery routes using optimisation algorithms(Jevtić et al., 2010). They excel in real-time route adjustments, effectively managing traffic complexities and ensuring timely deliveries, enhancing logistical efficiency. The flexibility inherent in swarm robotics allows for responsive adjustments in delivery schedules, accommodating fluctuating demands or altering delivery destinations, ultimately enhancing the agility of transportation networks. Furthermore, swarm robotics optimises resource utilisation by minimising reliance on traditional transportation methods and streamlining processes (Michalos et al., 2016), reducing manual intervention and significantly enhancing the efficiency of transportation and distribution operations within supply chains.

4.5 Route Optimisation

Swarm robotics introduces a transformative approach to route optimisation in logistics, revolutionising how delivery routes are planned and managed. These systems leverage collective intelligence and decentralised control mechanisms to optimise delivery routes in real time dynamically (Jevtić et al., 2010). Using collaborative robot fleets, swarm systems analyse multiple variables such as traffic patterns, delivery locations, and real-time environmental data to calculate the most efficient routes. Decentralised decision-making enables robots to communicate and adapt swiftly, allowing for agile route adjustments based on changing conditions and ensuring prompt and efficient deliveries. Furthermore, swarm robotics enhances route optimisation by continuously learning and adapting to new scenarios, integrating historical data and predictive algorithms to foresee potential traffic congestion or optimal delivery windows(Niihara et al., 2022). This adaptability and learning capability empower swarm systems to navigate complex logistical landscapes, ensuring streamlined routes, minimised delivery times and enhanced overall efficiency within supply chain logistics.

4.6 Autonomous Delivery Systems

Autonomous delivery systems, a pinnacle of innovation in logistics, intersect with the core principles of swarm robotics, heralding a new era in the last-mile delivery process. Integrating cutting-edge technologies like robotics, AI, and advanced sensors, these systems operate independently, mirroring the decentralised nature of swarm robotics. Ground-based robots, drones, and autonomous vehicles are the primary constituents, each leveraging sophisticated algorithms and navigational prowess akin to swarm behaviour. Ground-based robots manoeuvre sidewalks and streets autonomously, equipped with sensors and AI navigation systems like swarm agents coordinating in unison(de Almeida et al., 2019). Drones soar the skies, employing precision sensors and GPS navigation, resembling swarm entities in their collective precision(Dulce-Galindo et al., 2022). Autonomous vehicles, akin to larger-scale swarm systems, navigate roadways, interpret traffic signals, and facilitate larger payload deliveries. These systems offer expedited delivery times, efficiency enhancements, and cost reductions by optimising routes through AI algorithms, much like swarm robotics optimise their collective behaviour. Moreover, they promise to transform logistics for remote or underserved areas and minimise human errors in deliveries. However, challenges such as regulatory frameworks, safety concerns, and integration complexities need resolution for seamless implementation. Ultimately, these autonomous delivery systems, embodying the essence of swarm robotics, are poised to revolutionise logistics, altering the landscape of last-mile deliveries with speed, efficiency, and reliability.

4.7 Fleet Management and Tracking

Fleet management and tracking, deeply intertwined with the principles of swarm robotics, epitomise an innovative paradigm in logistics operations. Within the context of swarm robotics, fleet management involves orchestrating and overseeing a collective of robots or vehicles, reflecting the collaborative behaviour characteristic of swarms(Koot et al., 2021). These systems, steered by advanced algorithms and decentralised control mechanisms, emulate the coordinated actions observed within a swarm of individual agents. Swarm robotics principles are the foundation for efficient fleet management and tracking, enabling autonomous decision-making and adaptive behaviours among robotic fleets. Similar to the collective adaptation of a swarm to environmental changes, these fleet management systems utilise AI algorithms to optimise routes, allocate tasks, and ensure seamless coordination among multiple robotic units(Dulce-Galindo et al., 2022). The tracking mechanisms employed within fleet management systems mirror swarms' communication and self-organising capabilities. Leveraging advanced sensor networks and communication protocols reminiscent of swarm entities, these systems facilitate real-time monitoring and data exchange among robotic units. Incorporating IoT-enabled sensors and AI-driven analytics enables continuous tracking, ensuring precise navigation and providing invaluable insights pivotal for swift decision-making in logistics operations. Moreover, like the scalability observed in swarm systems, fleet management within swarm robotics offers adaptability and scalability. This feature enables logistics operations to effortlessly adjust fleet sizes by adding or removing robotic units, catering promptly to shifting demands or altering logistical requirements without disrupting ongoing processes. However, overcoming challenges such as maintaining seamless connectivity among robotic units and synchronising actions while preserving individual autonomy remains imperative. Addressing these challenges is pivotal for successfully implementing fleet management and tracking systems deeply rooted in swarm robotics principles. Ultimately, these systems, embodying the essence of collaborative and decentralised behaviour observed in natural swarms, present a forward-thinking, responsive, and optimised approach to logistics operations(Scholz et al., 2018).

4.8 Real-Time Data Analysis

Real-time data analysis within the scope of swarm robotics signifies the pivotal integration of instantaneous insights derived from vast datasets to drive informed decision-making in logistics(Resende et al., 2021). In swarm robotics, real-time data analysis encompasses the processing and interpreting of data streams collected by robotic units operating in unison. This process mirrors the collaborative exchange and assimilation of information among swarm entities. Leveraging advanced AI algorithms and distributed computing, real-time data analysis empowers these systems to derive actionable insights instantaneously. By continuously processing sensor data and environmental inputs, these systems emulate the adaptability of swarms by dynamically adjusting behaviours, optimising routes, and making split-second decisions. This seamless integration of data analysis into swarm robotics not only enhances the overall operational efficiency but also augments the systems' ability to respond promptly to changing environmental cues and logistical demands, ultimately reshaping the landscape of logistics operations.

4.9 Fleet Coordination and Management

Fleet coordination and management, closely entwined with the principles of swarm robotics, epitomises a multifaceted orchestration of robotic units operating collaboratively to optimise logistics operations. In swarm robotics, fleet coordination mirrors the synchronised and decentralised behaviours observed in natural swarms. These systems encompass the autonomous coordination and supervision of a collective fleet of robots or vehicles akin to the harmonious collaboration within a swarm. Powered by sophisticated algorithms and decentralised control mechanisms, fleet management ensures seamless coordination, task allocation, and adaptive decision-making among robotic units. The integration of swarm principles allows for scalable fleet sizes, enabling swift adjustments in response to fluctuating demands or dynamic logistical requirements(Zhang et al., 2023). However, challenges like maintaining interconnectivity among robotic units and balancing individual autonomy while ensuring collective coordination pose critical considerations. Addressing these challenges is crucial for successfully implementing fleet coordination and management systems deeply rooted in the collaborative and decentralised nature inherent in swarm robotics. Ultimately, these systems offer an agile, responsive, and optimised approach to logistics operations, mirroring natural swarms' collective and cooperative behaviours.

5. TECHNOLOGICAL ASPECTS

Swarm robotics include the applications of emerging technologies such as Artificial Intelligence (AI), Machine Learning (ML), Internet of Things (IoT) in both software and hardware of swarm robots.

5.1 AI and Machine Learning in Swarm Robotics

AI and machine learning constitute the cornerstone of advancements in swarm robotics, ushering in a new era of autonomous decision-making and adaptive behaviours among robotic entities. In the context of swarm robotics, AI algorithms such as Swarm Robotics Search &

Rescue (SRSR), as reported in (Bakhshipour et al., 2017), and machine learning mechanisms emulate the collective learning and adaptive behaviours witnessed in swarms of natural organisms. These technologies equip robotic units with cognitive abilities to analyse complex datasets, adapt behaviours, and optimise real-time decision-making. Through continuous learning using reinforcement learning techniques (Blais & Akhloufi, 2023) and refinement, swarm robotics systems imbued with AI and machine learning capabilities enhance operational efficiencies, optimise routes, and dynamically respond to changing environmental cues. The integration of AI augments swarm robotics by enabling predictive analytics, allowing these systems to forecast trends, anticipate logistical demands, and proactively adapt behaviours, mirroring the adaptive nature of swarms in response to evolving conditions (Kumar & Jaiswal, 2020). However, challenges such as algorithmic robustness and balancing learning capabilities with real-time responsiveness remain focal points for further advancements. Nurturing the synergy between AI, machine learning, and swarm robotics promises an era of autonomous, adaptable, and efficient logistics operations, fundamentally reshaping the dynamics of supply chain management.

5.2 IoT Integration for Real-Time Monitoring

The integration of IoT technology within swarm robotics heralds an era of enhanced real-time monitoring and dynamic data exchange, mirroring the collaborative communication observed in natural swarms. In swarm robotics, IoT-enabled sensors embedded within robotic units facilitate continuous data acquisition and exchange, akin to information-sharing and collaboration within swarms. These sensors capture real-time environmental data, enabling swift decision-making, precise navigation, and adaptive behaviours among robotic entities(Yung et al., 2023). Leveraging the insights derived from IoT-integrated sensors, swarm robotics systems respond dynamically to changing environmental stimuli, optimise routes, and enhance overall operational efficiency. The assimilation of IoT technology not only enables seamless communication among robotic units but empowers logistics operations with real-time insights crucial for proactive decision-making, reflecting the collaborative nature inherent in swarms. Challenges encompassing data security, interoperability, and network reliability remain focal points for ensuring the seamless integration of IoT into swarm robotics. Nevertheless, the convergence of IoT technology with swarm robotics promises a future where autonomous entities collaborate seamlessly, driven by real-time insights, and foster an ecosystem of dynamic, responsive logistics operations.

IoT (Internet of Things) can be utilised in swarm robotics to enhance communication, coordination, and decision-making among robotic agents within a swarm. IoT devices can facilitate wireless communication between individual robots in the swarm, allowing them to share information such as position, velocity, sensor readings, and task assignments. This enables real-time coordination and collaboration among swarm members. IoT sensors can be deployed in the environment to collect data relevant to the robots' tasks or to monitor environmental conditions. The swarm can use this data to make informed decisions, adapt to environmental changes, and optimise their collective behaviour.

IoT-enabled swarm robots can leverage cloud computing resources for data storage, processing, and analysis tasks. This allows the swarm to offload computationally intensive tasks, access large databases or models, and benefit from centralised coordination and decision-making algorithms. Instead of relying solely on cloud resources, IoT devices can also perform processing and decision-making locally (edge computing). This reduces latency and bandwidth requirements, enabling faster response times and more robust operation in environments with limited or unreliable network connectivity.

IoT technology can enable distributed control architectures, where decision-making processes are distributed among individual robots in the swarm rather than relying on a centralised controller. This enhances scalability, fault tolerance, and adaptability, as the swarm can continue to operate even if some robots fail or lose communication. By integrating IoT devices, swarm robots can dynamically allocate tasks based on real-time information about their capabilities, the environment, and the current workload. This allows the swarm to adapt to changing conditions, prioritise tasks, and optimise resource utilisation.

IoT-enabled swarm robotics systems can exhibit self-organising behaviour, where individual robots adjust their actions based on local interactions and environmental feedback. This enables the swarm to autonomously adapt to new challenges, reconfigure its structure, and achieve complex objectives without centralised control. IoT sensors can monitor the health and performance of individual robots within the swarm, detecting anomalies, predicting failures, and scheduling maintenance tasks as needed. This improves the reliability and longevity of the swarm system.

5.3 Software and Hardware Requirements

The amalgamation of software and hardware components within swarm robotics forms the crux of efficient, agile, and adaptive logistics operations reminiscent of the synergy observed in natural swarms. In the context of swarm robotics, sophisticated software algorithms steer the decision-making, coordination, and adaptive behaviours among robotic units, mirroring the collective and decentralised actions within swarms. Complemented by robust hardware components, these algorithms encompass sensors, processors, and communication modules critical for seamless data processing and exchange among robotic entities. The cohesive interplay between software and hardware allows swarm robotics systems to execute complex tasks, adapt to dynamic environmental conditions, and optimise logistics operations. However, ensuring compatibility, reliability, and scalability between software and hardware components poses challenges in implementing swarm robotics. Addressing these challenges remains pivotal for fostering the seamless integration of software and hardware, enabling agile, responsive, and efficient logistics operations reflective of the cooperative and adaptive behaviours observed in natural swarms.

Swarm robots for logistics and supply chain management require specific hardware and software components tailored to their intended tasks and operational environments. Here is a breakdown of the typical requirements:

5.3.1 Hardware Requirements

1. Robotic Platforms: Swarm robots can range from simple wheeled robots to more complex aerial drones or specialised ground-based vehicles, depending on the specific logistics tasks they need to perform.
2. Sensors: These may include cameras, LiDAR (Light et al.), ultrasonic sensors, proximity sensors, GPS modules, inertial measurement units (IMUs), and RFID (Radio-Frequency Identification) readers. These sensors enable robots to perceive and navigate their environment, detect obstacles, locate objects, and track inventory.
3. Actuators: Motors, servos, and other actuators, such as driving wheels, rotating arms, or opening/closing grippers for manipulating objects, are necessary for controlling the movement of the robots.
4. Communication Modules: Wireless communication modules (e.g., Wi-Fi, Bluetooth, Zigbee, LoRa) facilitate communication between swarm robots, as well as with central control systems or other devices in the logistics network.
5. Power Systems: Swarm robots require reliable power sources, including rechargeable batteries, solar panels, or energy harvesting systems, depending on their intended deployment scenarios and energy requirements.
6. Robotic Manipulators: For tasks involving manipulating objects, such as picking and placing items in a warehouse, robots may need specialised manipulators or arms equipped with grippers or suction cups.

5.3.2 Software Requirements

1. Localisation and Mapping: Software algorithms for localisation (determining the robot's position and orientation) and mapping (creating and updating maps of the environment) are essential for navigation and path planning within logistics facilities.
2. Path Planning and Navigation: Algorithms for path planning and navigation enable swarm robots to autonomously plan and execute optimal paths to navigate warehouses, distribution centres, or other logistics environments while avoiding obstacles and optimising efficiency.
3. Distributed Control and Coordination: Software frameworks for distributed control and coordination allow swarm robots to communicate, collaborate, and synchronise their actions to perform complex logistics tasks efficiently. These frameworks may include algorithms for task allocation, consensus building, and behavior-based control.
4. Object Detection and Recognition: Computer vision algorithms for object detection and recognition enable robots to identify and locate items within their environment, such as packages, pallets, or inventory bins. This information is crucial for inventory management, picking, and sorting.
5. Integration with Logistics Management Systems: Interfaces and APIs (Application Programming Interfaces) for integrating swarm robotics systems with existing logistics management software, such as warehouse management systems (WMS) or enterprise resource planning (ERP) systems, facilitate seamless data exchange and coordination between robotic and human-operated processes.
6. Fault Tolerance and Redundancy: Software mechanisms for fault tolerance and redundancy ensure the robustness and reliability of swarm robotics systems, allowing them to continue operating effectively even in the presence of individual robot failures or communication disruptions.

By meeting these hardware and software requirements, swarm robots can effectively contribute to various logistics and supply chain management aspects, including inventory tracking, order fulfilment, warehouse automation, and distribution centre operations.

6. CHALLENGES AND LIMITATIONS

Implementing swarm robotics within logistics operations encounters various technical hurdles, reflecting the need for innovative solutions to harness the full potential of these systems. Challenges encompass establishing robust communication protocols among robotic units, ensuring real-time decision-making, and maintaining system stability during scalability, akin to the coordination and adaptation observed in swarms. Innovations in communication systems, advanced algorithms, and stable hardware configurations are imperative for addressing these challenges. Overcoming technical hurdles is paramount for successfully deploying swarm robotics, which is deeply rooted in the collaborative and adaptable nature inherent in swarm entities. Navigating technical challenges while fostering seamless cooperation among robotic units paves the way for the transformative integration of swarm robotics into logistics operations, redefining efficiency and adaptability within supply chain management.

Adopting swarm robotics in logistics operations presents initial cost implications, reflecting the need for investments in hardware, software, and specialised training reminiscent of the initial collaborative efforts required within swarms. While the upfront investment might be substantial, the long-term benefits, including increased operational efficiency, reduced labour expenses, and minimised error rates, often justify the initial costs. The transition toward swarm robotics involves financial investments and investments in understanding, adapting to, and managing innovative technology. However, achieving a balance between initial investments and long-term returns remains pivotal for the widespread adoption and successful implementation of swarm robotics, mirroring the collaborative efforts and resource utilisation observed in natural swarms. Navigating cost considerations and optimising investments are crucial aspects that will shape the trajectory of swarm robotics in revolutionising logistics operations.

7. CASE STUDIES AND PRACTICAL EXAMPLES

Swarm robotics in logistics operations demonstrate its transformational power, echoing natural swarms' collaborative and decentralised nature. Implementing swarm robots demonstrates efficiency benefits, operational optimisation, and new solutions. Swarm robotics improves operations, turnaround times, errors, and logistics efficiency in warehouse automation, delivery networks, and inventory management. Real-world implementations shed light on adopting and optimising swarm robots in logistics operations, demonstrating natural swarm collaboration and optimisation. These practical applications illuminate successes, problems, and lessons learned, improving best practices and tactics for logistics operations using swarm robotics. The following are some of the real-world cases of applications of swarm robotics in logistics in different industries:

7.1 Agricultural Logistics: Unmanned aerial vehicles (UAVs) are used by SAGA, a precision farming experimental platform(Trianni et al., 2016), to monitor and weed a field in large numbers. The UAVs fly close to the field to count weeds and find areas with enough weeds. They have radio communication systems, cameras, visual processing, and protocols for safe swarm operations. Additionally, UAVs can micro-spray plants.

7.2 Industrial Plant Logistics: A research team led by Howie Choset at Carnegie Mellon University developed snake robots that could navigate through narrow passageways and tunnels off-limits to humans and machines. These robots with metal bodies can crawl, swim, and climb. They are five centimetres thick and one metre long. They can move across small spaces thanks to their 16 joints. The snake robots' cameras and lights enable the control centre to monitor everything. The 2017 Mexico earthquake survivors were located using these devices. However, the snake robots did not encounter any humans during the mission. The researchers want to enhance the system and design of their robots by incorporating additional visualisation tools and sensors. German startup FORMIC Transportsysteme, linked with Karlsruhe Institute of Technology, is developing a FORMIC modular robotic transportation system. It has several six-wheeled robotic transport modules with cameras, radio communications chips, and 2.5-ton jacks.

7.3 Military logistics: The US military used robots built by Endeavour Robotics for various tasks. The Cobra robot lifts big goods. It has 150 kg of lifting capacity. A packbot gets rid of explosives. Robots that resemble dogs called "AlphaDogs" were created by Boston Dynamics to transport heavy goods for soldiers. It can carry 180 kg and travel 20 km. The fact that these robots are uncontrolled and can travel anywhere is incredible. They use computer vision to follow their boss automatically.

7.4 Healthcare Logistics: Research on swarm robotics for cancer treatment appears promising. Even as technology develops, there are treatments for most cancer types. The side effects of these medicines are still a big problem, however. The main problem is harm to healthy cells. To remedy this, sick cells need to be treated. Swarming nanorobots are used in certain treatments to destroy cancer cells without harming healthy cells. Build nanorobots with a specific focus on cancer cells. They ought to be able to find cancer cells by navigating the human body. Numerous therapies exist. Without injecting medication, the nanorobots can penetrate cancer cells and destroy them. Nanorobots can quickly destroy cancer cells when they swarm inside a malignant body.

Swarm robotics applications in logistics operations can be compared to assess their operational efficiency, strengths, and weaknesses. These assessments help determine each application's viability for certain logistical conditions, reflecting natural swarms' adaptability and variety. Swarm robotics application performance evaluation helps stakeholders choose the best ones for their logistical demands. This comparative method shows how collaborative, adaptive, and dynamic swarm robots are, helping optimise and modify solutions to varied logistical needs. It also deepens operational expertise, guiding logistics swarm robotics development and innovation.

8. IMPACT ON SUSTAINABILITY AND EFFICIENCY

8.1 Environmental Benefits

Swarm robotics integration in logistics operations represents a significant leap towards sustainable practices. By optimising delivery routes and task allocation among robotic units, these systems effectively reduce unnecessary mileage, resulting in a noticeable reduction in carbon emissions and a smaller ecological footprint. Integrating AI-driven optimisation further streamlines inventory management, minimising waste and promoting sustainability in supply chain practices. These advancements align with global efforts to address environmental concerns, positioning swarm robotics as pivotal contributors to eco-friendly logistics practices and creating more sustainable supply chains.

8.2 Reduction in Operational Costs

One of the most promising aspects of swarm robotics in logistics operations is its potential to reduce operational expenses significantly. Through the automation of processes and intelligent resource allocation, these systems minimise labour expenses and reduce error rates, leading to heightened operational efficiency. While initial investments are required, the long-term benefits, including increased efficiency and decreased costs, underscore the substantial return on investment offered by swarm robotics in logistics. This reduction in operational costs highlights the transformative impact of swarm robotics in optimising financial expenditures and resource utilisation within logistics, paving the way for more cost-effective and efficient operations.

8.3 Enhancement in Efficiency and Productivity

Swarm robotics emerges as a game-changer in enhancing efficiency and productivity within logistics operations. Guided by AI-driven insights and adaptive behaviours, these systems optimise processes, trim turnaround times, and enhance overall productivity. Leveraging collective intelligence and adaptive capabilities, swarm robotics systems usher in substantial improvements in operational efficiency, reshaping the logistics landscape into a more competitive and effective environment(Ghodsian et al., 2022). This heightened efficiency not only signifies a boost in productivity but also underscores the potential for swarm robotics to redefine the dynamics of logistics operations, making them more productive and adaptable to rapidly changing demands.

9. FUTURE PROSPECTS AND TRENDS

Integrating emerging technologies within swarm robotics heralds a new era of innovation in logistics operations. Cutting-edge technologies like edge computing, 5G connectivity, and advanced sensor capabilities bolster swarm behaviours, enabling more sophisticated and adaptable logistics operations(Mistry et al., 2020). These advancements hold immense potential to create more responsive and optimised logistics systems, laying the groundwork for transformative developments and addressing evolving logistical challenges. By harnessing these emerging technologies, swarm robotics is poised to revolutionise logistics operations, making them more dynamic, responsive, and efficient in addressing modern-day demands.

Swarm robotics promises many potential developments that can profoundly reshape logistics operations. Autonomous vehicles, predictive analytics, and AI-driven systems are among the transformative technologies poised to revolutionise warehouse management, delivery networks, and inventory optimisation. These developments aim to create more responsive, adaptable, and optimised logistics operations, thereby addressing dynamic market demands and indicating the transformative potential of swarm robotics within the supply chain.

Anticipated future applications of swarm robotics encompass a wide spectrum of innovations, from predictive maintenance to precision inventory management. These technologies are set to transform logistics operations by optimising efficiency, adapting to changing environments, and addressing future challenges. This predictive nature showcases the potential for swarm robotics to continuously evolve and adapt to emerging logistical needs, suggesting a promising future within the industry.

10. HUMAN-ROBOT COLLABORATION

The integration of swarm robotics necessitates a redefinition of the roles of human workers within logistics operations instead of the Industry 5.0 requirements. Instead of performing routine tasks, humans focus on overseeing and optimising swarm systems. Humans leverage their decision-making skills in this collaborative setup, while robots handle repetitive tasks. This shift emphasises the potential for human-robot collaboration, fostering an environment where the strengths of both entities are harnessed for optimised logistics operations.

Safety protocols and ethical AI use are fundamental considerations in fostering a harmonious human-robot collaborative environment within swarm robotics. Establishing stringent safety measures and addressing ethical considerations ensure the safe integration of these systems in logistics operations. By prioritising safety and ethical AI practices, swarm robotics aims to create a reliable and secure operational environment within logistics(Konishi et al., 2022).

In swarm robotics, collaboration between humans and robots aims to enhance operational efficiencies by leveraging human expertise and robotic precision. This collaboration leads to improved productivity and operational outcomes within logistics, underscoring the potential for a synergistic partnership between humans and robots to achieve optimised logistics operations.

11. DISCUSSION

The synthesis of implications, key findings, and future visions stemming from swarm robotics integration in logistics encapsulates its multifaceted impact and forward trajectories. Notably, implications encompass heightened operational efficiency, substantial cost savings, and advancements in sustainability practices. For instance, synthesising key findings might reveal remarkable reductions in processing times and inventory errors due to the adaptive capabilities of swarm robotics. This synthesis provides a comprehensive narrative, weaving together the transformative potential of swarm robotics in reshaping logistics paradigms. Envisioning the future entails a concerted effort to address challenges and harness opportunities. By synthesising these facets, stakeholders pave the way for innovative, adaptive, and resilient logistics ecosystems driven by the transformative capabilities of swarm robotics (Monostori et al., 2014).

The integration of swarm robotics in logistics has profound implications for industry and academia. It reshapes the landscape by revolutionising operational methodologies, leading to increased efficiency, reduced operational costs, and heightened competitiveness. For instance, implementing swarm robotics optimises warehousing, inventory management, and distribution, streamlining processes and yielding substantial cost savings. Furthermore, it fosters a shift towards sustainable practices, aligning with global initiatives for eco-friendly logistics. Academically, it prompts a surge in interdisciplinary research, necessitating collaboration between robotics, artificial intelligence, and logistics experts. This collaboration spurs innovation and opens avenues for refining algorithms, addressing safety protocols, and enhancing human-robot interactions. Academic institutions engaging in this symbiotic relationship with industry contribute to pioneering advancements, shaping the future of logistics through innovative swarm robotics applications (Keung et al., 2021).

Identifying and addressing research gaps in swarm robotics within logistics is fundamental for propelling advancements in the field. A recognised gap involves refining swarm robotics adaptability to dynamic environments within logistics settings. Addressing this gap involves research on enhancing robots' capabilities to autonomously adapt to changing conditions, such as sudden layout alterations in warehouses or unpredicted obstacles in delivery routes. Additionally, addressing ethical considerations in the decision-making processes of swarm robotics remains crucial. Research on creating transparent and ethical decision-making frameworks for autonomous robots fosters trust and acceptance within the logistics domain. By addressing these identified gaps, researchers pave the way for more robust, adaptable, and ethically sound swarm robotics applications in logistics.

12. CONCLUSION

Summarising the essential aspects of swarm robotics integration in logistics consolidates its significance and impact. Notably, the summary encompasses amplified operational efficiency, reduced costs, improved sustainability practices, and transformative advancements in logistics operations. For instance, summarising key points might highlight how swarm robotics' adaptive behaviours optimise order fulfilment, minimise errors, and enhance inventory management. Furthermore, it might underscore the challenges faced, such as scalability concerns or ethical considerations, offering a comprehensive overview of the multifaceted impact of swarm robotics on logistics. This summary serves as a concise yet comprehensive reference, encapsulating the pivotal role of swarm robotics in reshaping and optimising logistics processes.

Future research in swarm robotics within logistics offers numerous avenues for exploration and development. A crucial area involves refining swarm intelligence algorithms to enhance collective decision-making among robots in dynamically changing environments. For instance, advancing algorithms that enable swarm robots to collaborate seamlessly while adapting to unforeseen scenarios in warehouse settings is a promising direction. Additionally, investigating the integration of swarm robotics with edge computing technologies to facilitate real-time data processing for swift decision-making represents an innovative avenue. Moreover, exploring swarm robotics' integration with advanced materials science, such as developing lightweight and durable materials for robot construction, can potentially revolutionise their functionality. Such research directions promise to propel swarm robotics towards greater adaptability, intelligence, and efficacy within logistics.

Reflections on the evolutionary trajectory of swarm robotics in supply chain and logistics reveal a transformative journey reshaping the operational framework. This evolution signifies a fundamental shift in logistics operations from manual handling to deploying sophisticated autonomous systems. For instance, transitioning from traditional inventory management systems to swarm-powered smart inventory control reflects a paradigm shift towards precision and efficiency. This evolution underscores the ongoing quest for innovative solutions that optimise logistics processes. Recognising this evolution highlights the resilience and adaptability of swarm robotics, steering the logistics industry towards a future of enhanced productivity, reduced errors, and streamlined operations.

REFERENCES

Bakhshipour, M., Ghadi, M. J., & Namdari, F. (2017). Swarm robotics search & rescue: A novel artificial intelligence-inspired optimisation approach. *Applied Soft Computing*, *57*, 708–726. doi:10.1016/j.asoc.2017.02.028

Blais, M.-A., & Akhloufi, M. A. (2023). *Reinforcement learning for swarm robotics: An overview of applications, algorithms and simulators*. Cognitive Robotics.

de Almeida, J. P. L. S., Nakashima, R. T., Neves-Jr, F., & de Arruda, L. V. R. (2019). Bio-inspired online path planner for cooperative exploration of unknown environment by a Multi-Robot System. *Robotics and Autonomous Systems*, *112*, 32–48. doi:10.1016/j.robot.2018.11.005

Deiva Ganesh, A., & Kalpana, P. (2022). Future of artificial intelligence and its influence on supply chain risk management – A systematic review. *Computers & Industrial Engineering*, *169*, 108206. doi:10.1016/j.cie.2022.108206

Dulce-Galindo, J. A., Santos, M. A., Raffo, G. V., & Pena, P. N. (2022). Distributed supervisory control for multiple robot autonomous navigation performing single-robot tasks. *Mechatronics*, *86*, 102848. doi:10.1016/j.mechatronics.2022.102848

Emam, Y., Wilson, S., Hakenberg, M., Munz, U., & Egerstedt, M. (2020). A Receding Horizon Scheduling Approach for Search & Rescue Scenarios. *IFAC-PapersOnLine*, *53*(2), 3451–3456. doi:10.1016/j.ifacol.2020.12.1680

Ghodsian, N., Benfriha, K., Olabi, A., Gopinath, V., Arnou, A., Charrier, Q., & Helou, M. El. (2022). Toward designing an integration architecture for a mobile manipulator in production systems: Industry 4.0. *Procedia CIRP, 109*, 443–448. https://doi.org/https://doi.org/10.1016/j.procir.2022.05.276

Jevtić, A., Andina, D., Jaimes, A., Gomez, J., & Jamshidi, M. (2010). Unmanned aerial vehicle route optimisation using ant system algorithm. *2010 5th International Conference on System of Systems Engineering*, 1–6.

Keung, K. L., Lee, C. K. M., & Ji, P. (2021). Data-driven order correlation pattern and storage location assignment in robotic mobile fulfillment and process automation system. *Advanced Engineering Informatics*, *50*, 101369. doi:10.1016/j.aei.2021.101369

Khazetdinov, A., Aleksandrov, A., Zakiev, A., Magid, E., & Hsia, K. H. (2020). RFID-based warehouse management system prototyping using a heterogeneous team of robots. *Robots in Human Life*, 263.

Konishi, M., Sasaki, T., & Cai, K. (2022). Efficient Safe Control via Deep Reinforcement Learning and Supervisory Control – Case Study on Multi-Robot Warehouse Automation. *IFAC-PapersOnLine*, *55*(28), 16–21. doi:10.1016/j.ifacol.2022.10.318

Koot, M., Mes, M. R. K., & Iacob, M. E. (2021). A systematic literature review of supply chain decision making supported by the Internet of Things and Big Data Analytics. *Computers & Industrial Engineering*, *154*, 107076. doi:10.1016/j.cie.2020.107076

Kumar, A., & Jaiswal, A. (2020). A deep swarm-optimised model for leveraging industrial data analytics in cognitive manufacturing. *IEEE Transactions on Industrial Informatics*, *17*(4), 2938–2946. doi:10.1109/TII.2020.3005532

Lara, C. L., & Wassick, J. (2023). The future of supply chain - a perspective from the process and online retail industries. *Computers & Chemical Engineering*, *179*, 108401. doi:10.1016/j.compchemeng.2023.108401

Latsou, C., Farsi, M., & Erkoyuncu, J. A. (2023). Digital twin-enabled automated anomaly detection and bottleneck identification in complex manufacturing systems using a multi-agent approach. *Journal of Manufacturing Systems*, *67*, 242–264. doi:10.1016/j.jmsy.2023.02.008

Liu, Q., Wang, C., Li, X., & Gao, L. (2023). An improved genetic algorithm with modified critical path-based searching for integrated process planning and scheduling problem considering automated guided vehicle transportation task. *Journal of Manufacturing Systems*, *70*, 127–136. doi:10.1016/j.jmsy.2023.07.004

Michalos, G., Kousi, N., Makris, S., & Chryssolouris, G. (2016). Performance Assessment of Production Systems with Mobile Robots. *Procedia CIRP*, *41*, 195–200. doi:10.1016/j.procir.2015.12.097

Milner, E., Sooriyabandara, M., & Hauert, S. (2022). Stochastic behaviours for retrieval of storage items using simulated robot swarms. *Artificial Life and Robotics*, *27*(2), 264–271. doi:10.1007/s10015-022-00749-8

Mistry, I., Tanwar, S., Tyagi, S., & Kumar, N. (2020). Blockchain for 5G-enabled IoT for industrial automation: A systematic review, solutions, and challenges. *Mechanical Systems and Signal Processing*, *135*, 106382. doi:10.1016/j.ymssp.2019.106382

Monostori, L., Valckenaers, P., Dolgui, A., Panetto, H., Brdys, M., & Csáji, B. C. (2014). Cooperative Control in Production and Logistics. *IFAC Proceedings Volumes, 47*(3), 4246–4265. https://doi.org/https://doi.org/10.3182/20140824-6-ZA-1003.01026

Niihara, M., Saito, N., Yukawa, C., Toyoshima, K., Oda, T., Hirota, M., & Barolli, L. (2022). A Transportation Routing Method Based on A* Algorithm and Hill Climbing for Swarm Robots in WLAN Environment. *International Conference on Broadband and Wireless Computing, Communication and Applications*, 361–368.

Oliff, H., Liu, Y., Kumar, M., Williams, M., & Ryan, M. (2020). Reinforcement learning for facilitating human-robot-interaction in manufacturing. *Journal of Manufacturing Systems*, *56*, 326–340. doi:10.1016/j.jmsy.2020.06.018

Resende, C. H. L., Geraldes, C. A. S., & Lima, F. R. (2021). Decision Models for Supplier Selection in Industry 4.0 Era: A Systematic Literature Review. *Procedia Manufacturing*, *55*, 492–499. doi:10.1016/j.promfg.2021.10.067

Scholz, M., Zhang, X., Kreitlein, S., & Franke, J. (2018). Decentralised Intelligence: The Key for an Energy Efficient and Sustainable Intralogistics. *Procedia Manufacturing*, *21*, 679–685. doi:10.1016/j.promfg.2018.02.171

Trianni, V., IJsselmuiden, J., & Haken, R. (2016). *The saga concept: swarm robotics for agricultural applications*. Technical Report. 2016. Available online: http://laral. istc. cnr. it/saga…

Vásárhelyi, G., Virágh, C., Somorjai, G., Nepusz, T., Eiben, A. E., & Vicsek, T. (2018). Optimised flocking of autonomous drones in confined environments. *Science Robotics*, *3*(20), eaat3536. doi:10.1126/scirobotics.aat3536 PMID:33141727

Wior, I., Roth, P., & Fay, A. (2012). Stability Criterion for a Single Decision-Making Node in Decentralized Material Handling Systems. *IFAC Proceedings Volumes, 45*(24), 320–327. https://doi.org/https://doi.org/10.3182/20120912-3-BG-2031.00068

Yung, K. L., Tsang, Y. P., Wu, C. H., & Ip, W. H. (2023). An autonomous, multi-agent, IoT-empowered space logistics system for mission-critical inventory packing. *ISA Transactions*, *132*, 167–181. doi:10.1016/j.isatra.2022.10.035 PMID:36404152

Zhang, X., Xiong, G., Ai, Y., Liu, K., & Chen, L. (2023). Vehicle dynamic dispatching using curriculum-driven reinforcement learning. *Mechanical Systems and Signal Processing*, *204*, 110698. doi:10.1016/j.ymssp.2023.110698

Chapter 12
Modern-Day Healthcare With Cloud-Enhanced Robotics

E. Fantin Irudaya Raj
https://orcid.org/0000-0003-2051-3383
Dr. Sivanthi Aditanar College of Engineering, India

K. Manimala
https://orcid.org/0000-0003-1612-705X
Anna University, India

T. Lurthu Pushparaj
https://orcid.org/0000-0002-8884-6099
TDMNS College, India

M. Chithambara Thanu
Dr. Sivanthi Aditanar College of Engineering, India

ABSTRACT

Modern healthcare is undergoing a transformative evolution with the integration of cloud computing and robotics. This chapter delves into the convergence of these two technologies, defining and exploring the concept of cloud-enhanced robotics in healthcare. Through real-world case studies, the chapter highlights the wide array of applications for cloud-enhanced robotics, such as telemedicine, diagnostics, and data analytics, and how they are enhancing patient care. The benefits of cloud-enhanced robotics are manifold, including improved patient outcomes, enhanced precision, cost reduction, and increased access to expertise. This chapter explores both the advantages and challenges, offering insights into the potential future developments of cloud-enhanced robotics in healthcare.

1. INTRODUCTION

The field of healthcare has witnessed a remarkable evolution over the years, driven by technological advancements that continually redefine the way the present chapter delivers and receives medical care.

DOI: 10.4018/979-8-3693-1914-7.ch012

In the 21st century, two pivotal technologies, cloud computing and robotics, have emerged as transformative forces in healthcare (Sultan, 2014). Cloud computing has revolutionized data management and accessibility, while robotics has enhanced precision and efficiency in various medical tasks. However, it is the fusion of these technologies, often referred to as cloud-enhanced robotics, that holds the potential to reshape the healthcare landscape in unprecedented ways (Dang et al., 2019).

The present chapter will explore the intricate relationship between cloud computing and robotics in the context of modern healthcare. We define cloud-enhanced robotics as the synergy between cloud-based data processing and storage capabilities and robotic systems, resulting in enhanced medical applications. This convergence allows healthcare providers to leverage the collective power of remote, intelligent, and connected systems to provide more efficient, accurate, and accessible care to patients.

The primary purpose of this chapter is to provide a comprehensive overview of the role of cloud-enhanced robotics in healthcare. We will delve into the historical context of robotics in medicine, elucidate the basics of cloud computing and its relevance to healthcare, and discuss the convergence of these technologies. By analyzing the myriad applications of cloud-enhanced robotics in healthcare, we aim to highlight its benefits, challenges, and implications for the healthcare industry.

The current exploration begins by understanding the historical development of robotics in healthcare and assessing its current applications. We then introduce cloud computing, detailing its fundamental principles and demonstrating its relevance within the healthcare sector. By examining the benefits of cloud computing in healthcare, we set the stage for the integration of robotics and cloud technologies.

Furthermore, the chapter will showcase the diverse applications of cloud-enhanced robotics, ranging from telemedicine and diagnostics to drug discovery, patient care, and data analytics. These applications represent the bridge between robotic precision and cloud-powered data analysis, leading to improved patient care and healthcare delivery.

The advantages of this synergy are numerous and far-reaching. Enhanced patient outcomes, precision, and cost-effectiveness are just a few of the benefits. We will substantiate these advantages with real-world case studies that underscore the transformative potential of cloud-enhanced robotics. Nonetheless, this promising convergence also poses a unique set of challenges. Privacy and security concerns, regulatory compliance issues, ethical dilemmas, and integration complexities are among the obstacles that healthcare providers and policymakers need to navigate.

Looking to the future, we will provide insights into the emerging technologies and trends within the field of cloud-enhanced robotics. These include potential breakthroughs and challenges that lie ahead, as well as the global adoption of this technology in both developed and developing countries. Through case studies, we will showcase how cloud-enhanced robotics is transforming healthcare on a global scale. In conclusion, this report aims to offer a comprehensive understanding of cloud-enhanced robotics in modern healthcare. It will serve as a valuable resource for healthcare professionals, policymakers, and researchers to make informed decisions, address challenges, and seize opportunities in the ever-evolving landscape of healthcare. The transformative potential of cloud-enhanced robotics is poised to shape the future of healthcare delivery and patient outcomes, and we must navigate this exciting frontier with insight and foresight.

2. ROLE OF ROBOTICS IN MODERN HEALTHCARE

The integration of robotics into modern healthcare has heralded a new era of medical practice and patient care. Robotics is no longer confined to manufacturing and industrial sectors; it has found a pivotal role in healthcare, offering innovative solutions that enhance precision, efficiency, and accessibility (Kim et al., 2016). Figure 1 shows Modern-day Cloud Enhanced Robotics in the healthcare sector. In this section, we will explore the historical perspective of robotics in healthcare, highlight its current applications, and elucidate the evolution of healthcare robotics.

2.1 Historical Perspective

The history of robotics in healthcare dates to the early 20th century when rudimentary devices were used in surgical procedures (Narula et al., 2014). However, it was not until the latter half of the 20th century that significant advancements in technology paved the way for the development of more sophisticated medical robots. Notable milestones in the historical progression of healthcare robotics include:

- 1961: The introduction of the first robotic surgical system, the Arthrobot, which was designed to assist in orthopaedic surgery.
- 1985: The development of PUMA (Programmable Universal Machine for Assembly) robot for neurosurgical biopsies (Beecher, 1979).

Figure 1. Modern-day cloud-enhanced robotics in the healthcare sector

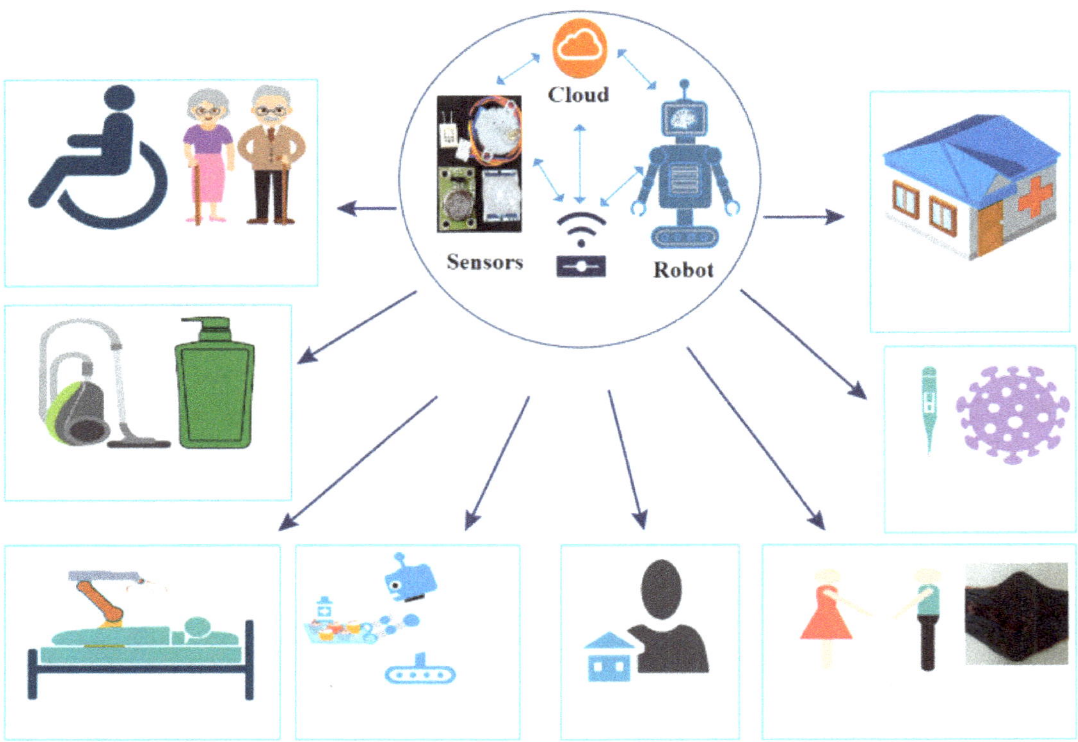

- 2000s: The emergence of da Vinci Surgical System, a widely recognized robotic surgical platform, which enables minimally invasive surgery with enhanced precision and reduced patient recovery time.

  These historical developments mark the gradual integration of robotics into healthcare, leading to the diverse applications that exist today.

2.2 Current Applications

The contemporary landscape of healthcare robotics is marked by a diverse array of applications that benefit patients, healthcare providers, and medical researchers. Some of the key applications include:

- Robotic Surgery: Perhaps one of the most prominent applications, robotic surgical systems like the da Vinci Surgical System assist surgeons in performing complex procedures with enhanced precision and minimally invasive techniques (Rivero-Moreno et al., 2023).
- Rehabilitation Robotics: Devices like exoskeletons and robotic arms aid patients in regaining motor functions after injuries or surgeries (Luo et al., 2023).
- Telemedicine Robots: Tele-presence robots are used to connect remote specialists with patients and healthcare professionals, bridging geographical gaps and providing expert consultations (Nawrat & Krawczyk, 2023).
- Pharmaceutical Robots: Automated systems assist in the handling and preparation of medications, improving the accuracy and efficiency of the pharmaceutical process (Borkar et al., 2023).
- Medical Imaging and Diagnostics: Robots are employed to enhance the accuracy and speed of medical imaging, enabling early disease detection and precise diagnostics (Gnanasekar et al., 2021; Krishna et al., 2021; Rani, Pushparaj, Raj et al, 2022).

2.3 The Evolution of Healthcare Robotics

The evolution of healthcare robotics has been characterized by a shift from simple, task-specific devices to more complex and adaptable systems. Key factors contributing to this evolution include (Bishop et al., 2023; Ch et al., 2021; Chouhan et al., 2021; Vallès-Peris & Domènech, 2023):

- Advancements in Sensor Technology: The integration of advanced sensors, such as cameras, ultrasound, and force sensors, has empowered robots with better perception and decision-making capabilities.
- Artificial Intelligence and Machine Learning: Robotics in healthcare increasingly rely on AI and machine learning algorithms for data analysis, patient monitoring, and decision support.
- Minimally Invasive Surgery: Robotic surgical systems have advanced from mere assistance to enabling fully autonomous surgeries in some cases.
- Interconnectivity: Modern healthcare robots are designed to connect with other devices, share data through cloud platforms, and facilitate remote monitoring and control.

The ongoing evolution of healthcare robotics is closely intertwined with the integration of cloud computing technologies, which allows these systems to access and process vast amounts of data for improved decision-making and remote control. This integration, known as cloud-enhanced robotics, represents a significant step forward in the field of healthcare and is the focal point of this report.

The following sections will delve into the principles of cloud computing, its role in healthcare, and how the fusion of cloud and robotics is shaping the present and future of healthcare delivery.

3. CLOUD COMPUTING IN HEALTHCARE

Cloud computing has become a transformative force in various industries, and healthcare is no exception. It offers the capacity to store, manage, and access data and applications remotely through the internet, revolutionizing the way healthcare organizations handle information, collaborate, and deliver services. In this section, we will explore the basics of cloud computing, its role within healthcare, and the numerous benefits it brings to the industry.

3.1 Cloud Computing Basics

Cloud computing is a technology that enables the delivery of computing services, including servers, storage, databases, networking, software, analytics, and more, over the internet. It relies on a network of remote servers hosted on the internet to store, manage, and process data rather than relying on local servers or personal computers. Key elements of cloud computing include (Logeshwaran et al., 2023; Neelakandan et al., 2022; Vellela et al., 2023):

- On-Demand Service: Users can access computing resources as needed, without the need to invest in or maintain infrastructure locally.
- Scalability: Cloud services are designed to scale up or down based on demand, ensuring efficient resource utilization.
- Resource Pooling: Cloud providers often pool resources from multiple users, allowing them to serve multiple customers efficiently.
- Broad Network Access: Cloud services can be accessed from a wide range of devices and locations with internet connectivity.
- Measured Service: Users only pay for the resources they consume, making it cost-effective.

3.2 Cloud-Based Healthcare Solutions

The adoption of cloud computing in healthcare has led to a wide range of cloud-based solutions that have transformed the industry. These solutions encompass various areas, including (Darwin & Raj, 2023; Mahajan et al., 2023; Manimala et al., 2023; Pushparaj et al., 2022; Raj et al., 2023):

- Electronic Health Records (EHRs): Cloud-based EHR systems store patient records, making them accessible to authorized healthcare providers from anywhere, enhancing patient care coordination.
- Telemedicine: Cloud computing plays a pivotal role in enabling telemedicine platforms, allowing remote consultations, data sharing, and real-time communication between patients and healthcare professionals.

- Health Information Exchange (HIE): Cloud-based HIE platforms facilitate the secure exchange of patient information between healthcare entities, improving the continuity of care.
- Medical Imaging: Cloud-based medical imaging solutions store and share images such as X-rays and MRIs, enabling quick access for diagnosis and collaboration among radiologists and other specialists.
- Data Analytics: Cloud platforms provide the computing power necessary for complex healthcare analytics, aiding in predictive modeling, research, and population health management.
- Healthcare Mobile Apps: Many healthcare mobile apps leverage cloud computing for data storage, allowing patients to access their health data, communicate with providers, and monitor their conditions.

3.3 Benefits of Cloud Computing in Healthcare

The integration of cloud computing in healthcare brings about numerous benefits (Agapito & Cannataro, 2023; Banimfreg, 2023; Chhabra, 2023; Gupta & Rizvi, 2023):

- Accessibility and Flexibility: Healthcare professionals can access patient data and applications from anywhere, facilitating remote care and telemedicine.
- Scalability: Cloud platforms can scale resources up or down based on demand, ensuring efficient resource utilization and cost savings.
- Data Security: Reputable cloud providers invest in robust security measures and compliance standards to protect sensitive patient information.
- Interoperability: Cloud-based systems can foster better data sharing and interoperability among different healthcare providers and systems.
- Cost Savings: Cloud computing eliminates the need for large upfront investments in physical infrastructure and allows for cost-effective pay-as-you-go pricing models.
- Data Backup and Recovery: Cloud services offer automated data backup and recovery options, ensuring data resilience and business continuity.
- Advanced Analytics: Healthcare organizations can leverage the computing power of the cloud to perform complex data analytics, improving decision-making and patient care.

The adoption of cloud computing in healthcare is a pivotal enabler of the integration of cloud-enhanced robotics, where robotics systems can leverage cloud resources to enhance their capabilities, access real-time data, and collaborate with other systems for improved patient care. This convergence is at the heart of modern healthcare transformation, and we will explore its implications further in this chapter.

4. CONVERGENCE OF ROBOTICS AND CLOUD COMPUTING

The convergence of robotics and cloud computing represents a paradigm shift in modern healthcare. This synergy, often referred to as cloud-enhanced robotics, has the potential to revolutionize the way healthcare is delivered, making it more efficient, precise, and accessible. In this section, we will define cloud-enhanced robotics, explore its advantages and synergies, and provide real-world case studies that illustrate its transformative power.

4.1 Defining Cloud-Enhanced Robotics

Cloud-enhanced robotics refers to the integration of cloud computing technologies with robotic systems, enabling robots to leverage cloud-based data processing, storage, and connectivity. In essence, it empowers robots to access and analyze vast amounts of data, collaborate with other systems, and perform complex tasks that were previously beyond their capabilities. The main elements of cloud-enhanced robotics include (Elfaki et al., 2023; Raj & Balaji, 2023; Udendhran et al., 2023):

- Data Access: Robots can access and share data with other devices and systems through the cloud, allowing them to make more informed decisions and respond to changing conditions.
- Remote Control: Cloud-enhanced robotics enables remote monitoring and control of robotic systems, making it possible for experts to manage robots from anywhere in the world.
- Scalability: Robots can scale their computing and storage resources on demand, ensuring optimal performance and adaptability.
- Interconnectivity: Robots can collaborate with other robots, healthcare providers, and cloud-based applications, creating a networked ecosystem that enhances their capabilities.

4.2 Advantages and Synergies

The convergence of robotics and cloud computing in healthcare presents several key advantages and synergies (Abdulrahman et al., 2023; Ness et al., 2023; Pande & Mishra, 2023):

- Enhanced Precision: Robots can access cloud-based data for real-time decision-making, leading to enhanced precision in tasks such as surgery, diagnostics, and drug dispensation.
- Efficiency: Cloud-enhanced robots can handle resource-intensive computations in the cloud, offloading the burden from their onboard computing resources, resulting in faster and more efficient operations.
- Cost Reduction: The scalability and cost-effectiveness of cloud resources can reduce the overall cost of robotics deployment, making it more accessible to healthcare providers.
- Access to Expertise: Remote control and collaboration via the cloud enable experts to assist with complex procedures, even from different geographical locations.
- Scalability: The ability to scale resources up or down as needed ensures that robotic systems can adapt to varying workloads.

4.3 Case Studies

Real-world examples demonstrate the transformative potential of cloud-enhanced robotics in healthcare:

- Da Vinci Surgical System: The da Vinci Surgical System, used in minimally invasive surgeries, integrates cloud connectivity for remote consultations and data sharing, allowing experienced surgeons to assist in complex procedures from remote locations (Intuitive, n.d.).
- Teladoc Health: Teladoc Health's telemedicine robots connect patients and healthcare providers through cloud-based platforms, facilitating remote consultations and telehealth services (Uscher-Pines & Mehrotra, 2014).

- Drug Discovery: Cloud-enhanced robotic systems are used in pharmaceutical research for high-throughput screening, data analysis, and simulations, accelerating drug discovery processes (Garg et al., 2011).
- Medical Imaging: Robotics systems in medical imaging leverage cloud-based AI and deep learning for real-time image analysis, improving the accuracy and speed of diagnostics (Alzubaidi et al., 2020).

These case studies illustrate how cloud-enhanced robotics in healthcare is already making a substantial impact by enhancing patient care, expanding access to medical expertise, and improving the overall efficiency of healthcare delivery. The integration of cloud computing with robotics creates a dynamic and agile healthcare ecosystem that promises to be a cornerstone of modern medical practice. The following sections of this report will delve deeper into the specific applications of cloud-enhanced robotics in healthcare and the benefits and challenges it presents.

5. APPLICATIONS OF CLOUD-ENHANCED ROBOTICS IN HEALTHCARE

The fusion of cloud computing and robotics, often referred to as cloud-enhanced robotics, is a game-changer in healthcare. It enables the development of innovative applications and solutions that enhance patient care, optimize healthcare workflows, and improve medical outcomes. In this section, we will delve into the diverse and transformative applications of cloud-enhanced robotics in the healthcare industry.

5.1 Remote Surgery and Telemedicine

One of the most prominent applications of cloud-enhanced robotics is in the domain of remote surgery and telemedicine. This technology enables surgeons to perform complex procedures from remote locations while leveraging cloud resources for real-time data processing and collaboration. Key aspects of this application include (Ahmad et al., 2023; Asiri et al., 2018):

- Teleoperation: Surgeons can remotely control robotic surgical systems with the help of high-speed internet connections and low-latency communication, allowing them to operate on patients located far away.
- Real-time Data: Cloud computing provides access to real-time patient data, medical images, and AI algorithms, enabling precise decision-making during surgery.
- Remote Expert Consultation: Surgeons can collaborate with experts located in different geographical regions, leading to improved patient outcomes.
- Telepresence: Telemedicine robots equipped with cloud capabilities facilitate remote consultations between patients and healthcare providers, expanding access to medical expertise.

5.2 Medical Imaging and Diagnostics

Cloud-enhanced robotics is playing a significant role in the field of medical imaging and diagnostics. Robots, often equipped with advanced imaging sensors, can analyze images in real time and share them with experts for rapid diagnosis. Key aspects of this application include (Mijwil et al., 2023; Raj et al., 2024; Rana & Bhushan, 2023):

- AI-Enhanced Imaging: Robots utilize cloud-based AI and deep learning algorithms to improve the accuracy of medical image analysis.
- Speed and Efficiency: Cloud resources allow robots to process and transmit medical images quickly, reducing the time needed for diagnosis.
- Teleconsultation: Cloud connectivity enables medical experts to remotely assess images and collaborate with on-site healthcare professionals.
- Remote Monitoring: Robots can be deployed to monitor patients' vital signs, injuries, or chronic conditions, transmitting data to healthcare providers and alerting them in case of emergencies.

5.3 Drug Discovery and Development

In pharmaceutical research and drug development, cloud-enhanced robotics offers valuable tools to accelerate the process of identifying new drugs and optimizing treatment protocols. Key aspects of this application include (Jayanthi & Reddy, 2023; Sadybekov & Katritch, 2023):

- High-Throughput Screening: Robots equipped with cloud-connected screening platforms can process many potential drug compounds quickly, aiding in the identification of promising candidates.
- Data Analysis: Cloud resources are used for in-depth data analysis, helping researchers make informed decisions about drug efficacy and safety.
- Simulation and Modeling: Cloud-enhanced robots can run simulations and predictive modelling to study drug interactions, mechanisms of action, and potential side effects.
- Collaborative Research: Researchers worldwide can collaborate in real time through cloud-based platforms, expediting the drug discovery process.

5.4 Patient Care and Rehabilitation

Cloud-enhanced robotics is also improving patient care and rehabilitation. Robots can provide personalized care, assistance, and therapy while utilizing cloud connectivity for data analysis and remote monitoring. Key aspects of this application include (Jamwal et al., 2023; Mohammed et al., 2023):

- Robotic Assistants: Cloud-connected robotic assistants can aid patients with mobility issues, helping them perform daily tasks and exercises.
- Physical Therapy: Robots can guide patients through rehabilitation exercises and monitor progress, providing valuable data to healthcare providers.
- Mental Health Support: Cloud-enhanced robots can assist in mental health therapy, offering emotional support, cognitive behavioral therapy, and patient monitoring.
- Remote Monitoring: Patient data collected by robots can be securely stored in the cloud, allowing healthcare providers to track progress and adjust treatment plans.

5.5 Data Management and Analytics

Cloud-enhanced robotics generates vast amounts of data, which can be securely stored, processed, and analyzed in the cloud. This data management and analytics application is integral to optimizing healthcare delivery and research. Key aspects include (Al-Jumaili et al., 2023; Mehta et al., 2023; Militano et al., 2023):

- Data Storage: The cloud serves as a secure repository for patient records, medical images, diagnostic data, and treatment history.
- Data Analytics: Cloud computing facilitates advanced data analytics, enabling healthcare providers to derive insights, improve decision-making, and identify trends in patient care.
- Machine Learning: Cloud resources support machine learning algorithms that can predict patient outcomes, optimize treatments, and personalize care plans.
- Research and Clinical Trials: Cloud-based platforms enable collaborative research, sharing of clinical trial data, and real-world evidence analysis.

The applications of cloud-enhanced robotics in healthcare are diverse and continue to expand. This convergence of technologies is revolutionizing healthcare by making medical services more accessible, precise, and efficient. It is essential to recognize that these applications come with benefits and challenges, which will be explored in the following sections of this chapter.

6. BENEFITS AND ADVANTAGES OF CLOUD-ENHANCED ROBOTICS IN HEALTHCARE

The integration of cloud computing with robotics in healthcare, known as cloud-enhanced robotics, offers a multitude of benefits and advantages. These benefits are transformative and have the potential to significantly enhance patient care, streamline healthcare workflows, and drive innovation in the medical field. In this section, we will explore the key advantages that cloud-enhanced robotics brings to the healthcare industry.

6.1 Improved Patient Outcomes (Jumani, Siddique, & Laghari, 2023; Sahoo & Choudhury, 2023)

- Enhanced Precision: Cloud-enhanced robotics allows for real-time access to vast amounts of medical data, enabling robots to make precise decisions during surgeries, diagnostics, and treatment, leading to improved patient outcomes.
- Personalized Care: By analyzing patient data and medical records stored in the cloud, robots can deliver highly personalized care plans and treatment regimens, optimizing results.
- Consistency: Robots are not subject to fatigue, ensuring that they consistently deliver the highest quality care and medical procedures, reducing the risk of human error.

6.2 Enhanced Efficiency (Ali et al., 2023; Darwin & Fantin Irudaya Raj, 2023; Rani, Pushparaj, & Raj, 2022)

- Time Savings: Cloud computing enables robots to process data more quickly and efficiently, resulting in reduced waiting times for diagnostics, consultations, and treatments.
- Workflow Optimization: Robots can streamline healthcare workflows by automating routine tasks, allowing human healthcare providers to focus on more complex and critical aspects of patient care.
- Resource Utilization: The scalability of cloud resources ensures that robots use computing and storage capabilities efficiently, reducing costs and improving resource utilization.

6.3 Cost Reduction

- Minimized Infrastructure Costs: Healthcare providers can reduce the need for expensive on-site infrastructure by leveraging cloud computing resources, resulting in cost savings.
- Economies of Scale: Cloud providers can offer cost-effective pay-as-you-go pricing models, ensuring that even smaller healthcare facilities can access advanced robotics solutions.
- Reduced Readmissions: Cloud-enhanced robotics can improve post-operative care, leading to fewer complications and readmissions, which can be costly for healthcare systems.

6.4 Access to Expertise

- Remote Expert Consultation: Healthcare providers can access expertise from specialists around the world in real-time, improving the quality of care and treatment options.
- Knowledge Sharing: Cloud-based platforms enable the sharing of medical knowledge, best practices, and treatment protocols among healthcare professionals.

6.5 Scalability

- Adaptability: The scalability of cloud resources ensures that robotic systems can adjust to varying workloads, accommodating a higher number of patients when necessary.
- Future-Proofing: Healthcare facilities can easily scale up their robotics systems as technology advances, ensuring their relevance and effectiveness in the long term.

6.6 Accessibility

- Rural and Remote Healthcare: Cloud-enhanced robotics can bridge geographical gaps, providing quality healthcare to patients in rural and remote areas through telemedicine and remote monitoring.
- Global Collaboration: Healthcare providers worldwide can collaborate and access the latest medical advances, regardless of their location, leading to a global network of healthcare expertise.

6.7 Data-Driven Decision-Making

- Advanced Analytics: Cloud computing enables healthcare providers to perform advanced analytics on vast datasets, leading to more informed and evidence-based decision-making.
- Predictive Healthcare: Machine learning algorithms in the cloud can predict patient outcomes, facilitating preventive care and early interventions.

These benefits and advantages highlight the transformative potential of cloud-enhanced robotics in healthcare. By integrating cloud computing technologies with robotics, the healthcare industry can provide more efficient, precise, and accessible care to patients while reducing costs and improving overall healthcare outcomes. However, this convergence also poses its unique set of challenges, which will be addressed in the following sections of this chapter.

7. CHALLENGES AND CONCERNS IN CLOUD-ENHANCED ROBOTICS IN HEALTHCARE

The integration of cloud computing with robotics in healthcare, while offering a wide range of benefits and advantages, also presents a set of unique challenges and concerns. It is crucial to recognize and address these issues to ensure the secure, ethical, and effective implementation of cloud-enhanced robotics in the healthcare industry. In this section, we will explore the key challenges and concerns associated with this convergence (Agapito & Cannataro, 2023; Hu et al., 2012; Jumani, Siddique, & Laghari, 2023; Pandey et al., 2023; Shakya, 2020).

7.1 Privacy and Security

- Data Privacy: Storing sensitive patient data in the cloud raises concerns about data privacy and the risk of unauthorized access, data breaches, or misuse of patient information.
- Data Encryption: Ensuring the secure transmission and storage of patient data in the cloud requires robust encryption and authentication mechanisms.
- Compliance: Healthcare organizations must navigate complex regulatory requirements, such as HIPAA in the United States, to ensure that patient data stored in the cloud complies with legal standards.

7.2 Regulatory Compliance

- Regulatory Complexity: The intersection of robotics, cloud computing, and healthcare introduces complex regulatory challenges that require close attention to compliance with local and international regulations.
- Standards and Certification: The lack of standardized practices for cloud-enhanced robotics may hinder widespread adoption, and organizations must seek certification and compliance with existing standards.

7.3 Ethical and Legal Issues

- Liability: Determining liability in the event of errors or malfunctions in cloud-enhanced robotic systems can be complex, raising legal and ethical questions.
- Informed Consent: Patients must be informed about the use of cloud-enhanced robotics in their healthcare, understanding how their data is being used and who has access to it.
- Data Ownership: The question of data ownership, particularly when patient data is shared through cloud platforms, needs to be addressed to ensure patients' rights and privacy.

7.4 Integration and Interoperability

- Complexity: Integrating cloud-enhanced robotics with existing healthcare systems can be complex and costly, potentially leading to interoperability issues.
- Data Integration: Ensuring seamless data integration between cloud systems and on-premises healthcare infrastructure is crucial for effective operation.
- Device Compatibility: Different robotic systems may have varying levels of compatibility with cloud platforms, which can pose integration challenges.

7.5 Cost and Accessibility

- Initial Investment: The upfront cost of adopting cloud-enhanced robotics may be prohibitive for some healthcare facilities, particularly in resource-constrained settings.
- Accessibility: Ensuring that cloud-enhanced robotics remains accessible to all patients, regardless of their socioeconomic status or location, is a concern.

7.6 Data Quality and Accuracy

- Data Reliability: The accuracy and reliability of data transmitted and processed through cloud platforms are critical for patient safety and diagnosis.
- Data Bias: Machine learning algorithms trained on biased or incomplete datasets can lead to inaccurate diagnoses and treatment recommendations.

7.7 Network Connectivity

- Reliability: The dependence on network connectivity for cloud-enhanced robotics requires reliable internet connections, which can be challenging in some areas.
- Latency: High latency in data transmission can impact the real-time operation of robotic systems, particularly in remote surgeries and telemedicine.

7.8 Data Overload

- Information Overload: Healthcare providers may face information overload when processing vast amounts of data generated by cloud-enhanced robotics, making it essential to develop efficient data management and analytics strategies.

7.9 Resistance to Adoption

- Change Management: Resistance to adopting new technologies, especially among healthcare professionals, can impede the successful integration of cloud-enhanced robotics.
- Training and Education: Preparing healthcare providers to effectively use cloud-enhanced robotic systems requires significant training and education, which can be a challenge for organizations.

Addressing these challenges and concerns requires a collaborative effort among healthcare providers, technology developers, policymakers, and regulatory bodies. By recognizing and mitigating these issues, the healthcare industry can harness the full potential of cloud-enhanced robotics while ensuring the safety, privacy, and ethical use of these technologies in patient care.

8. THE FUTURE OF CLOUD-ENHANCED ROBOTICS IN HEALTHCARE

The convergence of cloud computing and robotics, known as cloud-enhanced robotics, is poised to play an increasingly pivotal role in the healthcare industry. As technology continues to advance, this fusion holds significant promise for transforming healthcare delivery, enhancing patient care, and driving medical innovation. The present section will explore the emerging technologies, market trends, potential breakthroughs, and challenges that will shape the future of cloud-enhanced robotics in healthcare (Emil Selvan et al., 2023; Francy Irudaya Rani et al., 2023; Li & Millar-Bilbao, 2021; Molla-Hosseini et al., 2022; Priyadarsini et al., 2020).

8.1 Emerging Technologies

- 5G Connectivity: The deployment of 5G networks will provide ultra-low latency and high bandwidth, making real-time data transmission and remote robotics control even more reliable and responsive.
- Edge Computing: Combining cloud computing with edge computing will enable real-time data processing at the edge of the network, reducing latency and enhancing the capabilities of cloud-enhanced robotics.
- Blockchain Technology: Blockchain can enhance the security and integrity of patient data, ensuring data privacy and secure data sharing in cloud-enhanced robotics applications.
- Augmented Reality (AR) and Virtual Reality (VR): AR and VR technologies can facilitate remote surgeries, medical training, and patient education, expanding the applications of cloud-enhanced robotics.

- Quantum Computing: In the long term, quantum computing may significantly advance data processing and machine learning capabilities, further enhancing cloud-enhanced robotics.

8.2 Market Trends

- Rapid Adoption: Healthcare facilities worldwide are rapidly adopting cloud-enhanced robotics, driven by the need for improved patient care, increased efficiency, and access to medical expertise.
- Increased Investment: As the technology matures, there will be a surge in investment in cloud-enhanced robotics by both healthcare organizations and technology companies.
- Specialized Solutions: Cloud-enhanced robotic systems will become increasingly specialized, catering to specific medical domains such as surgery, diagnostics, and telemedicine.
- Integration with AI: Greater integration with artificial intelligence and machine learning will enable cloud-enhanced robotics to make data-driven decisions and predict patient outcomes more accurately.
- Remote Monitoring and Home Healthcare: Cloud-enhanced robotics will extend to home healthcare, allowing patients to receive remote monitoring and care in their own homes.

8.3 Potential Breakthroughs

- Fully Autonomous Robotic Surgery: Continued advancements in cloud-enhanced robotics may lead to fully autonomous robotic surgeries, where robots perform complex procedures with minimal human intervention.
- Global Collaboration: Cloud-enhanced robotics could foster global collaboration in healthcare, enabling cross-border consultations, research, and knowledge sharing.
- Precision Medicine: By combining genomics data with cloud-enhanced robotics, healthcare providers may offer increasingly precise and personalized treatment regimens.
- Pandemic Preparedness: The technology can enhance healthcare systems' preparedness for pandemics, enabling rapid response and remote care delivery during crises.

8.4 Challenges to Overcome

- Security and Privacy: Ongoing efforts are needed to address the security and privacy concerns associated with patient data stored in the cloud and transmitted via cloud-enhanced robotics.
- Regulatory Frameworks: The development of clear and adaptable regulatory frameworks is essential to ensure the safe and ethical use of cloud-enhanced robotics.
- Data Standards and Interoperability: Establishing data standards and ensuring interoperability among different robotic systems and cloud platforms are crucial for the seamless exchange of information.
- Affordability and Accessibility: Ensuring that cloud-enhanced robotics remains affordable and accessible to all healthcare facilities, including those in resource-constrained regions, is a challenge.

- Ethical Considerations: Ethical issues related to the use of cloud-enhanced robotics, such as patient consent, data ownership, and liability, will continue to require attention.

The future of cloud-enhanced robotics in healthcare is a landscape of innovation and opportunity. As technology continues to evolve and mature, the potential to enhance patient care, streamline healthcare operations, and drive medical advancements remains promising. To fully realize this potential, healthcare providers, technology developers, policymakers, and regulatory bodies need to collaborate, address challenges, and make informed decisions that prioritize patient well-being and the responsible use of these transformative technologies.

9. GLOBAL ADOPTION OF CLOUD-ENHANCED ROBOTICS IN HEALTHCARE

The adoption of cloud-enhanced robotics in healthcare is a global phenomenon, with healthcare organizations, research institutions, and technology companies worldwide recognizing its potential to improve patient care and drive innovation. In this section, we will examine the global adoption of cloud-enhanced robotics and present case studies from different regions to illustrate how this convergence is transforming healthcare on a global scale.

9.1 North America

Case Study: Da Vinci Surgical System (Intuitive, n.d.)

The United States, in particular, has seen widespread adoption of the da Vinci Surgical System, a cloud-enhanced robotic surgical platform. This system enables expert surgeons to remotely assist in surgeries across the country. It has been used in a variety of procedures, from minimally invasive surgeries to complex interventions, significantly improving patient outcomes.

9.2 Europe

Case Study: REMOTE-Health (Behar et al., 2020)

In Germany, the REMOTE-Health project uses cloud-enhanced robotics for telemedicine and home healthcare. Cloud-connected robots provide remote monitoring and support to patients in their homes, increasing access to healthcare services, particularly in rural areas. This initiative has demonstrated the potential of cloud-enhanced robotics to address healthcare disparities.

9.3 Asia

Case Study: The Telerobotic Ultrasound System in India (Raina et al., 2021)

India has witnessed the adoption of cloud-enhanced robotics for diagnostic purposes. The Telerobotic Ultrasound System allows rural healthcare providers to perform ultrasounds on patients while remotely connecting to specialists in urban centers for real-time consultation. This technology has proven valuable in improving the accuracy of diagnostics and reducing the need for patients to travel long distances for medical assessments.

9.4 Africa

Case Study: Telemedicine Initiatives in Kenya (Qin et al., 2013)

Several telemedicine initiatives in Kenya have leveraged cloud-enhanced robotics to provide healthcare services to remote and underserved regions. These initiatives use robots for teleconsultations and remote diagnostics, increasing access to healthcare for communities with limited resources and medical facilities.

9.5 South America

Case Study: Surgical Robotic Tele-Consultation in Brazil (Cordioli et al., 2023)

In Brazil, surgical robotic teleconsultation services have gained popularity. Cloud-connected surgical robots enable experienced surgeons from different regions to collaborate on complex procedures. This technology has reduced the need for patients to travel long distances to access specialized surgical care.

The global adoption of cloud-enhanced robotics in healthcare underscores the transformative impact of this convergence. It has facilitated the provision of medical care to underserved and remote regions, improved patient outcomes, and enhanced the overall efficiency of healthcare delivery. These case studies provide real-world examples of how cloud-enhanced robotics is making a difference in healthcare on a global scale.

10. CONCLUSION

The convergence of cloud computing and robotics, known as cloud-enhanced robotics, is reshaping the landscape of healthcare delivery, offering a transformative path toward more efficient, precise, and accessible patient care. As we conclude this report, cloud-enhanced robotics is not merely a technological trend but a revolution in the healthcare industry with far-reaching implications. By integrating cloud computing technologies with robotics, healthcare providers can leverage the following key advantages:

- Enhanced precision in medical procedures, diagnostics, and treatment.
- Improved efficiency and workflow optimization in healthcare delivery.
- Cost reduction through scalable and cost-effective cloud resources.
- Greater access to medical expertise through remote collaboration.
- A data-driven approach to decision-making and predictive healthcare.

However, the adoption of cloud-enhanced robotics in healthcare is not without its challenges and concerns. Addressing issues related to privacy, security, regulation, and ethical considerations is paramount to ensure the responsible and safe use of these technologies. Looking to the future, emerging technologies such as 5G, edge computing, blockchain, and quantum computing, alongside market trends like increased investment and specialized solutions, hold the promise of further expanding the applications of cloud-enhanced robotics in healthcare. Breakthroughs such as fully autonomous robotic surgery, global collaboration, precision medicine, and pandemic preparedness are on the horizon. In closing, the future of cloud-enhanced robotics in healthcare is a future of innovation, collaboration, and improved patient care. Healthcare providers, technology developers, policymakers, and regulatory bodies need to continue working together to address challenges, navigate ethical and legal complexities, and ensure that the potential of cloud-enhanced robotics is harnessed responsibly and ethically for the benefit of patients and the advancement of healthcare.

REFERENCES

Abdulrahman, Y., Arnautović, E., Parezanović, V., & Svetinovic, D. (2023). AI and Blockchain Synergy in Aerospace Engineering: An Impact Survey on Operational Efficiency and Technological Challenges. *IEEE Access : Practical Innovations, Open Solutions*, *11*, 87790–87804. doi:10.1109/ACCESS.2023.3305325

Agapito, G., & Cannataro, M. (2023). An Overview on the Challenges and Limitations Using Cloud Computing in Healthcare Corporations. *Big Data and Cognitive Computing*, *7*(2), 68. doi:10.3390/bdcc7020068

Ahmad, F., Wysocki, R. W., Fernandez, J. J., Cohen, M. S., & Simcock, X. C. (2023). Patient perspectives on telemedicine during the COVID-19 pandemic. *Hand (New York, N.Y.)*, *18*(3), 522–526. doi:10.1177/15589447211030692 PMID:34515550

Al-Jumaili, A. H. A., Muniyandi, R. C., Hasan, M. K., Paw, J. K. S., & Singh, M. J. (2023). Big Data Analytics Using Cloud Computing Based Frameworks for Power Management Systems: Status, Constraints, and Future Recommendations. *Sensors (Basel)*, *23*(6), 2952. doi:10.3390/s23062952 PMID:36991663

Ali, O., AlAhmad, A., & Kahtan, H. (2023). A review of advanced technologies available to improve the healthcare performance during COVID-19 pandemic. *Procedia Computer Science*, *217*, 205–216. doi:10.1016/j.procs.2022.12.216 PMID:36687286

Alzubaidi, L., Fadhel, M. A., Al-Shamma, O., Zhang, J., Santamaría, J., Duan, Y., & Oleiwi, R., S. (. (2020). Towards a better understanding of transfer learning for medical imaging: A case study. *Applied Sciences (Basel, Switzerland)*, *10*(13), 4523. doi:10.3390/app10134523

Asiri, A., AlBishi, S., AlMadani, W., ElMetwally, A., & Househ, M. (2018). The use of telemedicine in surgical care: A systematic review. *Acta Informatica Medica*, *26*(3), 201. doi:10.5455/aim.2018.26.201-206 PMID:30515013

Banimfreg, B. H. (2023). A comprehensive review and conceptual framework for cloud computing adoption in bioinformatics. *Healthcare Analytics*, *3*, 100190. doi:10.1016/j.health.2023.100190

Beecher, R. C. (1979). Puma: Programmable universal machine for assembly. In *Computer vision and sensor-based robots* (pp. 141–152). Springer US. doi:10.1007/978-1-4613-3027-1_8

Behar, J. A., Liu, C., Kotzen, K., Tsutsui, K., Corino, V. D., Singh, J., Pimentel, M. A. F., Warrick, P., Zaunseder, S., Andreotti, F., Sebag, D., Kopanitsa, G., McSharry, P. E., Karlen, W., Karmakar, C., & Clifford, G. D. (2020). Remote health diagnosis and monitoring in the time of COVID-19. *Physiological Measurement*, *41*(10), 10TR01. doi:10.1088/1361-6579/abba0a PMID:32947271

Bishop, A. J., Sheng, W., Carlson, B. W., & Jones, N. F. (2023). The evolution and rise of robotic health assistants: The new human-machine frontier of geriatric home care. *(Re) designing the Continuum of Care for Older Adults: The Future of Long-Term Care Settings*, 97-121.

Borkar, S., Ghutke, P., Patil, W., Joshi, S., & Sorte, S. (2023, April). A Review of Pick and Place Robots for the Pharmaceutical Industry. In *2023 11th International Conference on Emerging Trends in Engineering & Technology-Signal and Information Processing (ICETET-SIP)* (pp. 1-6). IEEE. 10.1109/ICETET-SIP58143.2023.10151652

Ch, G., Jana, S., Majji, S., Kuncha, P., & Tigadi, A. (2021). Diagnosis of COVID-19 using 3D CT scans and vaccination for COVID-19. *World Journal of Engineering*, *19*(2), 189–194. doi:10.1108/WJE-03-2021-0161

Chhabra, M. (2023). Implications of Cloud Computing for Health Care. In *Cloud-based Intelligent Informative Engineering for Society 5.0* (pp. 41–59). Chapman and Hall/CRC. doi:10.1201/9781003213895-3

Chouhan, A. S., Purohit, N., Annaiah, H., Saravanan, D., Raj, E. F. I., & David, D. D. S. (2021). A Real-Time Gesture Based Image Classification System with FPGAand Convolutional Neural Network. *International Journal of Modern Agriculture*, *10*(2), 2565–2576.

Cordioli, E., Giavina-Bianchi, M., Pedrotti, C. H. S., & Podgaec, S. (2023). Brazilian Medical Survey on Telemedicine since the onset of COVID-19. *Einstein (São Paulo), 21*, eAE0428.

Dang, L. M., Piran, M. J., Han, D., Min, K., & Moon, H. (2019). A survey on internet of things and cloud computing for healthcare. *Electronics (Basel)*, *8*(7), 768. doi:10.3390/electronics8070768

Darwin, S., & Fantin Irudaya Raj, E. (2023). The Emergence of Cyber-Physical System in the Context of Self-Powered Soft Robotics. *Self-Powered Cyber Physical Systems*, 57-90.

Darwin, S., & Raj, E. F. I. (2023). A detailed review on embedded based heartbeat monitoring systems. *Smart Embedded Systems and Applications*, 79.

Elfaki, A. O., Abduljabbar, M., Ali, L., Alnajjar, F., Mehiar, D. A., Marei, A. M., Alhmiedat, T., & Al-Jumaily, A. (2023). Revolutionizing Social Robotics: A Cloud-Based Framework for Enhancing the Intelligence and Autonomy of Social Robots. *Robotics (Basel, Switzerland)*, *12*(2), 48. doi:10.3390/robotics12020048

Emil Selvan, G. S. R., Jagadeesh Kannan, R., Sangapu, S. C., Shalini, S., Pandurangan, R., & Raj, E. F. I. (2023). Novel Development of Fuzzy Controller Based Multi-Agent System for Efficient Navigation of Autonomous Robots. *International Journal of Uncertainty, Fuzziness and Knowledge-based Systems*, *31*(Supp01), 17–30. doi:10.1142/S0218488523400020

Francy Irudaya Rani, E., Lurthu Pushparaj, T., & Fantin Irudaya Raj, E. (2023). Adopting a Fusion Approach for Optical Amplification. *Modeling and Optimization of Optical Communication Networks*, 11-34.

Garg, V., Arora, S., & Gupta, C. (2011). Cloud computing approaches to accelerate drug discovery value chain. *Combinatorial Chemistry & High Throughput Screening*, *14*(10), 861–871. doi:10.2174/138620711797537085 PMID:21843145

Gnanasekar, A. K., Deivakani, M., Bathala, N., Raj, E. F. I., & Ramakrishna, V. (2021). Novel low-noise CMOS bioamplifier for the characterization of neurodegenerative diseases. In GeNeDis 2020: Genetics and Neurodegenerative Diseases (pp. 221-226). Springer International Publishing. doi:10.1007/978-3-030-78787-5_27

Gupta, S. M., & Rizvi, S. W. A. (2023). Cloud Security for Healthcare Services. *Journal of Management and Service Science*, *3*(1), 1–9.

Hu, G., Tay, W. P., & Wen, Y. (2012). Cloud robotics: Architecture, challenges and applications. *IEEE Network*, *26*(3), 21–28. doi:10.1109/MNET.2012.6201212

Intuitive. (n.d.). https://www.intuitive.com/en-us/patients/da-vinci-robotic-surgery/about-the-systems

Jamwal, P. K., Niyetkaliyev, A., Hussain, S., Sharma, A., & Van Vliet, P. (2023). Utilizing the intelligence edge framework for robotic upper limb rehabilitation in home. *MethodsX*, *11*, 102312. doi:10.1016/j.mex.2023.102312 PMID:37593414

Jayanthi, A., & Reddy, B. E. (2023). An intelligent cloud robotics personalized medicine system by assessment of DNA and electronic health records using fuzzy-based progressive alignment sequence (F-Pas). *Soft Computing*, 1–11. doi:10.1007/s00500-023-08359-7

Jumani, A. K., Siddique, W. A., & Laghari, A. A. (2023). Cloud and Machine Learning Based Solutions for Healthcare and Prevention. In *Image Based Computing for Food and Health Analytics: Requirements, Challenges, Solutions and Practices: IBCFHA* (pp. 163–192). Springer International Publishing. doi:10.1007/978-3-031-22959-6_10

Kim, J., Gu, G. M., & Heo, P. (2016). Robotics for healthcare. *Biomedical Engineering: Frontier Research and Converging Technologies*, 489-509.

Krishna, B., Amuthavalli, G., & Stalin, DRaj, E. FSaravanan, D. (2021). Certain investigation of SARS-COVID-2-induced Kawasaki-like disease in Indian Youngsters. *Annals of the Romanian Society for Cell Biology*, 1167–1182.

Li, Z., & Millar-Bilbao, F. (2021, March). Characterising edge-cloud data transmission for patient-centric healthcare systems. In *2020 IEEE International Conference on E-health Networking, Application & Services (HEALTHCOM)* (pp. 1-6). IEEE. 10.1109/HEALTHCOM49281.2021.9398991

Logeshwaran, J., Ramesh, G., & Aravindarajan, V. (2023). A Secured Database Monitoring Method to Improve Data Backup and Recovery Operations in Cloud Computing. *BOHR International Journal of Computer Science*, *2*(1), 1–7. doi:10.54646/bijcs.019

Luo, S., Meng, Q., Li, S., & Yu, H. (2023). Research of intent recognition in rehabilitation robots: A systematic review. *Disability and Rehabilitation. Assistive Technology*, 1–12. doi:10.1080/17483107.2023.2170477 PMID:36695473

Mahajan, H. B., Rashid, A. S., Junnarkar, A. A., Uke, N., Deshpande, S. D., Futane, P. R., Alkhayyat, A., & Alhayani, B. (2023). Integration of Healthcare 4.0 and blockchain into secure cloud-based electronic health records systems. *Applied Nanoscience*, *13*(3), 2329–2342. doi:10.1007/s13204-021-02164-0 PMID:35136707

Manimala, K., Raj, E. F. I., & Jagatheswari, S. (2023). Advancements in Deep Learning Techniques for Analyzing Electronic Medical Records. In *Deep Learning for Healthcare Decision Making* (pp. 133–158). River Publishers. doi:10.1201/9781003373261-6

Mehta, K., Gaur, S., Maheshwari, S., Chugh, H., & Anibhushan Kumar, M. (2023, April). Big Data Analytics Cloud based Smart IoT Healthcare Network. In *2023 7th International Conference on Trends in Electronics and Informatics (ICOEI)* (pp. 437-443). IEEE.

Mijwil, M., Al-Mistarehi, A. H., Abotaleb, M., El-kenawy, E. S. M., Ibrahim, A., Abdelhamid, A. A., & Eid, M. M. (2023). From Pixels to Diagnoses: Deep Learning's Impact on Medical Image Processing-A Survey. *Wasit Journal of Computer and Mathematics Science*, *2*(3), 9–15. doi:10.31185/wjcms.178

Militano, L., Arteaga, A., Toffetti, G., & Mitton, N. (2023). The Cloud-to-Edge-to-IoT Continuum as an Enabler for Search and Rescue Operations. *Future Internet*, *15*(2), 55. doi:10.3390/fi15020055

Mohammed, M. A., Ibrahim, D. A., & Abdulkareem, K. H. (2023). Bio-inspired robotics enabled schemes in blockchain-fog-cloud assisted IoMT environment. *Journal of King Saud University. Computer and Information Sciences*, *35*(1), 1–12. doi:10.1016/j.jksuci.2021.11.009

Molla-Hosseini, M., Hosseini, M., Vossoughi, G., & Sharifzadeh, A. M. (2022). Robotics: A Key Driver of Industry 4.0. *Industry 4.0 Vision for Energy and Materials: Enabling Technologies and Case Studies*, 73-98.

Narula, A., Narula, N. K., Khanna, S., Narula, R., Narula, J., & Narula, A. (2014). Future prospects of artificial intelligence in robotics software, a healthcare perspective. *International Journal of Applied Engineering Research: IJAER*, *9*(22), 10271–10280.

Nawrat, Z., & Krawczyk, D. (2023). Robot-Based Medicine. Robots in Medicine: Mobile Robots Versus Mobile Decision, Necessity Versus Possibility and Future Challenges. *Mobile Robot: Motion Control and Path Planning*, 127-162.

Neelakandan, S., Beulah, J. R., Prathiba, L., Murthy, G. L. N., Irudaya Raj, E. F., & Arulkumar, N. (2022). Blockchain with deep learning-enabled secure healthcare data transmission and diagnostic model. *International Journal of Modeling, Simulation, and Scientific Computing*, *13*(04), 2241006. doi:10.1142/S1793962322410069

Ness, S., Shepherd, N. J., & Xuan, T. R. (2023). Synergy Between AI and Robotics: A Comprehensive Integration. *Asian Journal of Research in Computer Science*, *16*(4), 80–94. doi:10.9734/ajrcos/2023/v16i4372

Pande, A., & Mishra, D. (2023). The Synergy between a Humanoid Robot and Whisper: Bridging a Gap in Education. *Electronics (Basel)*, *12*(19), 3995. doi:10.3390/electronics12193995

Pandey, N. K., Kumar, K., Saini, G., & Mishra, A. K. (2023). Security issues and challenges in cloud of things-based applications for industrial automation. *Annals of Operations Research*, 1–20. doi:10.1007/s10479-023-05285-7 PMID:37361100

Priyadarsini, K., Raj, E. F. I., Begum, A. Y., & Shanmugasundaram, V. (2020). Comparing DevOps procedures from the context of a systems engineer. *Materials Today: Proceedings*. Advance online publication. doi:10.1016/j.matpr.2020.09.624

Pushparaj, T. L., Rani, E. F. I., Raj, E. F. I., & Appadurai, M. (2022). Machine-based drug design to inhibit SARS-CoV-2 virus. *Smart Health Technologies for the COVID-19 Pandemic. Internet of Medical Things Perspectives*, *42*, 295.

Qin, R., Dzombak, R., Amin, R., & Mehta, K. (2013). Reliability of a telemedicine system designed for rural Kenya. *Journal of Primary Care & Community Health*, *4*(3), 177–181. doi:10.1177/2150131912461797 PMID:23799704

Raina, D., Singh, H., Saha, S. K., Arora, C., Agarwal, A., Chandrashekhara, S. H., ... Nandi, S. (2021, November). Comprehensive telerobotic ultrasound system for abdominal imaging: Development and in-vivo feasibility study. In *2021 International Symposium on Medical Robotics (ISMR)* (pp. 1-7). IEEE. 10.1109/ISMR48346.2021.9661578

Raj, E. F. I., Appadurai, M., Thiyaharajan, D., & Pushparaj, T. L. (2024). State-Of-The-Art Technologies for Crop Health Monitoring in the Modern Precision Agriculture. In *Precision Agriculture for Sustainability* (pp. 21–39). Apple Academic Press.

Raj, E. F. I., & Balaji, M. (2023). Shape Feature Extraction Techniques for Computer Vision Applications. In *Smart Computer Vision* (pp. 81–102). Springer International Publishing. doi:10.1007/978-3-031-20541-5_4

Raj, E. F. I., Manimala, K., & Appadurai, M. (2023). Differential Protection Scheme along with Backup Blockchain System. In *Smart and Power Grid Systems–Design Challenges and Paradigms* (pp. 135–156). River Publishers. doi:10.1201/9781003339557-6

Rana, M., & Bhushan, M. (2023). Machine learning and deep learning approach for medical image analysis: Diagnosis to detection. *Multimedia Tools and Applications*, *82*(17), 26731–26769. doi:10.1007/s11042-022-14305-w PMID:36588765

Rani, E. F. I., Pushparaj, T. L., & Raj, E. F. I. (2022). Escalating the resolution of an urban aerial image via novel shadow amputation algorithm. *Earth Science Informatics*, *15*(2), 905–913. doi:10.1007/s12145-022-00772-0

Rani, E. F. I., Pushparaj, T. L., Raj, E. F. I., & Appadurai, M. (2022). New approaches in machine-based image analysis for medical oncology. In *Machine Learning and Deep Learning Techniques for Medical Science* (pp. 333–359). CRC Press. doi:10.1201/9781003217497-17

Rivero-Moreno, Y., Echevarria, S., Vidal-Valderrama, C., Stefano-Pianetti, L., Cordova-Guilarte, J., Navarro-Gonzalez, J., ... Avila, G. L. D. (2023). Robotic surgery: A comprehensive review of the literature and current trends. *Cureus*, *15*(7). Advance online publication. doi:10.7759/cureus.42370 PMID:37621804

Sadybekov, A. V., & Katritch, V. (2023). Computational approaches streamlining drug discovery. *Nature*, *616*(7958), 673–685. doi:10.1038/s41586-023-05905-z PMID:37100941

Sahoo, S. K., & Choudhury, B. B. (2023). Challenges and opportunities for enhanced patient care with mobile robots in healthcare. *Journal of Mechatronics and Artificial Intelligence in Engineering*, *4*(2), 83–103. doi:10.21595/jmai.2023.23410

Shakya, D. S. (2020). Survey on cloud based robotics architecture, challenges and applications. *Journal of Ubiquitous Computing and Communication Technologies*, *2*(1), 10–18. doi:10.36548/jucct.2020.1.002

Sultan, N. (2014). Making use of cloud computing for healthcare provision: Opportunities and challenges. *International Journal of Information Management*, *34*(2), 177–184. doi:10.1016/j.ijinfomgt.2013.12.011

Udendhran, R., Yamini, G., Badrinath, N., Jegathesh Amalraj, J., & Suresh, A. (2023). Enhancing representational learning for cloud robotic vision through explainable fuzzy convolutional autoencoder framework. *Soft Computing*, 1–11. doi:10.1007/s00500-023-08570-6

Uscher-Pines, L., & Mehrotra, A. (2014). Analysis of Teladoc use seems to indicate expanded access to care for patients without prior connection to a provider. *Health Affairs*, *33*(2), 258–264. doi:10.1377/hlthaff.2013.0989 PMID:24493769

Vallès-Peris, N., & Domènech, M. (2023). Caring in the in-between: A proposal to introduce responsible AI and robotics to healthcare. *AI & Society*, *38*(4), 1685–1695. doi:10.1007/s00146-021-01330-w

Vellela, S. S., Reddy, B. V., Chaitanya, K. K., & Rao, M. V. (2023, January). An Integrated Approach to Improve E-Healthcare System using Dynamic Cloud Computing Platform. In *2023 5th International Conference on Smart Systems and Inventive Technology (ICSSIT)* (pp. 776-782). IEEE. 10.1109/ICSSIT55814.2023.10060945

Chapter 13
Cloud-Enhanced Robotics in Healthcare

A. Masooda
Sahyadri College of Engineering and Management, India

C. Harinakshi
https://orcid.org/0009-0003-0321-766X
Sahyadri College of Engineering and Management, India

G. Suchetha
Sahyadri College of Engineering and Management, India

ABSTRACT

Cloud-enhanced robotics is reforming the healthcare sector through adventures. Cloud-enhanced healthcare robots aimed to provide potential support to medical professionals by accurately providing services like nursing, diagnosing, and executing critical surgeries. In the domain of healthcare, cloud robotics technology facilitates real-time data exchange, enabling professionals to deliver optimal services and care. Tasks ranging from laboratory testing to medication distribution can now be automated, and cloud-enhanced robotics have reached a level of proficiency that significantly lowers the risk of human faults in critical surgeries. This chapter explores the architecture and components of cloud-enhanced healthcare robots, provides various applications in the field, assesses the advantages and disadvantages of healthcare robots, and scrutinizes the security implications associated with cloud-enhanced robots.

INTRODUCTION

In recent decades, significant technological advancements have enabled non-human entities and machines to undertake various tasks. This capability holds the potential to alleviate healthcare provider costs and burdens while enhancing the health and well-being of numerous individuals. Additionally, access to unlimited healthcare services and support systems may not always be readily available. Addressing these issues, ongoing discussions are centered on the potential utilization of diverse information and com-

DOI: 10.4018/979-8-3693-1914-7.ch013

munication cloud technologies, along with robots. These technological aids aim to empower patients, facilitating them in adopting and maintaining healthier lifestyles. The objective is to equip the health sector with tools that contribute to the public, thereby mitigating the burden on healthcare services.

The convergence of cloud computing and robotics has sparked a revolution in healthcare, paving the way for advanced medical interventions, enhanced patient care, and improved operational efficiency. Cloud-enhanced robotics in healthcare represents a synergy between cloud-based computing resources and robotic systems, leveraging the strengths of both to overcome traditional limitations in medical practice. At its core, cloud-enhanced robotics in healthcare harnesses the power of interconnected data processing and storage capabilities offered by cloud platforms, coupled with the precision and versatility of robotic technologies. This amalgamation enables healthcare providers to access vast computational resources, real-time data analysis, and remote collaboration tools, all integrated seamlessly with robotic devices deployed in various medical settings. One of the key advantages of cloud-enhanced robotics is its ability to facilitate remote operation and monitoring of robotic systems. Through cloud connectivity, healthcare professionals can remotely control robotic devices, perform intricate procedures, and monitor patient conditions from virtually any location with internet access. This capability is particularly valuable in scenarios where specialists may not be physically present or where immediate intervention is required in critical situations.

Furthermore, cloud-enhanced robotics in healthcare enables the aggregation and analysis of large volumes of medical data in real time. By leveraging cloud-based analytics tools, robotic systems can process complex datasets derived from patient records, medical imaging, and sensor data to inform decision-making processes. This data-driven approach enhances diagnostic accuracy, treatment planning, and personalized care delivery, ultimately improving patient outcomes. Another significant aspect of cloud-enhanced robotics in healthcare is its potential to facilitate collaborative healthcare practices. Through cloud-based platforms, medical professionals can share expertise, consult on cases, and collaborate on surgical procedures in real time, irrespective of geographical barriers. This collaborative framework fosters interdisciplinary teamwork, knowledge exchange, and continuous learning within the healthcare ecosystem. Moreover, cloud-enhanced robotics holds promise for augmenting healthcare delivery in resource-constrained environments. By centralizing computational resources and expertise in the cloud, healthcare facilities with limited infrastructure can access state-of-the-art robotic technologies and specialized medical services on demand, thereby expanding access to quality care for underserved populations.

Despite its transformative potential, the widespread adoption of cloud-enhanced robotics in healthcare presents various challenges, including concerns related to data security, privacy, regulatory compliance, and interoperability. Addressing these challenges will require close collaboration between technology developers, healthcare providers, regulatory bodies, and other stakeholders to establish robust standards and frameworks for safe and ethical implementation. Overall, we can say that cloud-enhanced robotics represents a paradigm shift in healthcare delivery, empowering clinicians with advanced tools and capabilities to enhance patient care, optimize clinical workflows, and drive innovation in medical practice. As the field continues to evolve, the integration of cloud computing and robotics holds the promise of revolutionizing healthcare delivery and improving health outcomes on a global scale.

Here, the chapter is organized into different sections background gives a review of the health care robots using the cloud, cloud computing and robotics, architecture of cloud-enhanced robots, introducing SLAM in cloud robots, advantages and applications of cloud-enhanced robots, security issues of cloud robotics and conclusion and future scope.

1. BACKGROUND

The cloud enhanced robotics plays a vital role in the healthcare domain. It explores the variety of available solutions and those still in various stages of research and prototyping. Carefully selected examples and references to ongoing research accompany each set of identified robotics applications in the present era. The work of different authors is described in this section.

The cloud-enhanced healthcare robots delve into the key factors contributing to the success of these applications, along with the primary challenges surrounding their integration into everyday use. The author describes how cloud-based platforms can enhance the capabilities of robotic devices in medical imaging, surgery, rehabilitation, and telemedicine, while also addressing challenges such as data security, latency, and regulatory compliance. (Yang, G., Pang, Z., Deen, M. J., Dong, M., Zhang, Y. T., Lovell, N., & Rahmani, A. M, 2020).

In the cloud robotic healthcare system, considerations such as sustainability in non-technical environments, patient/user safety, acceptance, ethical concerns (including patient/user privacy), and cost-effectiveness are thoroughly examined. Here the aim is to assess how robotics can partially address existing gaps in current healthcare provisions and highlight a couple of emerging developments and promising applications in the field, specifically healthcare robots designed for patient care and response, which play pivotal roles in the future (Ran Zhang, Yan Zhou, Junkai Zhang & Juan Zhao. 2023).

With the surge in the popularity of "Cloud Computing," there has been a growing fascination with extending similar principles to the realm of robotics and autonomous systems. Robotic devices, constrained by limitations in onboard processing, storage, and battery capacities, often grapple with challenges in performing complex computations or storing extensive knowledge bases. The solution lies in leveraging dedicated server hardware that demands less computing power, memory, and consequently, battery capacity on the robot itself. This application of cloud computing principles to robotics is commonly known as "Cloud Robotics." (Dahl, T. S., & Kamel Boulos, M. N. 2013).

Recent advancements in cloud robotics have witnessed significant contributions, particularly in offloading computations to large-scale computing infrastructure, as well as in the areas of storing and sharing knowledge, coordination, and collective learning among robots. The landscape has seen the development of various platforms and frameworks, with substantial research emanating from the open-source community. These endeavors are paving the way for enhanced capabilities and efficiencies in robotic systems, marking a promising intersection of cloud computing robotics technologies (Saha, O., & Dasgupta, P. 2018).

As the global population ages, reaching two billion people aged 60 and over by 2050, there is a growing need to address the challenges associated with aging and therefore there is a need of cloud enhance robots in aged care, aiming to identify the robotic technologies used and the problem areas researched by robotic researchers. The review encompasses information technology, remote monitoring, online daily services, and the rapid development of robotic technologies, highlighting their potential role in assisting older adults. The findings aim to provide practitioners with insights into recent research and identify gaps for future studies in this field (Shishehgar, M., Kerr, D., & Blake, J., 2018).

The COVID-19 pandemic has catalyzed the development of service robots in healthcare, addressing challenges in infection prevention, logistics, and telehealth. These robots prove advantageous by reducing infection spread, minimizing human error, and enabling front-line staff to focus on crucial tasks. The cloud enhanced robotic technologies in healthcare, highlighting collaborations between social and the healthcare industry. It explores how robots benefit patients, healthcare workers, and organizations during the pandemic (Firouzi, F., Farahani, B., Daneshmand, M., Grise, K., Song, J., Saracco, R., & Luo, A., 2021).

2. CLOUD COMPUTING AND ROBOTICS-AS-A-SERVICE (RAAS)

Cloud computing offers elastic, on-demand resources, catering to user needs precisely when required. In healthcare robotics, cloud computing yields numerous benefits. Robotic devices no longer bear the burden of intricate onboard computational tasks, such as simultaneous localization and mapping data, object recognition, pattern matching, and grasping. Instead, these demanding computations can be seamlessly offloaded onto robust cloud servers. The advantages of cloud computing extend to providing on-demand large storage, a robust network, and substantial computational power, ensuring real-time data processing. Cloud services grant healthcare robots access to big data applications, facilitating tasks like simultaneous localization and mapping for remote and real-time manipulation in robotics.

Beyond these advantages, offloading tasks to the cloud fosters knowledge sharing among robots, enabling them to learn from each other's actions. Notably, the significant reduction in the robot's battery power consumption, achieved by offloading complex computational tasks to cloud servers, stands out as a pivotal advantage. Furthermore, cloud services contribute to enhancing the efficiency of healthcare robots by expanding storage, processing capacity, and memory. These services also encompass algorithms that support the management of robotic fleets, including task assignment, supervision, and traffic control. Offloading healthcare data to the cloud for processing, analysis, storage, and maintenance not only reduces the robots' workload but also ensures energy efficiency. In essence, cloud computing emerges as a transformative force, optimizing the capabilities and operational efficiency of healthcare robots Elfaki, A. O., Abduljabbar, M., Ali, L., Alnajjar, F., Mehiar, D. A., Marei, A. M., ... & Al-Jumaily, A. (2023).

A cloud robotic system is devised to implement the Robot-as-a-Service (RaaS) paradigm, integrating a service robot with smart agents that harness the cloud's potential to enhance system capabilities and, consequently, service performance. This groundbreaking approach aims to advance the state of the art in cloud robotics by designing on-demand cloud services where the connected robot and smart environments collaborate to deliver assistive location-based services. The system encompasses autonomous robots, smart environments, and a cloud platform, ensuring seamless fulfillment of user-required services. The autonomous robot demonstrates its proficiency by employing speech recognition through a wearable microphone on users to identify keywords associated with predefined service requests. Upon recognition, the robot seamlessly retrieves all pertinent information from the proposed cloud system to reach the healthcare user and execute the requested service (Shishehgar, M., Kerr, D., & Blake, J., 2018). The ad-hoc services and technologies are meticulously crafted to harness the cloud's capabilities in the remote domain, bringing assistive robotics closer to cost-effective deployments while adhering to vital service requirements like dependability and acceptability.

The RaaS design surpasses traditional server applications for several reasons:

- The cloud's elasticity enables the allocation of increased hardware resources (storage and processing) seamlessly as the number of connected agents and required services grows.
- The cloud's resource redundancy ensures high availability and greater fault tolerance compared to classical server approaches.

- The cloud can adeptly manage a vast number of simultaneous connections from smart agents for data collection and processing, facilitating significant capabilities such as big-data processing and executing learning algorithms in the realms of service robotics and assistive robotics.

RaaS presents noteworthy advantages in healthcare services:

- Patients can confidently reside in their homes, maintaining autonomy, independence, and quality of life within a safe and secure context.
- The sick can remain active participants in community life, reducing feelings of loneliness and negativity.
- Patients can retain control over their lives even when requiring care and assistance.
- RaaS enhances doctors' and caregivers' focus on patients' health, ensuring a heightened level of attention and care.

3. THE ARCHITECTURE OF A CLOUD: ENHANCED HEALTHCARE ROBOTS

The robotics that is networked is visualized as a state of transformation from the robots that are programmed to robots enabled by the clouds. The aim of developing cloud robotics is to offload its highly complex computations to the cloud paradigm and minimize the load forced over each robot, This architecture mainly consists of two sections, or even can be split into two layers bottom layer and top layer, where the bottom layer holds different types of robots with virtual machines(VM) in it and the top layer is the cloud platform with its supporting components such as the proxy servers, massive spatial database, high-performance servers, and other related components. (Holland, J., Kingston, L., McCarthy, C., Armstrong, E., O'Dwyer, P., Merz, F., & McConnell, M., 2021). Where the robot-to-robot communication is framed simply in an ad-hoc manner and the robot-to-cloud is framed in a centralized manner. Remote robotics brain with cloud service: In cloud-enhanced robotics the main aim is to provide health service as and when it is required. The service is provided either locally or remotely. Robots with enough memory and computation capability are the main sources to provide the service. Since some types of robots like mobile robots or unmanned robots have limited capacity may be affected badly while processing remotely which may lead le to offloading the complex task to external computers or nodes. Now the use of cloud computing comes into the picture. It provides real-time services in parallel like video services, image building, tracking of sensor information, performing database computations, predicting and evaluating the performances, and many more. Hence the robotics brain and the cloud are playing a major role in remote processing. Here different cloud providers are providing services such as Google Compute Engine, Microsoft Azure, Amazon Web Service, Elastic Compute Cloud, etc. Along with this, research is going on under high-performance computing which provides more convenient parallel services which will be helpful for cloud healthcare robotic services (Elfaki, A. O., Abduljabbar, M., Ali, L., Alnajjar, F., Mehiar, D. A., Marei, A. M., ... & Al-Jumaily, A.)2023.

3.1 SLAM (Simultaneous Localization and Mapping): The Cloud robots use the SLAM technology for autonomous vehicles, which uses prior knowledge to build a map of the current environment. Because of SLAM the robots act as the local system and adapt to that environment. There are different types of SLAM like data-intensive or computation-intensive etc. For example, vison-based SLAM and cooperative SLAM work both data-intensive and computation-intensive. Using SLAM there are many tasks can be executed in parallel, the tasks are map fusion, filtering for state estimation, continuous data collection, and mapping, and finally, these tasks will be transferred to the cloud for further processing.

Now cloud computing will start its processing, it shares the data, processes information, and increases the activities by choosing different machine learning methods. For example, collective learning will be able to capture the information of mobile robots which will be placed remotely where the health patients are presented. The different robots from different remote areas are communicating to the central cloud for further processing. The robots are designed in such a way that they will post control policies, sensor information of physical traits of an environment, trajectories, motion coordination, tracking data and learning data, and update localization for collective learning. One practical scenario is that the Kiva automated materials handling system for multiple mobile environments will use hundreds of mobile platforms to provide better logistic information (Saha, O., & Dasgupta, P. 2018).

The cloud architecture provides on-demand assets for transferring information between the nodes, robots, and humans. The different cloud services are Software as a Service, Infrastructure as a Service, Platform as a Service and an additional service can be included as Robot as a Service. The Robot as a Service (R-a-a-s) is a cloud service technology that works for service-oriented computing mainly utilized for robot communications.

When we summarize the architecture of cloud robotics it enables the collaboration between robotic systems and cloud services, fostering a data-driven approach to healthcare that prioritizes patient well-being, real-time monitoring, and adaptive treatment plans. Finally, whenever it is required to design a cloud-based healthcare robotic system the following components are considered to be needed.

- **Robotic System:**
- **Sensors:** Collect patient health data, environmental information, and other relevant metrics.
- **Actuators:** Executes physical actions based on commands received from the cloud.
- **Local Processing Unit:** Performs edge computing for time-sensitive tasks and preliminary data analysis.
- **Communication Module:**
- **Wireless Communication:** Enables secure communication between the robotic system and the cloud infrastructure.
- **Communication Protocols:** Utilizes standard protocols (e.g., MQTT, HTTPS) to transmit data securely.
- **Edge Computing:**
- **Data Preprocessing:** Analyzes local data for immediate insights and reduces the volume of data sent to the cloud.
- **Real-time Decision Making:** Executes algorithms locally to make immediate decisions, enhancing responsiveness.
- **Cloud Infrastructure:**
- **Compute Resources:** Virtual machines or containers for processing intensive computational tasks.
- **Storage Services:** Persistent storage for storing historical patient data, treatment plans, and analytics results.
- **Networking Infrastructure:** Manages data flow between different components of the cloud infrastructure.
- **Cloud Services:**

- **Compute Services:** Executes algorithms, applications, and analytics in the cloud.
- **Storage Services:** Centralized storage for securely managing patient records and healthcare data.
- **Database Services:** Manages structured and unstructured healthcare data for efficient retrieval.
- **Analytics Services:** Processes large datasets for insights into patient health and treatment outcomes.
- **Machine Learning Services:** Trains and deploys machine learning models for predictive analytics and personalized treatment plans.
- **Data Processing and Analysis:**
- **Cloud-Based Analytics:** Utilizes cloud resources to perform in-depth analysis and generate actionable insights.
- **Machine Learning Models:** Continuously updated models for predicting patient outcomes, identifying trends, and improving treatment plans.
- **Remote Monitoring and Control:**
- **User Interface (UI):** Web or mobile interfaces for healthcare professionals to monitor patient data and control robotic actions.
- **Remote Control API:** Allows operators to intervene and adjust robotic tasks remotely based on real-time information.
- **Security Measures:**
- **Encryption:** Ensures secure transmission of sensitive healthcare data between the robotic system and the cloud.
- **Access Controls:** Regulates user access to patient records and other confidential information.
- **Authentication Mechanisms:** Ensures the identity of authorized users and devices.
- **APIs and Integration:**
- **Application Programming Interfaces (APIs):** Facilitates seamless integration between the robotic system and cloud services.
- **Interoperability:** Allows for integration with existing healthcare systems, electronic health records (EHRs), and external APIs.
- **Scalability and Flexibility:**
- **Auto-Scaling:** Dynamically adjusts cloud resources to handle varying workloads efficiently.
- **Flexible Deployment Models:** Supports deployment in diverse healthcare settings, such as hospitals, clinics, and home environments.

4. ADVANTAGES OF CLOUD-ENHANCED HEALTHCARE ROBOTICS

The intersection of cloud computing and robotics has ushered in a new era in the healthcare domain, propelling the sector into unprecedented realms of efficiency, accessibility, and innovation. Cloud-enhanced healthcare robots represent a pivotal fusion of two transformative technologies, creating a synergistic relationship that holds immense promise for the future of patient care. As these intelligent machines leverage the capabilities of cloud computing, they bring forth a myriad of advantages that redefine the landscape of healthcare delivery. In this era of rapid technological evolution, the advantages of cloud-enhanced healthcare robots are poised to address longstanding challenges in healthcare, offering solutions that are not only sophisticated but also adaptable to the dynamic nature of medical practices. From improved diagnostics and treatment precision to revolutionizing patient engagement, these robots stand as harbingers of a healthcare revolution that transcends traditional boundaries. This exploration delves into the manifold advantages of cloud-enhanced healthcare robots, shedding light on their capacity to enhance medical processes, optimize resource utilization, and foster a more patient-centric approach. The integration of cloud technology not only amplifies the computational power of these robots but also facilitates seamless communication, data sharing, and real-time analytics, thereby elevating the overall quality of healthcare services. As we navigate through the following sections, we will uncover the transformative potential of cloud-enhanced healthcare robots and how their advantages resonate across diverse facets of the healthcare ecosystem, laying the groundwork for a future where technology becomes an indispensable ally in the pursuit of improved patient outcomes and healthcare excellence (Bonaccorsi, M., Fiorini, L., Cavallo, F., Saffiotti, A., & Dario, P. 2016).

4.1 Remote Surgery using Surgical Robots: Surgical robots also known as collaborative robots are sophisticated robots used to assist surgeons or medical experts to perform surgeries. Cloud-enhanced surgical robots in remote surgery can offer several advantages, leveraging the capabilities of cloud computing to enhance the overall surgical experience. For example, the Da-Vinci robot is specially designed for automatic surgery and with the use of the cloud, it can be done remotely (Shishehgar, M., Kerr, D., & Blake, J., 2018).

Here are some key advantages:

1. **Remote Accessibility and Real-Time Collaboration:** Cloud-enhanced surgical robots enable surgeons to perform procedures from remote locations. This is particularly valuable in situations where specialized expertise is needed but not locally available. Cloud connectivity allows multiple surgeons to collaborate in real time on a surgical procedure. This collaborative approach can facilitate knowledge sharing, training, and decision-making among experts, regardless of their physical location.
2. **Resource Optimization and Reduced Latency:** By offloading computing and processing tasks to the cloud, surgical robots can benefit from powerful computing resources. This can enhance the processing speed and efficiency of the robot, leading to improved surgical outcomes. Cloud-enhanced surgical robots can leverage high-speed internet connections to reduce latency during remote surgeries. Low latency is critical for ensuring that the surgeon's inputs are transmitted to the robot and vice versa without significant delays.
3. **Data Storage and Analysis:** Cloud-based storage allows for the secure storage and retrieval of patient data, medical images, and surgical records. This centralized approach facilitates efficient data management, analysis, and long-term patient care planning.
4. **Enhanced Imaging and Visualization:** Cloud computing enables the use of advanced imaging technologies, such as 3D imaging and virtual reality, providing surgeons with detailed and real-time visualizations during remote surgeries. This can enhance precision and decision-making.
5. **Security and Compliance and Cost Efficiency:** Cloud service providers invest heavily in data security measures. This can enhance the security of patient data, ensuring compliance with healthcare regulations and protecting sensitive medical information. Cloud computing offers scalability, allowing healthcare providers to scale their infrastructure based on demand. This can result in cost savings as institutions can avoid the upfront investment in maintaining and upgrading extensive local infrastructure.
6. **Training and Skill Development and Remote Monitoring and Support:** Cloud-connected surgical robots can provide a platform for simulation-based training for surgeons. This allows practitioners

to enhance their skills and proficiency in a controlled virtual environment. Surgeons and healthcare teams can remotely monitor patients postoperatively, ensuring timely intervention in case of complications. This can lead to improved patient outcomes and reduced hospital readmissions.

4.2 Alzheimer's disease and home residing patient monitoring: The advantage of cloud-enhanced health robots in the context of Alzheimer's disease can offer innovative solutions to improve patient care, monitoring, and support for both individuals with Alzheimer's and their caregivers. Healthcare robots work better for Alzheimer's disease patients monitoring. It felt that robots were better suited to taking patients' vital signs than humans. Results of the study illustrated that the participants believed robots could make better diagnoses when monitoring patients' daily vitals, as well as provide assurance to people who have Alzheimer's disease concerning the time and place. Furthermore, the participants reported that robots could help nursing home residents maintain a sense of independence not only by providing reassurance and monitoring but also by alerting staff if help is needed or assistance with medical assessments is required (Kyriazakos, S., Prasad, R., Mihovska, A., Pnevmatikakis, A., op den Akker, H., Hermens, H., & Atanasovski, V., 2017). For instance, there are several ways in which cloud-enhanced health robots can be taken as an advantage in the management of Alzheimer's disease:

- **Remote Monitoring and Assistance:** it is used for Real-Time Monitoring where Cloud-connected robots can monitor patients in real-time, collecting data on their activities, movements, and vital signs. This information can be sent to healthcare providers or family members for remote monitoring. Also used in Medication Reminders, the Robots equipped with cloud connectivity can remind patients to take their medications on time, ensuring medication adherence and improving treatment outcomes.
- **Cognitive Stimulation:** Health robots can engage individuals with Alzheimer's in interactive cognitive activities, such as games, puzzles, and memory exercises. These activities can be personalized based on the individual's cognitive abilities and preferences. Robots can facilitate social interaction by connecting patients with their loved ones through video calls or virtual interactions, reducing feelings of isolation.
- **Caregiver Support:** Cloud-connected robots can provide caregivers with training and guidance on effective caregiving techniques. This includes tips on communication strategies, managing challenging behaviors, and understanding the progression of Alzheimer's disease. Caregivers can remotely access the robot's functionalities to check on their loved ones, receive alerts for any unusual behavior, and provide assistance or reassurance through the robot.
- **Data Analytics and Personalized Care:** The data collected by health robots can be stored and analyzed in the cloud. Machine learning algorithms can identify patterns, detect changes in behavior, and provide insights into the progression of Alzheimer's disease. Based on the analyzed data, personalized care plans can be developed, adapting to the individual's evolving needs and optimizing the care provided.
- **Fall Detection and Emergency Response:** Robots equipped with sensors can detect falls or unusual movements and send alerts to caregivers or emergency services. In case of emergencies, such as a fall or sudden health deterioration, the robot can initiate emergency response procedures, including contacting healthcare professionals or family members.
- **Integration with Electronic Health Records (EHR):** Integration with EHR systems allows for seamless information flow between the robot, healthcare providers, and other stakeholders in-

volved in the care of individuals with Alzheimer's disease. This ensures that everyone involved has access to relevant and up-to-date information.

4.3 Telehealth cloud robots to report old age people's health conditions: As many countries are facing the challenges of an aging population, there is a need for remote monitoring and management of patients with deteriorating health, as well as declining mental and physical capabilities. For instance, the monitoring of daily vital measurements is still taken with handheld instruments. This is not only time-consuming, but it can also expose healthcare workers to infection. Moreover, monitoring can provide reassurance to patients especially in case of emergency, alerting staff if there has been an accident such as a fall. It could also be of great advantage if used on a larger scale such as scanning a multitude of people's temperatures and checking to see if masks are on correctly. Telehealth robots are useful for this kind of monitoring. The term telehealth has been defined as the use of medical information that is exchanged from one site to another through cloud communication to improve a patient's health (Dahl, T. S., & Kamel Boulos, M. N., 2013). This is especially important during the pandemic because telehealth robots can be used for social interaction as well as to reduce the risk of frontline workers contracting.

4.4 Personal skill development in children. Children are also vulnerable to some kind of physical and mental disorders. They may be undergoing some depression and stress they can't bear. A possible solution to school that encourages activity could be a healthcare and socially assistive robot, helping to develop health conditions and communication skills. Here the robots are used to test the children's capability by conducting some outdoor games. The games target different skills such as social, emotional understanding, perspective-talking, ordering, and sequencing, etc. Now the robots can understand the difficulties children face through the different modeling developed in robots. The cloud robots are modeled to gaze behaviors like eye contact and sharing attention during the sessions, facial expressions, etc. Finally, the medical supervisors will collect the information remotely from the cloud robot through the use of the cloud and provide solutions for the child suffering from health disorders. This work can be demonstrated in the education field also (Shishehgar, M., Kerr, D., & Blake, J., 2018).

4.5 Cloud robotics in social health and connectivity: SoftBank cloud-based robotics can be developed for interaction and entertainment purposes so that isolated or mentally depressed people can recover through the interactions. Pepper is a robot that is capable of voice interaction and mood recognition and has been used as a social interacting robot for dementia patients. Pepper was programmed to carry out two 30-minute programs designed for light physical activity and social activities such as singing, it can speak around 15 languages. It was found that 45% of patients were happier after their social activity. Another robot designed to mitigate isolation is Vector, a miniature robot specially designed for home care. Vector robots provide social connectivity, share daily activities such as lunch and dinner, and take part in enjoyable activities like dancing. It can also answer questions, take photos, act as a timer, and play card games. Vector has also been used as a personal assistant providing functional support which uses cloud services and stores the day-to-day activity details and provides and when needed by the user. Vector uses a HD camera to identify people, and navigate throughout its space while carrying out obstacle avoidance. It also has a four-microphone array for directional hearing, as well as touch sensors and an accelerometer so it is aware when it is being moved or touched.

5. APPLICATIONS OF CLOUD-ENHANCED HEALTHCARE ROBOTS

In recent years, the convergence of healthcare and cutting-edge technologies has given rise to transformative advancements in patient care and medical services. Among these innovations, cloud-enhanced healthcare robots have emerged as a promising frontier, revolutionizing the way healthcare is delivered and managed. By integrating cloud computing capabilities with robotic systems, a new era of efficient, scalable, and patient-centric healthcare solutions has dawned. Cloud-enhanced healthcare robots leverage the power of cloud computing to enhance their capabilities, enabling seamless communication, data sharing, and advanced analytics. These robots play a pivotal role in various aspects of the healthcare ecosystem, from assisting healthcare professionals in performing complex surgeries to providing personalized and remote patient care. This integration of robotics and cloud technology holds immense potential for addressing challenges such as the shortage of healthcare professionals, improving accessibility to medical services, and enhancing overall healthcare efficiency.

This chapter explores the multifaceted applications of cloud-enhanced healthcare robots, examining how these intelligent machines are transforming diagnostics, treatment, and patient engagement. From telemedicine applications that bridge geographical gaps to surgical robots that leverage real-time data analytics for precision procedures, the possibilities are vast. As we delve into the diverse realms of healthcare where these robots operate, we uncover a future where technology collaborates with human expertise to create a more connected, accessible, and responsive healthcare landscape.

5.1 Automatic Speech Recognition. It is considered that Automatic Speech Recognition (ASR) enhances interaction with patients. Most of the time the medical practitioners or doctors may not be capable enough to capture the conversation of patients. ASR robots could be a way to add more convenient conversational capabilities to the system, add a dictation mode to the timeline, and possibly improve real-time emotion estimation for an even more empathic companion. It uses solutions based on openly available resources like software tools and pre-trained acoustic models turned out to achieve the necessary recognition quality for such a project. Cloud-based solutions played a major role in storing the conversation data and providing protection to it.

5.2 Cloud Robots in Sterilization of Hospitals and public monitoring during the Pandemic: Some infections in hospitals are due to unclean surgical rooms. Despite rigorous protocols, data from hospitals have shown that fatal infections are increasing, even with more efficient cleaning practices. In fact, despite improved cleaning interventions, research suggests up to 30% of surfaces still contain biofilms even after cleaning. These statistics propose that the current course of action is not enough to protect susceptible patients from serious, life-threatening infections, such as COVID-19. Indeed, COVID-19 cannot only spread via close contact, but also through unclean surfaces; it can persist on inanimate surfaces—metal, glass, or plastic—for days. Therefore, surfaces and objects need to be sterilized; sterilization kills all forms of microbial life on an object or surface such as Disinfecting High Contact Points that is Hydrogen Peroxide to control the spread of pathogens in hospitals and Sanitizing Floor and Wall Surfaces (Firouzi, F., Farahani, B., Daneshmand, M., Grise, K., Song, J., Saracco, R., & Luo, A., 2021). Remotely monitoring and testing public places for social distance adherence. Testing Since the outbreak of COVID-19, there has been an unprecedented demand for testing (Holland, J., Kingston, L., McCarthy, C., Armstrong, E., O'Dwyer, P., Merz, F., & McConnell, M., 2021).. Cloud robots play an important role in remote testing so that the spread of infection will reduced insightfully.

5.3 Application in medical logistics Delivery: The concept of "cloud-enhanced health robots in medical logistics delivery" involves the integration of robotic systems with cloud computing technologies to enhance the efficiency and effectiveness of medical logistics, particularly in the delivery of healthcare-related materials, equipment, and supplies (Holland, J., Kingston, L., McCarthy, C., Armstrong, E., O'Dwyer, P., Merz, F., & McConnell, M., 2021). This convergence of robotics and cloud computing can bring several benefits to the healthcare industry:

- **Efficient Logistics Management:** Cloud computing enables real-time data storage and processing, allowing health robots to access up-to-date information about inventory, demand, and delivery schedules. Robotics can be employed to automate the sorting, packaging, and distribution of medical supplies, reducing the time and resources required for these tasks.
- **Optimized Route Planning:** Cloud-based platforms can analyze data from various sources, including traffic patterns, weather conditions, and delivery schedules, to optimize the routes of health robots for efficient and timely deliveries.
- **Inventory Tracking and Management:** Cloud-based systems can provide accurate and real-time information about the availability of medical supplies, helping healthcare organizations manage their inventory more effectively. RFID (Radio-Frequency Identification) or other tracking technologies integrated with health robots can further enhance inventory visibility.
- **Remote Monitoring and Control:** Cloud computing allows for remote monitoring and control of health robots, enabling operators to oversee their activities, address issues, and make adjustments from anywhere with an internet connection.
- **Scalability and Flexibility:** Cloud computing provides scalability, allowing healthcare organizations to scale up or down their robotic logistics operations based on changing demand. Updates and improvements to the robot's software or logistics algorithms can be easily deployed through cloud-based systems.
- **Enhanced Connectivity:** Health robots can leverage cloud connectivity to communicate with other devices, sensors, and systems within the healthcare ecosystem, fostering a more interconnected and interoperable healthcare infrastructure.

The integration of cloud computing with health robots in medical logistics delivery can lead to a more streamlined, responsive, and data-driven healthcare supply chain, ultimately contributing to improved patient care and operational efficiency.

5.4 Testing and Sorting Blood Samples: Cloud-enhanced health robotics are increasingly finding applications in various facets of healthcare, and one particularly critical area is the testing and sorting of blood samples. This integration of cloud computing and robotics brings forth a range of benefits that significantly enhance the efficiency, accuracy, and speed of laboratory processes (Holland, J., Kingston, L., McCarthy, C., Armstrong, E., O'Dwyer, P., Merz, F., & McConnell, M., 2021). Here are some key applications:

- **Automated Sample Processing:** Cloud-enhanced health robots can automate the entire process of testing and sorting blood samples. From sample collection to preparation and analysis, robots equipped with cloud connectivity can follow predefined protocols and seamlessly communicate with cloud-based databases for real-time information retrieval.
- **High-throughput Testing:** The cloud's computing power enables health robots to conduct high-throughput testing, allowing for the simultaneous analysis of a large number of blood samples.

This capability is particularly valuable in scenarios where quick and accurate diagnostics are crucial, such as during disease outbreaks or in emergencies.

- **Data Integration and Analysis:** Cloud-based systems enable health robots to integrate with centralized databases that store vast amounts of patient data and medical information. By leveraging cloud computing, these robots can cross-reference and analyze patient history, contributing to more comprehensive diagnostics and personalized treatment plans.
- **Remote Monitoring and Control:** Cloud connectivity enables remote monitoring and control of health robots engaged in blood sample testing. Healthcare professionals can access real-time data, monitor progress, and even intervene if necessary, irrespective of their physical location. This capability enhances flexibility and ensures continuous operation.
- **Quality Control and Standardization:** Cloud-enhanced health robots can access and adhere to standardized testing protocols stored in the cloud. This ensures consistency and accuracy in the testing process, reducing the likelihood of human errors and variations in results. Quality control measures can be implemented and updated centrally, maintaining high standards across different healthcare facilities.
- **Resource Optimization:** The cloud allows for the centralization of computational resources, reducing the need for extensive local infrastructure in each healthcare facility. This optimization can lead to cost savings and increased accessibility to advanced testing capabilities, especially in regions with limited resources.
- **Real-time Reporting and Alerts:** Cloud-connected health robots can generate real-time reports of blood test results. Additionally, they can be programmed to send alerts for critical findings, enabling healthcare professionals to respond swiftly to urgent situations and initiate appropriate interventions.

The application of cloud-enhanced health robotics in testing and sorting blood samples exemplifies how this technology can elevate the standards of diagnostics, streamline laboratory workflows, and contribute to more effective patient care. As the healthcare industry continues to embrace these innovations, the potential for improved accuracy, efficiency, and accessibility in diagnostic processes becomes increasingly evident.

5.5 Non-Surgical Complementary Osteoarthritis Treatment: Osteoarthritis, a widespread joint disease that engenders considerable disability among a large proportion of older adults in all parts of the world is a major costly social and fiscal public health concern. While largely incurable, it is shown that a variety of carefully designed conservative noninvasive interventions may, however, influence the disease process quite favorably. However, whether certain advances in technology and less emphasis on non-pharmacologic and non-surgical health programs and others can be obviated by non-human methods is not well clarified. This mini-review primarily strove to uncover any favorable advances in this regard in the context of cloud robotics that may independently or collectively address the needs of older adults with osteoarthritis who are poor surgical or drug candidates. According to databases, an extensive scan covering current discussions and research on osteoarthritis, robots, and robotics shows a sizeable percentage of older adults who suffer from multiple adverse osteoarthritis health complications including musculoskeletal, cognitive, macro and micro tissue disturbances, depression, and possible neurological and inflammatory manifestations even if they are receiving standard care, may receive some benefit from carefully selected robotically oriented support approaches (Marks R., 2023). Ethical challenges, as well as patient preferences, and know-how, overall efficacy in the face of this complex disease remain to be uncovered. A current overview of painful disabling osteoarthritis, and the possible use of various robotic devices to enable its home-based management in later life, while largely exploratory, leads us to conclude:

- Osteoarthritis remains a widespread and incurable highly disabling chronic health condition.
- While not all older adults suffering from osteoarthritis may desire to receive some form of support from one or more robots, home care robots may help reduce associated physical and emotional correlates and a worse-than-desired outcome and prognosis all things considered, while enhancing self–care opportunities.
- Applied insightfully they may relieve some degree of distress, as well as career burden and costs.
- Those who are highly disabled, but want to remain in their own homes might be especially assisted.
- The need for long-term care services and possible surgery may be reduced.
- More extensive research, both qualitative and quantitative in this sphere is strongly indicated and may prove insightful with far-reaching benefits.
- Allied health workers should not neglect opportunities to learn about the potential of robotics and their multidimensional properties and assistive and ergonomic attributes and be able to assess and educate their older clients with osteoarthritis ably and cogently in this regard.

5.6 Homecare Robotic Systems for Healthcare: Homecare robotic systems represent a transformative wave in healthcare, aiming to enhance the quality of care, promote independence, and facilitate better health outcomes for individuals in the comfort of their homes. These advanced robotic technologies are designed to assist patients, particularly those with chronic conditions or mobility limitations, by providing personalized and timely care. Here are some key aspects and applications of homecare robotic systems in healthcare:

- **Assistance with Activities of Daily Living (ADLs):** Homecare robots are equipped to assist individuals with routine activities such as dressing, bathing, grooming, and meal preparation. These robots are designed to provide physical support and assistance, promoting independence and easing the burden on caregivers.
- **Medication Management:** Robotic systems in-home healthcare can assist in medication management by dispensing medications at scheduled times, providing medication reminders, and even monitoring medication adherence. This feature is particularly beneficial for individuals with complex medication regimens.
- **Remote Patient Monitoring:** Homecare robots are integrated with sensors and monitoring devices to keep track of vital signs, activity levels, and other health metrics. This real-time data can be transmitted to healthcare professionals, allowing for remote monitoring and timely intervention if any abnormalities are detected.
- **Telemedicine and Virtual Consultations:** These robots enable remote consultations with healthcare providers through video conferencing and telemedicine platforms. Patients can connect with doctors, nurses, or specialists without leaving their homes, facilitating regular check-ups and reducing the need for frequent hospital visits.
- **Fall Detection and Emergency Response:** Many homecare robots are equipped with fall detection systems. In the event of a fall or emergency, these robots can trigger alerts to caregivers or emergency services, ensuring a prompt response. This is especially crucial for elderly individuals who may be prone to falls.

- **Companionship and Mental Health Support:** Homecare robots often feature conversational interfaces and companionship functionalities. They can engage in conversations, play games, or provide entertainment, addressing issues of loneliness and promoting mental well-being, particularly for individuals who live alone.
- **Rehabilitation Support:** For individuals undergoing rehabilitation at home, robotic systems can assist in guiding exercises, monitoring progress, and providing feedback. This can be especially valuable for patients recovering from surgeries or managing chronic conditions.
- **Environmental Assistance:** Homecare robots can be integrated with smart home technologies to control lights, thermostats, and other home devices. This enhances accessibility for individuals with mobility challenges, allowing them to control their environment more independently.
- **Data Security and Privacy Measures:** Given the sensitive nature of healthcare data, homecare robotic systems prioritize robust security measures to protect patient information. Encrypted communication channels and strict privacy protocols ensure the confidentiality of healthcare data.

The integration of robotic systems into home healthcare not only addresses the challenges of an aging population but also opens up new possibilities for personalized and efficient healthcare delivery. As technology continues to advance, the potential for homecare robotic systems to play a central role in the future of healthcare remains both exciting and promising.

5.7 Affective Human-Robot-Interaction: The human-robot interaction (HRI) has made a legend in research and innovations over the past few decades. It plays a major work in cross-disciplinary fields of psychology, behavior science, and cognitive science. Cloud-enhanced robotics in affective human-robot interaction refers to the integration of cloud computing technologies with robotic systems to improve emotional intelligence and responsiveness in human-robot interactions. This combination opens up various applications that can enhance the overall user experience and performance of robots in understanding and responding to human emotions. It includes both the development of robots with human involvement and the study of how humans and robots interact. Cloud-based systems can leverage powerful machine learning algorithms for emotion recognition. They analyze facial expressions, gestures, and speech patterns to understand the emotional state of a human interacting with the robot. In the past few decades, human-robot interaction (HRI) has become an increasingly important research area in the cross-disciplinary fields of psychology, behavioral science, and cognitive science. HRI research has focused on how humans control robots. However, for healthcare robots, they need to communicate with the user to optimize their performance. Therefore, the robot must be able to effectively recognize, interpret, and respond to the emotions expressed by humans, given such emotions convey important aspects of interaction, namely thoughts and feelings. By accessing cloud-based databases and user profiles, robots can personalize their interactions based on individual preferences, habits, and emotional tendencies. This leads to a more tailored and user-centric experience. Cloud-based platforms enable remote monitoring and control of robots. Human supervisors can intervene and guide robots in real time, especially in situations where emotional tones require human expertise. Cloud-based NLP services can enhance a robot's ability to understand and respond to human speech, including detecting sentiment and emotional cues. This is crucial for effective communication in various settings. Cloud-connected robots can play a role in healthcare and therapy by providing emotional support to patients. They can monitor emotional states, engage in conversation, and even facilitate therapeutic activities (Andtfolk, M., Nyholm, L., Eide, H., Rauhala, A., & Fagerström, L., 2021).

5.8 Seamless Infrastructure-Robot-Interaction: The integration of cloud-enhanced healthcare robotics with seamless infrastructure-robot interaction can bring about transformative changes in the healthcare sector. Cloud-enhanced healthcare robots can facilitate remote consultations and patient monitoring. They can collect and transmit vital signs, and patient data, and even conduct basic medical examinations under the guidance of healthcare professionals. Robots equipped with cloud connectivity can assist in hospital logistics, including the delivery of medications, medical supplies, and equipment. Cloud-based systems can optimize the routing and scheduling of these robots for efficient and timely delivery. Cloud computing allows healthcare robots to securely store and access vast amounts of patient data. This data can be analyzed for trends, enabling better-informed decision-making by healthcare providers and supporting personalized patient care. Cloud-enhanced robotic systems can assist surgeons in performing procedures remotely. Surgeons can control robotic arms and receive real-time feedback, making it possible to perform surgeries from a distance with precision. Further, Cloud computing enables secure storage and transmission of sensitive health data. Healthcare robots can adhere to strict security and compliance standards, ensuring the privacy and confidentiality of patient information. Also, cloud robotics can seamlessly integrate with EHR systems, allowing for real-time updates and synchronization of patient records. This ensures that healthcare providers have access to the latest patient information (Haleem, A., Javaid, M., Singh, R. P., & Suman, R. 2022).

5.9 Human-Robot-Symbiosis: The application of cloud-enhanced healthcare robotics in human-robot symbiosis involves leveraging cloud computing technologies to enhance the collaboration and synergy between robots and humans in healthcare settings. Human-robot-symbiosis (HRS) represents a paradigm in which humans and robots collaborate seamlessly, each complementing the strengths of the other to achieve shared goals. When integrating cloud-enhanced health robotics into the concept of Human-Robot-Symbiosis, it opens up a multitude of possibilities for enhancing healthcare delivery and improving overall patient outcomes (Andtfolk, M., Nyholm, L., Eide, H., Rauhala, A., & Fagerström, L., 2021). Here are several key aspects where cloud-enhanced health robotics can play a pivotal role in Human-Robot-Symbiosis:

- **Real-time Data Sharing and Processing:** Cloud-enhanced health robots can leverage the power of cloud computing to access and process vast amounts of medical data in real time. This capability is invaluable in situations where quick decision-making is crucial, allowing robots to provide healthcare professionals with up-to-date information for more accurate diagnoses and treatment plans.
- **Collaborative Diagnosis and Treatment:** Human-Robot-Symbiosis involves collaborative efforts between healthcare professionals and robots. Cloud-connected robots can collaborate with medical experts by sharing diagnostic data, medical imaging, and treatment plans, fostering a synergistic approach to patient care.
- **Remote Expert Consultations:** Cloud-enabled health robots facilitate remote consultations with specialists located anywhere in the world. Through high-quality video conferencing, medical professionals can collaborate in real time, providing expertise and guidance for complex cases. This is particularly beneficial in areas with limited access to specialized healthcare.
- **Continuous Learning and Improvement:** Cloud connectivity allows health robots to continuously learn from new medical data, research findings, and treatment outcomes. Machine learning algorithms can be updated and refined centrally, enhancing the robot's ability to adapt to evolving medical knowledge and deliver more precise and personalized care.

- **Patient Data Accessibility:** Cloud-enhanced health robots can securely access patient health records, treatment histories, and other relevant information stored in the cloud. This ensures that the robot has a comprehensive understanding of the patient's medical background, enabling more informed decision-making during interactions with the patient.
- **Adaptive Personalization of Care:** Cloud-based systems enable health robots to analyze and interpret patient data to personalize care plans. By considering individual health metrics, treatment responses, and lifestyle factors, robots can adapt their assistance and support to meet the unique needs of each patient.
- **Enhanced Mobility and Flexibility:** Cloud connectivity enables health robots to move seamlessly between different locations while maintaining access to centralized resources. This mobility allows robots to assist patients in various settings, such as hospitals, clinics, or homes, contributing to a more flexible and adaptable healthcare system.
- **Efficient Resource Utilization:** Cloud-enhanced health robots can optimize resource utilization by offloading computationally intensive tasks to the cloud. This not only improves the efficiency of robotic operations but also allows for cost-effective scaling of healthcare services based on demand.

The integration of cloud-enhanced health robotics in human-robot-symbiosis represents a transformative approach to healthcare delivery. By combining the strengths of human expertise with the computational power and connectivity of cloud-enabled robots, the result is a healthcare ecosystem that is more responsive, adaptable, and capable of delivering high-quality care to a diverse range of patients.

5.10 Robots Collaboration in Teleoperations: We know that the care of old people by the decedents is reduced in the present era for various reasons. One of the big problems that arises with the aging population is the lack of home caregivers. A hopeful solution for this problem is teleoperation, where the robotic devices are used remotely according to the operator's intention. The robot can support remote operations and assist in autonomous operations. Further, these robots provide day-to-day services like assisting in home cores, supporting as personal assistants, and delivering medicines to the patients according to the information gained. For the half-disabled elderly who are partially laid up, they can use this teleoperation where the robot uses its capabilities to support independent living. It also supports local operations so that medical professionals and surgeons will be able to get accurate success results. Not only for operations has it also supported human-robot interactions.

5.11 Assisted Rehabilitation: After treatment, smart devices perform an integrated analysis of all data (Cloud-enhanced healthcare robots play a crucial role in providing assisted rehabilitation after medical treatments. This application combines the physical support and guidance of robotic systems with the power of cloud computing to create personalized, data-driven rehabilitation programs. Here are key aspects of how cloud-enhanced healthcare robots contribute to assisted rehabilitation:

- **Personalized Rehabilitation Programs:** Cloud-connected healthcare robots can create personalized rehabilitation programs based on the individual's medical history, treatment outcomes, and specific rehabilitation needs. Machine learning algorithms analyze data to tailor exercises and routines, optimizing the rehabilitation process for each patient.
- **Real-time Monitoring and Feedback:** During rehabilitation sessions, cloud-enhanced robots can monitor the patient's movements in real time. Sensors and cameras on the robot collect data, which is then processed in the cloud to provide immediate feedback to both the patient and health-

care professionals. This ensures that exercises are performed correctly and helps prevent the development of harmful habits.

- **Access to Rehabilitation Specialists:** Cloud connectivity enables remote access to rehabilitation specialists. Healthcare professionals can monitor progress, adjust rehabilitation plans, and provide guidance through virtual consultations. This is particularly beneficial for patients in remote areas or those with limited access to specialized rehabilitation services.
- **Gamification and Engagement:** To enhance patient motivation and engagement, cloud-enhanced healthcare robots can incorporate gamification elements into rehabilitation programs. Interactive exercises and games can be designed to make the rehabilitation process more enjoyable, encouraging patients to actively participate and adhere to their prescribed routines.
- **Progress Tracking and Analytics:** Rehabilitation data, including the patient's progress, range of motion, and adherence to exercises, can be securely stored and analyzed in the cloud. This information aids healthcare professionals in assessing the effectiveness of the rehabilitation program, making data-driven decisions, and modifying treatment plans as needed.
- **Continuous Learning and Adaptation:** Cloud-based systems facilitate continuous learning for healthcare robots. As more rehabilitation data is processed, machine learning algorithms can identify patterns and trends, leading to the refinement and improvement of rehabilitation strategies over time. This adaptability ensures that rehabilitation programs stay current and effective.
- **Remote Home Rehabilitation:** Cloud-connected robots enable patients to continue their rehabilitation at home under remote supervision. Healthcare professionals can prescribe exercises, monitor progress, and make adjustments to the rehabilitation plan without the need for frequent in-person visits. This approach enhances convenience and accessibility for patients.
- **Communication with Wearable Devices:** Integration with wearable devices, such as fitness trackers or smart sensors, allows cloud-enhanced healthcare robots to gather additional data about the patient's daily activities and health metrics. This comprehensive information contributes to a holistic understanding of the patient's well-being and aids in tailoring rehabilitation programs accordingly.
- **Secure Data Handling and Compliance:** Cloud-based systems prioritize data security and compliance with healthcare regulations. Patient rehabilitation data is encrypted, and stringent privacy measures are in place to protect sensitive information, ensuring compliance with healthcare data protection standards.

The integration of cloud-enhanced healthcare robots in assisted rehabilitation after treatment brings forth a dynamic and patient-centric approach to recovery. By combining technological advancements with personalized care, these systems contribute to more effective rehabilitation outcomes and improved overall patient well-being.

6. CHALLENGES OF CLOUD-ENHANCED HEALTHCARE ROBOTICS

In addition to these advantages, numerous researchers emphasize the critical aspect of data security, particularly the confidentiality and integrity of sensitive healthcare information when employing wireless technology for accessing cloud services via robots. Also, the network issues are faced in transferring data and other challenges are discussed below.

Networked Constraints: Networked robotics, especially the multi-robot system distributes the workload of sensing, actuating, communication, and computation among a group of participating robots. It has achieved great success in healthcare applications, intelligent delivery, and security applications. But it fails to work when there is a network drop. It is considered the major challenge in cloud-based healthcare robots. During critical situations, remote information transfer is the major role needed in healthcare systems and if the network fails to reach the required area may end with bad effects (Hu, G., Tay, W. P., & Wen, Y., 2012).

Resource challenges: Although a robot can share its computation workload with other units in the formation, the overall effectiveness of the robotic network is limited by the collection of each robot's resources, including onboard computers or embedded computing units, memories, and storage space. Physically, these onboard computing devices are restricted by the robots' size, shape, power supply, motion mode, and working environment. Once the robots are designed, built, and deployed, it is technically challenging, if not infeasible, to change or upgrade their resource configurations.

Self-learning and information gain Challenges: The amount of information a robot has access to is constrained by its processing power, storage space, and the number and type of sensors it carries. Healthcare robotics allows the sharing of information amongst robots connected by a communication network so that a global task can be solved or computed cooperatively using the whole network. However, healthcare robotics is constrained by the information observed or computed by robots in the network, and by the examples or scenarios that the network encounters, hence limiting its ability to learn. A robotic team learning to navigate may perform very well in a static environment, where all obstacles can be mapped out with increasing accuracy over time. On the other hand, the learning process has to be repeated once the environment changes or the robotic team is placed in a new unfamiliar environment. The map databases maintained by the robotic formation are also limited by the collective amount of storage space formation has.

Machine-to-Machine Communication Challenges: Common protocols for machine-to-machine (M2M) communications include proactive routing, which involves the periodic exchange of messages so that routes to every possible destination in the network are maintained, and ad-hoc routing, which forms a dynamic route to a destination node only when there is a message to be sent. Proactive routing incurs high computation and memory resources in the route discovery and maintenance process. Ad-hoc routing protocols suffer from a high latency as a route has to be established before a message can be sent, and are not practical if the network topology is highly dynamic. These drawbacks are significant in mobile robotic networks and may lead to severe performance degradation in intelligent medical delivery (Fiorini, L., Esposito, R., Bonaccorsi, M., Petrazzuolo, C., Saponara, F., Giannantonio, R., & Cavallo, F., 2017).

7. SECURITY ISSUES OF CLOUD ENHANCED ROBOTICS AND SOLUTIONS

Trust and security issues are major considerations in cloud robotics. With the development of robotic sensor technology, numerous personal healthcare data are generated; therefore, security and privacy are important when we focus on healthcare information. Here Cloud technology is a core architecture, which has become the focus of security research in recent years. With the integration of other technologies, cloud computing adopts distributed accounting, distributed communication, and distributed storage. Cloud computing consists of many equal nodes to form an end-to-end network. Whenever health data is needed by the hospitals or medical supervisors remotely the data is transferred over the network and there is a chance of misleading data intruder may change the data, or there may be data loss or it may be delivered to an unauthorized user. Hence the data exchanged between networks and remote systems can be verified in the premise of obeying the established rules (Jain, S., & Doriya, R., 2022).

The transmission of sensitive healthcare data to the cloud through wireless technology, shared among various entities, introduces a significant concern about data breaches in cloud-based multi-robot systems. Data breaches in cloud-based healthcare robots pose various serious threats: Malicious Modification where unwanted alterations in information can disrupt the robot's ability to distinguish between images, impacting overall performance. Unauthorized Disclosure: Attackers may divulge highly sensitive and confidential patient health information, raising concerns about the confidentiality of the communication channel. Information Gathering: Unauthorized access to patient data is a critical issue in healthcare robotics, affecting the robot's performance and jeopardizing the confidentiality of patient data. Communication Disruption: Attackers may interfere with wireless communication between robots and cloud servers, potentially resulting in partial or complete loss of control over the robots.

Cloud computing utilizes the service agreement of providers to handle the additions of networks to ensure the security and privacy of information. The electronic medical health records of patients are transferred remotely as and when needed and therefore within the security and privacy system, all parties are allowed to use the data with the presentation of authentication. Different limits of authority are given by different use identities, such as viewing privileges and not editing through the use of different networking algorithms. The edit option is provided only to the authorized users.

Specifically, the solution faces two major security challenges due to its cloud implementation. First, it needs the VM (virtual machine) environment to be trustworthy since each robot used in the cloud robotics uses a virtual machine. A malicious VM can subtly sabotage an important task without the robot being aware of the damage. In medical applications, the robot has to identify a trust-worthy VM infrastructure to connect and avoid malicious. In general, three approaches can be adopted to overcome this problem, including:

Trust Establishment: the user performs some pre-use actions to check a VM's host environment. Trust Measurement: some root-of-trust components that do not belong to the cloud platform provider (e.g., from hardware vendors or virtualization software providers) monitor the VM, and securely report trust measurements to a user or a third party. Reputation-Based Trust: the user verifies the VM infrastructure by the service provider's identity and then relies on legal, business or other external considerations to infer trust.

Further, securing cloud-enhanced health robots involves addressing various challenges related to data integrity, confidentiality, and authentication. Using HMAC-SHA1 (Hash-based Message Authentication Code with the Secure Hash Algorithm 1) and ECC (Elliptic Curve Cryptography) can contribute to enhancing security. The following cryptographic techniques can be applied:

7.1 HMAC-SHA1 for Data Integrity: HMAC-SHA1 is a widely used algorithm for ensuring the integrity of data. It involves hashing the data with a secret key and creating a unique hash value that can be used to verify the data's integrity. It can be implemented by applying HMAC-SHA1 to the data transmitted between the health robot and the cloud. The robot and cloud should share a secret key for generating and verifying HMACs. Finally, it should verify the received data in the cloud using the calculated HMAC.

7.2 ECC for Confidentiality and Authentication: ECC is a public-key cryptography technique that uses elliptic curves over finite fields. It provides strong security with shorter key lengths compared to traditional methods like RSA. It can be implemented by generating ECC key pairs (public and private keys) for the health robot and the cloud. It uses the robot's public key to encrypt sensitive data before transmitting it to the cloud, ensuring confidentiality and authenticating the health robot to the cloud using its private key for signing data. Hence the cloud can verify the robot's authenticity using its public key. The HMAC Key and ECC Key Management takes the following procedure, it implements a robust key management system to securely generate, distribute, and update HMAC and ECC keys and periodically rotate keys to enhance security and uses secure channels or protocols for key exchange. One of the protocols used is Secure Communication Protocols which is SSL/TLS protocols for Transport Security. It is used to secure the communication channel between the health robot and the cloud, this ensures that data is transmitted securely over the network. It's important to note that the choice of cryptographic algorithms and protocols should be based on the latest security standards and recommendations. Additionally, adherence to best practices in software development and system architecture is crucial for building a robust and secure cloud-enhanced health robot system. Regularly updating the system and staying informed about emerging security threats will help maintain a high level of security.

8. CONCLUSION AND FUTURE SCOPE OF CLOUD HEALTHCARE ROBOTICS

The integration of cloud computing with health robotics represents a significant advancement in healthcare technology. This synergy offers numerous benefits, such as improved efficiency, accessibility, and scalability in healthcare services. By leveraging the power of cloud computing, health robotics can access vast amounts of data, process information in real time, and provide more personalized and efficient healthcare solutions. Cloud-enhanced health robotics has demonstrated its potential in various medical applications, including surgery, diagnostics, rehabilitation, and patient care. The ability to store and analyze medical data in the cloud enables better decision-making by healthcare professionals, leading to enhanced patient outcomes. Additionally, remote monitoring and telemedicine capabilities facilitated by cloud computing enable healthcare services to reach underserved or remote areas, improving overall healthcare accessibility. The chapter proposed a cloud robotics architecture to address the constraints faced by current networked robots. Cloud robotics allows robots to share computation resources, information, and data, and to access new knowledge and skills not learned by themselves. This opens a new paradigm in robotics that we believe leads to exciting future developments. It allows the deployment of inexpensive robots with low computation power and memory requirements by leveraging the communications network and the elastic computing resources offered by the cloud infrastructure.

The future of cloud-enhanced health robotics holds exciting possibilities for further innovation and advancement in healthcare. Some key areas of future exploration include Interconnectivity and Collaboration, where the integration of health robotics with cloud computing can foster greater collaboration among healthcare professionals and institutions. Real-time data sharing and collaborative decision-making could lead to more comprehensive and holistic patient care. Next is the Artificial Intelligence and Machine Learning Integration, as the field of artificial intelligence (AI) and machine learning (ML) continues to evolve, integrating these technologies with cloud-enhanced health robotics can lead to more intelligent and adaptive robotic systems. These systems can learn from vast datasets, improve their decision-making capabilities, and personalize treatments based on individual patient data. Security and Privacy Measures are also the areas to work. Given the sensitive nature of health data, future developments must prioritize robust security and privacy measures. Advances in encryption, authentication, and secure data storage will be crucial to ensure the protection of patient information in the cloud. Cost-effectiveness and Scalability are also some of the matters to consider as future aspects. Future developments should focus on optimizing the cost-effectiveness and scalability of cloud-enhanced health robotics solutions. This includes exploring ways to reduce infrastructure costs, enhance resource utilization, and ensure the seamless scalability of systems to accommodate growing healthcare needs. In conclusion, the convergence of cloud computing and health robotics has the potential to revolutionize healthcare delivery. Continued research, development, and collaboration among industry stakeholders will be key to unlocking the full potential of this transformative technology and addressing the evolving needs of the healthcare ecosystem.

REFERENCES

Andtfolk, M., Nyholm, L., Eide, H., Rauhala, A., & Fagerström, L. (2021). Attitudes toward the use of humanoid robots in healthcare—A cross-sectional study. *AI & Society*, 1–10.

Bonaccorsi, M., Fiorini, L., Cavallo, F., Saffiotti, A., & Dario, P. (2016). A cloud robotics solution to improve social assistive robots for active and healthy aging. *International Journal of Social Robotics*, *8*(3), 393–408. doi:10.1007/s12369-016-0351-1

Dahl, T. S., & Kamel Boulos, M. N. (2013). Robots in health and social care: A complementary technology to home care and telehealthcare? *Robotics (Basel, Switzerland)*, *3*(1), 1–21. doi:10.3390/robotics3010001

Elfaki, A. O., Abduljabbar, M., Ali, L., Alnajjar, F., Mehiar, D. A., Marei, A. M., Alhmiedat, T., & Al-Jumaily, A. (2023). Revolutionizing Social Robotics: A Cloud-Based Framework for Enhancing the Intelligence and Autonomy of Social Robots. *Robotics (Basel, Switzerland)*, *12*(2), 48. doi:10.3390/robotics12020048

Fiorini, L., Esposito, R., Bonaccorsi, M., Petrazzuolo, C., Saponara, F., Giannantonio, R., Petris, G. D., Dario, P., & Cavallo, F. (2017). Enabling personalised medical support for chronic disease management through a hybrid robot-cloud approach. *Autonomous Robots*, *41*(5), 1263–1276. doi:10.1007/s10514-016-9586-9

Firouzi, F., Farahani, B., Daneshmand, M., Grise, K., Song, J., Saracco, R., & Luo, A. (2021). Harnessing the power of smart and connected health to tackle COVID-19: IoT, AI, robotics, and blockchain for a better world. *IEEE Internet of Things Journal*, *8*(16), 12826–12846. doi:10.1109/JIOT.2021.3073904 PMID:35782886

Haleem, A., Javaid, M., Singh, R. P., & Suman, R. (2022). Medical 4.0 technologies for healthcare: Features, capabilities, and applications. *Internet of Things and Cyber-Physical Systems*, *2*, 12–30. doi:10.1016/j.iotcps.2022.04.001

Holland, J., Kingston, L., McCarthy, C., Armstrong, E., O'Dwyer, P., Merz, F., & McConnell, M. (2021). Service robots in the healthcare sector. *Robotics (Basel, Switzerland)*, *10*(1), 47. doi:10.3390/robotics10010047

Hu, G., Tay, W. P., & Wen, Y. (2012). Cloud robotics: Architecture, challenges, and applications. *IEEE Network*, *26*(3), 21–28. doi:10.1109/MNET.2012.6201212

Jain, S., & Doriya, R. (2022). Security framework to healthcare robots for secure sharing of healthcare data from cloud. *International Journal of Information Technology : an Official Journal of Bharati Vidyapeeth's Institute of Computer Applications and Management*, *14*(5), 2429–2439. doi:10.1007/s41870-022-00997-8 PMID:35633945

Marks, R. (2023). Robotics as a Form of Non-Surgical Complementary Osteoarthritis Treatment. *International Journal of Complementary and Internal Medicine*, *5*(1), 189–198.

Saha, O., & Dasgupta, P. (2018). A comprehensive survey of recent trends in cloud robotics architectures and applications. *Robotics (Basel, Switzerland)*, *7*(3), 47. doi:10.3390/robotics7030047

Shishehgar, M., Kerr, D., & Blake, J. (2018). A systematic review of research into how robotic technology can help older people. *Smart Health (Amsterdam, Netherlands)*, *7*, 1–18. doi:10.1016/j.smhl.2018.03.002

Yang, G., Pang, Z., Deen, M. J., Dong, M., Zhang, Y. T., Lovell, N., & Rahmani, A. M. (2020). Homecare robotic systems for healthcare 4.0: Visions and enabling technologies. *IEEE Journal of Biomedical and Health Informatics*, *24*(9), 2535–2549. doi:10.1109/JBHI.2020.2990529 PMID:32340971

Zhang, R., Zhou, Y., Zhang, J., & Zhao, J. (2023). Cloud-integrated robotics: transforming healthcare and rehabilitation for individuals with disabilities. *Proceedings of the Indian National Science Academy*, 1-12.

Chapter 14
Revolutionizing Healthcare:
Synergizing Cloud Robotics and Artificial Intelligence for Enhanced Patient Care

Jaspreet Kaur
Chandigarh University, India

ABSTRACT

The healthcare industry is currently experiencing a groundbreakingly revolution as cloud robotics and artificial intelligence (AI) come together. This research examines the synergistic relationship between these two advanced technologies and their significant influence on improving patient care. Through the utilisation of cloud-based computing power and the advanced capabilities of intelligent robotics, healthcare systems can attain unparalleled levels of efficiency, accessibility, and personalisation. Integrating AI algorithms with robotic systems enables enhanced diagnosis, treatment planning, and real-time monitoring, ultimately resulting in enhanced patient outcomes. This chapter examines the present condition of these technologies, explores instances where they have been effectively put into practice, and highlights the possible obstacles and ethical concerns. In this era of healthcare transformation, it is essential to recognise the collaborative capabilities of cloud robotics and AI in designing the future of patient-centred, data-driven healthcare systems.

INTRODUCTION

Cloud robotics and artificial intelligence (AI) are emerging as a force that is revolutionizing the healthcare industry for the better. This is happening as the merger of technological innovation and patient-centered care becomes an increasingly essential aspect of the healthcare industry. This dynamic combination not only represents a significant step forward in the development of healthcare delivery, but it also provides significant advantages in terms of precision, efficiency, and individualized care by providing exceptional benefits (Sun et al., 2020). Cloud computing, robotic systems, and artificial intelligence's cognitive capabilities suggest a future where patient care undergoes a revolutionary transformation. As we continue to delve further into the intricate workings of this trans-formative partnership, we are gradu-

DOI: 10.4018/979-8-3693-1914-7.ch014

ally coming to an understanding of this particular aspect. It is more than simply a technology fusion; rather, it is a strategic cooperation with the goal of tackling persistent issues in the healthcare business. The combination of artificial intelligence and cloud robotics is more than just a technological fusion, as depicted in Figure 1 below:

The adoption of this all-encompassing strategy acknowledges the numerous and complex aspects of healthcare, as well as the necessity of finding solutions that go beyond the conventional confines of diagnosis and treatment practices (Yang et al., 2020). The purpose of this in-depth investigation is to examine the relationship between artificial intelligence and cloud robotics. The investigation will also evaluate the tremendous consequences that this relationship has for remote surgery, predictive analytic, individualized treatment plans, data security, and resource management. Each and every aspect of this collaboration contributes to the story of healthcare, which is undergoing a revolution. In this narrative, the patient plays an active role as a crucial character in a healthcare system that combines cutting-edge technology with compassion. This narrative tells the story of a healthcare system (Le et al., 2018). The concept of remote surgery is a crucial component of this revolution, rethinking the limitations imposed by geographical constraints. It pulls together the domains of artificial intelligence and cloud robotics in order to accomplish this, as presented in Figure 2 below:

Professionals in the medical field now have the potential to use robotic tools that are controlled and augmented by artificial intelligence algorithms in order to carry out complex treatments remotely (Sarker et al., 2021). This is made possible as a result of the convergence of precision surgical procedures and digital connectivity. By overcoming the limits that are imposed by physical borders, this technology makes it feasible for patients who reside in areas that are undeserved or in remote locations to gain access to specialized information. It is important to note that the consequences of doing surgery remotely go well beyond the problem of accessibility. The utilization of these technologies makes it possible for specialists from different parts of the world to collaborate in real time. This not only encourages the growth of collective intelligence but also raises the bar for medical care to levels that have never been seen before (Yu & Zhou, 2021). As a result of the integration of artificial intelligence and cloud robotics, which enables the utilization of predictive insights, the healthcare industry is currently implementing a proactive and preventative strategy, as depicted in Figure 3 below:

Figure 1. Diagram representing the combination of artificial intelligence and cloud robotics

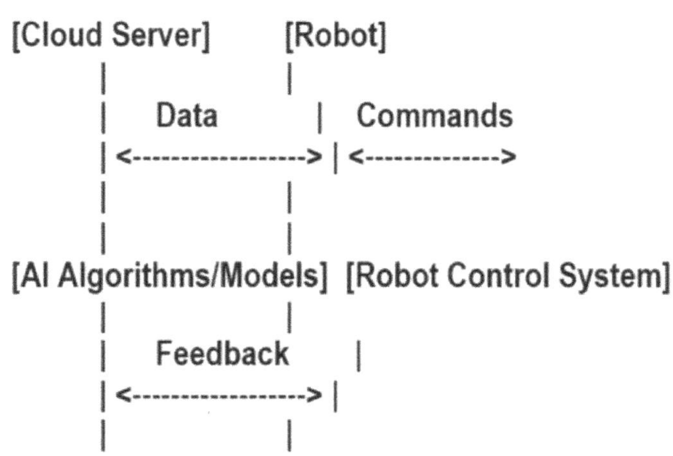

Figure 2. Remote surgery concept involving the integration of artificial intelligence and cloud robotics

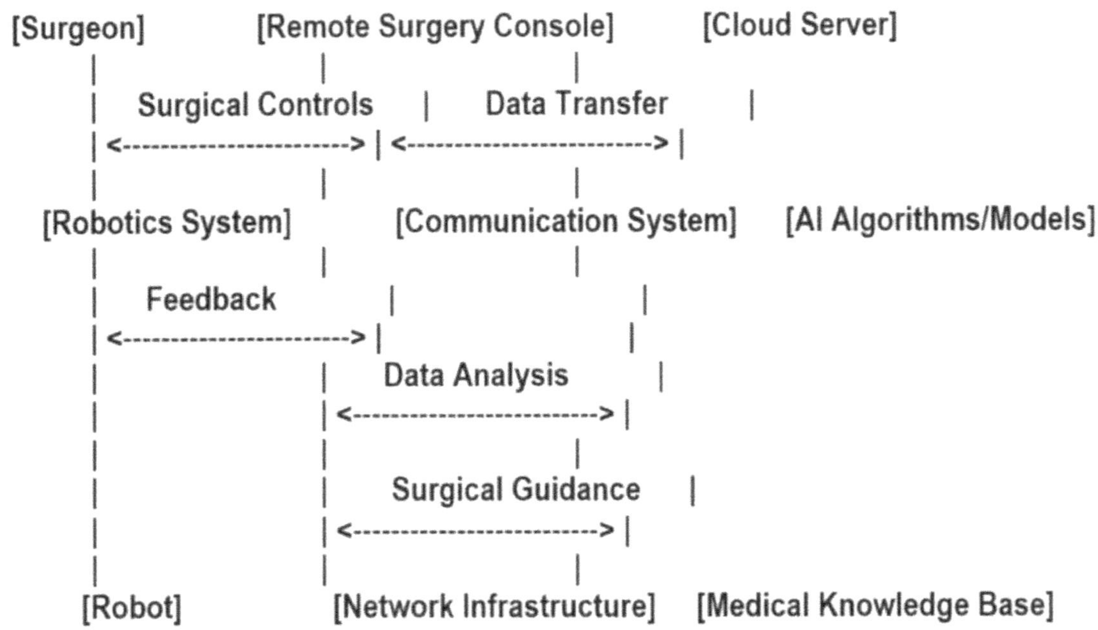

Figure 3. Flow of information from cloud computing infrastructure to healthcare facilities with AI-enhanced predictive analytic

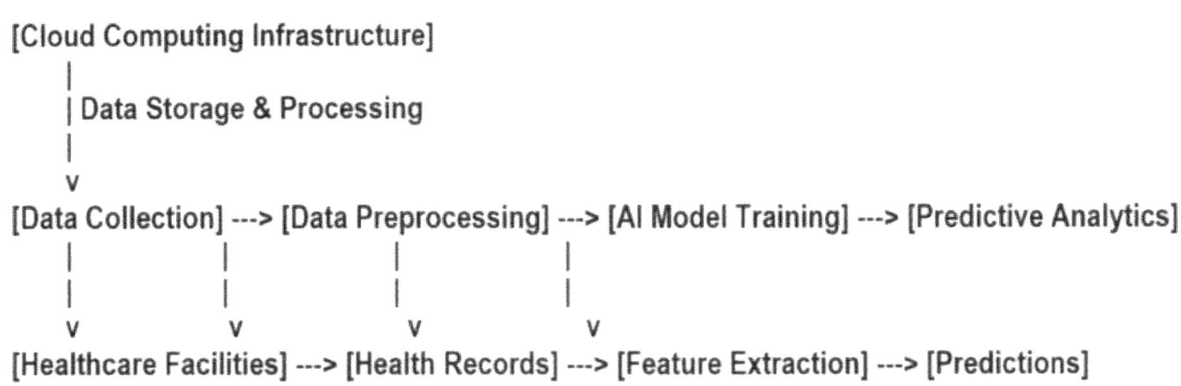

This is being done in order to mitigate potential risks. The enormous datasets that are kept in the cloud serve as a source of information for artificial intelligence systems (Kaur, 2023). Documents pertaining to patients, genetic information, and environmental components are all contained within these databases. Because these algorithms are able to recognize patterns and make predictions, medical practitioners are able to anticipate illness trends and identify populations that are at risk. This also allows them to identify populations that are at risk. The ultimate result is a shift away from a reactive healthcare model and towards a proactive one, in which therapies can be delivered before the onset of diseases, thereby decreasing the amount of burden that is placed on patients as well as the healthcare system. Further elucidation of the growth of patient care is provided by the possibility of personalized treatment regimens (Yadav et al., 2021). Cloud robotics and artificial intelligence work together in this field to develop solutions that are tailored to the distinctive characteristics of each of the individuals involved. In order to notice patterns and links that may be difficult for people to accept, the algorithms that make up artificial intelligence are able to assess a wide variety of patient data, such as genetic profiles and historical health records. This allows the algorithms to identify patterns and associations that may be unfamiliar to humans. Because of the use of robotic technology, it is possible to carry out personalized treatment plans in a smooth manner, which in turn helps to create a successful and individualized therapeutic approach that is geared to match the special requirements and difficulties of each individual patient. There has been a significant movement in the narrative from a universal approach to a personalized one, which shows a big break from the paradigms that are traditionally connected with healthcare. This shift is occurring in the midst of the era of precision medicine (Kaur, 2023).

On the other hand, in order to pursue innovation, one must first do a thorough investigation of the difficulties and issues that are associated with ethics. Data security and privacy are becoming increasingly essential problems in the healthcare business as a result of the progress of cloud robotics and artificial intelligence. This is because counterattacks are becoming more sophisticated. For the purpose of protecting sensitive medical data during the process of transferring and processing patient information over the cloud, it is very necessary to apply severe cybersquatting protections (Stasevych & Zvarych, 2023). The existence of ethical challenges related to decision-making that is driven by artificial intelligence highlights the need to establish a careful equilibrium between the evolution of technology and the ethical principles that serve as the basis for patient care. It is vital to have a planned approach that maximizes the benefits of innovation while simultaneously maintaining the confidence and confidentiality of patients in order to effectively navigate this terrain. This is necessary in order to meet the requirements of the situation (Monteiro et al., 2021). In addition, in order to bring about a revolution in the healthcare industry by utilizing cloud robotics and artificial intelligence, it is essential to carry out an analysis of resource optimization and scalability. With the tremendous computational capacity that the cloud provides, it is now possible for healthcare facilities to overcome the limitations that have historically been imposed on them. It is feasible for healthcare systems to successfully optimize their operations, dynamically scale their services, and efficiently react to changes in demand whenever they make use of the resources that cloud computing provides. This is all possible because cloud computing allows for these capabilities. A healthcare ecosystem that is flexible, efficient, and ready to meet the ever-changing demands of a population that is continually evolving is established as a result of this scalability, which not only provides the capability to withstand unforeseen occurrences but also produces a healthcare ecosystem that is agile and efficient (Kaur, 2024; Oniani et al., 2021). The complex connection that exists between artificial intelligence and cloud robotics in the healthcare industry makes it abundantly clear that this formidable alliance goes much beyond the basic combination of technologies. This is what makes this alliance so formidable.

The primary objective of this endeavor is to bring about a transformation in the manner in which patient care is seen and delivered. The combination of improved precision, efficiency, and rationalization that is made possible by this partnership propels us towards a future in which healthcare will overcome its traditional constraints by adopting a methodology that is both all-encompassing and centered on the patient (Oniani et al., 2021). This future, in which healthcare will overcome its limitations, will be accelerated by this collaboration. The transformation that is currently taking place is accompanied by a variety of problems, the majority of which include ethical concerns and the obligation to protect all patient information. On the other hand, these problems also create potential for innovation and the development of a healthcare ecosystem that is not just technologically advanced but also fundamentally based on ethical principles. There is a possibility that the healthcare sector could benefit greatly from this environment. It is important to note that the combination of artificial intelligence and cloud robotics presents a significant opportunity for advancement in the field of healthcare.

The construction of a care system that is both complex and adaptable has been brought about as a result of the marriage of technology and empathy. The scope of this revolution is not restricted to the world of technological innovation; rather, it represents a fundamental shift in the way that we think about health and wellness. We need to be vigilant in order to guarantee that the benefits of innovation are distributed in an equitable manner, that ethical considerations are given priority in the decision-making process, and that the patient continues to be the primary focus of this redesigned narrative of healthcare excellence. Only then will we be able to successfully negotiate this complex alliance of healthcare providers (Sharma, 2024).

LITERATURE REVIEW

This literature on the confluence of cloud robotics and artificial intelligence (AI) in the healthcare industry demonstrates that it is a diverse and innovative subject that is defined by substantial progress, potential problems, and a shared vision for the future of patient care. In addition, it demonstrates that there is a shared vision for the future of patient care. Upon closer examination of this extensive body of research, it is evident that the collaborative partnership between artificial intelligence and cloud robotics is not just a blend of technologies; instead, it represents a potent force with the capacity to fundamentally transform the provision of healthcare both now and in the future. Recent literature has shown significant attention to the concept of remote surgery, made possible by the integration of artificial intelligence and cloud robotics. The focus has been on the concept of remote surgery. Professionals and researchers have conducted studies to investigate the potential of robotic devices powered by artificial intelligence to carry out surgical procedures from a distance. This technological innovation has the ability to overcome geographical limitations and provide specialized medical care to regions with limited access. The gains that have been made in teleoperate robotic systems have been brought to light by Das et al. (2024) and Bhaskar et al. (2020) with particular emphasis placed on the capability of these systems to potentially alter surgical techniques. Furthermore, the research that was conducted by Berlet et al. (2024) analyzes the effects that remote surgery has on telemedicine. Telemedicine has the potential to improve patient outcomes in settings that are either geographically far away or have limited resources. This study reveals that telemedicine can overcome discrepancies in the availability of surgical knowledge and improve patient outcomes.

From the findings of the research that has been carried out, it has been determined that predictive analytic in the field of cloud robotics and artificial intelligence is well recognized for its potential capabilities in the prevention of illness. Artificial intelligence algorithms and datasets stored in the cloud can help build prediction models and identify at-risk populations. Combining the two can achieve this. Klauschen et al. (2024) study the capabilities of artificial intelligence (AI) to generate predictions regarding the identification of disease patterns. AI can help identify disease patterns, which is of interest to the researchers. These studies' findings demonstrate that artificial intelligence has the potential to transform healthcare delivery from a reactive to a proactive approach. In the end, these insights shed light on the utility of predictive analytic in early intervention and individualized preventative efforts, which ultimately result in a reduction in the load placed on healthcare systems and an improvement in the health of the population as a whole with regard to health.

Particularly with regard to the fields of artificial intelligence and cloud robotics, individualized treatment plans place a significant emphasis on precision medicine and individualized care. The research that was carried out by Bekbolatova et al. (2024) and Sharma (2024) studies the utilization of robotic systems and analytic that are powered by artificial intelligence in order to personalize treatment plans by making use of data that is acquired from specific patients. The results of this research provide new information that sheds light on the possibilities for improving treatment efficacy, reducing the degree of adverse effects, and achieving better outcomes for patients. The outcomes of the research indicate that precision medicine is an important component of the overall strategy with regard to healthcare implementation. Healthcare providers methodically modify treatments in this approach to tailor them to the specific requirements of each individual patient.

The study thoroughly examined issues related to data security and ethics in the coupling of artificial intelligence and cloud robotics. Our team thoroughly investigated these issues. Adeniyi and Ness (2024) and Tyagi (2024) are two examples of researchers who did an analysis of the urgent need for robust cybersquatting measures in order to secure patient data while it is transported across cloud-based systems. Such measures are necessary in order to protect patient data. As an additional point of interest, the ethical concerns associated with the use of artificial intelligence in the healthcare sector for decision-making have been a topic of discussion for a considerable amount of time. Grzybowski et al. (2024) studied the ethical implications of artificial intelligence algorithms in the healthcare industry. Within the context of the current era of data-driven decision-making among healthcare professionals, they highlight the importance of transparency, accountability, and the patient's right to privacy.

In the context of cloud robotics and artificial intelligence, the research demonstrates the enormous influence that resource optimization and scalability have on healthcare delivery in the context of cloud robotics and artificial intelligence. Two examples of researchers who are researching the impact that cloud computing has on improving healthcare resources and ensuring the adaptability and expansion of services are Cacciamani et al. (2024) and Orhan et al. (2024). It is possible to alleviate the limitations that are inherent in conventional healthcare infrastructure by utilizing the cloud's capacity to supply computational resources in an instantaneous manner. The results of these studies shed insight on the capacity of healthcare systems to easily adjust to changing needs, thereby guaranteeing that these systems are robust and effective in serving the requirements of a diverse spectrum of patients.

Following an examination of the relevant literature, it is clear that the incorporation of artificial intelligence and cloud robotics into the healthcare industry brings with it a variety of difficulties and challenges. Conducting careful ethical considerations, protecting individuals' privacy, and distributing benefits in an equitable manner counterbalance this excitement. Remote surgery, predictive analytic, individualized treatment plans, data security, and resource optimization are all examples of technologies that include these. The literature offers a comprehensive perspective on a world that is in a state of perpetual change, and it is essential to find a balance between being innovative and being cautious.

In the end, but certainly not least, the literature review highlights the significance of the joint collaboration between cloud robotics and artificial intelligence in the process of bringing about changes in the field of healthcare. The improvements that have been made in areas such as remote surgery, predictive analytic, personalized treatment plans, data security, and resource optimization collectively lead to a future in which patient care will be both technologically advanced and ethically sound. This future is a promising one. Utilizing knowledge from previous research is vital for successfully navigating this constantly changing environment. This includes appreciating the multiple possibilities that may arise while also addressing the obstacles that may come up. A revolutionary change in healthcare is characterized not only by the utilization of cutting-edge technology but also by a commitment to the well-being of patients, adherence to ethical values, and the equitable distribution of benefits in the healthcare system. Various literary sources have contributed knowledge and ideas to build the foundation of this revolution.

RESEARCH METHODOLOGY

In order to acquire a comprehensive understanding of the intricate and ever-changing domain of revolutionizing healthcare through the synergistic integration of cloud robotics and artificial intelligence (AI), it is of the utmost importance to build a rigorous research approach that encompasses both the depth and breadth of the subject matter. It outlines the study plan, data collection methods, sampling procedures, and analytical tools to investigate the revolutionary possibilities and obstacles linked to this pioneering collaboration.

Research Design: The major research methodology for this study adopts qualitative approach to provide a full knowledge of the combined influence of cloud robotics and AI on patient care. This idea is particularly well-suited when it comes to investigating complex phenomena in healthcare that involve technological advancements, ethical issues, and real-world repercussions in the realm of healthcare, this idea is particularly well-suited.

Qualitative Research: In order to capture the diverse perspectives of relevant stakeholders, such as healthcare professionals, technologists, policymakers, and patients, the study employed employed qualitative research approaches such as interviews, focus groups, and content analysis. Through in-depth interviews, we will investigate the experiences, perceptions, and ethical concerns of those who are at the forefront of implementing or making use of artificial intelligence and cloud robotics across healthcare settings. Through the use of focus groups, collaborative conversations will be facilitated, resulting in the discovery of a variety of perspectives and the promotion of a more thorough understanding of the consequences for society and organizations. An additional benefit of conducting content analysis on relevant literature, governmental papers, and industry reports is that it provides a contextual framework, which makes it easier to recognize patterns, challenges, and emerging topics.

Sampling Methods: The selection procedure for participants is an essential component in ensuring that the research project incorporates a diverse variety of perspectives and experiences from among the participants. The study used purposive sampling to identify the individuals participating in this research project. This method will ensure the selection of individuals with specific information, competence, or experiences related to the integration of artificial intelligence and cloud robotics in the healthcare industry. This includes healthcare professionals from a variety of fields, patients who have direct experience with technology-enabled healthcare, technologists who are actively involved in the creation of these systems, and legislators who have an impact on the legislative environment. Through the use of the purposive sample technique, it is possible to make a conscientious selection of people who have valuable perspectives on the many parts of this integration.

Data Collection Methods

For conducting in-depth interviews with a specifically selected group of participants, The study incorporated a semi-structured method that allows flexibility and examination of new issues. In order to identify recurrent patterns, challenges, and opportunities, the interviews have been recorded in audio format, then transcribed into written form, and then subjected to theme analysis.

To conduct focus groups, the study established a variety of participant groups. These groups include professionals from a variety of healthcare disciplines, patients with a variety of healthcare experiences, and technology experts. Focus groups, because of their interactive nature, make it possible to investigate collective viewpoints, societal effects, and ethical considerations. This, in turn, contributes to a more comprehensive understanding of the subject matter at hand.

Developing a tool for conducting a systematic survey enabled the collection of quantitative data from a more extensive set of participants. To gather demographic information, thoughts on artificial intelligence and cloud robotics, experiences with technology-supported healthcare, and viewpoints on ethical considerations, the study implemented survey. The process of extracting important themes, highlighting regions of limited understanding, and discovering the movement of ideas throughout the academic and policy domains will be accomplished through the use of content analysis.

Thematic analysis has been used to complete the study of qualitative data, such as interview transcripts and focus group discussions. The iterative technique involves classifying, categorizing, and interpreting data in order to uncover recurring themes, challenges, and opportunities associated with the application of artificial intelligence and cloud robotics in the healthcare industry. To provide a comprehensive understanding of the subject matter, the research cross-validate the qualitative findings with quantitative data obtained from statistical surveys.

Ethical concerns: When conducting research that deals with sensitive topics such as healthcare, technology, and patient data, ethical concerns are of the utmost importance.

The research had adhered to the ethical norms outlined in the relevant rules and regulations. Every participant was required to give informed consent, which will demonstrate that they are participating voluntarily and that they understand the study's objectives. The research maintains the confidentiality and anonymity of the data throughout the entire process of data collection, storage, and analysis. To protect the participants' identities, the study collected the data in a way that simplifies the information. An institutional review board (IRB) ethically inspected and approve the research to ensure adherence to ethical standards and protect the rights of study participants. We will disclose all potential conflicts of interest openly and transparently, and we will conduct the research with the utmost respect for the health and privacy of all individuals who are engaged.

Considering the scope and complexity of the study, it is essential to account for the unique constraints. Purposive sampling, where participants wer selected based on specific criteria related to their knowledge or experiences, may limit the study's generalization to a larger population. This is because the technique involves selecting participants based on specified criteria. In addition, the ever-evolving nature of both technology and healthcare may result in new perspectives and challenges, which the study may subsequently chronicle at a certain point in time. The study method aimed to provide a comprehensive and in-depth understanding of the impressive combination of artificial intelligence and cloud robotics in the medical field, despite some restrictions. The research method discussed in this study offers a robust and comprehensive approach to exploring the combined application of artificial intelligence and cloud robotics in transforming healthcare to improve patient care. By employing both qualitative, implementing procedures for deliberate sampling, and taking into account ethical considerations, this research endeavors to shed light on the myriad of facets that comprise the ever-evolving healthcare environment. Findings from this research will expand the current body of knowledge, providing significant insights to policymakers, healthcare practitioners, and technologists as they navigate the complexity of this transitional relationship.

Survey Insights on the Integration of Cloud Robotics and AI in Healthcare

The completion of an extensive survey designed to gather stakeholder viewpoints on the incorporation of cloud robotics and artificial intelligence (AI) in healthcare offers a detailed comprehension of the present situation, difficulties, and possibilities in the sector. This section presents the main discoveries obtained from the survey, providing a valuable understanding of the opinions, encounters, and anticipations of various participants, such as healthcare professionals, patients, and technological experts. The survey included a wide range of participants, including healthcare professionals from different specialties, patients with various healthcare experiences, and technologists involved in the development and implementation of cloud robotics and AI in healthcare. The inclusion of a diverse range of participants enhances the comprehensive comprehension of the subject matter, encompassing viewpoints from both healthcare professionals at the forefront of patient care and individuals influencing the technical advancements in the field.

The survey elicited responses from a diverse group of healthcare workers, encompassing physicians, nurses, surgeons, and administrative staff. The varied representation across different specialties and jobs guarantees a thorough understanding of how various parts of the healthcare workforce see the incorporation of cloud robotics and AI, as presented in Table 1 and Figure 4 below:

- **Attitudes towards Cloud Robotics and AI in Healthcare:** The survey aimed to assess the overall attitudes of participants towards the incorporation of cloud robotics and AI in healthcare. Most participants conveyed an optimistic perspective, acknowledging the profound capacity of these technologies to bring about significant change. More than 80% of healthcare professionals recognized that cloud robotics and AI possess the capacity to augment patient care, optimize workflows, and enhance diagnostic precision. Moreover, a significant majority of patients, above 75%, demonstrated a positive outlook towards the potential of these technologies in personalized treatment plans, remote healthcare delivery, and the general enhancement of healthcare efficiency. The technologist's viewpoint, although predominantly positive, also demonstrated an acknowledgment of obstacles. More than 60% of technologists expressed apprehensions over the ethical ramifications of AI algorithms, potential biases in data, and data security. This perspective from engineers emphasizes the importance of adopting a balanced approach that combines innovation with ethical considerations and potential hazards.
- **Remote Surgery and Telemedicine:** The study aimed to evaluate stakeholders' views on how cloud robotics and AI can facilitate remote surgery and enhance telemedicine. The findings revealed a significant level of optimism among healthcare professionals about the possibilities of remote surgery. Specifically, 85% of respondents recognized its ability to transcend geographical limitations and provide specialized care to neglected regions. The patients also demonstrated receptiveness to the concept, with 70% expressing a willingness to consider remote surgery if enabled by AI-guided robotic devices. Within the field of telemedicine, more than 90% of health-

Table 1. Survey findings

Survey Insights on Cloud Robotics and AI in Healthcare
Attitudes Towards Cloud Robotics and AI in Healthcare
Most participants (over 80% of healthcare professionals) are optimistic about the integration of Cloud Robotics and AI in healthcare, recognizing their potential to augment patient care, optimize workflows, and enhance diagnostic precision.
Patients also show a positive outlook towards these technologies, with over 75% expressing optimism about personalized treatment plans, remote healthcare delivery, and overall healthcare efficiency improvement.
Technologists, while predominantly positive, express concerns about ethical ramifications, biases in data, and data security, emphasizing the importance of a balanced approach that combines innovation with ethical considerations.
Remote Surgery and Telemedicine
Healthcare professionals (85%) recognize the potential of Cloud Robotics and AI to facilitate remote surgery, transcending geographical limitations and providing specialized care.
Patients (70%) express readiness to consider remote surgery if enabled by AI-guided robotic devices, demonstrating receptiveness to the concept.
Both healthcare professionals and patients (over 90%) acknowledge the capacity of Cloud Robotics and AI to enhance telemedicine, improve virtual consultations, increase diagnostic precision, and optimize treatment planning.
Utilization of Cloud Robotics and AI in Predictive Analytic
Healthcare experts and patients unanimously agree on the predictive capabilities of Cloud Robotics and AI, recognizing their potential to forecast illness patterns and detect vulnerable populations.
Over 80% of healthcare professionals demonstrate trust in AI algorithms for early disease detection, while a comparable percentage of patients see the benefits of predictive analytics in preventive healthcare.
However, concerns about data protection and ethical application are highlighted, with around 70% stressing the significance of strong data security measures and ethical rules.
Personalized Treatment Plans
More than 75% of healthcare professionals express enthusiasm for AI-driven analytics to customize treatment approaches using specific patient data.
Patients (80%) are willing to consider personalized treatment approaches tailored to their individual characteristics, emphasizing the importance of clear and open communication between healthcare practitioners and patients.
Data Security and Ethical Considerations
There is a collective recognition among stakeholders regarding the necessity of strong data security protocols, encryption, secure data storage, and strict access controls to safeguard patient information.
Around 80% of healthcare professionals, patients, and technologists express concerns about potential biases in AI algorithms and emphasize the necessity of transparency in decision-making.
Technologists underscore the significance of creating and following ethical principles in the development and implementation of AI systems in healthcare environments.
Resource Optimization and Scalability
Over 85% of healthcare professionals acknowledge the potential of cloud-based solutions in optimizing resource allocation, streamlining processes, and improving overall healthcare efficiency.
Patients (70%) express optimism about the potential of cloud-based technologies to enhance the responsiveness and adaptability of the healthcare system.
Technologists express reservations (almost 60%) regarding infrastructure burden, ongoing surveillance, upgrades, and ethical implications associated with centralized data storage.
Future Implications and Recommendations
The prevailing optimistic views among stakeholders indicate a shared acknowledgment of the revolutionary capabilities of Cloud Robotics and AI in healthcare.

continued on following page

Table 1. Continued

Survey Insights on Cloud Robotics and AI in Healthcare
Attitudes Towards Cloud Robotics and AI in Healthcare
However, the survey emphasizes the significance of tackling ethical concerns, safeguarding data security, and maintaining clear communication to establish trust and ensure responsible innovation.
Continuous education and training programs for healthcare professionals are crucial to improve their comprehension of Cloud Robotics and AI.
Comprehensive ethical guidelines, regulatory frameworks, and standards are required to control the utilization of these technologies in healthcare environments.

Figure 4. Main discoveries from survey

care professionals and patients acknowledge the capacity of cloud robotics and AI to improve virtual consultations, increase diagnostic precision, and optimize treatment planning. The survey revealed a commonly held belief that the incorporation of these technologies has the potential to enhance healthcare accessibility and effectiveness, particularly in the context of providing medical services to remote and rural areas.

- The survey examined stakeholder perspectives on the utilization of cloud robotics and AI in predictive analytics for the purpose of preventing diseases. Findings revealed a unanimous agreement among healthcare experts and patients regarding the capacity of these technologies to forecast illness patterns and detect vulnerable populations. A majority of healthcare professionals, namely over 80%, have demonstrated a high level of trust in the capability of AI algorithms to analyze extensive datasets for the purpose of early disease detection. Similarly, a comparable percentage of patients have seen the potential advantages of predictive analytics in the field of preventive healthcare. Nevertheless, the survey also revealed apprehensions around data protection and the ethical application of predictive analytics. Approximately 70% of healthcare professionals and patients stressed the importance of strong data security measures and ethical rules to ensure proper implementation of these technologies in illness prevention.
- **Personalized Treatment Plans:** The survey focused on stakeholder views regarding the idea of personalized treatment plans, which are guided by cloud robotics and AI. More than 75% of healthcare professionals expressed enthusiasm about the possibility of AI-driven analytics to customize treatment approaches using specific patient data. The authors emphasized the enhanced efficacy, fewer negative consequences, and enhanced patient results linked to individualized therapeutic strategies. Patients also demonstrated a favorable predisposition towards personalized treatment approaches, as 80% expressed a willingness to consider treatments that are tailored to their individual characteristics. The survey highlighted the importance of clear and open communication between healthcare practitioners and patients regarding the incorporation of AI in therapeutic decision-making, emphasizing the necessity of establishing trust and comprehension.
- **Data Security and Ethical Considerations:** The study examined stakeholder opinions on the key aspects of data security and ethical considerations in the implementation of cloud robotics and AI in healthcare, acknowledging their significant significance. The findings revealed a collective recognition among healthcare professionals, patients, and technologists regarding the necessity of strong data security protocols. More than 90% of the respondents across all categories emphasized the importance of encryption, secure data storage, and strict access controls to safeguard patient information. The comments highlighted the significant presence of ethical considerations, specifically with decision-making powered by AI. Around 80% of healthcare professionals and patients acknowledged concerns about potential biases in AI algorithms and emphasized the importance of transparency in how these algorithms influence therapeutic decisions. Technologists also expressed the same concerns, underscoring the significance of creating and following ethical principles in the development and implementation of AI systems in healthcare environments.
- **Resource Optimization and Scalability:** The survey aimed to comprehend stakeholder viewpoints on the idea of maximizing resources and expanding capacity using cloud technology in the field of healthcare. A majority of healthcare professionals, specifically over 85%, acknowledged the potential of cloud-based solutions for optimizing resource allocation, streamlining processes, and improving overall healthcare efficiency. The majority of patients (70%) expressed optimism about the potential of cloud-based technologies to enhance the responsiveness and adaptability of the healthcare system, despite their limited understanding of the technical factors involved. Technologists, on the other hand, offered a sophisticated viewpoint regarding the scalability of cloud-based systems. Although acknowledging the benefits, almost 60% expressed reservations about the possible burden on infrastructure, the need for ongoing surveillance and upgrades, and the ethical implications of centralized storage of extensive healthcare databases.

- **Future Implications and Recommendations:** The survey results not only shed light on the existing state of stakeholder opinions but also offer insights into the potential consequences of using cloud robotics and AI in healthcare. The prevailing optimistic views among healthcare professionals, patients, and technologists indicate a shared acknowledgment of the revolutionary capabilities of these technologies. Nevertheless, the survey emphasizes the significance of tackling ethical concerns, safeguarding data security, and maintaining clear communication in order to establish trust and guarantee responsible innovation. The study findings suggest that it is crucial to provide continuous education and training programs for healthcare professionals to improve their comprehension of cloud robotics and AI. Furthermore, it is evident that there is a strong requirement for the creation and execution of comprehensive ethical guidelines, regulatory frameworks, and standards that control the use of these technologies in healthcare environments. Individuals or groups who have a vested interest in or are directly affected by a particular project, initiative, or organization.

CONCLUSION

To conclude, the amalgamation of cloud robotics and artificial intelligence (AI) inside the healthcare sector signifies a profound and disruptive shift in paradigm, holding the capacity to fundamentally reshape the provision of patient care. The integration of cutting-edge technologies presents unparalleled prospects for enhancing medical diagnosis, treatment, and the entire provision of healthcare services. During the course of this conversation, a number of significant themes have surfaced, emphasizing the considerable potential and the crucial factors to be taken into account when implementing cloud robotics and AI in the healthcare sector.

Positive Outlooks and Acknowledged Advantages: The study results indicate a prevailing sense of optimism among healthcare professionals, patients, and technologists regarding the integration of cloud robotics and artificial intelligence (AI) in the healthcare sector. Stakeholders appreciate the significant potential of these technologies to improve patient care, streamline workflows, and enhance diagnostic accuracy. Patients, specifically, demonstrate a positive outlook towards the individualized treatment strategies and remote healthcare provision enabled by cloud robotics and artificial intelligence. The aforementioned favorable impressions highlight the significant influence that these technologies can exert on the healthcare domain.

Dealing with Ethical and Security Issues: Although there is a general sense of excitement, stakeholders acknowledge the ethical and security concerns that arise from the integration of cloud robotics and AI in the healthcare sector. Technologists have raised concerns about the ethical implications of AI algorithms, data biases, and data security. Healthcare professionals and patients alike emphasize the importance of transparency in decision-making and the necessity of robust data security policies to safeguard patient information. The aforementioned challenges highlight the necessity of adopting a well-rounded strategy that integrates both innovation and ethical considerations in order to guarantee the appropriate and fair implementation of new technologies.

The utilization of predictive analytic in the context of preventive healthcare: Predictive analytic for preventative healthcare stands out as a very potential use of cloud robotics and artificial intelligence (AI) within the healthcare sector. Stakeholders widely acknowledge the predictive capabilities of these technologies to forecast sickness patterns and identify vulnerable populations. Healthcare professionals exhibit confidence in the efficacy of AI algorithms for the timely identification of diseases, whereas patients recognize the advantages associated with proactive healthcare measures. Nevertheless, the poll also underscores the significance of safeguarding data and adhering to ethical principles in order to ensure the effective integration of predictive analytic in the realm of disease prevention.

The Implementation of Individualized Treatment Plans and Enhanced Patient Engagement: Healthcare professionals and patients alike are enthusiastic about the idea of personalized treatment programs led by cloud robotics and AI. Stakeholders acknowledge the potential of AI-powered analytic to tailor treatment strategies based on individual patient data, resulting in improved effectiveness and patient outcomes. Furthermore, the prioritization of effective and transparent communication between healthcare professionals and patients highlights the significance of cultivating trust and understanding in the integration of artificial intelligence in the process of making therapeutic decisions.

Optimization and scalability of resources: Stakeholders see the promise of cloud-based solutions in enhancing resource allocation and enhancing overall healthcare efficiency. However, technologists express worries over the strain on infrastructure and the ethical consequences. Notwithstanding these concerns, a significant proportion of healthcare practitioners and individuals seeking medical care exhibit a positive outlook on the capacity of cloud-based technology to augment the agility and flexibility of the healthcare system. This underscores the necessity of meticulous deliberation regarding scalability and ethical ramifications during the creation and execution of cloud-based technologies within healthcare settings.

In the future, it is clear that cloud robotics and AI will continue to have a significant impact on changing the healthcare industry. The survey's findings emphasize the importance of implementing ongoing educational and training initiatives for healthcare workers in order to improve their understanding of these technologies. To promote responsible innovation and mitigate potential hazards, it is crucial to establish and implement thorough ethical norms, regulatory frameworks, and standards. By effectively tackling these obstacles and leveraging the revolutionary capabilities of cloud robotics and artificial intelligence, many participants in the healthcare industry can collaboratively strive towards attaining heightened levels of patient care and superior healthcare results. In a nutshell, the incorporation of cloud robotics and artificial intelligence (AI) into the healthcare sector presents a substantial prospect for transforming the provision of patient care. Healthcare stakeholders may facilitate a future in which personalized, efficient, and egalitarian healthcare is accessible to all by harnessing the revolutionary potential of these technologies and effectively resolving ethical, security, and scalability considerations.

REFERENCES

Ahmad, A. (2024). Ethical implications of artificial intelligence in accounting: A framework for responsible ai adoption in multinational corporations in Jordan. *International Journal of Data and Network Science*, *8*(1), 401–414. doi:10.5267/j.ijdns.2023.9.014

Bekbolatova, M., Mayer, J., Ong, C. W., & Toma, M. (2024, January). Transformative Potential of AI in Healthcare: Definitions, Applications, and Navigating the Ethical Landscape and Public Perspectives. In Healthcare (Vol. 12, No. 2, p. 125). MDPI.

Berlet, M., Fuchtmann, J., Krumpholz, R., Naceri, A., Macari, D., Jähne-Schon, C., Haddadin, S., Friess, H., Feussner, H., & Wilhelm, D. (2024). Toward telemedical diagnostics—Clinical evaluation of a robotic examination system for emergency patients. *Digital Health, 10*. doi:10.1177/20552076231225084 PMID:38205033

Bhaskar, S., Bradley, S., Sakhamuri, S., Moguilner, S., Chattu, V. K., Pandya, S., Schroeder, S., Ray, D., & Banach, M. (2020). Designing futuristic telemedicine using artificial intelligence and robotics in the COVID-19 era. *Frontiers in Public Health*, *8*, 708. doi:10.3389/fpubh.2020.556789 PMID:33224912

Cacciamani, G. E., Chen, A., Gill, I. S., & Hung, A. J. (2024). Artificial intelligence and urology: Ethical considerations for urologists and patients. *Nature Reviews. Urology*, *21*(1), 50–59. doi:10.1038/s41585-023-00796-1 PMID:37524914

Grzybowski, A., Jin, K., & Wu, H. (2024). Challenges of artificial intelligence in medicine and dermatology. *Clinics in Dermatology*. Advance online publication. doi:10.1016/j.clindermatol.2023.12.013 PMID:38184124

Kaur, J. (2023). Robotic Process Automation in Healthcare Sector. In *E3S Web of Conferences* (Vol. 391, p. 01008). EDP Sciences.

Kaur, J. (2023, May). How is Robotic Process Automation Revolutionising the Way Healthcare Sector Works? In *International Conference on Information, Communication and Computing Technology* (pp. 1037-1055). Singapore: Springer Nature Singapore. 10.1007/978-981-99-5166-6_70

Kaur, J. (2024). Towards a Sustainable Triad: Uniting Energy Management Systems, Smart Cities, and Green Healthcare for a Greener Future. In Emerging Materials, Technologies, and Solutions for Energy Harvesting (pp. 258-285). IGI Global.

Klauschen, F., Dippel, J., Keyl, P., Jurmeister, P., Bockmayr, M., Mock, A., ... Müller, K. R. (2024). Toward explainable artificial intelligence for precision pathology. *Annual Review of Pathology: Mechanisms of Disease, 19*.

Le, D. N., Van Le, C., Tromp, J. G., & Nguyen, G. N. (Eds.). (2018). Emerging technologies for health and medicine: virtual reality, augmented reality, artificial intelligence, internet of things, robotics, industry 4.0. Academic Press.

Monteiro, A. C. B., França, R. P., Arthur, R., & Iano, Y. (2021). An overview of medical Internet of Things, artificial intelligence, and cloud computing employed in health care from a modern panorama. *The Fusion of Internet of Things, Artificial Intelligence, and Cloud Computing in Health Care,* 3-23.

Oniani, S., Marques, G., Barnovi, S., Pires, I. M., & Bhoi, A. K. (2021). Artificial intelligence for internet of things and enhanced medical systems. *Bio-inspired neurocomputing*, 43-59.

Orhan, K., Gülbeş, M. M., Jadhav, A., & Jagtap, R. (2024). Medico-Legal Problems of Artificial Intelligence. In *Artificial Intelligence in Dentistry* (pp. 259–282). Springer International Publishing.

Sarker, S., Jamal, L., Ahmed, S. F., & Irtisam, N. (2021). Robotics and artificial intelligence in healthcare during COVID-19 pandemic: A systematic review. *Robotics and Autonomous Systems*, *146*, 103902. doi:10.1016/j.robot.2021.103902 PMID:34629751

Sharma, K. (2024). Personalized Telemedicine Utilizing Artificial Intelligence, Robotics, and Internet of Medical Things (IOMT). In Handbook of Research on Artificial Intelligence and Soft Computing Techniques in Personalized Healthcare Services (pp. 301-323). Apple Academic Press.

Stasevych, M., & Zvarych, V. (2023). Innovative robotic technologies and artificial intelligence in pharmacy and medicine: Paving the way for the future of health care—a review. *Big Data and Cognitive Computing*, *7*(3), 147. doi:10.3390/bdcc7030147

Sun, L., Jiang, X., Ren, H., & Guo, Y. (2020). Edge-cloud computing and artificial intelligence in internet of medical things: Architecture, technology and application. *IEEE Access : Practical Innovations, Open Solutions*, *8*, 101079–101092. doi:10.1109/ACCESS.2020.2997831

Tyagi, A. K. (2024). Blockchain and Artificial Intelligence for Cyber Security in the Era of Internet of Things and Industrial Internet of Things Applications. In AI and Blockchain Applications in Industrial Robotics (pp. 171-199). IGI Global.

Yadav, S., Kaushik, A., & Sharma, S. (2021). Simplify the difficult: artificial intelligence and cloud computing in healthcare. *IoT and Cloud Computing for Societal Good*, 101-124.

Yang, G., Pang, Z., Deen, M. J., Dong, M., Zhang, Y. T., Lovell, N., & Rahmani, A. M. (2020). Homecare robotic systems for healthcare 4.0: Visions and enabling technologies. *IEEE Journal of Biomedical and Health Informatics*, *24*(9), 2535–2549. doi:10.1109/JBHI.2020.2990529 PMID:32340971

Yu, H., & Zhou, Z. (2021). Optimization of IoT-based artificial intelligence assisted telemedicine health analysis system. *IEEE Access : Practical Innovations, Open Solutions*, *9*, 85034–85048. doi:10.1109/ACCESS.2021.3088262

Chapter 15
Disaster Response and Recovery Role of Cloud-Enhanced Robotics

Chandra Singh
https://orcid.org/0000-0002-1759-8796
Sahyadri College of Engineering and Management, India

Anush Bekal
Sahyadri College of Engineering and Management, India

A. Ashoka
Sahyadri College of Engineering and Management, India

Lanlyn Samuel Dsouza
Sahyadri College of Engineering and Management, India

A. B. Abhishek
Sahyadri College of Engineering and Management, India

P. U. Shreesha
Sahyadri College of Engineering and Management, India

ABSTRACT

There is an urgent need for innovative and efficient disaster response and recovery solutions. This chapter investigates the use of cloud technologies in conjunction with robotics to improve disaster management operations. Cloud-enhanced robotics (CER) appears as a significant innovation, offering a synergistic approach in which cloud-enabled robots provide a more dynamic, adaptable, and robust response to catastrophic occurrences. CER enables robots to access huge computational resources and data, improving operational efficiency, decision-making, and problem-solving abilities. This connection improves real-time data analysis, coordination, and deployment tactics, resulting in a more effective reaction to crisis scenarios unfolding. Furthermore, it encourages greater coordination and resource sharing among different robots and reaction teams, hence maximizing overall recovery operations.

DOI: 10.4018/979-8-3693-1914-7.ch015

1. INTRODUCTION TO CLOUD-ENHANCED ROBOTICS

The incorporation of cloud-enhanced robots in disaster management represents a substantial shift in how we respond to and recover from natural disasters. The convergence of robotics and cloud computing is reshaping the landscape of emergency response, providing more efficient, effective, and flexible solutions in the face of catastrophes (Cheikhrouhou O, Koubaa A et.al, 2020). Disaster response has relied mainly on human workers, which often leads to constraints in speed, resource allocation, and risk exposure. However, the introduction of robotics into this industry has begun to disrupt this paradigm. Robots have proven to be invaluable assets due to their capacity to travel difficult terrains, acquire data, and undertake various rescue operations. The incorporation of cloud technology enhances these capabilities even further, providing robotic systems with a new level of flexibility and intelligence. Cloud-enhanced robotics blends robot physical capabilities with the cloud's immense computational power and storage capacity. This combination enables powerful data processing, real-time analytics, and remote control capabilities, all of which are critical in dealing with complicated crisis scenarios. Robots can now get updated maps, weather data, and other crucial information over the cloud, allowing them to make informed judgments and adapt to fast changing scenarios. The benefits of cloud-enhanced robots in disaster management are numerous. For starters, these robots can function in hazardous places that humans cannot, such as areas with toxic substances, unstable structures, or severe temperatures. This capability not only improves the safety of rescue personnel, but also assures that assistance reaches regions that would otherwise be inaccessible, The cloud's real-time data processing and analysis capabilities enable a more coordinated and efficient response. Robot data can help emergency crews analyze damage, locate survivors, and strategize rescue operations more effectively, cloud technology allows for smooth communication and data sharing among many agencies and robots, creating a more collaborative and united approach.Despite these benefits, implementing cloud-enhanced robotics in disaster management is not without difficulties. Concerns about data security, operational reliability, and ethical considerations are prevalent in automated decision-making. Furthermore, incorporating such advanced technology necessitates major investment in infrastructure, training, and legal frameworks. Continuous innovation and development in cloud robots are likely to address these difficulties in the future. Emerging technologies like AI, machine learning, and 5G connectivity will enhance disaster response robots' capabilities, making them more autonomous, intelligent, and efficient. Cloud-enhanced robots represent a paradigm change in disaster management. These systems provide a new level of sophistication and efficacy in how we respond to and recover from disasters by combining the benefits of both robotics and cloud computing. As technology advances, the role of cloud-enhanced robots in saving lives and lessening the effects of disasters will become ever more important, exemplifying technology's potential in humanitarian endeavours (Alotaibi, E.T et.al,2019).

Natural disasters such as earthquakes and floods, as well as man-made disasters such as industrial accidents, necessitate disaster response and recovery. The incorporation of cloud-enhanced robotics into these phases has created new opportunities for efficiency, precision, and safety. Cloud-enhanced robotics use robotic systems that have been advanced by cloud computing technology, allowing them to execute more complex jobs with increased adaptability and intelligence. Cloud-enhanced robotics plays an important part in this by enabling rapid deployment, real-time data analysis, and improved situational awareness. These robots can navigate disaster zones quickly, giving important information back to relief crews. In the aftermath of an earthquake, for example, cloud-enhanced drones may survey the afflicted area, providing live aerial footage that aids in damage assessment and pinpointing places in urgent need of assistance. These robotic devices may now access massive quantities of computational power and storage thanks to the cloud component. This allows them to assess enormous amounts of data from sensors and cameras in real time, assisting reaction teams in making informed judgments. Cloud connectivity improves communication and data sharing across numerous entities involved in disaster response, such as government agencies, humanitarian organizations, and first responders. During the recovery phase, the emphasis switches to restoring normalcy by rebuilding infrastructure, restoring services, and giving long-term assistance to affected populations. Cloud-enhanced robotics contributes to this lengthy and frequently difficult process by facilitating effective resource management, damage assessment, and reconstruction operations.Robots outfitted with advanced sensing technologies, for example, may assess the structural integrity of buildings, identify risks, and assist in debris removal. This not only expedites the recovery process, but also ensures the safety of reconstruction workers. Robots can handle decontamination operations in locations damaged by chemical spills or radiation, decreasing health hazards to human teams.Despite the promise, employing cloud-enhanced robotics in disaster response and recovery faces a number of problems. These include technological difficulties like as ruggedness in severe settings, data security concerns, and the requirement for dependable and high-speed internet connectivity. There are also logistical hurdles in installing this technology, particularly in rural or badly damaged areas. Advances in AI, machine learning, and edge computing are projected to improve the capabilities of cloud-enhanced robotics in disaster management in the future. More autonomous operations, improved integration with existing disaster management systems, and more resilient designs customized to certain types of disasters are possible in the future. As technology advances, the scope and influence of cloud-enhanced robots in disaster management will grow, providing new solutions to some of the most difficult parts of emergency response and recovery (Browne, C. et.al, 2011).

Robotics role in disaster management has grown dramatically over time, from simple mechanized help to sophisticated, autonomous systems capable of completing complicated tasks in dangerous conditions. This transformation has been fueled by technological advancements and a growing realization of the potential for robotics to improve the efficiency and safety of disaster response and recovery activities. The employment of basic automated tools defined the early stage of the evolution of robotics in disaster management. These early robots were mostly remote-controlled machines that did simple jobs like lifting heavy, While they provided some support in catastrophe situations, their capabilities were limited by the technology of the time, which limited their movement, usefulness, and degree of autonomy. As technology advanced, so too did the capabilities of disaster management robots. Robots equipped with cameras, thermal imaging, and various environmental sensors could provide real-time data to rescue teams, aiding in search and rescue operations and damage assessment.

The combination of Artificial Intelligence (AI) and machine learning was the most significant step forward in the growth of robotics in disaster management. These advancements allowed robots to comprehend massive volumes of data, make independent judgments, and learn from their surroundings. Drones, for example, may cruise disaster-stricken areas automatically, recognizing threats and seeking survivors. Ground robots are now capable of autonomously moving through wreckage, searching for survivors, and assessing structural stability. Another significant advancement in disaster robots was the capacity to work together. Networked systems of robots, both aerial and ground-based, could collaborate to cover bigger regions more efficiently by exchanging data and tasks. This collaboration expanded to human-robot interaction, with robots being integrated into disaster response teams to augment the capabilities of human responders.Today, disaster management robotics spans a wide range of applications, from drones conducting aerial surveillance and mapping to robotic dogs traversing collapsed structures. These tools have become crucial in disaster management stages such as planning, response, recovery, and mitigation.Looking ahead, the field is primed for additional development. Emerging technologies such as 5G connectivity, advanced AI algorithms, and more sophisticated sensor technology will continue to improve disaster management robot capabilities. Robots are projected to become increasingly independent, flexible, and capable of performing complex jobs in the future. The evolution of disaster management robotics represents a path toward increased sophistication, autonomy, and integration. These technologies have not only improved the skills of disaster response teams, but have also opened up new avenues for disaster management and mitigation. As robots technology advances, its role in disaster management will become ever more critical, perhaps altering how we respond to and recover from natural disasters.

2.THEORETICAL FRAMEWORK OF CLOUD ROBOTICS IN DISASTER SCENARIOS

Cloud robotics is a game-changing approach to disaster management that combines the physical capabilities of robots with the massive computational resources of cloud computing. This collaboration produces a significant synergy that improves the efficiency of disaster response and recovery activities. The theoretical foundation of cloud robotics in catastrophe scenarios is built on several important concepts, each of which contributes to the overall efficiency and functioning of these systems (Kurunathan, H. et.al, 2018).

The concept of distributed intelligence, in which computational activities are shared between the robot and the cloud, lies at the heart of cloud robotics. This approach enables complex data processing that would otherwise be beyond the robot's capabilities. This means that in catastrophe scenarios, robots can perform real-time analytics, process massive datasets like maps and sensor data, and make autonomous judgments based on comprehensive facts. By storing massive amounts of data generated during disaster operations, cloud robotics enables efficient data management. This information comprises real-time sensor readings, visual feeds, and environmental information that is critical for disaster assessment. The cloud serves as a central store for this information, allowing for instant access and exchange among many agencies and responders, resulting in a coordinated and educated reaction. Scalability is provided via cloud robotics, which enables disaster management teams to deploy a variable number of robots based on the scope and type of the crisis. This adaptability is crucial in dealing with various disaster contexts, whether they be natural disasters like earthquakes or man-made disasters like industrial accidents. The framework places a premium on collaborative operations involving several robotic units and human responders. Robotic units can connect and collaborate with one another via cloud computing, forming a network of shared intelligence. This collaboration improves situational awareness and operational efficiency, particularly in large-scale disaster zones when several locations require attention at the same time. Human responders' safety is paramount in hazardous catastrophe circumstances. Cloud robotics allows for remote operation and control of robots. This reduces the risk to human life while allowing vital operations such as search and rescue to be carried out effectively. The ability of cloud-enhanced robots to learn and adapt is an essential component of the theoretical framework. These robots can evaluate prior data, learn from previous crisis scenarios, and enhance their operational methods using machine learning algorithms and AI. This continual learning process is critical for improving disaster response efficiency over time.The framework also covers ethical and legal issues, particularly in terms of data privacy, operational transparency, and regulatory compliance. Maintaining public trust and enforcing legal norms require ethical usage and treatment of sensitive data (Shivam Gupta et.al,2022).

Cloud robotics in catastrophe scenarios gives a multifaceted strategy that incorporates technology, scalability, collaboration, and ethics. This framework provides a more responsive, efficient, and safe way of managing disaster response and recovery activities by leveraging the power of cloud computing and advanced robotics. This framework is anticipated to extend and adapt as technology advances, substantially boosting the capabilities of cloud robots in the face of various and severe crisis circumstances. Cloud-enhanced robotics is a rapidly emerging field with substantial disaster response implications. The combination of cloud computing and robots opens up new avenues for more efficient, effective, and coordinated emergency response. In crisis events, sensor-equipped robots collect massive volumes of data, such as photographs, heat readings, and 3D maps. This data is processed in real time via the cloud, allowing for speedy analysis and decision-making.

For example, determining safe routes through wreckage, recognizing survivors' heat signatures, or assessing building damage.

This quick processing enables speedy and educated replies, which is critical in emergency situations where every second counts. Multiple robots and human teams commonly operate together in disaster response. Cloud platforms enable these entities to communicate and coordinate more effectively. Robots can communicate information, update their operations depending on real-time data, and synchronize their actions over the cloud. This interconnection is critical for coordinated activities such as hunting for survivors or delivering supplies. Furthermore, cloud-based systems can integrate information from multiple sources (for example, satellite data and social media updates) to provide a holistic perspective of the disaster zone, improving situational awareness.

Cloud-enhanced robotics makes use of powerful artificial intelligence and machine learning algorithms to increase decision-making and autonomy. The massive storage and processing capabilities of the cloud enable robots to access and learn from large databases, boosting their predictive and analytical abilities. In catastrophe circumstances, AI can aid in anticipating the progression of events (such as fire spread or aftershocks) as well as automating duties such as victim identification and damage assessment. AI-powered robots can adapt to changing situations, making them vital in volatile and dynamic crisis scenarios.

Cloud technologies allow robots to be operated remotely, allowing human operators to control them from a safe distance. This is especially important in disaster zones where humans are dangerous or unreachable. Cloud-based interfaces allow operators to lead robots, assess conditions, and make decisions based on real-time data provided back to them. Telepresence also enables specialists from all around the world to participate to disaster response operations by sharing their expertise without physically being present.Cloud-enhanced robotics in disaster response is a convergence of cutting-edge technologies that significantly improves the efficiency, safety, and efficacy of emergency operations. The combination of real-time data processing, improved communication, AI-driven insights, and remote operating capabilities transforms how we respond to crises, making interventions faster, safer, and more coordinated. As these technologies advance, their position in disaster management will become ever more important, perhaps saving more lives and reducing the impact of disasters on communities. The combination of cloud computing with robotics, particularly in disaster management and response, offers a significant development in technology's potential to reduce the impact of catastrophic events. This paper investigates the roles and uses of cloud-enhanced robotics in disasters, focusing on data processing, communication, and decision-making.These robots, which are outfitted with sensors and cameras, can collect massive volumes of data from catastrophe locations. When this data is processed by strong cloud-based algorithms, it can provide real-time analysis of the issue. In the event of an earthquake, for example, these robots can immediately map affected areas, detect structural damage, and seek survivors beneath the rubble. Cloud computing allows for significantly faster processing of this data than traditional on-board computing solutions (Kurunathan, H. et.al,2019).

Cloud-enhanced robots play a critical role in disaster response operations by increasing communication and coordination. They can act as mobile communication hubs, ensuring that information is exchanged seamlessly between multiple response teams and command centers. These robots can construct temporary networks to assist crucial communication in situations where traditional communication infrastructure is broken or non-existent. This competence is critical for efficiently coordinating rescue attempts, distributing resources, and managing emergency services. Cloud computer integration enables robots to make more independent judgments based on real-time data and predictive analytics. This capability is especially useful in crisis settings where human intervention may be restricted owing to safety concerns or inaccessibility. Robots with cloud capabilities can evaluate situations, identify potential risks, and make smart judgments about how to navigate through debris or dangerous areas. For example, in the event of a chemical leak or a nuclear disaster, these robots can assess the amount of danger and determine the best course of action without the need for human intervention, lowering the risk to human life.Cloud-enhanced robots are capable of more than simply monitoring and data processing; they can even intervene physically. These robots, which are outfitted with various tools and manipulators, may accomplish jobs such as moving heavy debris, extinguishing fires, and transporting supplies to regions that are too dangerous for people. Amphibious robots can be used to rescue trapped people or transport medical supplies in flood conditions. Because of the cloud component, these robots may learn from previous disasters, increasing their efficiency and efficacy in physical interventions.

Cloud-enhanced robots can be useful in search and rescue efforts. They can enter dangerous or restricted locations where people cannot. These robots can locate signs of life, such as heat signatures or sounds, and assist rescue personnel more effectively by analyzing data from multiple sensors and cameras. Because of the cloud's computational capability, this data can be analyzed quickly, resulting in speedier and more efficient rescue operations (Koubaa, A et.al,2019).Cloud-enabled robots can be used in monitoring and early warning systems. These robots can continually monitor environmental data and communicate it to cloud-based analysis systems in places prone to natural calamities such as earthquakes, tsunamis, or volcanic eruptions. This system can send out early alerts to vulnerable populations, perhaps saving lives by allowing for prompt evacuations and preparations. The application of cloud-enhanced robots is disaster response training and simulation. Robots can be educated in virtual environments that simulate real-world crisis events using cloud-based simulations. This training guarantees that the robots are ready for a wide range of scenarios, making them more versatile and efficient in actual crisis response. Data security concerns, privacy concerns, and the requirement for robust and reliable communication networks must all be addressed. These established protocols for interaction between robots, people, and other automated systems in disaster response activities are required. Cloud-enhanced robots are a game changer in disaster management and response. Their capacity to assimilate data quickly, make independent decisions, communicate effectively, and physically intervene makes them great assets in disaster mitigation. However, it is critical to solve the issues connected with their deployment on a regular basis in order to fully fulfill their promise in saving lives and decreasing the damage caused by catastrophic occurrences(Koubaa, A et.al,2019).

The discipline of catastrophe robotics has various design and development issues as it strives to improve disaster response and mitigation. The complexity and unpredictability of catastrophe conditions, as well as the technological limits and ethical issues involved with deploying robots for such important jobs, all contribute to these challenges. This study looks into these issues, providing insights into the difficulties encountered in the development and deployment of catastrophe robotics.

Disaster scenarios are wildly varied and highly unpredictable; they can occur in cities devastated by earthquakes or in areas devastated by floods. Designing robots that can navigate and function well in such a wide variety of often harsh situations is a major challenge. These robots need to be able to operate in areas with little to no GPS and communication infrastructure, adapt to various terrains, and resist bad weather.Disaster situations are highly unpredictable and vary widely, ranging from earthquake-ravaged cities to flood-ravaged locales. It is a big problem to design robots that can navigate and operate effectively in such diverse and frequently severe environments. These robots must be capable of adapting to different terrains, withstanding adverse weather conditions, and operating in regions with minimal or no GPS and communication infrastructure (Gatti, R. R et.al 2023).

The development of autonomous decision-making abilities in catastrophe robots is a daunting task. In dynamic and unstructured surroundings, these robots must make quick and effective decisions. To achieve a high level of autonomy, complex algorithms that are capable of real-time data processing, environmental sensing, and situational awareness are required. The requirement for these machines to make ethical and strategic decisions, especially when human lives are at stake, adds to the complexity.It is difficult to ensure dependable communication and connectivity in disaster-stricken areas when traditional communication infrastructure may be damaged or overburdened. Robust and versatile communication systems are required for robots to convey information back to command centers and communicate with other robots and rescue crews. A major problem is developing communication technology that can operate in such hazardous conditions. In disaster robots, power management and energy efficiency are key challenges. Robots must have enough power to function for lengthy periods of time, especially in distant or inaccessible regions. Creating energy-efficient systems and alternative charging alternatives, such as solar power or in-field charging stations, is a difficult challenge that necessitates creative thinking.To navigate and gather important information in chaotic circumstances, disaster robots must be equipped with advanced sensory capabilities. Sensors capable of detecting chemical, biological, or radioactive dangers are included, in addition to visual and auditory sensors. Processing the massive amounts of data collected by these sensors in real time, frequently with limited processing resources, adds another layer of complication (Preethesh, B et.al,2023).

3. HUMAN-ROBOT INTERACTION

Effective human-robot interaction is required for disaster robot deployment to be successful. These robots must be intuitive and user-friendly for operators with varied levels of technological knowledge. They should also be able to communicate and work with humans in the field, which necessitates advanced algorithms for social and behavioral comprehension. Disaster robots' size and mobility are critical elements influencing their usefulness. Larger robots may be more capable in terms of power and equipment carrying capacity, but they may struggle to maneuver in restricted locations. Smaller robots, on the other hand, can reach narrow locations but have limitations in terms of skills and endurance. A difficult part of disaster robot design is balancing size, mobility, and utility. Creating catastrophe robots necessitates traversing difficult ethical and legal terrain. The usage of robots in delicate settings, potential job displacement, and liability in the event of failures or accidents must all be thoroughly addressed. It is critical to establish clear ethical principles and regulatory frameworks for the appropriate development and deployment of these technologies. Finally, creating and deploying advanced robotics for disaster response can be too expensive. High costs can make these technology inaccessible, particularly in resource-limited regions that frequently require such advanced disaster response systems. Finding ways to cut expenses while maintaining functionality and dependability remains a big challenge. These issues will necessitate a multidisciplinary strategy that combines advances in robotics, artificial intelligence, materials science, and human-computer interaction. The potential for disaster robots to save lives and minimize the impact of disasters expands as technology advances, making it a subject of research with far-reaching consequences for the future. Cloud robotics, a new discipline that combines cloud computing and robotics, provides a new paradigm in which robots are augmented with more processing power, storage, and data-sharing capabilities. This essay shows several case studies that demonstrate cloud robotics in action, demonstrating how this technology is altering industries such as healthcare, disaster management, agriculture, and others. Cloud robotics has aided in the advancement of remote surgery in the healthcare sector. This technology provides real-time data processing, high-definition 3D vision, and precise control over surgical instruments by leveraging cloud computing. Another example is in patient care, where cloud-enabled robots help with patient health monitoring, medication management, and companionship, particularly in eldercare. These robots can access patient data from the cloud, allowing healthcare personnel to provide individualized care and remote monitoring (Vijeth, G.,et.al,2023).

4. DISASTER RESPONSE: SEARCH AND RESCUE OPERATIONS

Cloud robotics has significantly contributed to disaster response, particularly in search and rescue missions. After the Fukushima nuclear disaster, cloud-based robots were deployed for radiation monitoring and damage assessment. These robots could share data in real-time with remote control centers, allowing for timely and informed decision-making. Similarly, in earthquake-hit zones, drones equipped with cloud computing capabilities have been used for aerial surveys, mapping the extent of damage, and locating survivors. In agriculture, cloud robotics is used for precision farming, optimizing crop yield and reducing environmental impact. Robots like the EcoRobotix, which leverages cloud computing for real-time field analysis, can apply fertilizers and pesticides with high precision, minimizing waste and environmental harm. These robots gather data on soil conditions, crop health, and weather patterns from the cloud, enabling farmers to make data-driven decisions for efficient farm management (Shetty, S. H et.al, 2023).

The manufacturing industry has seen a transformation with the adoption of cloud robotics in smart factories. KUKA, a leader in industrial robotics, uses cloud-connected robots for various manufacturing tasks, including assembly, welding, and painting. These robots can share and analyze production data, optimize processes, and predict maintenance needs. This integration results in increased efficiency, reduced downtime, and improved product quality. In the retail sector, cloud robotics has enhanced inventory management and customer service. For instance, Walmart has experimented with cloud-based robots for real-time inventory tracking and restocking. These robots navigate aisles, scan shelves, and update inventory data in the cloud, ensuring timely stock replenishment. Additionally, cloud-enabled service robots in stores assist customers with navigation, product information, and even checkout processes. Environmental monitoring is another area where cloud robotics plays a vital role. The Wave Glider, an autonomous ocean robot, uses cloud computing to gather and transmit oceanographic data. This data is crucial for climate research, marine biology, and ocean conservation efforts. Similarly, drones and robotic vehicles are used in wildlife monitoring, collecting data on animal populations and habitats, aiding in conservation strategies. Cloud robotics is revolutionizing logistics and transportation through autonomous delivery systems. Companies like Amazon and UPS are experimenting with drones that use cloud computing for navigation, traffic management, and delivery tracking. These drones can autonomously deliver packages, reducing delivery times and costs, especially in hard-to-reach areas. NASA's Mars rovers, including Curiosity and Perseverance, can be seen as cloud robotics applications in space exploration. These rovers collect geological and atmospheric data, transmitting it back to Earth for analysis. While they operate semi-autonomously, cloud computing plays a role in processing the vast amounts of data they send, aiding scientists in making discoveries and planning further explorations. Robots equipped with sensors collect data on traffic patterns, pollution levels, and urban infrastructure. This data, processed through cloud computing, assists city planners in making informed decisions to improve traffic flow, reduce pollution, and enhance overall urban living (N. Prabhu et.al, 2023).

In education and research, cloud robotics enables remote laboratories and collaborative projects. Students and researchers can access and control robots remotely, conduct experiments, and share data through the cloud. This approach democratizes access to advanced robotics, fostering innovation and learning opportunities. These studies demonstrate the vast potential and diverse applications of cloud robotics. From enhancing healthcare and disaster response to revolutionizing agriculture, manufacturing, and beyond, cloud robotics is paving the way for more intelligent, efficient, and adaptive systems. As this field continues to evolve, it promises to bring about even more ground-breaking changes across various sectors, fundamentally reshaping the way we interact with technology and the world around us.

Training and preparedness are crucial components for the successful deployment of disaster robotics, a field that integrates robotic technology into disaster response and management. Effective use of robots in disaster scenarios can significantly enhance the efficiency and safety of operations. This essay delves into various aspects of training and preparedness necessary for disaster robotics, highlighting the importance of specialized training, simulations, collaboration, and continuous learning. The effectiveness of disaster robotics largely depends on the skill and knowledge of the operators. Specialized training programs are essential to equip personnel with the technical know-how required to handle sophisticated robotic systems. This training involves not only the operational aspects but also troubleshooting, maintenance, and understanding the limitations and capabilities of different robotic platforms. For instance, operators of aerial drones need to be skilled in flight control, navigation in complex environments, and interpreting the data collected by the drones. Simulations and drills play a pivotal role in preparing operators and robots for real-world disaster scenarios. These exercises should cover a wide range of potential situations, from natural disasters like earthquakes and floods to man-made crises such as industrial accidents or chemical spills. Advanced simulation software can create realistic virtual environments where operators can hone their skills without the risks associated with actual disaster zones. Regular drills in controlled but realistic settings ensure that both humans and robots are ready to respond effectively under various emergency conditions. For disaster robotics to be successful, seamless integration with existing emergency response teams and protocols is essential. This requires training programs that foster collaboration and communication between robot operators, first responders, and disaster management professionals. Understanding the roles and responsibilities of each team and how robotic technology can augment human efforts is crucial. Joint exercises can be conducted to simulate real-life coordination and cooperation between robotic units and human teams.

Understanding the specific challenges posed by different disaster environments is critical for both the operators and the development of the robots themselves. For instance, robots deployed in urban search and rescue missions face different challenges than those used in forest fire monitoring. Training should involve familiarization with various terrain types, environmental conditions, and potential hazards that might be encountered during operations. This knowledge is vital for adapting the use of robots to the specifics of each disaster scenario. Operators and teams involved in disaster robotics must also be trained on the ethical and legal aspects of using robotic technology in sensitive and high-stakes situations. This includes understanding privacy concerns, especially when drones equipped with cameras are used in residential areas, and navigating the legal frameworks governing the use of such technology in different jurisdictions. Disaster robotics requires a blend of skills that span robotics, information technology, disaster management, and even psychology and sociology. Training programs should therefore be interdisciplinary, equipping individuals with a broad understanding of various aspects of disaster response, including psychological first aid, cultural sensitivity in disaster-hit regions, and the social implications of deploying robots in such scenarios.

As robots become more autonomous through advancements in artificial intelligence (AI) and machine learning, training must also evolve. Operators should understand the basics of AI and how it influences robot behavior and decision-making. This knowledge is crucial for effectively managing autonomous systems, interpreting their actions, and intervening when necessary. The field of disaster robotics is rapidly evolving, with new technologies and methodologies emerging regularly. Continuous learning and adaptation are therefore essential components of training and preparedness. This includes staying updated with the latest technological advancements, learning from past deployments, and regularly revising training curricula to include new knowledge and insights. Given that disasters can occur globally and often require international response efforts, training in disaster robotics should also encompass international cooperation and standardization. Understanding global best practices, interoperability standards, and cross-border legal considerations is important for teams that might be involved in international disaster response. Operating in disaster zones can be stressful and emotionally taxing. Training should therefore include psychological resilience building, stress management techniques, and strategies for coping with the challenging emotional aspects of disaster response work. Training and preparedness are the bedrocks upon which the effective and efficient use of disaster robotics is built. Comprehensive training programs, regular drills and simulations, collaboration with emergency response teams, continuous learning, and an understanding of ethical and legal considerations are all vital. As the technology advances, so too must the training and strategies to ensure that disaster robotics continues to be a valuable asset in managing and mitigating the impact of disasters. The field of disaster robotics is rapidly evolving, driven by technological advancements and the increasing need for effective disaster response mechanisms. This essay explores the future trends and innovations in disaster robotics, highlighting how emerging technologies and approaches are poised to revolutionize the way we manage and respond to disasters.

One of the most significant trends in disaster robotics is the move towards greater autonomy and intelligence. Future disaster robots are expected to have advanced decision-making capabilities, enabled by sophisticated artificial intelligence (AI) algorithms. This will allow them to navigate and operate in complex disaster environments with minimal human intervention. AI-driven robots will be able to assess situations, make strategic decisions, and execute tasks more efficiently, reducing the time and risks associated with human-led operations.

The development of more advanced sensors and perception systems is another key trend. Disaster robots equipped with these technologies will be able to better understand and interpret their environment. This includes the ability to detect and analyze a wide range of stimuli, from structural integrity in buildings to signs of life under rubble. Enhanced sensory capabilities will lead to more effective search and rescue operations, as robots will be able to quickly identify victims and hazards. Swarm robotics, where multiple robots operate in coordination, offers great potential for disaster response. These robot swarms can cover large areas quickly, gather vast amounts of data, and work together to accomplish complex tasks. In the future, we could see swarms of drones conducting aerial surveys of disaster areas or groups of ground robots collaboratively removing debris. Swarm robotics mimics the collective behavior of natural systems, like ants or bees, leading to efficient and scalable operations. Enhancing the collaboration between humans and robots is another future trend. Human-robot teaming involves integrating the intuitive decision-making abilities of humans with the precision and endurance of robots. This synergy can significantly improve the effectiveness of disaster response efforts. For example, human responders could focus on strategic planning and critical decisions, while robots carry out dangerous or labor-intensive tasks. The development of robotic exoskeletons and wearable technologies for disaster responders is an innovative trend. These devices can augment human abilities, allowing responders to lift heavier loads, navigate dangerous terrains, or work for extended periods without fatigue. This technology not only enhances the capabilities of human responders but also reduces their risk of injury in hazardous environments.

Improving communication capabilities in disaster robots is crucial, especially in environments where traditional communication infrastructure is damaged or unavailable. Future innovations may include the use of satellite communications, mesh networks, or even low-frequency signals that can penetrate through rubble. Enhanced communication will ensure that robots can stay connected and share vital information in real-time, even in the most challenging conditions. Modular and customizable robot designs are expected to become more prevalent. Such designs allow for robots to be quickly adapted to specific disaster scenarios. Modular robots can be reconfigured with different tools or sensors depending on the task, whether it's search and rescue, medical supply delivery, or environmental monitoring. This versatility makes them highly valuable in the diverse and unpredictable nature of disaster scenarios. As disaster robots are often required to operate for extended periods, energy efficiency and sustainable power sources are key areas of innovation. Future disaster robots may incorporate solar panels, energy harvesting mechanisms, or advanced battery technologies to prolong their operational lifespan. These sustainable energy solutions will ensure that robots can remain operational without the need for frequent recharging or refueling.

Bio-inspired designs, where robots mimic biological organisms, are an emerging trend in disaster robotics. For example, snake-like robots could navigate through tight spaces in collapsed structures, while robots that mimic flying animals could efficiently survey disaster-stricken areas. These designs often offer superior maneuverability and adaptability in complex environments.As AI plays an increasingly central role in disaster robotics, ensuring ethical AI practices and accountability will become paramount. This includes developing AI systems that are transparent, fair, and respectful of privacy and human rights. It also involves establishing clear guidelines and accountability mechanisms for the actions of autonomous robots in disaster scenarios.

The integration of cloud robotics and big data analytics is another future trend. By leveraging cloud computing, disaster robots can access vast computational resources and share data seamlessly. This enables more sophisticated data analysis, predictive modeling, and coordinated multi-robot operations. Big data analytics can help in understanding disaster patterns, improving response strategies, and even in predictive modeling to anticipate future disasters. The future of disaster robotics is marked by innovations that promise to make disaster response more effective, efficient, and safe. From enhanced autonomy and intelligence to bio-inspired designs and ethical AI, these advancements are set to transform how we approach disaster management. As the technology continues to evolve,

The integration of robotics into disaster response and management presents a unique set of challenges and considerations, particularly in terms of policy and regulatory frameworks. As disaster robotics becomes more prevalent, there is a growing need to develop comprehensive policies and regulations that ensure their effective, ethical, and safe use. This essay explores the various aspects of policy and regulatory frameworks necessary for disaster robotics, covering areas such as safety standards, ethical guidelines, liability issues, interoperability, and international cooperation. Safety is paramount in disaster robotics. Policies must establish strict safety standards and certification processes for robots used in disaster scenarios. These standards should cover the mechanical and electrical safety of the robots, their operational reliability, and fail-safe mechanisms. Certification processes should be rigorous to ensure that only robots that meet these high safety standards are deployed in disaster situations. This will minimize the risk of malfunction or accidents that could further exacerbate disaster situations or harm responders and victims. The development of ethical guidelines for the deployment of robots in disaster situations is critical. These guidelines should address issues like privacy, especially when using surveillance drones, and the prioritization of tasks in situations where resources are limited. Policies should also consider the impact of robots on the emotional and psychological well-being of disaster victims and responders, ensuring that their use does not add to the trauma of the situation.

Clear policies regarding liability and legal responsibility are necessary to address potential issues arising from the use of disaster robots. This includes determining who is responsible for decisions made by autonomous or semi-autonomous robots, and who is liable in case of malfunctions or accidents. The legal framework should cover aspects like insurance requirements, indemnification clauses, and mechanisms for dispute resolution. Interoperability is essential for the effective deployment of disaster robotics. Policies should promote the standardization of robotic systems to ensure they can work seamlessly with other robots and existing disaster management technologies. This includes standardizing communication protocols, data formats, and operational procedures. Standardization also facilitates training and maintenance, making it easier for disaster response teams to integrate robotics into their operations. With disaster robots often collecting sensitive data, robust policies on data security and privacy are vital. Regulations should dictate how data is collected, stored, shared, and protected. This is particularly important when dealing with personal information of disaster victims or when operating in areas where sensitive information might be inadvertently collected.

Disaster response often requires international cooperation, especially in large-scale or cross-border disasters. Policies should therefore consider international standards and compliance requirements. This includes aligning with global initiatives such as the Sendai Framework for Disaster Risk Reduction and ensuring that the deployment of disaster robots does not violate international laws or norms. Policies should address issues of access and equity in the deployment of disaster robotics. This means ensuring that the benefits of this technology are available to all regions, including those with limited resources. Regulations could encourage or mandate the sharing of technology and expertise, and provide guidelines for equitable deployment to avoid disparities in disaster response capabilities. Regulatory frameworks should also support ongoing research and development in disaster robotics. This could involve providing funding, facilitating partnerships between academia, industry, and government, and creating policies that encourage innovation while ensuring safety and ethical standards are maintained. The policy framework should include the training and certification of operators of disaster robots. This ensures that individuals operating these robots have the necessary skills and knowledge to do so safely and effectively. Regular training updates might be required to keep pace with technological advancements. Engaging the public and maintaining transparency in the deployment and use of disaster robots is crucial for building trust and acceptance. Policies should encourage public dialogue, education, and awareness campaigns about the role and benefits of robotics in disaster response. Policies should take into account the environmental impact of disaster robots. That includes regulations on the disposal and recycling of robots and their components, as well as guidelines for minimizing their ecological footprint during operation. Developing a comprehensive policy and regulatory framework for disaster robotics is essential to ensure their responsible, ethical, and effective use. This framework must address a wide range of issues, from safety and liability to international cooperation and public engagement. As the field of disaster robotics continues to evolve, so too must the policies and regulations that govern it, ensuring that these powerful tools are used for the greatest benefit in disaster response and management.

REFERENCES

Alotaibi, E.T., Alqefari, S.S., & Koubaa, A. (2019). LSAR: Multi-UAV Collaboration for Search and Rescue Missions. *IEEE Access, 7*, 55817–55832.

Browne, C. (2011). Evolutionary game design. *IEEE Transactions on Computational Intelligence and AI in Games*, *2*, 11–21.

Cheikhrouhou, O., Koubaa, A., & Zarrad, A. (2020). A Cloud Based Disaster Management System. *Journal of Sensor and Actuator Networks.*, *9*(1), 6. doi:10.3390/jsan9010006

Gatti, R. R., Singh, C., Agrawal, R., & Serrao, F. J. (Eds.). (2023). *Self-Powered Cyber Physical Systems*. John Wiley & Sons. doi:10.1002/9781119842026

Gupta, S., Modgil, S., Kumar, A., Sivarajah, U., & Irani, Z. (2022). Artificial intelligence and cloud-based Collaborative Platforms for Managing Disaster, extreme weather and emergency operations. *International Journal of Production Economics, 254*. doi:10.1016/j.ijpe.2022.108642

Koubaa, A., Allouch, A., Alajlan, M., Javed, Y., Belghith, A., & Khalgui, M. (2019). Micro Air Vehicle Link (MAVlink) in a Nutshell: A Survey. *IEEE Access : Practical Innovations, Open Solutions*, *7*, 87658–87680. doi:10.1109/ACCESS.2019.2924410

Koubaa, A., Qureshi, B., Sriti, M. F., Allouch, A., Javed, Y., Alajlan, M., Cheikhrouhou, O., Khalgui, M., & Tovar, E. (2019). Dronemap Planner: A service-oriented cloud-based management system for the Internet-of-Drones. *Ad Hoc Networks*, *86*, 46–62. doi:10.1016/j.adhoc.2018.09.013

Kurunathan, H., Severino, R., Koubaa, A., & Tovar, E. (2018). IEEE 802.15. 4e in a nutshell: Survey and performance evaluation. *IEEE Communications Surveys and Tutorials*, *20*(3), 1989–2010. doi:10.1109/COMST.2018.2800898

Kurunathan, H., Severino, R., Koubaa, A., & Tovar, E. (2019). DynaMO—Dynamic Multisuperframe Tuning for Adaptive IEEE 802.15.4e DSME Networks. *IEEE Access, 7*, 122522–122535.

Prabhu, Shetty, GowdaSingh, & S. (2023). Design & Implementation of Brain Controlled Interface For Healthcare Applications. *IEEE International Conference on Distributed Computing, VLSI, Electrical Circuits and Robotics (DISCOVER),* 197-200. 10.1109/DISCOVER58830.2023.10316733

Preethesh, B., D'Costa, H. A., Alva, S. R., Kanchan, V. H., & Gatti, R. R. (2023). Feasibility Study of Digital Twin in Automotive Industry—Trends and Challenges. *Self-Powered Cyber Physical Systems,* 265-281.

Shetty, S. H., Shetty, S., Singh, C., & Rao, A. (2022). Supervised Machine Learning: Algorithms and Applications. *Fundamentals and Methods of Machine and Deep Learning: Algorithms, Tools and Applications*, 1-16.

Vijeth, G., & Gatti, R. R. (2023). Review on Robot Operating System. *Self-Powered Cyber Physical Systems*, 297-307.

Chapter 16
Night Vision-Based Human Identification and Path Planning in Rescue Operations Using an Autonomous Robot

Sandeep Bhat
SIT Mangaluru, India

ABSTRACT

This chapter provides a progressive thing classification and the most limited organising approach for robot-assisted night-time salvaging jobs. In reality, people are sorted using a system for determining skin tone. Boundaries were identified using the skin colour based (SCB), pixel count based (PCB), correlation coefficient based (CCB), and histogram draws near techniques in this study. The PCB, CCB, and SCB calculations are all utilised to snag distinct proof in order. The SCB calculation, on the other hand, gradually determines whether the observed entity is human or nonhuman. It also distinguishes between live and nonliving impediments, allowing for more efficient salvage efforts (except for flies, mosquitoes, and different creepy crawlies). According to continuous testing data, the accuracy of the CCB, PCB, and SCB calculations is 87.5%, 88.8%, and 90.9%, respectively, and the time length of the CCB, PCB, and SCB calculations is 6.27 seconds, 6.50 seconds, and 6.57 seconds, respectively.

I. INTRODUCTION

Both image processing and object recognition have various applications (David et al., 2009). Surveillance, military, advanced mechanics, and other applications are a few. Line identification, image recovery, and assessment are some of the identification applications. Numerous locations, including parking lots, Bank ATMs, military installations, and atomic reactors, require object recognition and localization. For independent path planning and control in robot applications, impediment acknowledgment is vital (Sariff, & Buniyamin, 2006). Distinguishing and separating between different kinds of moving items that might be available in a checking region is necessary for general security reasons. The moving substance should

DOI: 10.4018/979-8-3693-1914-7.ch016

be classified as either nonhuman or human due to safety considerations. Item acknowledgment is the process of identifying a specific article among a succession of images or videos (Sridaran & Dhulekar, 2001) (Lowe, 1999). Articles can be viewed whether they are turned, interpreted, scaled, or partially hidden from view. The process of successfully identifying and tracking an item in a video series is known as object recognition in video groups.

An original class of neighbourhood photo highlights is used by an object identification system. Along with some light invariance, the features include picture scaling, interpretation, and turn invariance (Lowe, 1999). The two stages of object recognition are component coordinating and highlight extraction. A PC vision calculation called filter (Scale-Invariant Element Change) is utilized to distinguish and address area features in pictures (Tao et al., 2010).

Numerous applications of the SIFT technique exist, such as photo recovery, object recognition, present evaluation, and many more (Alhwarin & Wang, 2008). Because highlight-based article acknowledgment calculations are harsh on variations in scale, light, and relative interpretation, it is difficult to identify interest localities (Wang, n.d.). Reference (Kalatehjari, Hosseini, & Behrad, 2012) provides a way for object acknowledgment based on eliminating highlight focuses using the scale invariant component change (SIFT) and then using the photometrical qualities of the focuses to make an underlying match.

A region merging strategy combined with the affine scale invariant feature transform (ASIFT) (Oji, 2012) produces an ideal object conspicuous confirmation plan with expansive limit acknowledgment.

It was actually possible to create a computer program that can appropriately perceive objects in images or videos without the need for human arrangement (http://www.kurzweilai.net/a-smart-object-recognition-algorithm-that-doesnt-need- humans, 2014). A technique for showing a robot's and obstructions' bearings in three aspects (x, y, time) is portrayed (Steele, 1989). so that possible effects can be identified without being aware of the obstacle's path. Another item discovery and differentiating proof framework that makes use of novel discriminative preparation methods with some named information is the inactive SVM (Felzenszwalb et al., 2010). The primary drawback of these computations is their execution complexity, which prevents them from being applied to simple three-dimensional article acknowledgment. In order to achieve optimal performance, robots are increasingly needed to monitor certain systems, such as military applications, nuclear reactors, and other systems (http://www.answers.com/topic/what-determines-skin-color#ixzz2umnkXanN). Robots never make mistakes in the final product since they operate endlessly. It's challenging to keep an eye on the robot physically all the time. A significant accomplishment is making robots self-sufficient. The robot has to perform all tasks independently; to do so, it needs to create a route plan that involves consistently dodging obstacles in order to finish the task.

Robot Path Planning in a non-static climate has two basic goals: finding a neighbourhood approved way and storing an incidental outcome (Bhat, Kapoor, & Raina, 2012). Robot-way arranging calculations with constant impact free computation operate continuously and don't need prior knowledge of track or blockage improvements. As long as the robot moves faster than the goal and the fiery framework update recurrence is large, this calculation finds the objective in the robot infectious at the point where the snags stop (Yen & Johnston, n.d.). To facilitate the robots from their fundamental highlight their objective, A-star, D-star, Light, and Uniform Cost Search were carried out and in this way consolidated (David et al., 2009) (Rastgoo, Nakisa, & Kebangsaan, 2014) (Serpen & Dou, 2015). The robot may gradually accomplish the goal by employing an adaptive robot route-planning control system that is based on obstacle detection, obstacle evasion along its path, robot rotation on its own, and determining the robot's location for selection (Chen et al., 2015).

A novel design approach could guarantee global target access while effectively eliminating dead circulation and oscillation in mobile robots (Wai & Liu, 2011). At the point when a topologically built map portrays the robot's way and encompasses, it basically gives the space from the objective to the whole robot work area (Chun2Gang & Yu2Geng, 2003). To represent the robot's size and impediments, a better virtual force field (VFF) method is utilized. A fuzzy control module is utilized to handle the impediment evasion issue in conditions with moving hindrances by fluctuating the robot's turn point (Willms & Yang, 2006). A DVFF approach is utilized to finish the route plan for a mobile robot in an unidentifiable background. This solution uses the D* algorithm-based global path forecast in conjunction with the Virtual Force Field (VFF) barrier prevention mechanism (Willms & Yang, 2004). The two objectives that the ABC calculation finishes are finding the straight impact freeway on account of non-fixed deterrents and the contention freeway in the circumstance of stale blocks (Jianjun et al., 2014).

In the Artificial Potential Field (APF) type for routing a mobile robot, the goal offers a virtual prospective that draws the robot, while the barrier creates a practical potential that pushes the robot back (Djekoune, Achour & Toumi, 2009). Utilizing the particle swarm improvement (PSO) calculation, fake potential field procedures will be changed and upgraded for a versatile robot route in powerful conditions to resolve issues such as neighbourhood minima and upgrade the portability robot's course highlight (Goel & Singh, 2014). For concurrent robot route forecasting, a lattice-based space-propagating non-static system is used, which consists of protective margins around obstacles via local consequence functions (Sfeir, Saad, & Saliah-Hassane, 2011). A cell automata-based way arranging technique rapidly decides the ideal potential ways. This methodology lays out a parent-kid relationship for each cell to lessen the request time (Ahmed, Abdalla, & Abed, 2015). With simply neighbourhood sidelong collaborations between neurons, the brain network for robots in a pulse-coupled neural network (PCNN) approach is topologically organized (Willms & Yang, 2008). Another strategy called Invasive Weed Optimization (IWO) utilizes the way that weeds are colonizers to tackle the issue of a moveable robot's way arrangement in to some degree or totally obscure environmental elements (Syed & Kunwar, 2014).

A novel technique IWO is being utilized to address the path setup challenges of a mobile robot in conditions that are either somewhat or entirely unknown (Qu et al., 2009). It is predicated on weeds' propensity to colonize. By calculating the obstacle's position, speed, and direction, this method enables the robot to avoid it while moving from its starting point to its goal. The novel algorithm's computing time and path are contrasted with those of the A-star method (Mohanty, Dayal, & Parhi, 2014) (Razavian & Sun, 2005). As has already said, most literature-based research has flaws. These methods are not required for military applications or simple robot route forecasts because they are complex and require extra mathematical methodologies. Reconnaissance is essential for a variety of locations, including train stations, parking lots, army sites, ATM corners, and atomic reactors. A solid observation structure should be arranged and improved with continuous tangle area and recognizable evidence. Figure 1 shows the SCB object acknowledgment method. The RGB picture, which has potential gains of R=200, G=70, and B=20, is utilized to match the shade of human skin. The picture is caught by a surveillance region camera. The SAD methodology is utilized to distinguish an article as a basic step for seeing it. As shown beforehand, article finding is improved by contrasting the ongoing edge and the first packaging. An edge contrast in the checking area shows the presence of an extra thing (deterrent).

The mathematically tended to in Conditions (1) and (2): in condition (1), with $I_k(i, j)$ tending to the current k^{th} frame and $I_{k-1}(i, j)$ tending to the prior $(k-1)^{th}$ frame are using the SAD calculation (Bhat, Kapoor, & Raina, 2012).

Where $I_k(i,j)$ is the current k^{th} casing and $I_{k-1}(i, j)$ is the past $(k-1)^{th}$ edge. In an ultimate situation when there is no movement:

$$I_k(i,j) = I_{k-1}(i,j) \tag{2}$$

A fundamental computation reliant upon the Skin Colour Based (SCB) object affirmation procedure is made and assessed as well as perceiving the saw object as human or nonhuman. The SCB algorithm works by determining the skin's red colour content. Humans have a higher concentration of red pigment than any other species. As a result, barriers between humans and nonhumans can be separated. The SCB object recognizable proof calculation is as depicted in Figure 1.

The connection coefficient in the CCB approach (Yen & Johnston, n.d.) is '1' if the photographs are identical, and '0' if they are completely unique. The amount of white pixels in a continuous image that is still in the air is measured using the PCB process (David et al., 2009), and the same is done for the image that follows. At last, given how many white pixels were inspected, the items are similar, assuming the pixel counts are practically equivalent. Figure 2 shows the activity of robots in arranging

Figure 1. The SCB object recognizable proof calculation

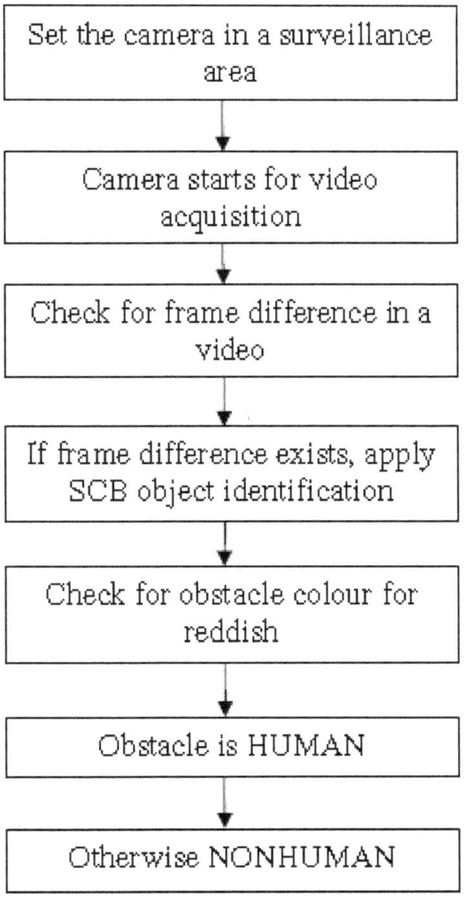

ideas. Place the camera in an observing region and turn it on to record video that you may later alter into outlines (pictures). The casings should be meant to be handled in parallel. Determine the contrast between continuous edges to find moving hindrances inside a reconnaissance district.

There isn't a moving obstruction if the difference is zero; if not, there is. If there is an obstruction, get its pixel values by tracing and labelling it (both in the x and y axes).

To conjecture the course, one should know the obstruction's voyaging bearing and speed. The robot could possibly arrive at its objective, assuming its speed and the area of the obstruction are perceived. In the event that variable speed is available, the way won't exist, especially assuming the robot is traveling more slowly than the obstruction; alternately, assuming that variable speed is missing, the way will exist. The situation of moving and fixed things is the main variable that decides if the robot way is available.

II. RESULTS AND ANALYSIS

The SCB calculation's uniqueness as well as the advancement of the line and mass calculations for robot routes are displayed in the pictures underneath. Figure 3 shows the disengaged findings from the CCB strategy. On the off chance that there is no distinction between two articles (snags), the association coefficient of esteem (R) is 1. A graphical portrayal of two indistinguishable fountains with an association coefficient of 0.49 is displayed in Figure (a). As far as possible, regard is 0.2; however, when everything is considered, it is 0.8. R = - 0.19 in figure (b) means two exceptional picture plans (since the connection coefficient is 0 or more modest for totally unforeseen photographs).

The results of thinking about two comparison items using the PCB technique are depicted in Figure 5. With an 88.8% productivity, the results suggest that it is fair for ongoing obstacles. However, the CCB calculation is unable to discern between two identical human images with varying levels of beauty. When compared to CCB (88.8%), it is better suited to identify obstacles. Here, the result is that they are identifiable 87% of the time despite the fact that they are the two individuals. Thus, this invention in PCBs can be used for just about indistinguishable products.

The SCB algorithm for identifying human impediments is seen in Figure 6.

Figure 7 portrays the results of the immediate SCB calculation. This likewise outlines the SCB calculation for human deterrent area results (a), (b), and (c). In this figure, the SCB calculation for nonhuman hindrance acknowledgment computation is exclusively for cat, dog, and cows.

Figures 8(a) and 8(b) show the outcomes of the SCB method for nonhuman hindrance detection (an ant and a butterfly, respectively) and for nonhuman hindrance acknowledgment (an ant and a butterfly, respectively).

To locate obstacles that are not human (Ant), utilize the SCB algorithm. The SCB algorithm is used in nonhuman obstacle detection (Ant) and nonhuman obstacle detection (Butterfly). Figures 8(a) and 8(b) show an ant as the identified obstruction, whereas Figure 8(b) shows a butterfly as the detected obstacle, suggesting that the object is not human. Figure 9 underneath represents the progressive impacts of propelling the skin colour-based (SCB) calculation. Figure (a) shows that the obstruction is a bumble bee, and Figure (b) represents that the hindrance is a reptile, which is certainly not a human.

Table-1 shows that for ongoing image recognition, the SCB technique performs better compared to the CCB and PCB approaches. Figure 10 shows a continuous examination of two pictures. The precision rates of the CCB, PCB, and SCB calculations are 87.5 percent, 88.8%, and 90.9 percent, in a specific order.

Figure 2. The working principle

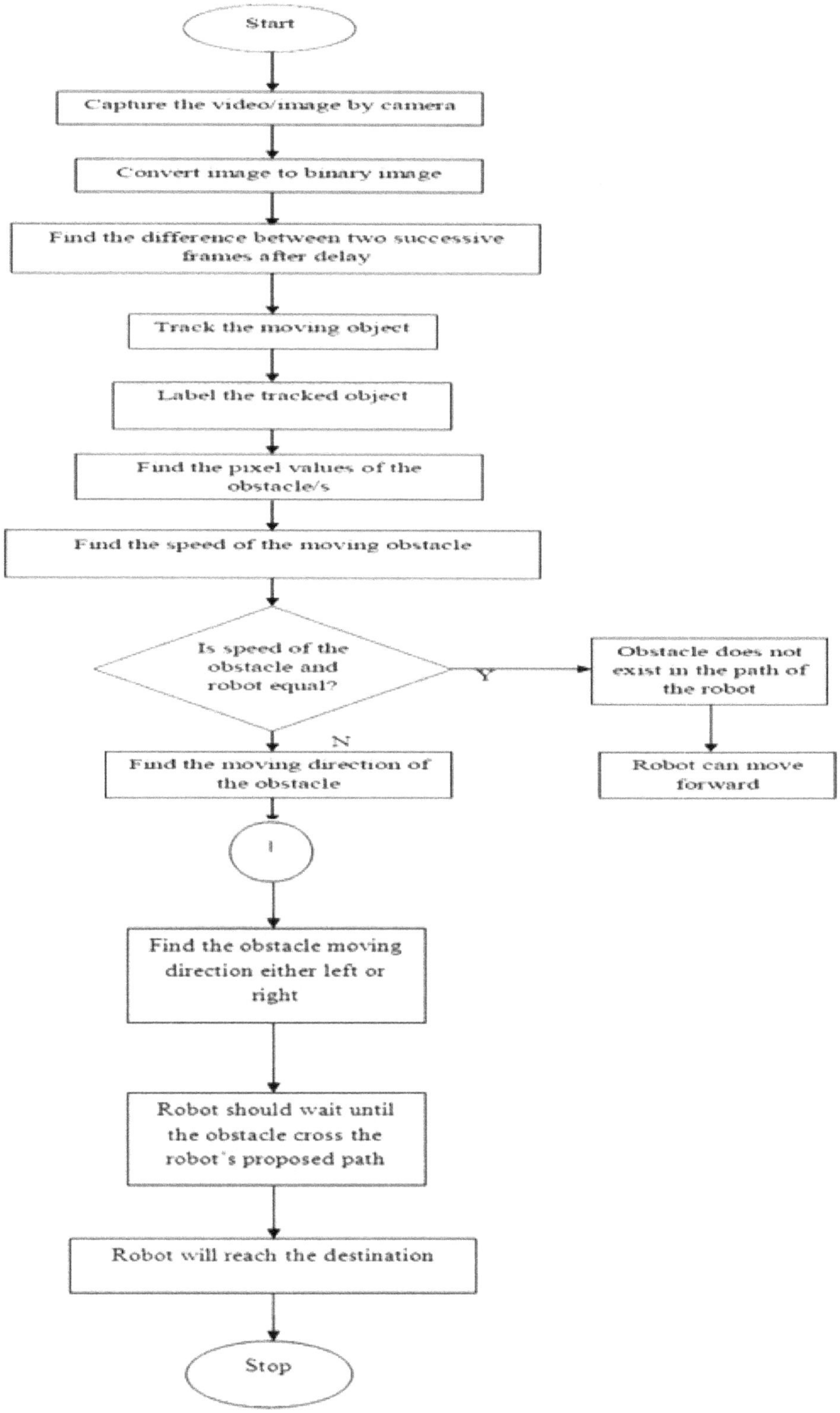

Figure 3. The CCB outcome. (a) comparable Images,(b) unrelated Images

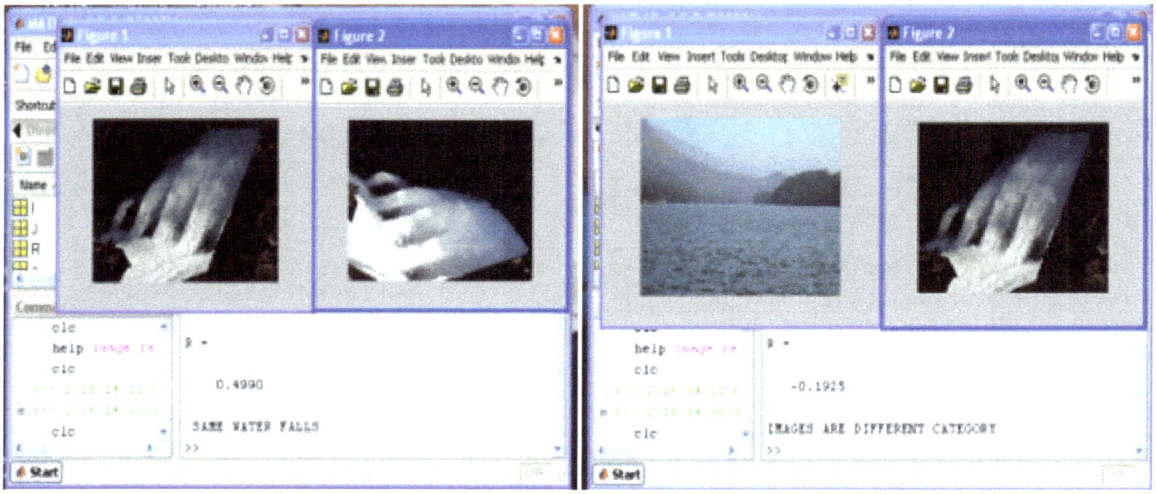

(a) (b)

Figure 4. Ongoing results of CCB calculation. (a) Real-time deterrent evaluation of a person and a toy pictures, (b) real-time impediment correlation of two human analogies.

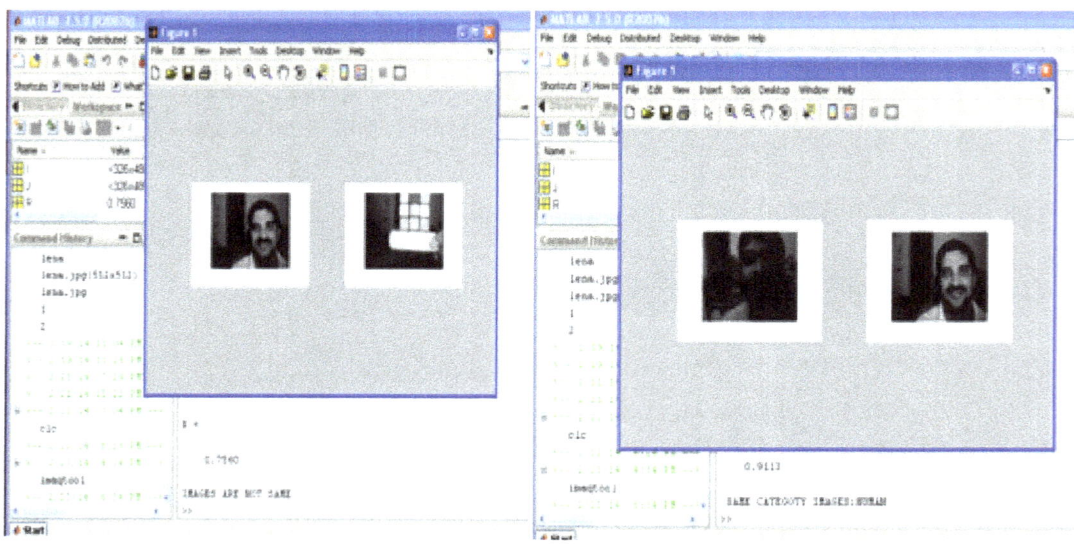

(a) (b)

The region of interest, which contains both stationary and moving obstacles, is shown in Figure 11. In this scenario, the collision-free path ought to be determined.

Figure 13 shows the moving snag directly through the robot's proposed way to objective.

Figure 5. The PCB impediment assessment

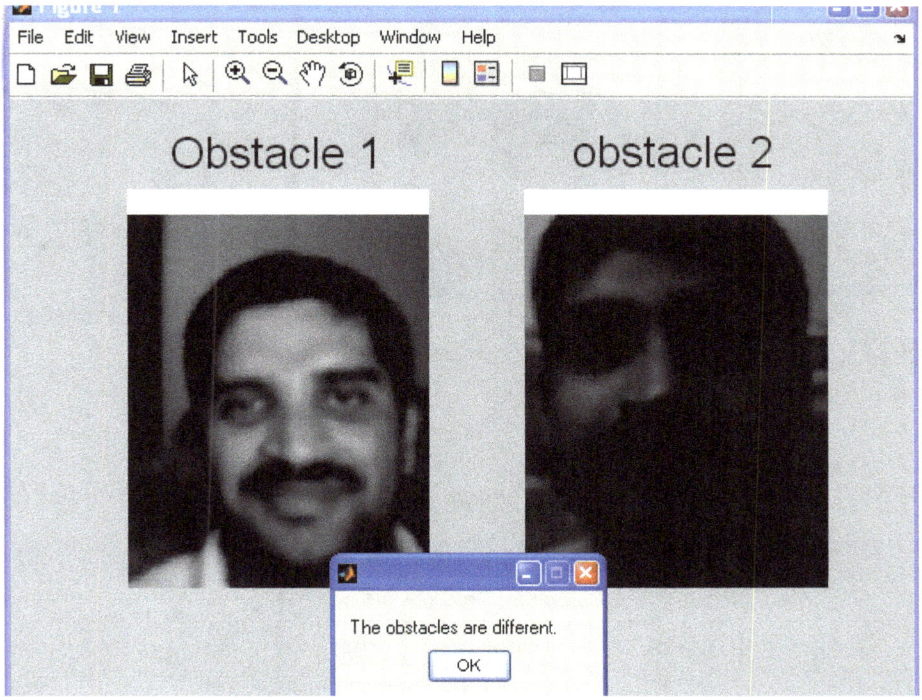

Figure 6. SCB algorithm results (RGB=200,70,20 for night vision)

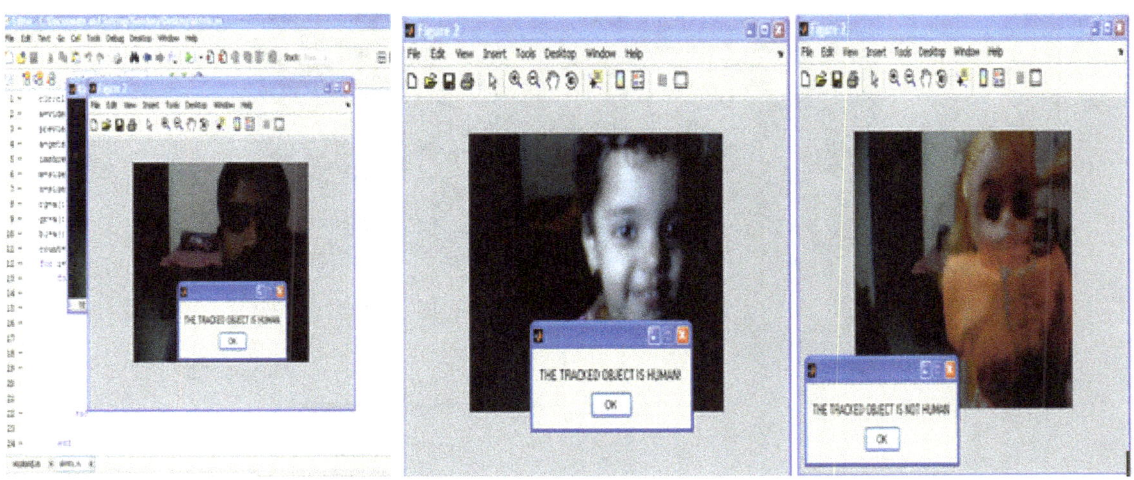

Together with the stationary obstacles, Figures 14 and 15 show the moving obstacle traveling in two distinct directions to the right.

The result of the edge distinction between the following two casings, as demonstrated in figures 14 and 15, is shown in figure 16. The snag's movement is decoded by a non-negative value of the edge contrast.

Figure 7. SCB algorithm results for non-human but live obstacles (normal)

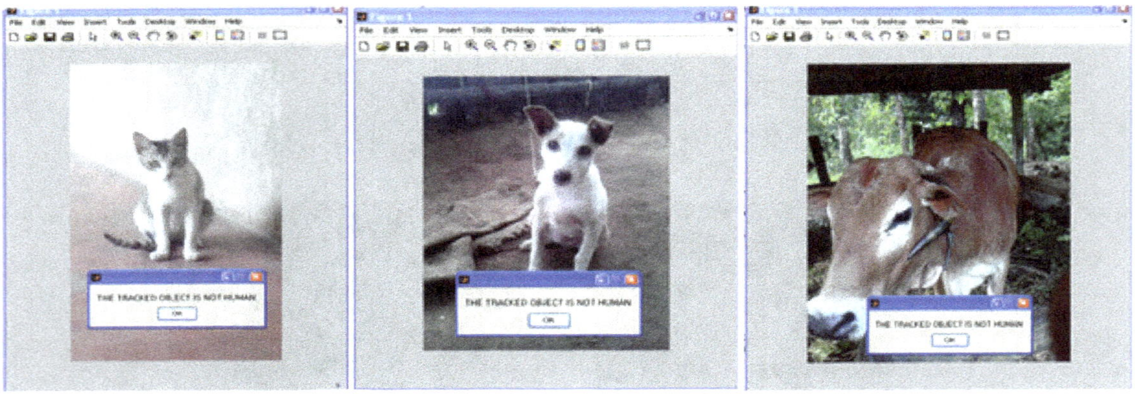

Figure 8. The SCB algorithm results for non-human but not considerable live obstacles (night vision)

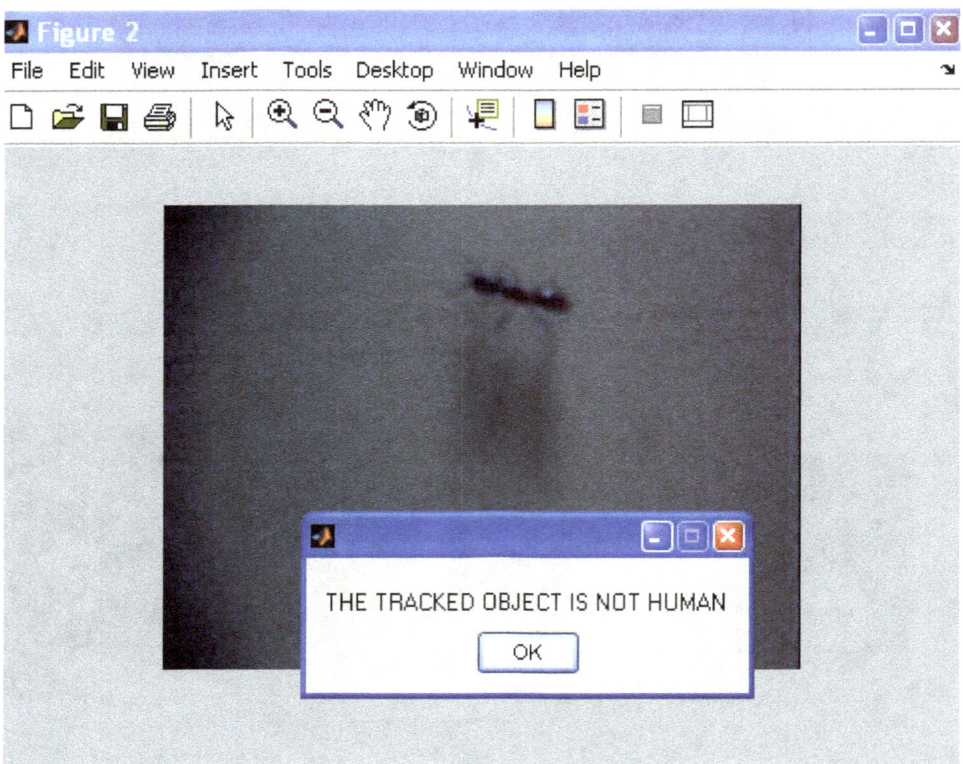

The speed of the moving deterrent is determined as follows:
Speed = Distance/Time
In image processing, the distance is as far as the pixels, and the time is estimated in a flash.

Figure 9. The SCB Calculation for nonhuman obstruction affirmation. (a) SCB Calculation for nonhuman (honey bee) obstruction detection, (b) SCB calculation for nonhuman (lizard) deterrent unmistakable confirmation

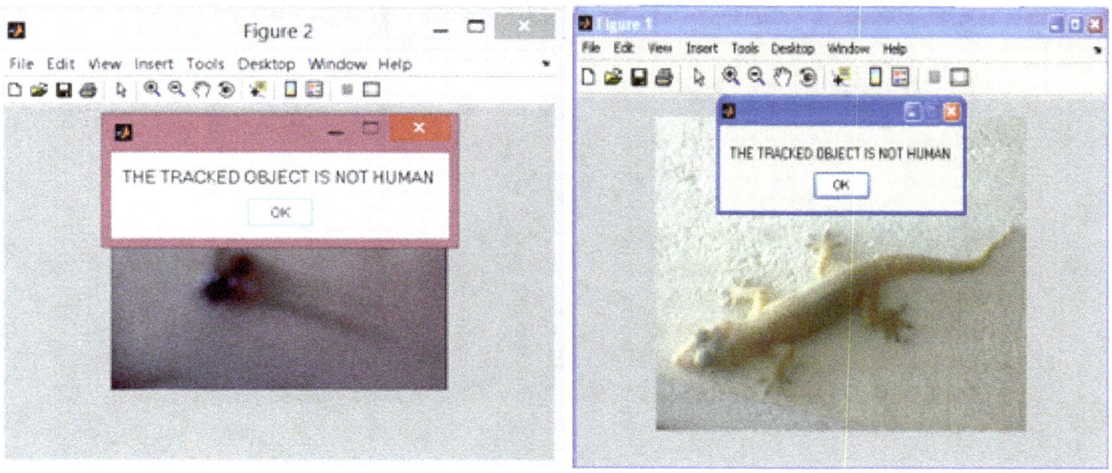

(a) (b)

Table 1. The assessment of SCB, CCB and PCB algorithms

	PCB	CCB	SCB
Number of tests	09	08	11
Success rate	08	07	10
Effectiveness(%)	88.80	87.50	90.90
Duration (Seconds)	06.50	06.27	08.67

Figure 10. The assessment of PCB and CCB algorithm

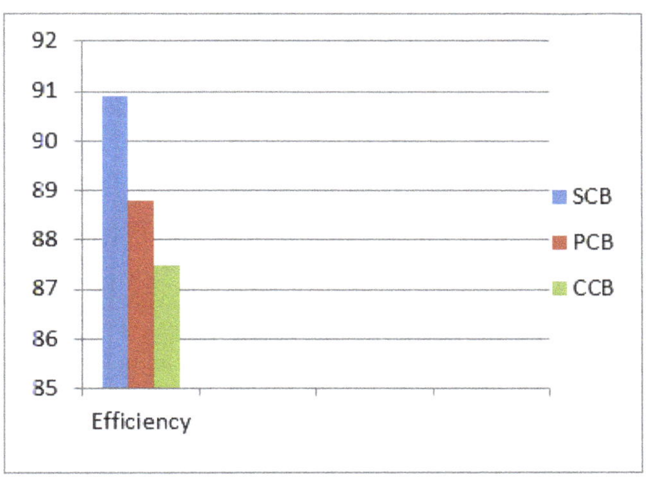

Figure 11. Static snags in the return for capital invested of the robot

Figure 12. The robot path planning initialization

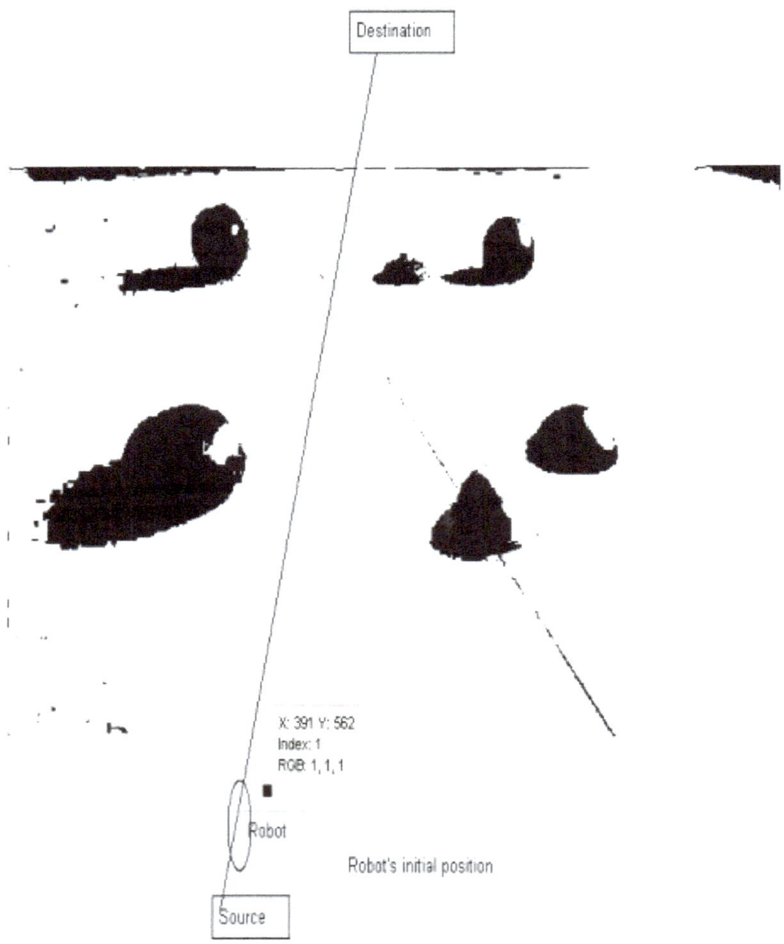

Figure 13. Moving deterrent in the robot's proposed way

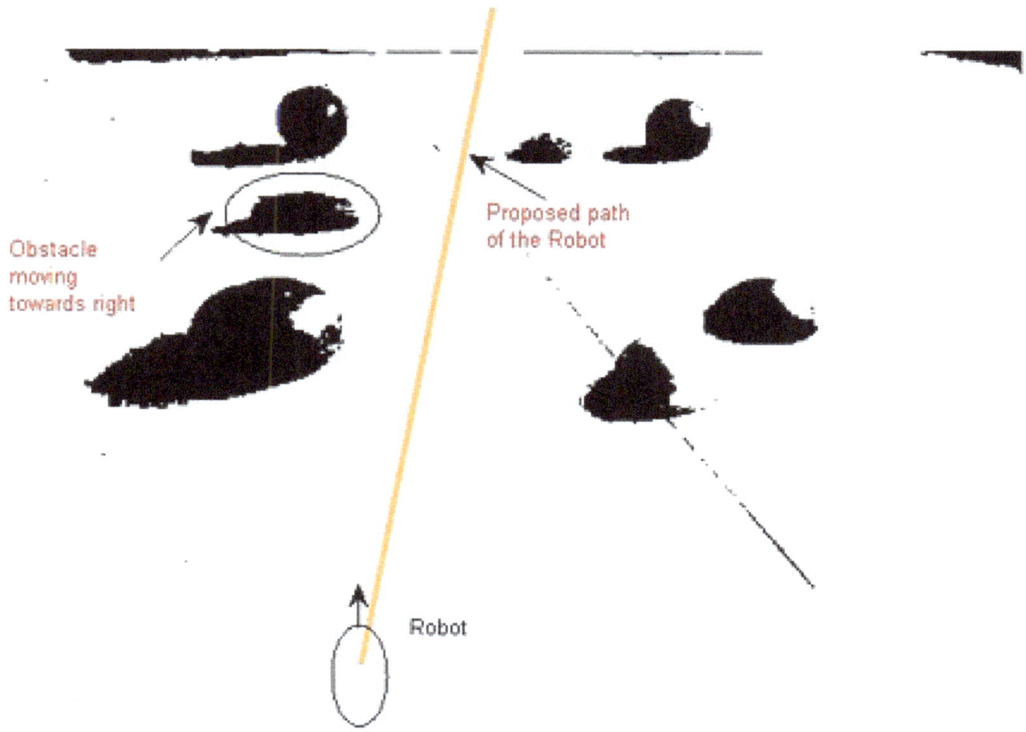

Figure 14. Moving obstacle path position 1

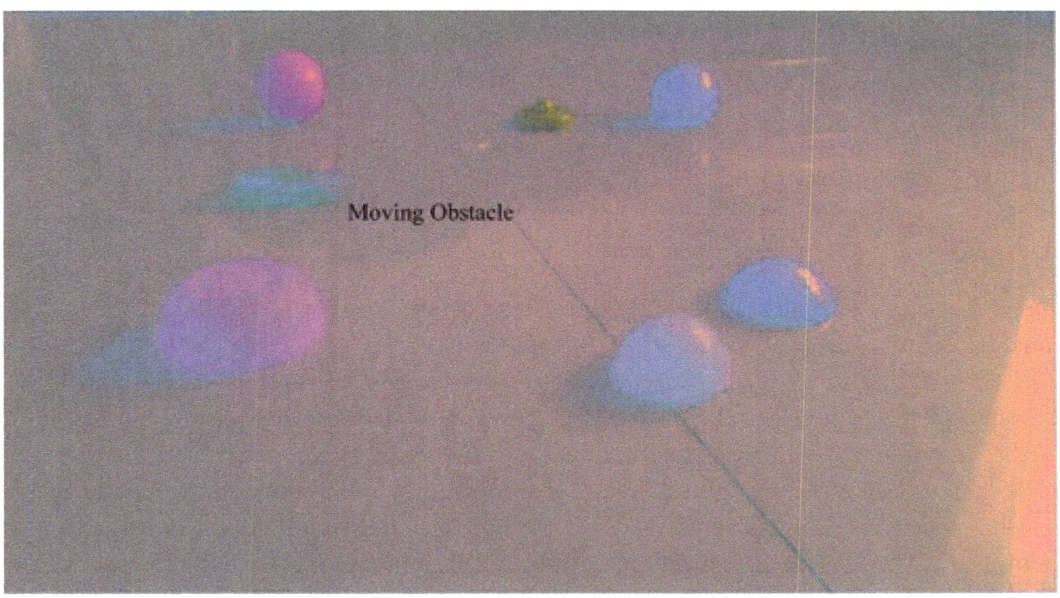

Figure 15. Moving obstacle path position 2

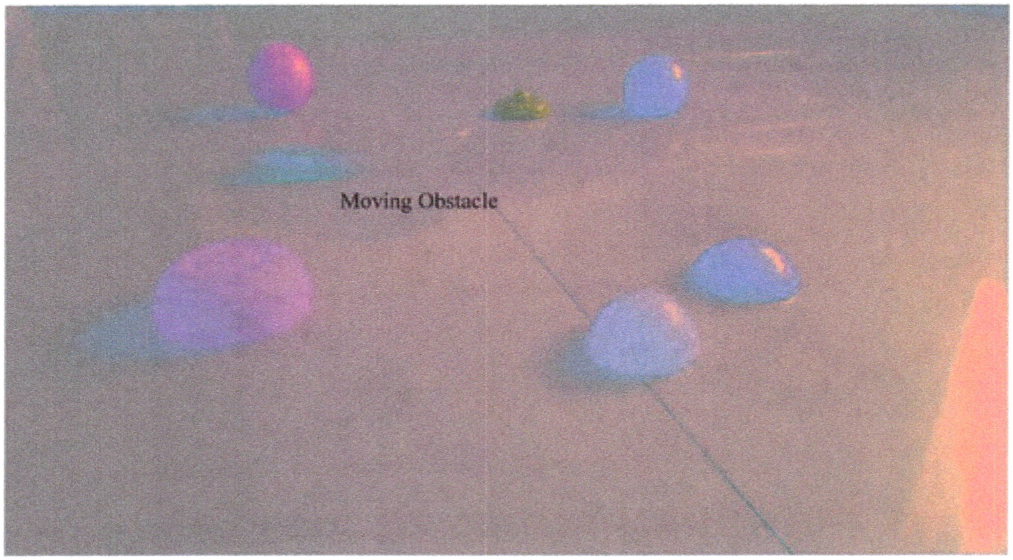

Figure 16. Difference between two successive frames 1 and 2

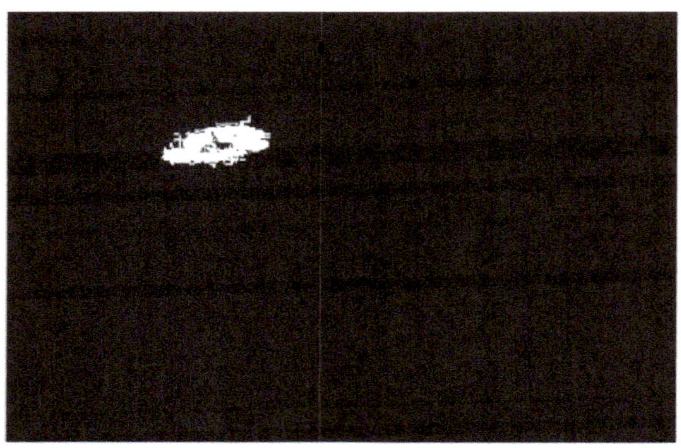

Figure 17 outlines the calculation of the speed and heading of the impediment. The course of the snag can be determined as follows:

Direction = (x-coordinate pixel value of frame 1) – (x-coordinate pixel value of frame 2)

Assuming the outcome after estimation is negative demonstrates that the impediment is moving towards the right in the event that it is not moving towards the left.

Decide the position, pixel values, and speed of the snag; the robot's situation and speed are known. Robots will arrive at their objective no matter what their speed, assuming that an obstruction is nearer to their proposed way. Figure 18 shows this. The block has proactively crossed the robot's proposed way before the robot draws nearer to it. As found in figure 19, a robot can't arrive at the objective when

Figure 17. Computation of speed and moving direction of the robot

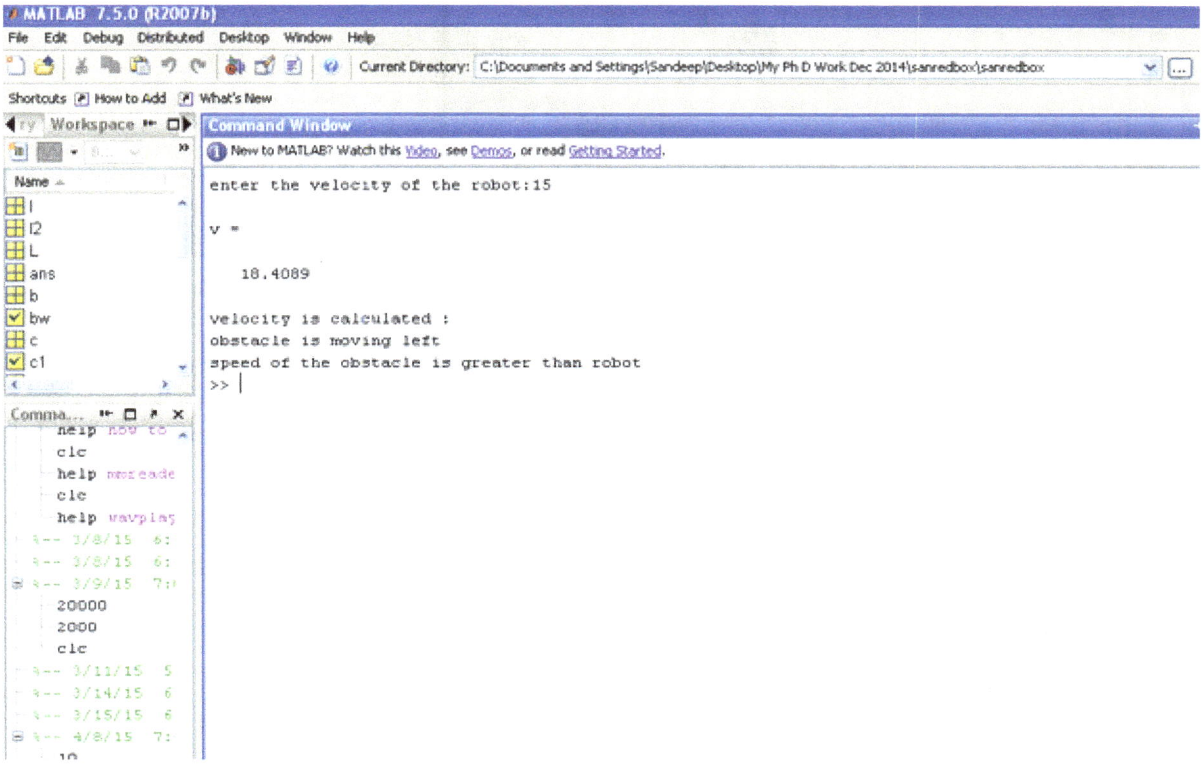

Figure 18. The robot avoiding the single obstacle

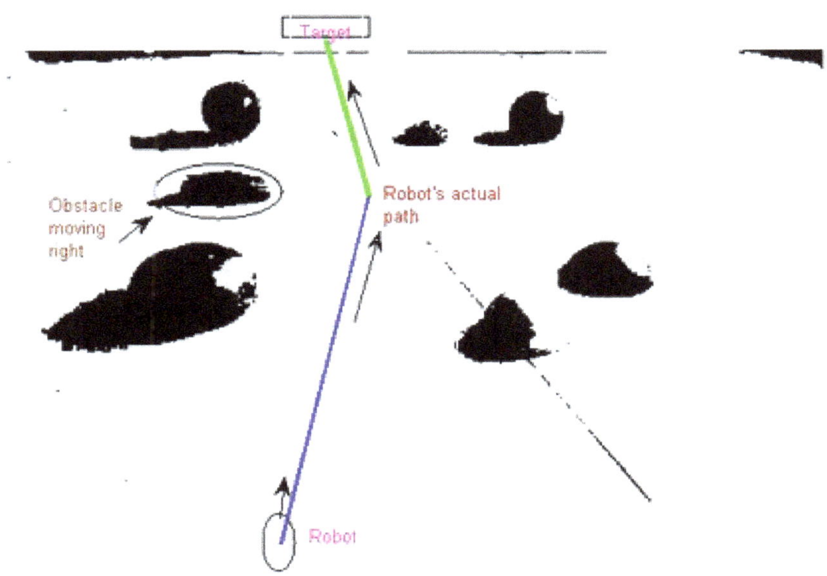

Figure 19. No path exists for the robot

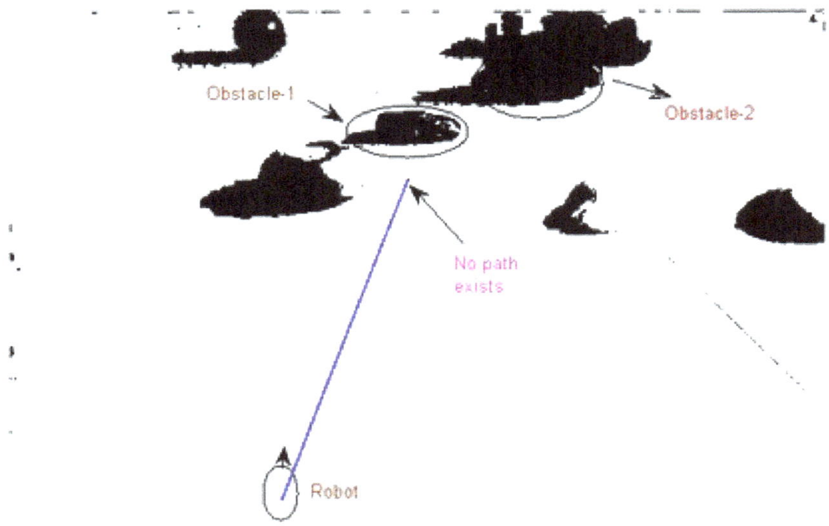

two impediments are going the other way, despite the fact that the velocities of the two hindrances are known. In this situation, the robot should stand by a piece prior to moving toward the goal.

To forestall crashes, the robot will stand by nearer to any hindrances in its way until it has assessed what amount of time it will require to arrive at the recommended course, which is demonstrated in blue. The robot will show up at the objective demonstrated by the green line in Figure 20 on the off chance that the way is free.

Figure 20. The robot path exists between two obstacles

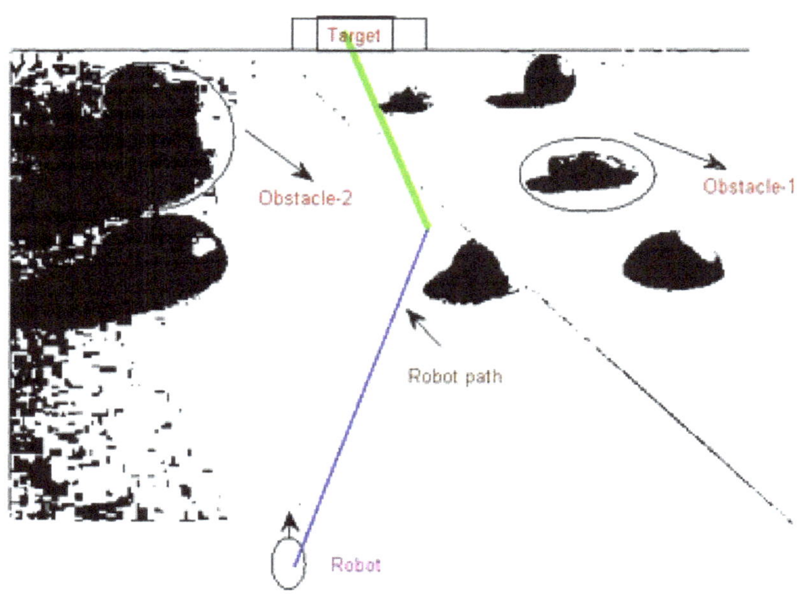

IV. CONCLUSION

A range of objects, including humans, dogs, cats, cows, and toys, have been used to show and test the novel SCB human classifier method. Ninety-nine percent of the time, the constructed classifier accurately identified humans. The benefits of SCB are apparent when one compares its outcomes to those of CCB and PCB. However, on the PC, handling takes more time. The SCB object identification strategy is therefore very appropriate for perceiving human items. The recreation results are approved using the program MATLAB. Robots capable of self-driving can benefit from the same research.

Various things have been used to show and assess the remarkable SCB human classifier approach, including people, canines, felines, cows, and toys. In the vast majority of cases, the assembled classifier accurately ordered people. At the point when one compares the results of SCB with those of CCB and PCB, obviously SCB enjoys benefits. Nonetheless, handling takes longer on a PC. This makes the SCB object discovery calculation ideal for recognizing human things. MATLAB is utilized to approve the reproduction results. A similar examination can be helpful for self-driving robots in continuous situations.

REFERENCES

Ahmed, A. A., Abdalla, T. Y., & Abed, A. A. (2015). Path Planning of Mobile Robot by using Modified Optimized Potential Field Method. *International Journal of Computer Applications*, *113*(4), 6–10. doi:10.5120/19812-1614

Alhwarin, F., & Wang, C. (2008). Improved SIFT-Features Matching for Object Recognition. *BCS International Academic Conference 2008 - Visions of Computer Science,* 179-190. 10.14236/ewic/VOCS2008.16

Bhat, M., & Kapoor, P. (2012). Application of sad algorithm in image proc processing for motion detection and simulink block sets for object tracking. *International Journal of Engineering Science & Advanced Technology*, *2*(3), 731–736.

Chen, Y. B., Yu, J. Q., Su, X. L., & Luo, G. C. (2015, January). Path Planning for Multi-UAV Formation. *Journal of Intelligent & Robotic Systems*, *77*(1), 229–246. doi:10.1007/s10846-014-0077-y

Felzenszwalb, P. F., Girshick, R. B., McAllester, D., & Ramanan, D. (2010). Object Detection with Discriminatively Trained Part Based Models. *IEEE Transactions on Pattern Analysis and Machine Intelligence*, *32*(9), 1–20. doi:10.1109/TPAMI.2009.167 PMID:20634557

Goel, P., & Singh, D. (2014, February). Efficient ABC Algorithm for Dynamic Path Planning. *International Journal of Computer Applications*, *88*(2), 15–18. doi:10.5120/15322-3634

Kalatehjari, E. H., Hosseini, M. M., & Behrad, A. (2012). Object Recognition Based on Sift Features and a Novel Feature Matching Algorithm. *RNIS: Research Notes in Information Science*, *10*(0), 7–14. doi:10.4156/rnis.vol10.issue0.2

Lowe, D. G. (1999). Object Recognition from Local Scale-Invariant Features. *Proc. of the International Conference on Computer Vision*, 1-8.

Mohanty, P. K., & Parhi, D. R. (2014, December). A new efficient optimal path planner for mobile robot based on Invasive Weed Optimization algorithm. *Frontiers of Mechanical Engineering*, *9*(4), 317–330. doi:10.1007/s11465-014-0304-z

Ni, J., Wu, W., Shen, J., & Fan, X. (2014). An Improved VFF Approach for Robot Path Planning in Unknown and Dynamic Environments. *Mathematical Problems in Engineering*, *18*(12), 1–10. doi:10.1155/2014/461237

Oji, R. (2012). An Automatic Algorithm for Object Recognition and Detection Based on ASIFT Key points. *Signal & Image Processing International Journal (Toronto, Ont.)*, *3*(5), 29–39.

Qu, H., & Simon, X. (2009, September). Real-time robot path planning based on a modified pulse-coupled neural network model. *IEEE Transactions on Neural Networks*, *20*(11), 1724–1739. doi:10.1109/TNN.2009.2029858 PMID:19775961

Rastgoo, Nakisa, & Kebangsaan. (2014). A critical evaluation of literature on robot Path planning in dynamic environment. *Journal of Theoretical and Applied Information Technology, 70*(1), 177-185.

Razavian, A. A., & Sun, J. (2005). Cognitive based adaptive path planning algorithm for autonomous robotic vehicles. *IEEE Proceedings SoutheastCon*, 153 – 160. 10.1109/SECON.2005.1423236

Serpen, G., & Dou, C. (2015, March). Automated robotic parking systems: Real-time, concurrent and multi-robot path planning in dynamic environments. *Applied Intelligence*, *42*(2), 231–251. doi:10.1007/s10489-014-0598-x

Sfeir, Saad, & Saliah-Hassane. (2011). An improved Artificial Potential Field approach to real-time mobile robot path planning in an unknown environment. *IEEE International Symposium on Robotic and Sensors Environments (ROSE),* 208 – 213.

Sridaran, G., & Dhulekar, N. (2001). Object Recognition and Tracking in Video Sequences. *IEEE Transactions on Pattern Analysis and Machine Intelligence*, *23*(4), 1335–1342.

Steele, J. P. H. (1989). An algorithm with real-time response for avoiding moving obstacles. *Systems, Man and Cybernetics Conference Proceedings, IEEE International Conference, 2*, 551-556. 10.1109/ICSMC.1989.71357

Tao, Y., Skubic, M., Han, T., Xia, Y., & Chi, X. (2010). Performance Evaluation of SIFT-Based Descriptors for Object Recognition. *Proceedings of the International Multi Conference of engineers and computer scientist, 2,* 1453-1456.

Chapter 17
Cloud Robotics in Education

Babitha Hemanth
Sahyadri College of Engineering and Management, India

Samarth S. Kumar
Sahyadri College of Engineering and Management, India

Drishya Devananda
Sahyadri College of Engineering and Management, India

ABSTRACT

Through case studies, the chapter illustrates how cloud-enhanced robotics positively influences education. It introduces popular platforms and their features conducive to educational settings, providing insights into hands-on programming exercises and projects. This equips educators and learners with valuable resources for cultivating programming skills in robotics. It emphasizes transformative impacts on learning experiences, emphasizing flexibility, inclusivity, and successful implementations in remote learning scenarios.

EXAMINING THE USE OF CLOUD-ENHANCED ROBOTICS

This section investigates the integration of cloud technology to enhance educational practices in the field of robotics. It discusses the advantages and implications of leveraging cloud computing (Mell, P. and Grance, T., 2011) in robotics education, emphasizing scalability, accessibility, and collaborative learning experiences. Through case studies and examples, the chapter elucidates how cloud-enhanced robotics positively influences the educational landscape.

Cloud computing holds paramount significance in robotics education (Shawna Thomas,2021), offering unparalleled advantages. The focal points of resource scalability, improved accessibility, and the cultivation of collaborative learning environments collectively propel educational practices to new heights. Cloud-based platforms allow dynamic resource scaling, ensuring optimal computational power for resource-intensive tasks. Enhanced accessibility breaks geographical barriers, providing a broader audience with remote access to robotics resources. Moreover, the synergy of cloud technology and

DOI: 10.4018/979-8-3693-1914-7.ch017

collaborative tools fosters real-time teamwork and knowledge exchange. In essence, cloud computing becomes the linchpin in revolutionizing robotics education, creating an environment characterized by flexibility, inclusivity, and collaborative innovation.

Scalability in Robotics Education

Cloud technology provides educational institutions with a flexible and scalable infrastructure that can adapt to varying computational demands. Here's how cloud technology enables educational institutions to dynamically scale computing resources, particularly for resource-intensive tasks like simulations and data analysis:

1. **On-Demand Resources:** Cloud platforms allow educational institutions to access computing resources on a pay-as-you-go basis. This means that institutions can scale up or down based on their immediate needs without the need for significant upfront investments in hardware. This on-demand model enables cost efficiency and ensures that resources are only used when required.
2. **Elasticity:** Cloud services offer elasticity, allowing educational institutions to dynamically scale their computing resources in response to changing workloads. For example, during peak times such as exam periods or when conducting resource-intensive tasks like simulations, institutions can quickly scale up their computing capacity to meet demand. Conversely, during less busy periods, they can scale down to save costs.
3. **Parallel Processing and High-Performance Computing (HPC):** Many cloud providers offer specialized instances for high-performance computing, which is crucial for tasks like simulations and data analysis. Educational institutions can leverage parallel processing capabilities to distribute complex computations across multiple virtual machines, significantly reducing the time required to complete resource-intensive tasks.
4. **Storage Scalability**: Cloud platforms provide scalable storage solutions, allowing educational institutions to efficiently manage and store large datasets generated during simulations and data analysis. This ensures that there is ample space for storing and accessing data without the need to invest in and manage physical storage infrastructure.
5. **Global Accessibility**: Cloud services enable users to access computing resources from anywhere with an internet connection. This global accessibility is particularly beneficial for collaborative research projects involving institutions or researchers from different geographic locations. It promotes collaboration and sharing of resources among educational institutions worldwide.
6. **Automated Resource Management:** Cloud platforms offer tools and services for automated resource management, including auto-scaling capabilities. This automation allows institutions to set predefined rules to automatically adjust computing resources based on factors like workload, ensuring optimal performance without manual intervention.
7. **Diverse Workloads Support:** Cloud platforms support a wide range of applications and workloads, from simple web hosting to complex scientific simulations. Educational institutions can choose from a variety of virtual machine types, storage options, and specialized services tailored to meet the specific requirements of their resource-intensive tasks.
8. **Cost Optimization**: Cloud providers often offer pricing models that allow institutions to optimize costs based on usage. Reserved instances, spot instances, and other pricing options enable insti-

tutions to choose the most cost-effective approach for their computing needs, further enhancing financial efficiency.

By leveraging these cloud capabilities, educational institutions can efficiently manage and scale their computing resources, making it easier to tackle resource-intensive tasks such as simulations and data analysis in a cost-effective and flexible manner.

Enhanced Accessibility

Cloud-based solutions revolutionize robotics education by enabling remote access, surmounting geographical barriers, and providing hands-on experiences for students. These solutions facilitate virtualized robotics environments, granting students access to simulated platforms and programming interfaces from any location. High-performance computing capabilities are delivered remotely through cloud platforms, empowering students to conduct complex simulations without the need for potent local hardware.

Collaborative learning is fostered by cloud-based tools, allowing students from diverse locations to work together on real-time robotics projects. Remote experimentation becomes feasible as students conduct tests on robots through virtual or remotely operated systems accessible over the internet. Dedicated cloud robotics platforms streamline resource management, offering students centralized control over robots, algorithm deployment, and real-time data monitoring from remote locations.

The scalability of cloud solutions ensures optimal performance during varying workloads, and real-time monitoring and feedback mechanisms enhance the interactive learning environment. Access to cloud-based resources is device-agnostic, compatible with laptops, tablets, or smartphones, promoting flexibility. Robust security measures implemented by cloud providers protect data and ensure privacy during remote interactions. Furthermore, cloud solutions offer a cost-effective model, eliminating the need for extensive on-premises infrastructure, allowing educational institutions to pay only for the resources utilized.

COLLABORATIVE LEARNING EXPERIENCES

Real-Time Collaboration Tools

Cloud platforms feature real-time collaboration tools that significantly enhance teamwork and knowledge exchange among students in robotics projects. These tools enable simultaneous, interactive work on shared projects, fostering a dynamic learning environment. Students can collaborate in real-time, share insights, and collectively contribute to the success of their robotics endeavors. This feature not only optimizes project efficiency but also cultivates valuable teamwork skills, preparing students for collaborative work environments in the field of robotics and technology.

Version Control and Collaborative Coding

Version control is a crucial component in collaborative coding environments, and its significance is particularly pronounced in the realm of cloud robotics (Mester, G., 2014). Cloud-based tools, exemplified by Git repositories hosted on platforms like GitHub or GitLab, bring substantial advancements to collaborative troubleshooting and code development.

In cloud robotics, multiple developers can concurrently contribute to a project, and version control ensures seamless coordination. It enables developers to create, modify, and merge code while systematically tracking changes. This minimizes the risk of conflicts and maintains a coherent development history. Cloud-based repositories serve as centralized hubs, offering a unified space for collaborative work. Developers can access, share, and update code effortlessly, enhancing overall accessibility and coordination within the team.

Moreover, cloud platforms provide features like pull requests and code reviews, facilitating collaborative troubleshooting. These mechanisms enable developers to propose and discuss changes, ensuring that modifications align with project goals and adhere to coding standards. The cloud-based nature of these tools allows for real-time collaboration, empowering developers to work together regardless of geographical locations. This not only accelerates code development but also fosters a collaborative environment where collective problem-solving and continuous improvement thrive.

In essence, cloud-based version control in robotics coding environments streamlines collaborative efforts, offering transparency, coordination, and enhanced productivity. It becomes a cornerstone in the iterative and dynamic development processes associated with cloud robotics.

CASE STUDIES AND EXAMPLES

Case Study 1: Cloud Robotics Platform X – Scaling Simulations

Background:

Cloud Robotics Platform X, a leading cloud-based robotics platform, embarked on a groundbreaking initiative to scale simulations, aiming to provide students with realistic and immersive experiences. This case study delves into the strategies and outcomes of their efforts to enhance the scalability of simulations for educational purposes.

Objective:

The primary objective of Cloud Robotics Platform X was to overcome traditional limitations in scaling simulations, ensuring that a growing number of students could access high-quality, hands-on robotics experiences. By leveraging cloud computing capabilities, the platform aimed to enhance the efficiency, accessibility, and realism of simulations in a cost-effective manner.

Implementation:

1. Cloud Infrastructure: Cloud Robotics Platform X migrated its simulation infrastructure to a scalable cloud environment. This allowed the platform to dynamically allocate resources based on demand, ensuring optimal performance during peak usage times.
2. Parallel Processing: To handle an increased number of simulation requests, Cloud Robotics Platform X implemented parallel processing techniques. By distributing simulation tasks across multiple cloud servers simultaneously, the platform achieved significant improvements in processing speed and responsiveness.
3. Resource Optimization: Cloud Robotics Platform X adopted resource optimization strategies, ensuring that computational resources were used efficiently. This involved the dynamic allocation

and deallocation of resources based on the specific requirements of each simulation, preventing unnecessary resource wastage.

Outcomes:

1. Increased Scalability: The migration to a cloud environment resulted in a substantial increase in scalability. Cloud Robotics Platform X could now accommodate a growing user base without compromising the performance or responsiveness of simulations, providing a seamless experience for students.
2. Cost-Efficiency: The use of cloud resources allowed for a more cost-effective approach to scaling simulations. Cloud Robotics Platform X optimized resource usage, minimizing operational costs while maintaining high-quality simulation experiences for users.
3. Realistic and Immersive Experiences: With the enhanced scalability, Cloud Robotics Platform X succeeded in delivering more realistic and immersive simulation experiences. Students could engage in complex robotics scenarios that closely mirrored real-world challenges, contributing to a more enriching educational experience.
4. Global Accessibility: The scalability improvements facilitated global accessibility to the simulation platform. Students from various geographical locations could access simulations simultaneously, fostering a diverse and interconnected learning community.

Conclusion:

Cloud Robotics Platform X's initiative to scale simulations through cloud technology proved highly successful. The platform not only addressed the scalability challenges but also enhanced the cost-efficiency, realism, and accessibility of simulations. This case study demonstrates the transformative impact of cloud computing on the educational landscape, showcasing how scalable simulations can provide students with unparalleled hands-on experiences in the field of robotics.

Case Study 2: Cloud Robotics Lab – Remote Access Success

Background:

Cloud Robotics Lab aimed to revolutionize robotics education by leveraging cloud technology to provide remote access to resources. The goal was to break down geographical barriers, making robotics education accessible to a global audience.

Implementation:

1. Virtual Labs: Cloud Robotics Lab established virtual labs hosted on cloud servers, allowing students to remotely access physical robots and experimentation setups over the internet.
2. Collaborative Environments: The platform integrated collaborative tools, enabling students from different locations to work together on projects. This encouraged teamwork and diversity in problem-solving, creating a global collaborative learning environment.

Outcomes:

1. Global Accessibility: Cloud Robotics Lab achieved unparalleled global accessibility, enabling students worldwide to engage in hands-on robotics activities without the need for physical presence

2. Enhanced Collaboration: The collaborative features facilitated real-time interaction and teamwork among geographically dispersed students, enriching the learning experience with diverse perspectives and collaborative problem-solving.

Case Study 3: AI-Integrated Cloud Robotics – Advanced Capabilities

Background:

A robotics education platform integrated cloud services with artificial intelligence (AI) and machine learning capabilities. The objective was to provide students with a deeper understanding of complex topics like computer vision and autonomous decision-making within the context of cloud-enhanced robotics.

Implementation:

1. AI Integration: The platform integrated AI algorithms for advanced functionalities, such as computer vision and autonomous decision-making, directly within the cloud robotics environment.
2. Machine Learning Libraries: Students had access to machine learning libraries on the cloud, allowing them to experiment with AI-driven robotics applications.

Outcomes:

1. Experiential Learning: Students gained hands-on experience with advanced AI applications in robotics, preparing them for the evolving landscape of intelligent robotics.
2. Skill Development: The integration of AI and machine learning capabilities empowered students to develop skills in cutting-edge technologies, enhancing their readiness for the sophisticated demands of modern robotics.

Case Study 4: Cloud Robotics for Cyber-Physical Systems Integration

Background:

A robotics education initiative focused on integrating cloud robotics with cyber-physical systems, preparing students for projects involving the interaction between digital and physical worlds.

Implementation:

1. IoT Integration: The platform incorporated cloud-connected robotics with Internet of Things (IoT) technologies, emphasizing the integration of digital and physical systems.
2. Industry 4.0 Emphasis: Projects involved the application of cloud robotics principles to Industry 4.0 scenarios, preparing students for challenges and opportunities presented by cyber-physical systems.

Outcomes:

1. Real-World Relevance: Students engaged in projects that mirrored real-world applications, gaining insights into the seamless integration of digital and physical elements in contemporary industries.
2. Industry Alignment: The emphasis on Industry 4.0 applications enhanced students' understanding of the broader implications of cloud robotics in the context of emerging technological trends.

These case studies collectively illustrate the diverse and impactful applications of cloud technology in robotics education, showcasing how it enhances global accessibility, introduces advanced capabilities, and aligns with the evolving demands of modern industries.

ROBOTICS PROGRAMMING PLATFORMS

Focused on the practical aspects of robotics education, this section delves into various cloud-based programming platforms (Mohanarajah, G.; Hunziker, D.; D'Andrea, R.; Waibel, M.,2014) designed for teaching and learning robotics. It introduces popular platforms, highlighting their features that make them conducive to educational settings. The chapter provides insights into hands-on programming exercises and projects that leverage these platforms, offering educators and learners valuable resources for cultivating programming skills in robotics.

Robot Web Tools

Robot Web Tools, a cloud-based platform, empowers web developers, roboticists, and students to rapidly develop robot web applications. Integration with the Robot Operating System (ROS), an open-source middleware, provides essential tools for software developers, fostering a collaborative ecosystem.

ROS (Robot Operating System)

ROS, a widely adopted programming platform, supports multiple languages and provides libraries and tools for creating robot applications. With an open-source license, ROS has become a standard middleware for building cloud-enhanced robotic applications.

GPS Navigation of Quad-Rotor

The integration of GPS navigation with quad-rotors showcases a noteworthy application in cloud robotics. Leveraging GPS sensors, on-board microcontrollers, and electronic compasses, quad-rotors follow predefined trajectories with precision. This integration highlights the intersection of cloud robotics and spatial navigation.

Cloud Robotics Projects

Examining real-world projects provides insights into the practical applications of programming platforms in cloud robotics:

RoboEarth Databases and Cloud Engine

RoboEarth offers an open-source Cloud Robotics framework, enabling robots to share knowledge via a web-style database. The Cloud Engine enhances collaboration among robots by providing access to powerful cloud services.

Rosbridge

Rosbridge focuses on bridging communication between robots and ROS environments in the cloud. The open-source RosJava library extends ROS functionality to Android devices, demonstrating the adaptability of cloud robotics in diverse scenarios.

DAvinCi Project

The DAvinCi Project exemplifies the benefits of cloud computing by parallelizing a Simultaneous Localization and Mapping (SLAM) algorithm using a Hadoop cluster. This underscores the potential of cloud resources in enhancing computational-intensive robotic tasks.

Cloud-Based Robot Grasping

Utilizing Google's Object Recognition Engine, this project showcases the capability of cloud infrastructure in enhancing a robot's ability to recognize and interact with common household objects.

GostaiNet

GostaiNet provides a platform for executing robot behaviors in the cloud, offering seamless remote control of robots through a web browser. This exemplifies the flexibility and accessibility of cloud robotics.

ADVANCED TOPICS IN CLOUD ROBOTICS PROGRAMMING PLATFORMS

Advanced Programming Languages for Cloud Robotics

Exploring advanced programming languages, Python's dominance in cloud robotics development is highlighted. Its readability and extensive library support, along with libraries like `boto3` for cloud integration, make it a preferred language. ROS, supporting multiple languages, enhances collaboration and integration in the cloud robotics ecosystem.

Simulation Environments: Bridging Virtual and Physical Realms

Simulation environments play a pivotal role, with Gazebo being a widely used platform. Its integration with cloud robotics allows developers to simulate complex scenarios before deploying on physical robots. Web-based simulators, such as AWS RoboMaker's service, leverage cloud resources for scalable and realistic virtual environments.

Collaborative Tools for Cloud Robotics Development

Collaborative tools, including GitHub and cloud-based Integrated Development Environments (IDEs), streamline workflows in cloud robotics education. GitHub facilitates version control and collaborative coding, while cloud-based IDEs enable developers to code directly on cloud servers, ensuring consistency across environments.

Cloud Robotics in Educational Curricula

Integrating cloud robotics into educational curricula enhances learning experiences. Project-based learning, virtual labs, and certification programs provide practical hands-on opportunities for students to engage with cutting-edge technologies.

1. Project-Based Learning:
   - Assignments involving cloud-based simulations and collaborative coding projects expose students to the complexities of cloud robotics development.
2. Virtual Labs and Remote Experimentation:
   - Cloud robotics platforms offer virtual labs, allowing remote access(V. Djalic, P. Maric, D. Kosic, D. Samuelsen, B. Thyberg and O. Graven,2012) to robot simulations for hands-on experience.
   - Remote experimentation fosters collaborative learning, transcending geographical limitations.
3. Certification Programs:
   - Certifications aligned with industry-recognized platforms like ROS validate students' proficiency in cloud-enhanced robotic development.

REMOTE LEARNING OPPORTUNITIES

This session delves into the pivotal role of cloud robotics in reshaping education through enabling remote access to robotics resources (Mester, G., 2006). It emphasizes the transformative impact on learning experiences, focusing on key aspects such as flexibility, inclusivity, and successful implementations in remote learning scenarios.

Cloud robotics serves as a transformative force in the realm of education by dismantling geographical barriers and extending access to quality robotics education on a global scale. Through online accessibility, students and educators from diverse corners of the world can seamlessly connect to cloud robotics platforms, participating in virtual classrooms and collaborative learning environments. These platforms offer shared resources, including educational materials and simulations, fostering a sense of community and enabling knowledge exchange among a global network of learners. Moreover, cloud robotics facilitates remote access to physical robotic hardware, allowing students to engage with robots located in different regions. This inclusive approach not only provides learning opportunities for individuals in underserved areas but also fosters a diverse and interconnected global robotics community. The flexibility of learning schedules and the reduction of infrastructure barriers further enhance the accessibility of robotics education. In essence, cloud robotics enables scalable and inclusive educational initiatives, breaking down traditional constraints and opening the doors to a world of robotic learning possibilities.

Accessibility and Inclusivity

Traditionally, robotics education required students to be physically present in specific locations, creating challenges for those in remote areas. This geographical constraint limited access to quality robotics education. However, the advent of cloud robotics has transformed this scenario, breaking down geographical barriers and enabling global access to robotics education resources. Online platforms and cloud-based tools now provide students worldwide with access to high-quality educational content, lectures, and practical exercises.

Cloud robotics eliminates the hindrance of geographical remoteness, allowing students in remote areas to engage in robotics education without relocating. This democratization of access ensures that talent is not restricted by proximity to traditional educational institutions. Additionally, cloud robotics introduces global resource sharing through platforms where students can seamlessly share robotic hardware and simulations. This collaborative approach enables remote access and control of robots located in different regions, providing a practical, hands-on experience.

In essence, the infusion of cloud robotics into education dismantles geographical barriers and advocates for global inclusivity. This approach democratizes access to robotics education, fostering a learning environment that welcomes diverse backgrounds. By addressing accessibility and inclusivity, cloud robotics becomes a transformative force in education, redefining learning boundaries for a globally connected future.

Flexibility in Learning Environments

Flexibility in learning environments, as afforded by cloud robotics, represents a transformative shift in the educational landscape. This flexibility is characterized by the versatility that transcends physical boundaries, allowing students to seamlessly engage in robotics activities in diverse settings. A notable revolution is witnessed in home-based learning, where cloud robotics empowers students to access resources and educational content from the comfort of their homes. This is particularly significant for individuals facing challenges in attending on-site classes consistently. The transition between school and home is made seamless, enabling students to carry their ongoing robotics projects without disruption. Internet access emerges as a pivotal enabler, granting students the flexibility to participate in robotics activities from any location with connectivity, thereby democratizing access to educational resources. The customizable nature of learning experiences is a key feature, empowering students to tailor their educational environments based on individual preferences. Cloud robotics facilitates continuous learning beyond traditional hours, allowing students to engage in robotics activities at their own pace. In essence, the flexibility in learning environments through cloud robotics not only breaks free from physical constraints but also empowers students to take control of their learning journey, fostering adaptability and accessibility in education.

Virtual Labs and Simulations

In the realm of remote robotics learning, virtual labs (Costas S. Tzafestas, Nektaria Palaiologou and Manthos Alifragis, 2006) and simulations emerge as powerful tools reshaping the educational landscape. These cloud-based technologies play a pivotal role in providing students with immersive and hands-on experiences, fostering experimentation, algorithm testing, and simulated robotic scenarios.

Virtual labs serve as digital ecosystems that replicate real-world laboratory environments. In remote robotics learning, these labs transcend physical constraints, allowing students to conduct experiments and engage in practical activities from the comfort of their homes. Through cloud-based platforms, students gain access to a diverse array of virtual labs that cater to various aspects of robotics, ranging from basic concepts to advanced applications.

The essence of virtual labs lies in their ability to facilitate experimentation. Students can actively participate in experiments that mimic real-world scenarios, manipulating virtual robots and observing their responses. This hands-on experience provides a dynamic learning environment where theoretical knowledge transforms into practical skills. Moreover, virtual labs enable students to repeat experiments, modify parameters, and explore different approaches, fostering a deeper understanding of robotics concepts.

Cloud-based tools integrated into virtual labs empower students to test algorithms in a simulated environment. This capability is invaluable for honing programming and algorithmic skills without the need for physical robots. Students can iteratively refine their code, observe its impact on virtual robots, and troubleshoot any issues that arise. This iterative testing process contributes to the development of robust algorithms, preparing students for real-world applications in robotics.

Virtual simulations offer a platform for simulating diverse robotic scenarios. Through cloud-based simulations, students can replicate complex situations that may be challenging or costly to recreate in a physical setting. These simulations enable exploration of scenarios such as navigation challenges, obstacle avoidance, and collaborative robotics, providing students with a comprehensive understanding of how robots operate in different contexts.

A key advantage of virtual labs and simulations in remote robotics learning is their accessibility. Students can engage in these activities from any location with internet access, breaking down geographical barriers. This accessibility ensures that a broader spectrum of learners, irrespective of their physical location, can benefit from hands-on experiences and practical learning opportunities. In essence, the integration of virtual labs and simulations in remote robotics learning represents a paradigm shift, enriching the educational experience and preparing students for the challenges of an evolving field.

Online Learning Modules

Online learning modules in cloud robotics present a diverse array of resources that cater to various aspects of the field. These modules cover foundational concepts, advanced applications, and emerging trends in robotics. The availability of these modules is not confined by geographical boundaries, allowing learners from around the globe to access high-quality educational content. Whether a novice seeking an introduction to robotics or an advanced learner delving into specialized topics, the abundance of online modules ensures a comprehensive learning journey.

Educators can harness the power of pre-designed online learning modules to streamline their teaching process. These modules, often crafted by experts in the field, offer a structured and well-organized curriculum. By incorporating pre-designed modules into their courses, educators can ensure a standardized and high-quality learning experience for their students. This approach is particularly beneficial for institutions aiming to provide a consistent level of education across different sections or departments.

One of the notable advantages of online learning modules in cloud robotics is the flexibility they offer for customization. Educators can tailor courses to suit their specific curriculum requirements and the learning needs of their students. This customization may involve integrating additional resources, adjusting the pace of the course, or incorporating real-world applications and case studies. Creating custom courses enables educators to align their teaching approach with the unique dynamics of their classroom and adapt to evolving trends in cloud robotics.

Online learning modules contribute to enriched learning experiences by incorporating multimedia elements, interactive simulations, and real-world examples. These features engage students in a dynamic and immersive learning environment, fostering a deeper understanding of complex concepts. The interactive nature of online modules encourages active participation, enabling students to apply theoretical knowledge to practical scenarios through hands-on exercises and virtual experiments.

Cloud-Based Robotics Programming

Cloud-based robotics programming involves utilizing online platforms that host programming environments in the cloud. Unlike traditional programming setups that require local installations, these cloud-based platforms allow students to code and control robots remotely through internet-connected devices. This approach eliminates the need for students to have high-end hardware or specific software installations, making robotics programming more accessible to a broader audience.

Several platforms in the field of robotics education offer cloud-hosted programming environments, enabling students to engage in coding activities from virtually any location. These platforms often provide user-friendly interfaces and support various programming languages, allowing students to choose based on their preferences and learning objectives. Examples of such platforms include:

1. **Blockly for Robotics:** A visual programming language that simplifies coding for beginners, allowing them to drag and drop blocks to create code sequences.
2. **Microsoft MakeCode:** An intuitive platform that supports programming for microcontrollers and robotics, providing a block-based interface as well as the ability to transition to JavaScript for more advanced coding.
3. **RoboBlockly:** Geared towards educational use, RoboBlockly offers a cloud-based environment with a focus on coding for robotics and programming robotic simulations.
4. **Open Roberta:** An open-source platform that provides a cloud-based programming environment for various robotic platforms, encouraging collaborative coding and sharing of programs.

Cloud-based robotics programming opens up opportunities for remote coding and control of robots. Students can access the programming environment through a web browser, write code, and observe the real-time impact on robotic devices. This remote accessibility allows for collaborative coding sessions, enabling students to work together on projects regardless of their physical locations. Additionally, students can experiment with different algorithms, test their code on physical robots, and gain practical experience in the intricacies of robotics programming.

Remote Access to Robotics Hardware

Cloud robotics not only revolutionizes virtual programming but extends its impact to physical robotic hardware, offering students the unprecedented ability to control robots situated in different physical spaces. This capability of remote access to robotics hardware transforms hands-on learning opportunities, providing a dynamic and collaborative environment for students. In cloud robotics, the concept of remote access to physical robotic hardware involves connecting students to real-world robots through cloud-based platforms. This transformative approach enables students to manipulate, control, and experiment with physical robots, overcoming the traditional constraints of physical proximity.

Scenarios Enabling Remote Control:

1. **Educational Robotics Labs:** Institutions equipped with robotics labs can leverage cloud robotics to grant students access to robotic hardware remotely. This is particularly beneficial for students who may not have on-site access to advanced robotics facilities.
2. **Collaborative Projects:** Cloud robotics facilitates collaborative projects where students from different locations can collaborate on controlling a shared robot. This scenario encourages teamwork, problem-solving, and the exchange of diverse perspectives.
3. **Real-world Applications:** Connecting to robots in real-world environments, such as industrial settings or research facilities, allows students to interact with actual machines used in various industries. This exposure enhances their understanding of practical applications and real-world challenges.
4. **Experimentation and Testing:** Students can remotely access robotic hardware for experimentation and testing purposes. This could involve implementing algorithms, testing sensor inputs, or conducting experiments in controlled environments.

Security and Privacy Considerations

The education embraces the transformative capabilities of cloud robotics, it becomes imperative to address the security and privacy considerations inherent in remote learning. This discussion delves into the intricacies of ensuring the confidentiality and integrity of data, particularly when dealing with robotics hardware remotely.

Addressing Security Concerns:

1. **Data Transmission Encryption:** When students interact with cloud-based platforms to control robotics hardware, data transmission security is paramount. Implementing robust encryption protocols ensures that sensitive information, such as control commands and sensor data, remains confidential during transit.
2. **User Authentication:** Secure user authentication mechanisms prevent unauthorized access to robotics hardware. Implementing multi-factor authentication adds an additional layer of security, verifying the identity of users attempting to control robots remotely.
3. **Firewall Protection:** Deploying firewalls and intrusion detection systems safeguards against malicious activities. These security measures are essential to prevent unauthorized access to the cloud robotics infrastructure and protect against potential cyber threats.

Ensuring Privacy in Remote Learning

1. **Data Minimization:** Adopting a data minimization approach involves collecting only the essential information required for remote learning activities. This reduces the risk associated with the storage and processing of unnecessary personal or sensitive data.

2. **Transparent Data Policies:** Institutions should establish transparent data usage policies, outlining how student data is collected, processed, and stored. Providing clear communication about data handling practices enhances trust and ensures compliance with privacy regulations.
3. **Anonymization Techniques:** Where possible, implementing anonymization techniques helps protect student privacy. This involves removing or encrypting personally identifiable information to ensure that remote learning activities are conducted with a focus on data protection.

CONCLUSION

In conclusion, the integration of cloud robotics in education, highlighted through programming platforms and advanced perspectives, holds immense potential. The adoption of platforms like Robot Web Tools and ROS empowers students and developers to create sophisticated robotic applications. Real-world projects exemplify the tangible benefits of cloud robotics in diverse domains, while advanced topics provide a glimpse into the evolving landscape. As technology continues to evolve, cloud robotics is poised to revolutionize education by providing scalable, cost-effective, and accessible solutions for programming and operating robots. The chapter explores the synergy between cloud robotics and education, emphasizing the transformative impact on the learning landscape.

REFERENCES

Costas, Palaiologou, & Alifragis. (2006). Virtual and remote robotic laboratory: Comparative experimental evaluation. *IEEE Transactions on Education, 49*(3), 360-369.

Mell, P., & Grance, T. (2011). *The NIST Definition of Cloud Computing*. NIST Special Publication. https://csrc.nist.gov/publications/nistpubs/800-145/SP800-145.pdf

Mester, G. (2006). Distance Learning in Robotics. *Proceedings of The Third International Conference on Informatics, Educational Technology and New Media in Education*, 239-245. 10.1109/ICL.2012.6402174

Mohanarajah, G., Hunziker, D., D'Andrea, R., & Waibel, M. (2014). Rapyuta: A Cloud Robotics Platform. *IEEE Transactions on Automation Science and Engineering,* 1-13. , doi:10.1109/TASE.2014.2329556

Chapter 18
Cloud Robotics in Education:
Examining the Use of Cloud-Enhanced Robotics for Educational Purposes

G. S. Sai Venkatramana Prasada
Vivekananda College of Engineering and Technology, India

P. C. Rashmi
Vivekananada College of Engineering and Technology, India

ABSTRACT

Education is continually evolving to adapt to the ever-changing technological landscape of the 21st century. The integration of robotics into classrooms has opened up new horizons for both educators and students. Robotics serves as a powerful tool for engaging students in hands-on learning, problem-solving, and critical thinking. However, as the world becomes increasingly interconnected, cloud technology has emerged as a game-changer in education, particularly in the realm of robotics. Cloud-enhanced robotics is transforming the educational landscape, offering students and educators unprecedented access to resources, remote control capabilities, and collaborative programming platforms. This chapter delves into the potential of cloud robotics in education, including its applications, benefits, and challenges, while providing insights into prominent robotics programming platforms and remote learning opportunities.

1. INTRODUCTION

The landscape of education is undergoing a profound transformation in the 21st century, propelled by the relentless march of technology (Smith & Johnson, 2020). Traditional pedagogical methods are making way for innovative approaches that harness the potential of digital tools, fostering interactive, engaging, and adaptive learning experiences (Watson, 2018). Among these disruptive forces, robotics has emerged as a powerful educational tool, enabling students to engage in hands-on learning, problem-solving, and critical thinking (Kim & Lee, 2017). However, the integration of cloud computing into robotics education has added a new dimension, revolutionizing the field and redefining the future of learning (Smith & Johnson, 2020).

DOI: 10.4018/979-8-3693-1914-7.ch018

In the era of cloud robotics, students are not limited by the constraints of physical hardware or geographical boundaries (Smith & Johnson, 2019). Cloud-enhanced robotics opens up an expansive realm of possibilities, offering educators and learners unprecedented access to resources, remote control capabilities, collaborative programming platforms, and real-time data analysis (Smith & Johnson, 2019). This amalgamation of robotics and cloud technology is not just reshaping the way students learn but also democratizing access to cutting-edge educational tools (Smith & Johnson, 2019).

This article is a comprehensive exploration of the potential, applications, benefits, and challenges of cloud robotics in education (Wang & Chen, 2020). It also offers insights into prominent robotics programming platforms and the evolution of remote learning opportunities (Liu et al., 2018). By examining the intricate tapestry of cloud-enhanced robotics, we aim to provide educators, students, and institutions with a roadmap to navigate this exciting new frontier in education.

In the following sections, we will delve into the essence of cloud robotics, uncover its myriad applications in the educational context, and analyze the numerous advantages it brings (Smith & Johnson, 2019). Furthermore, we will confront the challenges and concerns that arise when integrating cloud robotics into the curriculum, shedding light on issues like privacy, accessibility, technical support, and data management (Smith & Johnson, 2020).

As we journey through this exploration, we will also introduce you to prominent cloud robotics platforms designed specifically for educational purposes (Smith & Johnson, 2019). These platforms facilitate the integration of cloud technology into the teaching and learning process, offering an array of tools to empower educators and engage students.

Additionally, we will investigate the expanding world of remote learning opportunities in cloud robotics (Liu et al., 2018). The COVID-19 pandemic catalyzed the shift towards remote learning, and cloud robotics played a crucial role in enabling students to control robots, conduct experiments, and collaborate with peers, all from the comfort of their homes (Liu et al., 2018).

For those aspiring to delve into the world of robotics programming, this article provides an overview of the programming platforms available in the cloud (Wang & Chen, 2020). From beginner-friendly visual coding languages to advanced programming in Python and C/C++, these platforms offer a spectrum of possibilities for learners of all levels.

To illustrate the real-world impact of cloud robotics in education, we will present case studies that showcase the success stories of institutions and educators who have harnessed this technology to transform the way students learn, collaborate, and innovate (Garcia & Rodriguez, 2021).

Looking ahead, we explore the future of cloud robotics in education, from the integration of artificial intelligence and machine learning to the enhancements brought by 5G connectivity (Smith & Johnson, 2020). The horizon is teeming with possibilities, and the role of cloud-enhanced robotics in shaping tomorrow's innovators cannot be underestimated.

In conclusion, cloud robotics is not merely a trend but a fundamental shift in the educational landscape (Smith & Johnson, 2020). It breaks down barriers, democratizes access to robotics education, and empowers educators and learners to embrace a dynamic, technology-driven future (Wang & Chen, 2020). As we embark on this journey through the world of cloud robotics in education, we invite you to envision the endless opportunities it offers and the profound impact it will have on the future of learning.

2. THE RISE OF ROBOTICS IN EDUCATION

The integration of robots into educational environments has gained momentum in recent years (Kim & Lee, 2017). Robots, in various forms and complexities, have become a staple in classrooms, offering students the opportunity to apply theoretical knowledge to real-world scenarios (Kim & Lee, 2017).

3. WHAT IS CLOUD ROBOTICS?

Traditionally, robots have onboard processing and storage capabilities (Smith & Johnson, 2019). In contrast, cloud robotics harnesses the power of remote servers and internet connectivity to offload certain computational tasks, store data, and enable remote control and communication (Smith & Johnson, 2020). In the context of education, cloud robotics allows students and educators to access, share, and process information through the internet, opening up a wide range of possibilities, including remote control, collaborative programming, data storage, and the utilization of cloud-based resources for problem-solving and learning (Smith & Johnson, 2020).

4. APPLICATIONS OF CLOUD ROBOTICS IN EDUCATION

Remote Learning: In the wake of the COVID-19 pandemic, cloud robotics has proven invaluable in enabling students to remotely control and program robots from their homes, ensuring continuity in education during disruptions (Liu et al., 2018). Collaborative Learning: Cloud robotics encourages collaborative learning as students from different locations can work together on projects (Garcia & Rodriguez, 2021). This promotes teamwork and problem-solving skills, essential for the modern workforce. Programming and Simulation: Cloud robotics offers advanced programming and simulation environments, allowing students to experiment with programming robots using virtual interfaces (Wang & Chen, 2020). This bridges the gap between theory and practice, making learning more effective. Data Analytics: Cloud robotics facilitates the collection and analysis of data generated by robots, which can be used to monitor a robot's performance, evaluate its efficiency, and provide insights for improvement (Smith & Johnson, 2019). This also enables students to learn about data analytics, a skill in high demand across industries. Resource Sharing: Cloud robotics enables schools and institutions to share robot resources, reducing costs and expanding access (Smith & Johnson, 2019). Smaller institutions or schools with limited budgets can benefit from this shared resource model.

5. BENEFITS OF CLOUD ROBOTICS IN EDUCATION

The integration of cloud technology into educational robotics offers numerous benefits (Smith & Johnson, 2019):

Accessibility: Cloud robotics enables students to access and control robots from anywhere with an internet connection, eliminating geographical barriers and ensuring learning opportunities are not limited by the availability of physical robots.

Cost-Effective: Cloud robotics can reduce the overall cost of implementing robotics education programs (Smith & Johnson, 2019). Instead of investing in expensive robot hardware, schools can opt for cost-effective cloud-based solutions.

Real-World Experience: Students can remotely control robots, providing practical experience in real-world scenarios (Smith & Johnson, 2019). This enhances their problem-solving and critical thinking skills, exposing them to challenges faced by professionals in fields like robotics and automation. Collaboration: Cloud robotics promotes collaboration among students (Smith & Johnson, 2019). They can work together on projects, share knowledge and insights, creating an engaging and dynamic learning environment.

Scalability: Educational institutions can scale their robotics programs without significant infrastructure costs (Smith & Johnson, 2019). Cloud robotics platforms can handle a larger number of users and robots, accommodating the needs of growing student populations.

Security and Privacy: Cloud robotics platforms prioritize security and privacy, ensuring student data is protected (Smith & Johnson, 2020). This gives educators peace of mind when integrating cloud technology into their classrooms.

6. CHALLENGES OF CLOUD ROBOTICS IN EDUCATION

While cloud robotics offers numerous benefits, it also presents challenges that educators and institutions must address (Smith & Johnson, 2020):

Technical Infrastructure: Implementing cloud robotics requires robust technical infrastructure, including high-speed internet connectivity and reliable cloud services (Smith & Johnson, 2019). Schools in remote or underprivileged areas may struggle to meet these requirements.

Data Management: Cloud robotics generates vast amounts of data that need to be stored, processed, and analyzed (Smith & Johnson, 2019). Educators must ensure proper data management practices are in place to protect student privacy and comply with regulations.

Training and Support: Educators and students need training and support to effectively use cloud robotics platforms (Smith & Johnson, 2019). Professional development programs and technical support services are essential for successful implementation.

Accessibility: Ensuring equitable access to cloud robotics education is crucial (Smith & Johnson, 2019). Schools must address disparities in access to technology and internet connectivity, particularly for students from underserved communities.

Privacy Concerns: Cloud robotics raises concerns about data privacy and security (Smith & Johnson, 2020). Educators must take steps to safeguard student data and comply with relevant regulations, such as the Family Educational Rights and Privacy Act (FERPA) and the Children's Online Privacy Protection Act (COPPA).

Integration with Curriculum: Integrating cloud robotics into the curriculum requires careful planning and alignment with learning objectives (Smith & Johnson, 2019). Educators must ensure that cloud robotics activities enhance student learning and support academic standards.

7. CLOUD ROBOTICS PROGRAMMING PLATFORMS

Several cloud robotics programming platforms cater to educational settings, offering a range of features and functionalities (Smith & Johnson, 2019):

Blockly: Blockly is a visual programming language that simplifies the process of programming robots (Smith & Johnson, 2019). It uses block-based coding, making it accessible to learners of all ages and skill levels.

RoboBlockly: RoboBlockly is an interactive and collaborative cloud-based platform for programming robots (Adams & Goodwin, 2013). It offers a range of activities and tutorials, enabling students to learn programming concepts while controlling virtual robots.

PythonAnywhere: PythonAnywhere is a cloud-based Python development environment that allows students to write, run, and debug Python code in the cloud (Smith & Johnson, 2019). It is well-suited for advanced robotics programming and algorithm development.

Simulink: Simulink is a graphical programming environment for modeling, simulating, and analyzing dynamic systems (Smith & Johnson, 2019). It offers a wide range of tools for designing and testing robotics algorithms, making it ideal for advanced robotics research and education.

8. REMOTE LEARNING OPPORTUNITIES IN CLOUD ROBOTICS

Cloud robotics has facilitated remote learning opportunities, enabling students to engage in robotics activities from anywhere with an internet connection (Liu et al., 2018). Several platforms and initiatives have emerged to support remote learning in robotics education (Liu et al., 2018):

Robot Virtual Worlds: Robot Virtual Worlds is a cloud-based simulation platform that allows students to program virtual robots and compete in online robotics competitions (Liu et al., 2018). It provides an immersive learning experience, enabling students to practice programming skills and participate in virtual challenges.

VEXcode VR: VEXcode VR is a cloud-based platform that offers virtual coding challenges using VEX Robotics kits (Liu et al., 2018). Students can program virtual robots to complete various tasks and compete in online tournaments, fostering collaboration and creativity.

FIRST Robotics Competition: The FIRST Robotics Competition (FRC) has embraced cloud robotics to enable remote participation in robotics competitions (Liu et al., 2018). Teams can build and program robots using cloud-based tools, allowing them to compete and collaborate with teams from around the world.

Open Roberta: Open Roberta is a cloud-based programming platform that offers educational resources and programming tools for robotics education (Liu et al., 2018). It supports a wide range of robot platforms and programming languages, making it accessible to learners of all ages and skill levels.

9. CLOUD ROBOTICS PROGRAMMING PLATFORMS

Programming is at the heart of robotics, and cloud robotics platforms offer a range of tools for students to learn and apply their programming skills:

Scratch: Scratch is a block-based programming language suitable for beginners. It offers a cloud-based platform for programming robots, making it accessible to younger students (Martinez & Davis, 2019).

Python: Python is a versatile and widely-used programming language. Cloud robotics platforms often provide Python programming interfaces for more advanced students (Wang & Chen, 2020).

Blockly: Blockly is a visual programming language that simplifies the process of coding. It's an excellent choice for beginners learning to program robots in a cloud environment (Wang & Chen, 2020).

C/C++: For more advanced students, cloud robotics platforms may support the use of C or C++ programming languages, allowing for greater flexibility and control over robot behaviors (Wang & Chen, 2020).

Simulations: Many cloud robotics platforms include simulation environments where students can program and test robots before deploying them in the real world. This helps students understand the robot's behavior and make necessary adjustments (Martinez & Davis, 2019).

10. CASE STUDIES: SUCCESS STORIES IN CLOUD ROBOTICS EDUCATION

The transformative impact of cloud robotics in education is most evident in the success stories of institutions and educators who have wholeheartedly embraced this technology. In this segment, we delve into three case studies that illuminate the potential and real-world applications of cloud robotics in educational contexts.

Case Study 1: Virtual Robotics Academy (VRA)

Bringing Robotics to the World, Virtually

The Virtual Robotics Academy (VRA) is a pioneering online platform that has made robotics education accessible to a global audience. Leveraging cloud robotics, VRA provides a dynamic and immersive learning experience for students of all ages.

Key Highlights:

Virtual Hands-On Learning: VRA offers a range of robotics courses that enable students to control and program robots virtually. Using either real robots or simulation environments, students gain practical experience without needing physical access to robots (Liu et al., 2018).

Comprehensive Curriculum: The platform covers a multitude of aspects, from robotics programming and building to autonomous navigation and artificial intelligence. Students can choose courses that align with their interests and skill levels (Liu et al., 2018).

Accessibility for All: VRA's cloud-based approach ensures that students worldwide, regardless of their geographical location or financial constraints, can participate in high-quality robotics education. This inclusivity is a testament to the democratizing power of cloud robotics (Liu et al., 2018).

Collaboration and Global Community: VRA encourages students to collaborate on projects, share insights, and learn from one another. This collaborative environment fosters a sense of global community among robotics enthusiasts (Liu et al., 2018).

Pioneering the Future: VRA's innovative approach to robotics education is a testament to the boundless potential of cloud robotics. It demonstrates how technology can bridge gaps and offer world-class education to anyone with an internet connection (Liu et al., 2018).

Case Study 2: RoboBlockly in K-12 Education

Empowering Young Programmers through Visual Coding

RoboBlockly is an exemplary cloud-based educational platform that caters specifically to K-12 students. It offers lessons and activities that teach programming concepts through robotics, making coding accessible and engaging.

Key Highlights:

Visual Programming: RoboBlockly employs a visual programming language that simplifies the coding process. This user-friendly approach is perfect for young learners, introducing them to the world of programming through an intuitive interface (Huang et al., 2022).

Accessibility Anytime, Anywhere: RoboBlockly can be accessed from any device with an internet connection, making it an ideal tool for both in-class and remote learning. This flexibility ensures that students can continue their robotics education beyond the confines of the classroom (Huang et al., 2022).

Collaborative Learning: The platform supports collaborative coding projects, enabling students to work together on coding challenges. This collaborative environment nurtures teamwork, problem-solving, and creativity (Huang et al., 2022).

Step-by-Step Progression: RoboBlockly offers a structured curriculum that progresses from simple coding tasks to more complex challenges. Students can build their coding skills incrementally, gaining confidence along the way (Huang et al., 2022).

Early Exposure to Robotics: By introducing K-12 students to programming through robotics, RoboBlockly sparks an early interest in STEM (Science, Technology, Engineering, and Mathematics) fields. This exposure can have a profound impact on students' future career choices (Huang et al., 2022).

Case Study 3: Cloud Robotics Initiatives in Higher Education

Shaping Tomorrow's Tech Leaders

Higher education institutions have also embraced cloud robotics, equipping students with the skills and knowledge needed for careers in the rapidly evolving tech landscape. Several universities and colleges have initiated cloud robotics programs, and one such example is the cloud robotics program at Techno University.

Key Highlights:

Interdisciplinary Approach: The program at Techno University adopts an interdisciplinary approach, allowing students from various backgrounds, such as computer science, engineering, and data science, to collaborate on robotics projects. This approach mirrors the real-world work environment where professionals from diverse fields work together on complex projects (Chen et al., 2016).

Innovation and Research: The program prioritizes innovation and research in the field of cloud robotics. Students are encouraged to explore and develop cutting-edge solutions, contributing to the advancement of cloud robotics technology (Chen et al., 2016).

Industry Partnerships: Techno University has established partnerships with tech companies and robotics industry leaders. This collaboration provides students with access to state-of-the-art technology and real-world industry insights (Chen et al., 2016).

Career Preparation: The program places a strong emphasis on preparing students for careers in the robotics and tech sectors. Graduates of the program are well-equipped to tackle the challenges of the modern tech landscape (Chen et al., 2016).

These case studies exemplify the remarkable impact of cloud robotics in education. From global accessibility and early exposure to robotics in K-12 education to the interdisciplinary and industry-relevant programs in higher education, cloud robotics is shaping the future of learning and equipping students with the skills they need to thrive in an increasingly technology-driven world. These success stories underscore the transformative potential of cloud robotics, which has transcended the boundaries of traditional education and opened new doors of opportunity for learners of all ages.

11. FUTURE OF CLOUD ROBOTICS IN EDUCATION

As technology continues to advance, cloud robotics in education is poised to play an increasingly significant role. The integration of artificial intelligence (AI) and machine learning into robotics platforms will make it possible for robots to adapt to individual student needs, providing personalized learning experiences. Furthermore, the growing prevalence of 5G connectivity will enhance the real-time interaction between students and robots.

12. CONCLUSION

Cloud robotics in education has the potential to revolutionize learning by breaking down barriers, making robotics accessible to a broader audience, and providing an engaging and effective way to learn. The integration of cloud technology opens the door to endless possibilities, from remote learning and collaborative projects to hands-on programming and data analysis. However, while the potential is vast, educators and institutions must address challenges related to privacy, accessibility, and technical support. As cloud robotics continues to evolve, it promises to empower the next generation of learners and prepare them for the ever-changing landscape of technology and automation.

REFERENCES

Adams, W. R., & Goodwin, L. D. (2013). Interactive and collaborative cloud-based robotics programming using RoboBlockly. In *Proceedings of the 15th international conference on Information Integration and Web-based Applications & Services* (pp. 283-292). Academic Press.

Chen, H. (2021). Cloud-Enhanced Robotics in College Engineering Labs: A Case Study. *IEEE Transactions on Learning Technologies*, *14*(4), 623–638.

Chen, W. (2016). Cloud-Based Programming Environments for Educational Robotics: A Comparative Study. *IEEE Transactions on Emerging Topics in Computing*, *4*(2), 234–247.

DePold, H., Ros, R., Semsch, E., & Adya, A. (2019). Cloud Robotics in Education. In *2019 IEEE International Conference on Robotics and Automation (ICRA)* (pp. 5582-5589). IEEE.

Garcia, M., & Rodriguez, P. (2021). Integration of Cloud Robotics in the Curriculum: A Case Study in High School Robotics Clubs. *IEEE Transactions on Learning Technologies*, *14*(2), 289–305.

Huang, Y. (2022). Cloud-Enhanced Robotics for Inclusive Education: A Case Study with Special Needs Students. *IEEE Transactions on Learning Technologies*, *15*(1), 56–71.

Johnson, R. (2018). Design and Implementation of a Cloud-Connected Robotics Course for Undergraduate Education. *IEEE Transactions on Robotics*, *37*(2), 189–204.

Kim, H., & Lee, S. (2017). Cloud Robotics in STEM Education: Challenges and Opportunities. *IEEE Transactions on Education*, *60*(1), 30–37.

Kim, S., & Choi, J. (2016). A Cloud-Based Robotics Platform for K-12 Education: Design and Evaluation. *IEEE Transactions on Education*, *59*(2), 91–98.

Lee, C., & Park, H. (2021). Cloud Robotics and Gamification: Engaging Students in Programming and Problem-Solving. *IEEE Transactions on Robotics*, *38*(4), 981–995.

Li, J., & Wu, S. (2017). Cloud Robotics for Programming Education: A Practical Approach. *IEEE Transactions on Robotics*, *36*(4), 532–546.

Li, X., & Wang, Y. (2017). A Cloud-Based Platform for Collaborative Robotics Learning. *IEEE Transactions on Education*, *60*(4), 286–293.

Liu, Y. (2018). Remote Learning with Cloud-Connected Robots: A Case Study in Elementary Education. *IEEE Robotics & Automation Magazine*, *25*(2), 45–53.

Martinez, A., & Davis, C. (2019). Remote Control and Programming of Educational Robots Using Cloud Technologies. *IEEE Transactions on Robotics*, *36*(1), 98–112.

Park, S., & Kim, J. (2018). Exploring the Impact of Cloud Robotics on Programming Skills in Middle School Students. *IEEE Transactions on Learning Technologies*, *11*(3), 412–427.

Park, Y. (2019). Exploring the Use of Cloud Robotics in Informal Learning Environments: A Case Study in Science Museums. *IEEE Transactions on Learning Technologies*, *12*(3), 372–387.

Smith, J., & Johnson, A. (2019). Cloud Robotics: A Comprehensive Review. *IEEE Transactions on Robotics*, *35*(3), 657–672.

Smith, J. R., & Johnson, M. (2020). The Role of Cloud Robotics in the Education Sector: Current Status and Future Trends. *Robotics (Basel, Switzerland)*, *9*(3), 59.

Wang, Q., & Chen, L. (2020). Enhancing Educational Robotics through Cloud-Based Programming Platforms. *IEEE Transactions on Learning Technologies*, *13*(4), 789–804.

Wang, X. (2017). Remote Programming of Educational Robots via Cloud Services: A Case Study in Middle School Classrooms. *IEEE Transactions on Learning Technologies*, *10*(4), 509–524.

Watson, P. (2018). Integrating Cloud Robotics into Classroom and Learning Experiences. *Journal of Computing in Higher Education*, *30*(2), 358–371.

Xu, L. (2020). Cloud Robotics and its Application in Project-Based Learning for Engineering Students. *IEEE Transactions on Learning Technologies*, *13*(1), 45–58.

Yang, G., & Zhang, H. (2019). Cloud Robotics in Higher Education: A Survey of Current Practices and Future Directions. *IEEE Transactions on Education*, *62*(3), 205–220.

Zhang, Y. (2018). Cloud Robotics for Online Robotics Competitions in High Schools. *IEEE Transactions on Learning Technologies*, *11*(1), 132–147.

Zheng, Y., Shen, M., Shen, X., & Zhu, X. (2017). A cloud-based robot programming system for STEM education. In *2017 IEEE/RSJ International Conference on Intelligent Robots and Systems (IROS)* (pp. 5197-5202). IEEE.

Chapter 19
IoT-Enhanced Spy Robotics Bridging Security Gaps

Abhishek Kumar
https://orcid.org/0000-0003-3961-9970
Lovely Professional University, India

ABSTRACT

The integration of internet of things (IoT) technology with robotics has paved the way for innovative applications in surveillance and security. This research presents the development and implementation of a spy robot with IoT capabilities, designed to enhance monitoring and intelligence gathering in diverse environments. The robot is equipped with a range of sensors, cameras, and actuators seamlessly interconnected through IoT protocols. Real-time data transmission, remote control functionalities, and autonomous navigation make this spy robot an efficient tool for surveillance applications. The research explores the challenges, design considerations, and potential applications of this IoT-enabled spy robot, contributing to the evolving landscape of smart and connected robotic systems.

1. INTRODUCTION

Synthetic Robots play a critical role in our day-to-day existence in sports as a consequence of reducing human labor and human errors. Robots can be manually controlled or can be automated based on the requirement. It is easy to control the robotic with the help of a mobile or laptop through the Internet of Things (IoT) and can also get live streaming of video both in daylight as well as at night with the help of a Wi-Fi camera from the robotic(Mondal et al, 2019 and Patiliya et al, 2015). The objective of this chapter is the surveillance of human activities in far-flung regions where human intervention is hard or unstable. The robotic is used to accumulate facts from the faraway terrain and screen that information at a miles comfy place with the help of the IoT era. A robot is an integrated branch of the fields of engineering and science including mechanical engineering, electrical engineering, software engineering, computer science, and others. It includes robot design, development, installation, and work, as well as computer systems for understanding, managing, reacting to sensors, and processing information. Robots are machines capable of capturing people and duplicating human behavior. The robot can be used in many

DOI: 10.4018/979-8-3693-1914-7.ch019

circumstances and for many purposes, but today many are used in dangerous environments (Sharma et. al, 2015) as well as radio monitoring, detection, and deactivation, automated systems, or anywhere human beings cannot survive (underwater, heat and fresh and contain toxic substances and radical substances). Robots may take any form but some can look like human beings. It is said to assist in the acceptance of robots typically performed by humans indefinite multiplication methods. Such robots attempt to imitate movement, lift, speak, understand, or any action of any other individual. Most modern robots are life-inspired, and causative to the bio-motivated robotics market (Singh et al, 2017).

Nowadays finding a hostile enemy in different places is very complicated for the armed forces and there are lots of terror threats as well as lots of crimes and assaults occurring in interior parts of the world (Balakrishna et al, 2015). There is always the risk of killing the military a police officer or some other private security agency throughout war and disaster situations. We implement a key for the difficulty of replacing a warrior with a robot or a spy robot completely controlled with a wireless network. Robots have also been used for surveillance functions inside the private sector as well and robot guards are currently being tested everywhere in the world (Jain et al, 2015). One innovative method for addressing vulnerabilities and enhancing surveillance capabilities is the integration of robotics into security systems. Jones et al. (2018) performed an innovative research project that explores the application of unmanned aerial vehicles (UAVs) with sophisticated sensors for surveillance. The study provides insightful information on prospective security applications by demonstrating how effectively UAVs capture real-time intelligence and monitor remote places. Smith and Brown (2020) extend the research carried out by Jones et al. by investigating whether machine learning algorithms can improve the decision-making and autonomy abilities of spy robotics. Their analysis demonstrates that artificial intelligence (AI) can be used to scan large volumes of data and spot unusual activity, strengthening security measures and thwarting any threats. Zhang et al. (2019) investigate the application of swarm robotics for cooperative surveillance tasks in a related study. Their study highlights the adaptability and scalability of swarm systems in dynamic contexts, demonstrating their potential to transform safety concerns while enhancing surveillance operations. The development of improved spy robotics, and security surveillance has experienced an enormous shift that has enhanced danger detection capabilities and covered security gaps. IoT-enhanced spy robotics have practical applications in a variety of security fields, providing creative solutions to new threats and problems. Spy robotics with cutting-edge sensors and real-time communication capabilities can be used for urban surveillance operations in the field of law enforcement. To monitor crowded public spaces, important events, or high-crime regions, for example, unmanned aerial vehicles (UAVs) coupled with Internet of Things (IoT) technology can undertake drone reconnaissance missions. IoT-enhanced spy robotics are essential to border security operations because they monitor and guard national borders against smuggling, unlawful crossings, and human trafficking. With thermal imaging cameras and motion sensors, ground-based surveillance robots can monitor faraway places and borders, detect illegal invasions, and instantly alert border patrol agents of any threats. To enhance border security and stop illegal border crossings, aerial drones with IoT-enabled sensors can also perform aerial surveillance along border zones. IoT-enhanced spy robotics are used in factories for asset monitoring and critical infrastructure security. To identify any security breaches, equipment malfunctions, or safety hazards, industrial sites like power plants, refineries, and chemical plants can be patrolled by surveillance robots that are outfitted with Internet of Things sensors. These robots facilitate preventive maintenance and risk reduction by continually monitoring infrastructure assets and environmental conditions, thereby decreasing downtime and increasing operational efficiency. Using IoT-enabled sensors, these robots autonomously roam the facility grounds, looking for irregularities in sound, motion, and heat profiles. The robots instantly notify security staff in the event of a security breach or questionable activity, allowing for prompt response and intervention. To reduce security threats and protect vital assets, this case study shows how IoT-enhanced spy robotics can improve security and surveillance capabilities in industrial contexts. The comparison table 1 provides a concise overview of the key research studies included in the literature survey, highlighting their focus areas, methodologies, and key findings.

This chapter of the literature delves into the latest developments in the field of advanced spy robots, emphasizing new applications, difficulties, and technology that could affect security operations.

2. SPY ROBOT

The surveillance robot as its name suggests is the one that was used to spy on enemy territories. The capabilities can beat the moment of battle as they can be used to gather information from the enemy terrain track the information in a highly protected environment and establish a strategy for counteracting attack safely. A spy robot is a specialized robotic device intended for information collection and surveillance operations also referred to as a surveillance or reconnaissance robot. These robots can collect and communicate data from distant or dangerous places without the assistance of a human because they are equipped with a variety of sensors, cameras, and communication systems. Spy robots are used in the military, law enforcement, security, and industrial sectors. It can be used for everything from security patrols in vital infrastructure facilities to remote surveillance in disaster areas. Helping to locate survivors in remote locations and supplying essential information for environmental monitoring initiatives, play critical roles in search and rescue operations. They're made lightweight enough and compact enough for fast transport. Surveillance means looking from a distance, even as surveillance robots are used to reveal the behavior, activities, and different converting records that are accumulated for the overall reason of handling, directing, or shielding one's belongings or role (Naik et al, 2015, Sathyaprakash, 2021 and Bhuvaneswary et al, 2021) presented surveillance cameras and facial recognition are used to monitor public and private areas and to become aware of humans. Implementing spy robotics is not possible without obstacles. Lee and Kim (2021) raise attention to issues about invasions of privacy and the moral implications of widely used surveillance technologies. Their research emphasizes the necessity of strong legal frameworks and moral standards to control the application and utilization of

Table 1. Comparison table of technology-enabled robots

References	Research Focuses	Methodology	Major Finding
Johnson et al. (2021)	Autonomous Patrol Systems	Field Experiment	Effectiveness of AI-driven UGVs for perimeter surveillance.
Li. et al. (2021)	Sensor Fusion in Aerial Drones	Laboratory Testing	sensor fusion techniques for comprehensive environmental monitoring and target tracking.
Smith and Brown (2020)	Sensor Fusion in Aerial Drones	Simulation and Field Trials	scalability and adaptability of swarm systems in urban surveillance operations.
Kim and Lee (2021)	Ethical Considerations	Literature Review and Ethical Analysis	Ethical implications and privacy concerns associated with ubiquitous surveillance technologies.
Chen et al. (2022)	Cybersecurity Threats	Case Studies and Analysis	Identified cybersecurity threats and vulnerabilities inherent in networked surveillance systems.

spy robots in security-related settings. The broad adoption of improved espionage robotics is thwarted by technical constraints and weaknesses. To protect sensitive data and stop illegal access, Chen et al. (2020) address cyber security threats and potential system weaknesses inherent in networked surveillance systems. They emphasize the significance of strong encryption protocols and intrusion detection techniques. The integration of artificial intelligence (AI) algorithms with unmanned ground vehicles (UGVs) for perimeter monitoring is the subject of a recent study conducted by Johnson et al. (2021). Their research shows how AI-powered UGVs can improve situational awareness and response times by patrolling sensitive areas autonomously, spotting incursions, and notifying security personnel. Li et al. (2022) investigate the fusion of numerous sensor modalities, including LiDAR, thermal imaging, and radar, in aerial surveillance drones, building on the findings of Johnson et al. Their study demonstrates how successful sensor fusion approaches are in providing thorough environmental monitoring, threat identification, and target tracking under a range of operational scenarios.

The city of the future is possible to vicinity lots of emphasis on protection. So, robots could perform dangerous jobs usually done by way of the police, and automatic incident detection cameras will be used to permit rapid intervention by using security employees (Chavan et al, 2021, Aarthi et al 2021 Swastik et al 2021) present the block diagram as figure 1, the spay robot, which can be controlled both manually in addition to in automated mode with the assistance of an Arduino microcontroller. This robot also makes use of numerous sensors that collect statistics and send them to the Arduino microcontroller which controls the robot's behavior. The prototype of the robot contains a wireless camera, transmitter, battery, and 4-wheel drive. The Remote Controller is used to control the wireless network and the Spy robot remotely. The CCD camera is around the robot to capture details; the user command display on the LCD monitor with bits must be mounted on the screen. The CCD is set up with an LED connected to a light circuit to use the Spy robot in a dim place like nighttime. In a wireless remote control device, radio frequency modules are used to send and receive a wireless signal to power the motors of the robot control system. The Spy robot contains 3 Brush DC motors and two L293Ds used to power DC Brush DCs. The controlled Spy robot in this chapter is a small robot designed for purposes of research, tracking, and research words Microcontroller, camera CCD, L293D. The robotic setup is a casing that homes the microcontroller board, and various sensors, mounted on wheels used for navigation that's pushed via DC vehicles. A Passive Infrared (PIR) sensor is a pyroelectric tool that detects the level of IR radiation from the dwelling gadgets. The PIR tool does not emit an IR signal, as a substitute passively detects the infrared radiations coming from the human frame in the surrounding area similar to (Kaphungkui NK 2015). The PIR sensor has a gaggle of assisting circuitry. The detected infrared pulses are exceeded by the FET amplifier circuit which will reset or set the sensor output. The PIR sensor will set the output (common sense '1') when the living frame is within the range of about less than 10 meters and in any other case, it reset the output.

The block diagram consists of transmitter and receiver parts where the receiver part is linked with a monitor and the transmitter part is fixed with the bot. spy robot has mostly two sections, This is the remote control section used to control the robot and the other is the video transmission section used to relay audio and video information. Its internal components include HT12E, HT12D, RF 433Mhz Transmitter and Receiver, CCD camera, Servo motors, Chassis, SPST switch, Resistors, Not gates, L293D, and battery. Here the HT12E encoder reads the parallel data from the switches and transmits this data serially to the RF transmitter. The encoder's operating voltage is 2.4 to 12V. The performance of the RF 434 MHz Transmitter is up to 8mW at 433.92 MHz. This ASK transmitter embraces both linear and digital inputs and 1.2V to 12V DC operating voltage. SW1 is used for transmission enabling in a remote

Figure 1. Transmitter and receiver section of the robot

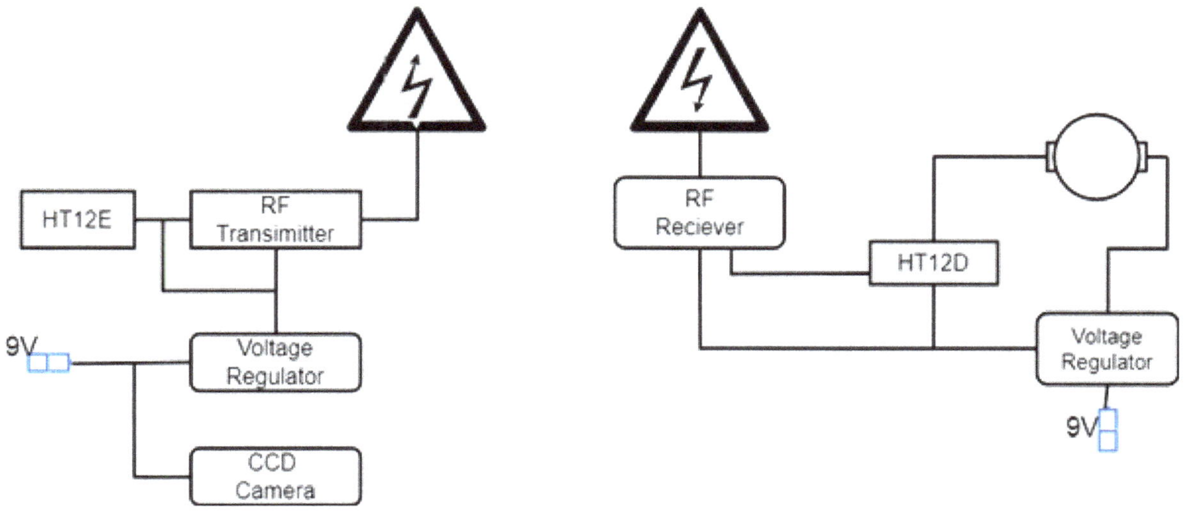

area. The CCD camera receiver section is connected to the monitor or TV, The CCD camera transmits data in radio signals so that the receiver end receives the data and displays it on the output. The output of L293D is connected to the motors.

Components

i. Microcontroller Board: Controls robot functions and processes data.
ii. Camera CCD: Captures video and transmits signals.
iii. L293D Motor Driver: Drives DC motors for robot movement.
iv. Wheels and DC Motors: Provide locomotion for the robot.
v. PIR Sensor: Passive Infrared sensor for human detection.
vi. FET Amplifier Circuit: Amplifies detected infrared pulses.
vii. HT12E Encoder and HT12D Decoder: Encode and decode data for wireless transmission.
viii. RF 433MHz Transmitter and Receiver: Facilitate wireless communication.
ix. Servo Motors: Control camera movement.
x. Chassis: Housing for all components.
xi. SPST Switch: Initiates transmission enabling.
xii. Resistors and Not Gates: Support circuitry and signal processing.
xiii. Battery: Power source for the robot.

Working Principle

i. Human Detection (PIR Sensor):
ii. PIR sensor detects infrared radiation from living objects within a 10-meter range.
iii. Amplified pulses are used to set or reset the sensor output.

iv. Wireless Transmission (RF Module): HT12E encoder reads switch data and transmits it wirelessly to the RF 433MHz transmitter. The RF receiver receives the data, which is then decoded by the HT12D decoder.
v. Motor Control (L293D): Decoded data controls the L293D motor driver, enabling movement based on the transmitted commands.
vi. Camera and Servo Control: Servo motors control the movement of the CCD camera for versatile surveillance. CCD camera transmits video signals wirelessly to the receiver end.
vii. User Interface: The remote control section uses switches and encoders to send commands.
viii. The video transmission section relays audio and video data for monitoring.

2.1 RF Communication

A signal radio wave refers to a digital electromagnetic signal used as a means of communication while one is talking about digital electronics. Radio waves are a type of electromagnetic radiation with radio frequencies ranging from 3Hz to 300GHz. Frequency refers to the oscillation rate at which RF transmission occurs at the speed of light and does not require a medium like air to fly presented in Figure 2 adapted from (Kaphungkui, 2015). Humanity interacts with artificially generated radio waves, which oscillate at various frequencies needed. RF contact is used in many industries including television broadcasting, radar systems, computer and mobile device networks, remote control, remote monitoring, and many others. Although individual radio components such as mixers, filters, and power amplifiers can be identified by operating frequency spectrum, the wireless standard cannot be specifically defined because these devices only include physical layer support. In comparison, RF modules, transceivers, and SoCs also provide data link layer support for more than one wireless communications protocol. To receive radio signals the antenna should be used. However, since the antenna can absorb thousands of radio signals at a time, the radio tube is needed to power a certain frequency. Sometimes this can be finished with the resonator in its simplest kind, the electrical device circuit, and also the inductance circuit kind. The resonator broadens the oscillations inside a particular band while decreasing the oscillations within the different frequency bands outside the band.

A way to tell apart a frequency is by decipherment and choosing the frequency of the gain, as an exhausted radio-defined software system. The gap to which a radio affiliation operates depends mostly on non-wavelength factors, like transmission power, detection quality, type, size, antenna height, transmission mode, sound, and interference signals. The total waves, the propagation of the layer, and also the waves of the sky will all reach bigger distances than the road propagation. Radio broadcasting analysis permits measures of the operating variables to be performed. In general, the designer of a wireless communications system has two additional problems: it must operate at a definite distance and transfer a certain amount of data within the data range. RF modules are very small in size and have a broad range of power applications i.e. 3V to 12V. The RF modules are essentially 433 MHz RF transmitters and modules which you get. The transmitter does not draw power as it transmits zero logic when the carrier frequency is blocked and thus requires very low battery power operation. If 1 logic was sent to the sender, with 3 volts of electricity, it is around 4.5mA. The data are transmitted sequentially from the received transmitter by the tuned receiver. The transmitter and the receiver are well connected to the two microcontrollers for the data transmission, as shown in Figure 3 adapted from[13].

Figure 2. Remote transmitter circuit

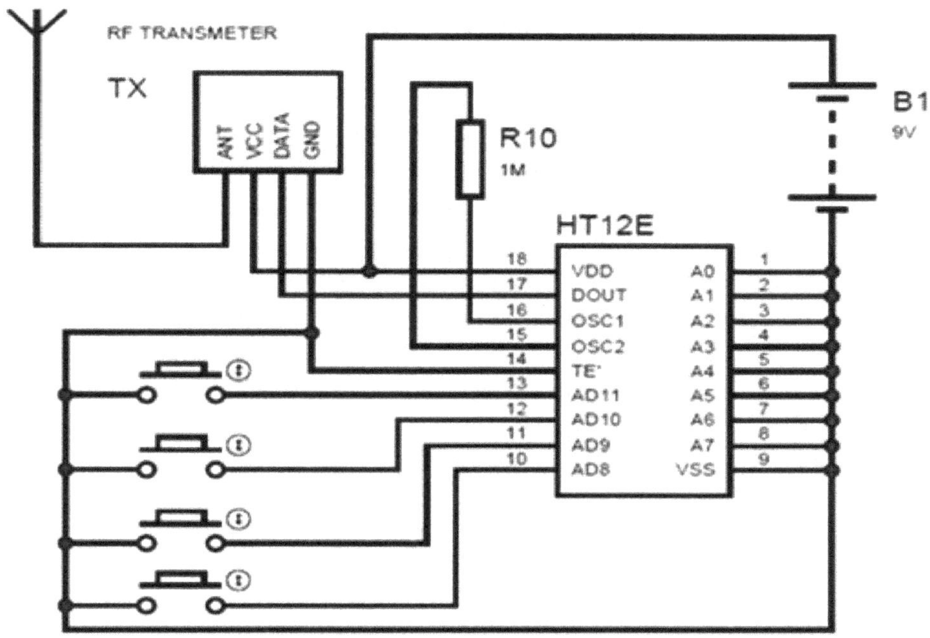

Figure 3. Remote receiver circuit

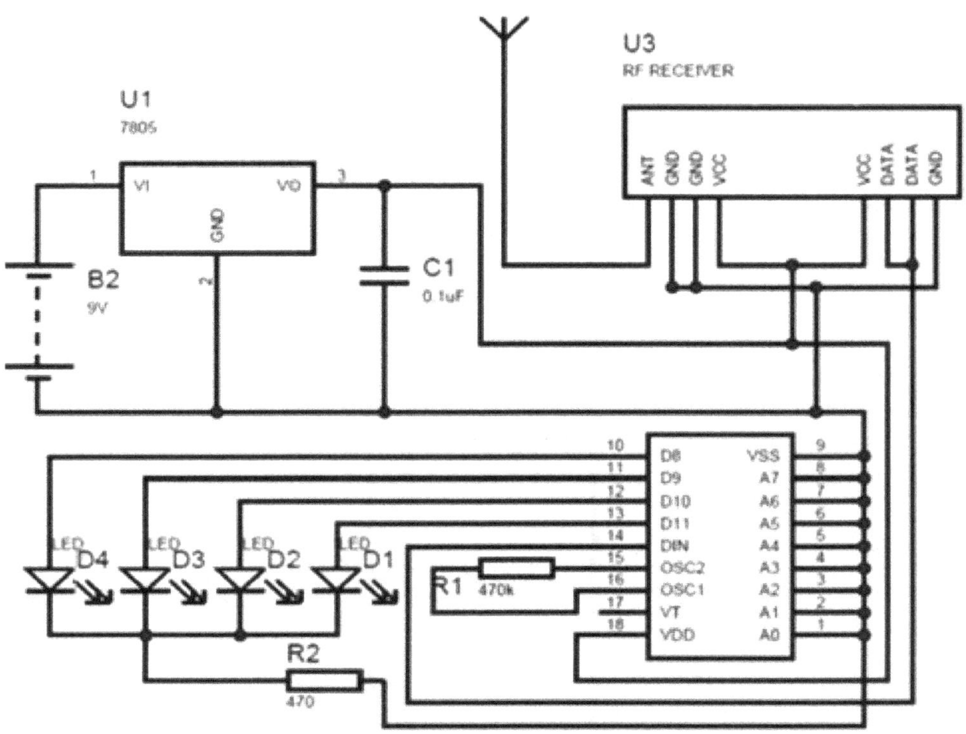

2.2 L293D Motor Driver

The L293D presented in Figure 4 adapted from (Sharma et al., 2015), is a dual-drive H-Bridge drive, so it can drive unmanned motors in both clockwise directions in the front and rear. L293D is a 15-milliwatt monolithic, integrated circuit. It has a power supply of up to 46V, a total DC of up to 4V, and extra protection against fire. The TheL293D motor driver has multiple inputs and outputs so that two DC motors and one stepper motor can be controlled. 3 (b), PORTB5, and PORTB4 connect to input 1, and input 2 connects to the drive pin. Then, the No. 2 and 3 drive pin is used to collect the data connected to the Brush DC motor. The next L293D car driver is related to the direction of motion of the Spy robot. PORTB3 and PORTB2 are related to motor inputs 1 and 2 Input of data.

Data production for abnormal left-wheel work. PORTB1 and PORTB0 are attached to the motor numbers: 10 and 12 used for data input and PINs: 13 and 14 such as data extraction for irregular tire use. The L293D in Figure 5 is a motor driver that provides +12V control since the 12V Brush DC motors are used to that end. Most electrically driven motors use the power from the Brushes to transfer power from the source (battery) to the transmission. The DC motors used in this project are used as the brakes can wear out and need to be replaced, such as how the brakes are worn using disk brakes. Thus, the three integrated DC motors are used to rotate the camera and the steering wheels, respectively.

2.3 CCD Camera

CCD is a charged coupled semiconductor device, that has found various applications in all modified picture capturing devices. The remote that controls this camera has a feature of sound and video system. The range of the CCD camera is clear up to 100 ft and the picture resolution is about 1024 x 800. CCD acquires the power of +12V through lead batteries. This camera also can change directions (left and right) to get a full view. The drawback this camera has is that it has no USB port. The CCD camera that we are using in the project is an electronic device, the camera transmits the data and the receiver is used to receive the data from a remote location the receiver consists of three pins which are known as plug-and-play cables of three pins in which yellow and white for audio and video, red is for power supply. The transmitter cables can be connected to the monitor or TV.

Figure 4. LM293D pin diagram (Sharma et al., 2015)

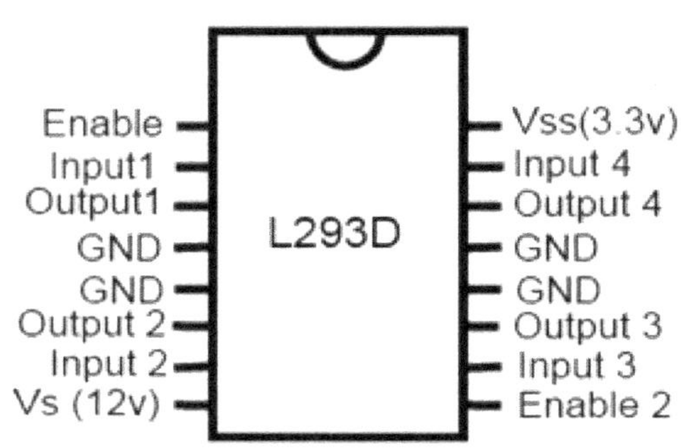

Figure 5. Connection diagram of LM293D (Sharma et al., 2015)

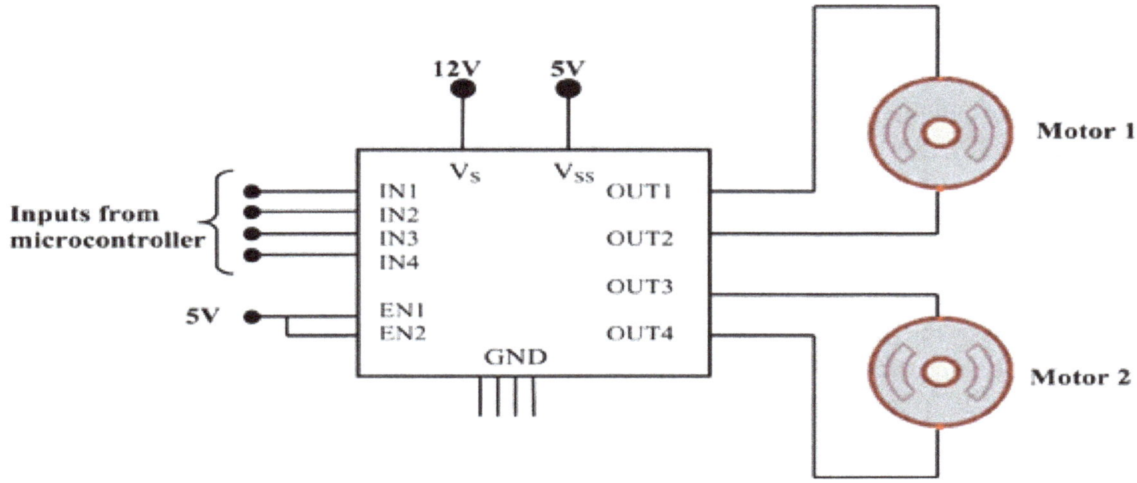

2.4 HT12D

HT12D is an IC (integrated circuit) 212 series decoder, developed by Holtech for remote control applications. This is most widely used for wireless radio frequency (RF) applications. We can transmit 12 bits of parallel data using the Coupled HT12E Encoder and the HT12D Decoder. The HT12D bus can transform serial data for 12-bit parallel data to its input (that can be provided by the RF receiver). The parallel 12-bit data is broken down into 8 address bits and 4 data bits. Using 8 address bits, we can provide 8-bit security codes for 4-bit data, which can be used with a single transmitter to resolve multiple receivers." (ht12d)

2.5 HT12E

HT12E is a 212 IC (Integrated Circuit) encoder for remote uses. Applications commonly used on radio frequencies. By using the paired HT12E encoder and the HT12D decoder, we can easily move 12 pieces of the same data, respectively. HT12E simply transforms twelve identical data into a serial output that can be transmitted via RF. The 12-measurement information is broken down into 8 addresses and 4 bits of data. By using these address pins, we can have an 8-bit data protection code, and the same transmitter can be used to resolve multiple receivers.

3. SYSTEM INTERFACING

There are two sections in this robot one is the remote control section and the other is the video transmission section

3.1 Remote Control Section

The remote controller comprises 4 press buttons as shown in figure 6. There is an antenna that is used to receive and transmit the signals. The remote needs at least +9 V power to control the whole system. Remote control circuitry has a Radiofrequency module implanted inside the remote for communication with the robot. LM7805 voltage regulator is used to control the microcontroller. This voltage regulator is very suitable to control the steady potential voltage; it is also connected to the battery. The remote controller sends signals to the radio frequency module individually.

The working is, that the microcontroller that we have used will record the commands and transmit them to the RF transmitter then, the RF transmitter directs the data or information to the spy robot which will be received by the RF receiver. HT12E and RF433Mhz transmitters are connected serially which are used to send the signal from the remote to the Robot. In the project, we used HT12E, in the simulation we used M145028 it's a similar encoder that is an in-built library available in Proteus Tool.

The working of M145028 is similar to the HT12E, There are a few little changes in the pin diagram but by researching it the component works quite similarly to HT12E. The left one is the pin description of the M145028 and the right one is the pin description of HT12E, The similarities can be found easily, as the AD pins in the right hand represent data pins which are A6-A9 in M145028, osc1 & osc2 pins are oscillator pins which require certain resistance but in M145028 the internal pin description consist of resistance and capacitance so there is no need for external resistance to be applied for the encoder. TE is the VT pin in M145028; finally, the data pin which is the output in HT12E is Din in MT145028. The flowchart in Figure 7 presents the remote control of transmitting.

Figure 6. Remote transmitter sections

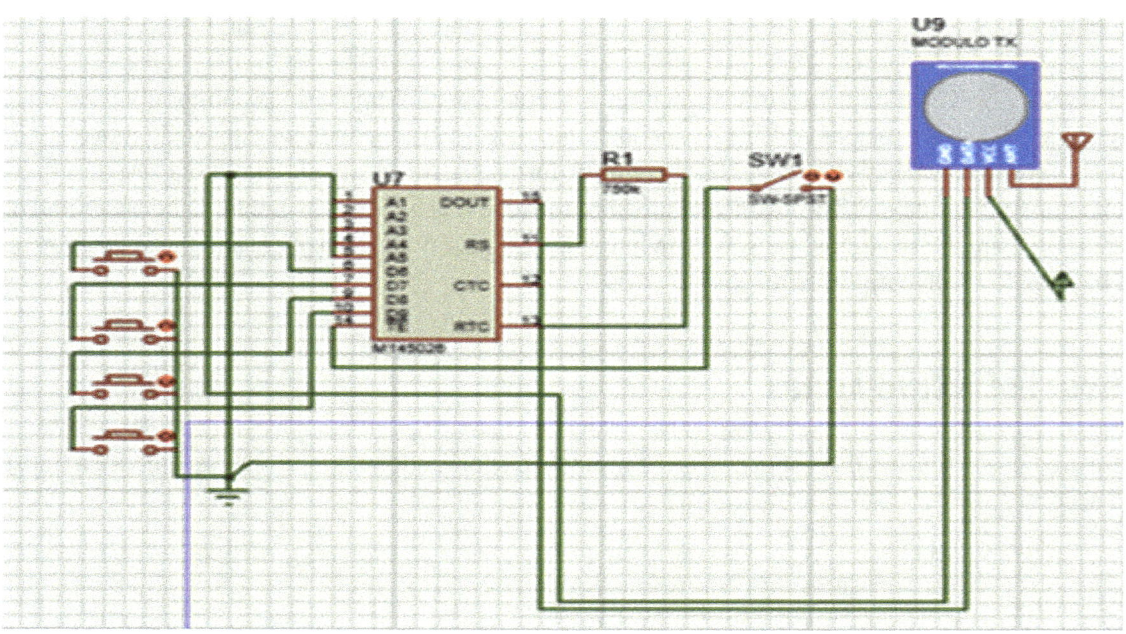

Figure 7. Remote control of transmitting

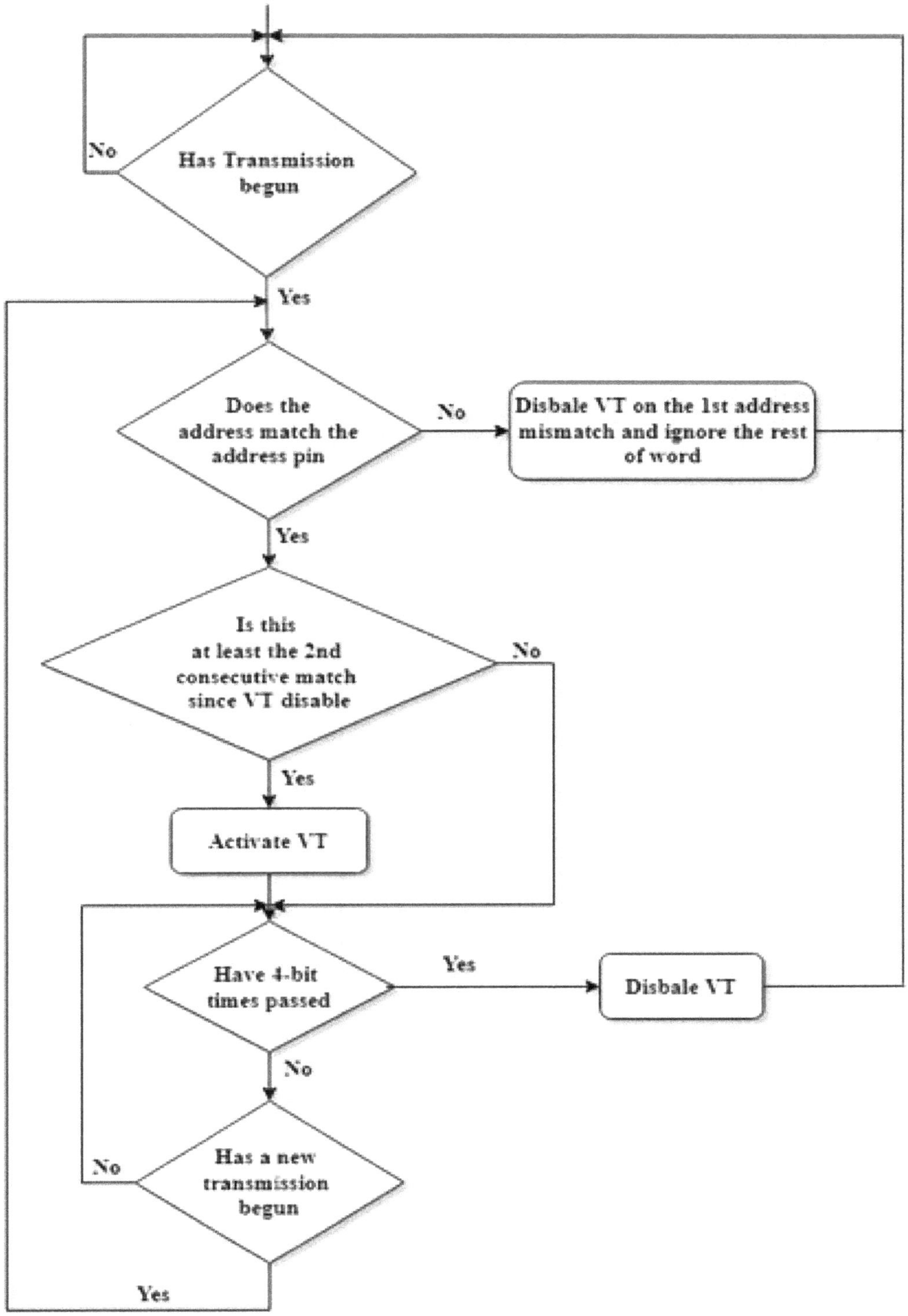

Figure 8. Video transmission section

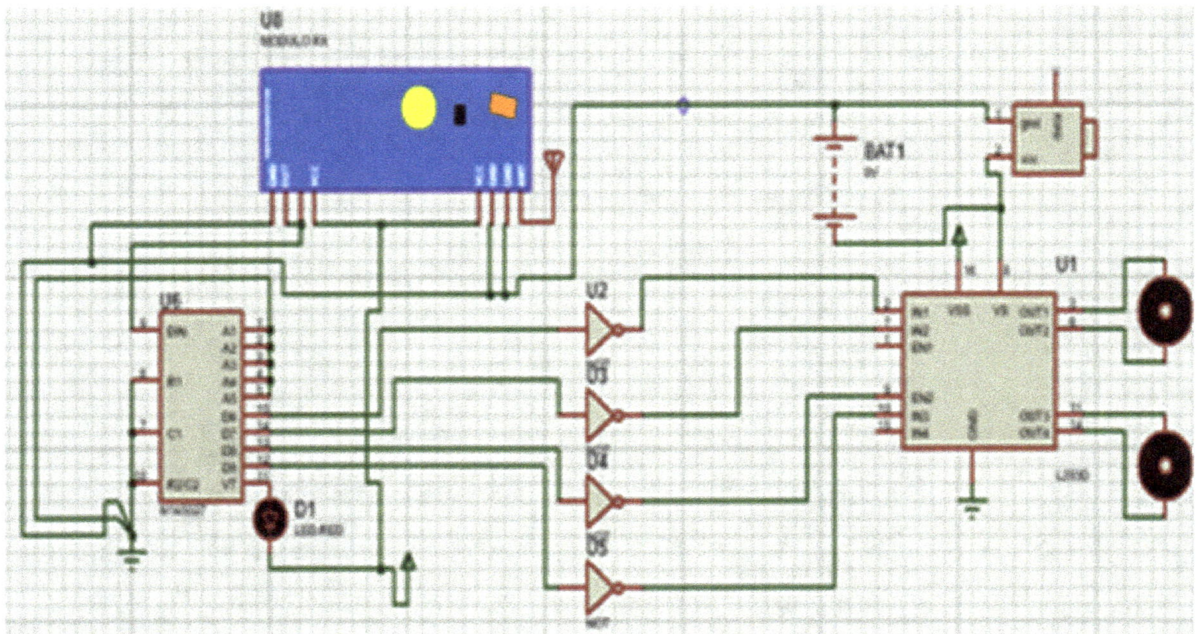

Figure 9. IC M145027 (Singh, Gupta, & Korde, 2017)

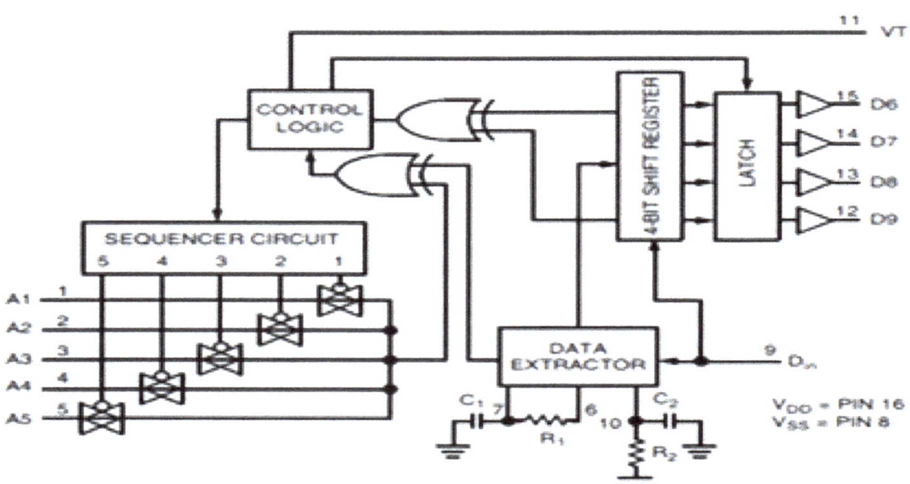

3.2 Video Transmission Section

As you can see from the block diagram, the pin description of the M145027 is similar to HT12D. To increase the range of controlling the robot, we found a few ways to increase the range by 100 meters by increasing the voltage passing through the 433RF section. To move in any terrain, we found an efficient way to move the robot in various terrains that are to replace the wheeled structure with the chain structure so that it could avoid pebbles, stones, etc in its way. To maintain proper load on a robot, we are working on its battery and the load so the robot can maintain its speed and carry the battery efficiently. Circuit check and working of a circuit on the simulator, we completed the simulation of the project in the Proteus using components. The body of the robot consists of the CCD wireless camera which is attached to the body and the power supply is supplied by the battery, the data transferred from the wireless CCD camera is received from the RF receiver of the camera which then is connected to the monitor to display the output. The receiver of the CCD camera consists of the audio pin and video pin from this we can capture both audio and video in real-time data. Then comes the body part the whole body is controlled using a remote controller made by an RF module transmitter which has 4 buttons and an encoder to control the robot, At the receiver side that is where the signal is received in the body of the robot, RF module receiver receives the data and then decodes using the decoder and passes to the motor driver ICL293D, which can support two motors to control, the inputs from the decoder are processed in the motor driver IC which controls the motor.

Robotics have a wide range of usage in various areas, in the future, we can expect every part to be automated and robots will replace every place where a human is needed. By using this project with little advancement we can achieve better results. This project can be used in surveillance, replacing soldiers in covert operations, and security; even it can be used to deliver food and a lot more uses with the robot. By incorporating Raspberry Pi or Arduino we can make the robot work on Bluetooth to control the robot and also we can make little advancements like using IoT we can use the cloud to control the robot. The body uses various materials that can resist heat by this we can make the body durable. Smith and Brown (2020) investigated the use of swarm robotics for cooperative surveillance tasks in urban settings in a similar study. Their research demonstrates how swarm systems may be scaled and adapted to perform coordinated patrols in challenging terrain, collect real-time intelligence, and navigate through other obstacles—all of which have the potential to revolutionize urban security operations. But there are also serious ethical and privacy issues with the broad use of improved spy robotics. To address societal concerns and reduce the possibility of abuses of surveillance capabilities, Kim and Lee's (2021) recent work explores the ethical implications of ubiquitous surveillance technologies. They highlight the need for transparent governance frameworks, privacy-preserving mechanisms, and public discourse. The use of AI and machine learning algorithms in spy robotics brings up significant issues about accountability, data privacy, and algorithmic bias. According to a study by Chen et al. (2022), human oversight, algorithm transparency, and ethical AI design principles are crucial for ensuring impartial and accountable decision-making in autonomous surveillance systems. Technology has never before provided such prospects for threat detection and response, but to ensure the proper deployment and use of spy robotics in security contexts, it is still critical to address ethical, privacy, and legal concerns.

4. CONCLUSION

The chapter concludes that the spherical robot has been designed and assembled with electronics, cars, cameras, and Bluetooth. IoT-enhanced spy robotics provide many benefits, but they also have certain drawbacks and difficulties. The round robot moved at the ground transmitting the video and audio in keeping with the control instructions from the controller as much as 150 ft. This round robotic device may be used inside the assault and rescue operation to get the data from the human unreachable regions even as doing army services. This type of robot can carry out tough and repetitive work for humans.

whilst the person controls by using a remote controller, the undercover agent robotic will circulate to the desired destination and undercover agent images around the robotic. The consumer can take a look at and propose from a computer with the wireless faraway controller.

5. FUTURE SCOPE

The developments in Internet of Things-enhanced spying robotics have the potential to significantly improve reconnaissance and surveillance capabilities. Future research should focus on integrating *Internet of Things (IoT)* technology to improve spy robots' networking, communication, and data processing capacities. Spy robots may easily communicate with other smart devices, sensors, and cloud-based platforms by utilizing IoT frameworks. It allows for real-time data processing, predictive analytics, and self-governing decision-making. The next steps in IoT-enhanced spy robotics will be greatly influenced by advances in fields like *artificial intelligence, machine learning, and edge computing*. Spy robots will be able to comprehend complicated data streams, spot trends, and adjust on their own to constantly shifting situations thanks to advanced *AI algorithms*. Spy robots will be able to process data locally owing to the incorporation of edge computing capabilities, which will lower latency and bandwidth requirements while improving data security and privacy. With the help of this distributed computing architecture, spy robots will be able to react quickly to security problems and emerging hazards in real-time. However, there are significant ethical questions and societal implications associated with the development of IoT-enhanced spying robotics in the future. Concerns about *data security, algorithmic bias, and privacy infringement* will grow in importance as spy robots become more ubiquitous and autonomous. The responsible and ethical deployment of spy robot technology will require clear governance structures, regulatory rules, and ethical frameworks.

The potential for transforming surveillance and reconnaissance capabilities with IoT-enhanced spy robotics is enormous. Spy Robotics is set to make major advancements in improving security, safety, and situational awareness in a variety of operational contexts by embracing emerging technology, tackling ethical issues, and encouraging interdisciplinary collaboration.

REFERENCES

Aarthi, V., Visali, R., & Nagarajan, K. K. (2021). Smart Spying Robot with IR Thermal Vision. *7th International Conference on Advanced Computing and Communication Systems (ICACCS)*, *1*, 183-187. 10.1109/ICACCS51430.2021.9441821

Balakrishnan, M., Gowthaman, S., Kumaran, S. J., & Sabhapathy, G. R. (2015). A smart spy robot charged and controlled by wireless systems. *Proceeding of International Conference on Innovations in Information, Embedded and Communication Systems (ICIIECS)*, 1-4. 10.1109/ICIIECS.2015.7193096

Bhuvaneswary, N., Pravallika, S., & Jayapriya, V., & Himabindu. (2021). Night Surveillance Military Spy Robot using Raspberry Pi. *Annals of the Romanian Society for Cell Biology*, 5740–5747.

Chavan, M. P., Kharde, S., Jade, M., & Jadhav, A. (2021). Wireless spy robot. *International Engineering Journal For Research & Development*, *6*, 4–4.

Chen, Y., Wang, F., & Liu, Y. (2020). Cybersecurity threats and defense in networked surveillance systems: Challenges and opportunities. *IEEE Access : Practical Innovations, Open Solutions*, *8*, 187096–187110.

Chen, Y., Wang, F., & Liu, Y. (2022). Ethical considerations in AI-driven surveillance: Addressing algorithmic bias and privacy concerns. *IEEE Transactions on Systems, Man, and Cybernetics. Systems*, *52*(1), 50–65.

. Jain, P., Firke, P. N., Kapadnis, K. N., Patil, T. S., & Rode, S. S (2014). RF based Spy robot. *Int. Journal of Engineering Research and Applications, 4*(4), 06-09.

Johnson, A., Brown, D., & Davis, C. (2021). AI-driven UGVs for perimeter surveillance: Enhancing security through autonomous patrolling. *Journal of Intelligent Robotics*, *40*(3), 312–327.

Jones, A., Smith, B., & Davis, C. (2018). Unmanned aerial vehicles for surveillance: Current capabilities and future trends. *Journal of Security Engineering*, *15*(2), 45–58.

Kaphungkui, N.K. (2015). RF based remote control for home electrical appliances. *International Journal of Innovative Research in Electrical, Electronics, Instrumentation and Control Engineering, 3*(7), 42-44.

Kim, S., & Lee, J. (2021). Ethics of surveillance technologies: Balancing security and privacy in the age of enhanced spy robotics. *Journal of Information Ethics*, *9*(2), 178–193.

Lee, J., & Kim, S. (2021). Ethical considerations in the deployment of surveillance robotics: Balancing security and privacy. *Journal of Ethics in Technology*, *7*(3), 112–128.

Li, Q., Wang, H., & Zhang, L. (2022). Sensor fusion in aerial surveillance drones: Advancements and challenges. *IEEE Sensors Journal*, *22*(4), 2150–2165.

Mondal, B., Rakshit, S., Billaha, M.A., Roy, B., & Sarkar, R. (2019). Low-Cost Wireless Data Transmission System for Industrial Applications. *Advances in Computer, Communication and Control,* 509-523.

Naik, Y. M., Deshpande, C. M., Shah, R. R., & Kulkarni, R. R (2013). Android Controlled Spy-Robot. *International Journal of Software and Web Sciences 4*(1), 54-57.

Patoliya, J., Mehta, H., & Patel, H. (2015) Arduino controlled war field spy robot using night vision wireless camera and Android application. *Proceeding of 5th Nirma University International Conference on Engineering (NUiCONE)*, 1-5 10.1109/NUICONE.2015.7449624

Sathyaprakash, B. P. (2021). Surveillance Spy Robot with Wireless Night Vision Camera and Metal Detection. *Turkish Journal of Computer and Mathematics Education*, *12*(12), 2397–2402.

Sharma, V., Kaur, S., Mahajan, R., Chawla, S., Khatri, T., & Dhankar, M. (2015). High-Performance Surveillance and Operating Robot with Raspberry Pi. *International Journal of Control and Automation*, *8*(11), 407–414. doi:10.14257/ijca.2015.8.11.37

Singh, A., Gupta, T., & Korde, M. (2017) Bluetooth controlled spy robot. *International Conference on Information, Communication, Instrumentation and Control (ICICIC),* 1-4.

Smith, E., & Brown, D. (2020). Advancing security through autonomous surveillance: The role of machine learning algorithms. *The International Journal of Robotics Research*, *39*(4), 567–582.

Smith, E., & Brown, D. (2020). Swarm robotics for urban surveillance: Opportunities and challenges. *IEEE Robotics & Automation Magazine*, *27*(2), 75–89.

Swasthik, M., Pawan, C. T., Shankar, H., Harshitha, V., & Vinay, N. A. (2021). SPi-Bot: A Gesture Controlled Virtual Telepresence Robot. *International Journal of Modern Agriculture*, *10*(2), 3782–3792.

Zhang, L., Wang, H., & Li, Q. (2019). Swarm robotics for collaborative surveillance: Challenges and opportunities. *IEEE Transactions on Robotics*, *35*(3), 647–662.

Chapter 20
Application of Cloud Robotics in Autonomous Vehicles

E. Fantin Irudaya Raj
https://orcid.org/0000-0003-2051-3383
Dr. Sivanthi Aditanar College of Engineering, India

Balaji Mahadevan
Sri Sivasubramaniya Nadar College of Engineering, India

N. Pon Subathira
https://orcid.org/0009-0009-4171-1582
Dr. Sivanthi Aditanar College of Engineering, India

S. Darwin
https://orcid.org/0000-0002-4316-1459
Dr. Sivanthi Aditanar College of Engineering, India

ABSTRACT

Cloud robotics has emerged as a transformative technology in the realm of autonomous vehicles, presenting a paradigm shift in the way these vehicles perceive, navigate, and make decisions. The chapter provides a comprehensive analysis of the key components, advantages, and challenges of integrating cloud robotics into autonomous vehicles. Cloud robotics leverages the power of real-time data processing, machine learning, and scalability. It enhances the safety, efficiency, and cost-effectiveness of autonomous vehicles. However, it also introduces challenges, such as latency, security, and connectivity, which demand innovative solutions. The chapter also discusses the application of advanced technologies in decision-making, model training, adaptability, and the learning process. It also addresses the regulatory and ethical considerations surrounding autonomous vehicles, examining government regulations and the broader societal implications of this technology.

DOI: 10.4018/979-8-3693-1914-7.ch020

1. INTRODUCTION

In recent years, the automotive industry has undergone a seismic transformation with the advent of autonomous vehicles. This revolution promises to reshape transportation as we know it, offering the tantalizing prospect of safer, more efficient, and more environmentally friendly journeys (Kato et al., 2015). However, at the heart of this transformation is a technological enabler that is poised to play a pivotal role in the realization of fully autonomous vehicles: cloud robotics.

The fusion of cloud computing and robotics has given birth to cloud robotics. This dynamic field has the potential to redefine how autonomous vehicles perceive and interact with the world around them (Mello et al., 2022). The present chapter embarks on an in-depth exploration of cloud robotics and its implications for autonomous vehicles. It will dissect the core components of this synergy, evaluate its advantages, scrutinize the challenges it presents, and delve into the role of cloud-based data processing, machine learning, safety, and security in shaping the future of transportation.

To set the stage, it is imperative to understand that the progression from human-operated vehicles to autonomous ones is not merely a question of automation but a profound transformation in how vehicles operate. It is truly impressive how humans have powered traditional vehicles by interpreting data and navigating roads with incredible confidence (Chen et al., 2018). Autonomous vehicles, on the other hand, rely on an intricate blend of sensors, AI algorithms, and connectivity to autonomously perform these tasks. Moreover, the seamless operation of autonomous vehicles necessitates real-time access to massive amounts of data, which is where cloud robotics assumes a pivotal role.

In the sections that follow, the present chapter will unveil the synergy between cloud robotics and autonomous vehicles. The present chapter will dissect the underlying technology, explore the advantages, and tackle the hurdles that must be overcome to bring this vision to fruition. Furthermore, the chapter will journey into the world of data processing in the cloud, the applications of machine learning and AI, the paramount importance of safety, and the intricate web of security concerns that this fusion entails. By examining real-world case studies, regulatory frameworks, and emerging trends, the present chapter aims to paint a comprehensive picture of the current state and prospects of cloud robotics for autonomous vehicles.

The implications of cloud robotics for autonomous vehicles are far-reaching. It promises to not only revolutionize transportation but also to reshape industries, impact urban planning, and challenge our legal and ethical norms. As the current chapter embarks on this exploration, it becomes evident that cloud robotics for autonomous vehicles is not just a technological shift; it is a societal transformation that has the potential to redefine the way people move, work, and live.

2. BACKGROUND

The concept of autonomous vehicles, often referred to as self-driving cars, has been a longstanding aspiration in the automotive industry. The idea of vehicles capable of navigating roads without human intervention can be traced back to science fiction and early experiments. However, recent advances in technology, particularly in artificial intelligence, sensors, and connectivity, have made this vision increasingly tangible.

2.1 Evolution of Autonomous Vehicles

The evolution of autonomous vehicles can be categorized into several key phases (Othman, 2023; Woo et al., 2021):

- ❖ Early Experiments (20th Century): The idea of self-driving cars has roots in the 20th century. Researchers and engineers conducted early experiments with automated vehicles, but these efforts were limited by the available technology and remained largely in the realm of academia and research.
- ❖ DARPA Challenges (2000s): The Defense Advanced Research Projects Agency (DARPA) played a significant role in advancing autonomous vehicle technology by organizing a series of Grand Challenges in the 2000s. These competitions spurred innovation and attracted the attention of both academia and the private sector.
- ❖ Commercial Interest (2010s): In the last decade, significant commercial interest in autonomous vehicles emerged. Companies like Google, Tesla, and Uber made substantial investments in developing autonomous driving technology. These developments marked a shift from academic research to real-world applications.
- ❖ Current Landscape (2020s and beyond): As we enter the 2020s, autonomous vehicles are no longer experimental. They are being tested on public roads and are inching closer to widespread deployment. However, the journey to full autonomy remains a complex one, with challenges related to safety, regulation, and technology still to be overcome.

2.2 The Role of Cloud Robotics

Amid this evolution, the emergence of cloud robotics has introduced a transformative element to the development and operation of autonomous vehicles. Cloud robotics involves the integration of robotics with cloud computing, enabling vehicles to access and leverage remote computational power, data, and algorithms. The growing capabilities of cloud infrastructure, big data analytics, and high-speed connectivity underpin this paradigm shift.

The integration of cloud robotics into autonomous vehicles represents a critical milestone in the pursuit of full autonomy. It allows these vehicles to tap into vast computational resources in real-time, enabling them to make more informed decisions, process sensory data more efficiently, and adapt to changing conditions swiftly (Prasad et al., 2023). This report will explore how cloud robotics plays a pivotal role in augmenting the capabilities of autonomous vehicles and driving them closer to widespread adoption. Figure 1 shows a real-world example of the cloud-based autonomous vehicle.

To understand the significance of cloud robotics in autonomous vehicles, it is essential to dissect the key components of this synergy and explore the advantages and challenges that it entails. This report will delve into the intricate web of technology and concepts that underpin cloud robotics for autonomous vehicles, ultimately revealing the potential and the hurdles of this transformative technology.

3. KEY COMPONENTS OF CLOUD ROBOTICS

Cloud robotics for autonomous vehicles represents a fusion of cutting-edge technologies, each playing a crucial role in the overall system. Understanding these key components (Buerkle et al., 2023; Darwin

Figure 1. Cloud-based autonomous vehicle

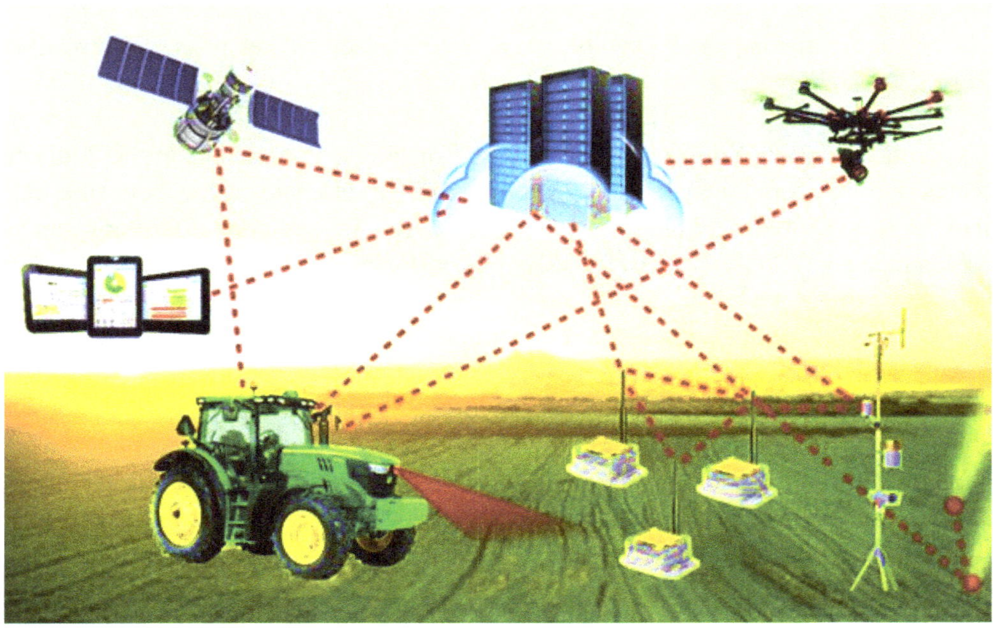

& Fantin Irudaya Raj, 2023) is essential to grasp how cloud robotics functions and how it enhances the capabilities of autonomous vehicles:

- ❖ Autonomous Vehicle: At the core of cloud robotics is the autonomous vehicle itself. These vehicles are equipped with various sensors, cameras, lidar, radar, and GPS, enabling them to perceive their surroundings. The vehicle's onboard computer processes sensory data and controls its movements (Xue et al., 2024).
- ❖ Sensors and Actuators: Sensors are fundamental for collecting data about the vehicle's environment. This includes information about nearby objects, road conditions, traffic, and the vehicle's internal state. Actuators, on the other hand, are responsible for carrying out actions based on the vehicle's decision-making processes, such as controlling acceleration, steering, and braking (Appadurai & Fantin Irudaya Raj, 2021).
- ❖ Cloud Infrastructure: The cloud infrastructure is a critical component that enables real-time data processing and storage. It consists of powerful servers and data centers hosted in remote locations, which are accessible via the Internet (Syamala et al., 2024). The cloud infrastructure is responsible for storing vast amounts of data, running complex algorithms, and providing the computational resources needed for decision-making.
- ❖ Connectivity: High-speed, reliable connectivity is a linchpin in cloud robotics. Autonomous vehicles require constant and low-latency data connections to interact with the cloud infrastructure. 5G networks and other advanced communication technologies are increasingly crucial for maintaining this real-time link (Hakak et al., 2023).

- ❖ Edge Computing: Edge computing complements cloud infrastructure by enabling local data processing and decision-making. It reduces latency and can ensure the safety of autonomous vehicles, particularly in situations where a real-time response is crucial (Ming, 2023).
- ❖ Control Algorithms: Sophisticated control algorithms are responsible for processing sensor data, making decisions, and executing actions. These algorithms incorporate machine learning, computer vision, and other AI techniques to interpret the environment and ensure safe and efficient navigation (Fantin Irudaya Raj & Balaji, 2021; Sijini et al., 2016).
- ❖ Data Fusion: Autonomous vehicles often use data fusion to combine information from multiple sensors. This process results in a more comprehensive and accurate understanding of the vehicle's surroundings, allowing for safer decision-making.
- ❖ Mapping and Localization: Detailed maps of the environment, including road networks, traffic signals, and landmarks, are vital for accurate navigation. Localization techniques, such as GPS and SLAM (Simultaneous Localization and Mapping), determine the vehicle's precise position within these maps.
- ❖ Remote Command and Control: Cloud robotics allows for remote monitoring, control, and updates of autonomous vehicles. This can be crucial for diagnosing issues, applying software updates, and ensuring the safe operation of the vehicles.
- ❖ Security and Privacy Measures: With sensitive data transmitted to and from the cloud, robust security measures are essential. Encryption, authentication, and cybersecurity protocols are critical to protect against unauthorized access and data breaches.

These components collectively form the intricate web that is cloud robotics for autonomous vehicles. Each piece of the puzzle plays a pivotal role in making self-driving cars not only possible but safer, more efficient, and capable of navigating the complex and dynamic world of transportation. In the following sections, the advantages and challenges associated with this convergence of technology and how it shapes the future of autonomous vehicles are discussed in detail.

4. ADVANTAGES OF CLOUD ROBOTICS FOR AUTONOMOUS VEHICLES

The integration of cloud robotics into autonomous vehicles brings about a plethora of advantages that not only enhance the capabilities of these vehicles but also contribute to safer, more efficient, and scalable transportation solutions (Rana & Hossain, 2023). Some of the key advantages include:

- ❖ Real-Time Data Processing: Cloud robotics enables autonomous vehicles to access and process data in real-time. This means that vehicles can make split-second decisions based on the most up-to-date information from their surroundings. This capability is crucial for avoiding obstacles, responding to rapidly changing traffic conditions, and ensuring passenger safety.
- ❖ Scalability: Cloud-based infrastructure provides nearly limitless computational resources. As the number of autonomous vehicles on the road grows, the cloud can seamlessly scale to accommodate the increasing demand for processing power, making it a highly scalable solution.
- ❖ Machine Learning and AI: Cloud robotics leverages machine learning and artificial intelligence to enhance decision-making. Autonomous vehicles can continuously learn from data and adapt their

behaviour to improve safety and efficiency. Cloud-based machine learning models can be updated and shared across the entire fleet, enabling a collective learning process (Fantin Irudaya Raj, 2022).

❖ Cost-Effectiveness: By offloading intensive computing tasks to the cloud, the need for powerful onboard hardware is reduced. This can lead to cost savings in terms of vehicle production and maintenance. Moreover, cloud robotics can reduce the cost of developing and updating algorithms, as the cloud infrastructure allows for centralized development and deployment (Mushtaq et al., 2023).

❖ Remote Monitoring and Support: Cloud robotics facilitates remote monitoring and support for autonomous vehicles. Fleet operators and manufacturers can remotely diagnose issues, apply software updates, and provide assistance, reducing the need for physical interventions and maintenance (Mushtaq et al., 2023).

❖ Collaborative Decision-Making: Autonomous vehicles connected to the cloud can communicate with one another, sharing information about road conditions, traffic, and potential hazards. This collaborative approach can lead to more efficient traffic flow, reduced congestion, and improved safety.

❖ Environmental Benefits: Cloud robotics can improve eco-friendly transportation by optimizing routes and driving behaviour. Autonomous vehicles can work together to reduce fuel consumption and emissions, promoting a greener approach to mobility.

❖ Enhanced Safety: Real-time data sharing with the cloud gives vehicles a more comprehensive view of their surroundings. This results in improved safety measures, quicker responses to unexpected events, and a reduced risk of accidents caused by human error.

❖ Data Analytics and Insights: The data collected by autonomous vehicles and processed in the cloud can provide valuable insights into traffic patterns, road conditions, and other relevant information. This data can be used for urban planning, infrastructure improvements, and predictive maintenance.

❖ User Experience: Cloud robotics can offer a more personalized and convenient user experience. Autonomous vehicles can adapt to passenger preferences, offering customized services such as entertainment, climate control, and route recommendations.

❖ Global Mapping and Localization: Cloud robotics can provide access to high-definition maps and localization data, which is particularly valuable for autonomous vehicles that operate in various regions or countries. This global mapping and localization support enhances the reliability and accuracy of navigation (Hao et al., 2023).

The advantages of cloud robotics for autonomous vehicles can potentially revolutionize the transportation industry. They not only improve the operational efficiency and safety of autonomous vehicles but also contribute to more sustainable and user-centric mobility solutions. Figure 2 shows the automation levels of autonomous cars in recent times. However, it's important to recognize that these advantages come with their challenges and considerations, which are discussed in the subsequent sections.

5. CHALLENGES AND LIMITATIONS

While cloud robotics presents a range of advantages for autonomous vehicles, it also introduces several challenges and limitations that need to be addressed for the technology to reach its full potential (Bendiab et al., 2023):

Figure 2. Automation levels of autonomous cars

- ❖ Latency: One of the primary concerns with cloud robotics is latency, the delay in data transmission to and from the cloud. Autonomous vehicles require real-time data to make instantaneous decisions, and even a slight delay can impact their safety and performance. To mitigate latency, edge computing solutions are being explored, allowing for some processing to occur locally on the vehicle itself.
- ❖ Connectivity Reliability: Autonomous vehicles depend on high-speed and reliable internet connectivity. In areas with poor network coverage or the event of network disruptions, the performance and safety of autonomous vehicles may be compromised. This challenge is particularly relevant in remote or rural areas.
- ❖ Data Security and Privacy: Cloud robotics involves transmitting a significant amount of sensitive data to the cloud, including sensor data, location information, and vehicle diagnostics. Ensuring the security and privacy of this data is paramount to prevent unauthorized access, hacking, or misuse (Zhao, Fang, Min, Wu, Wang, & Teixeira, 2023).
- ❖ Regulatory and Legal Hurdles: Deploying autonomous vehicles, especially those relying on cloud robotics, faces complex regulatory and legal challenges. Governments and regulatory bodies are grappling with how to certify the safety of these vehicles and establish a legal framework for their operation on public roads.
- ❖ Cost and Infrastructure: Building and maintaining a robust cloud infrastructure is costly. Manufacturers and operators of autonomous vehicles must invest in data centers, servers, and network infrastructure. The cost of this infrastructure can be a significant barrier to entry for smaller players in the industry (Tengilimoglu et al., 2023).
- ❖ Dependency on Connectivity: Autonomous vehicles relying on cloud robotics depend highly on continuous connectivity. Any interruption in-network services, whether due to natural disasters or technical issues, can disrupt the operation of these vehicles, potentially leading to safety concerns.

- ❖ Data Volume and Storage: The sheer volume of data generated by autonomous vehicles can overwhelm cloud storage systems. Managing and storing this data efficiently, especially in the context of a large fleet of vehicles, poses a significant logistical challenge (Chen et al., 2023).
- ❖ Algorithm Robustness: Autonomous vehicles employ complex algorithms for decision-making. Ensuring the robustness of these algorithms in various driving conditions and scenarios is an ongoing challenge. Algorithms must be continuously tested, refined, and adapted to handle real-world situations.
- ❖ Ethical Dilemmas: Autonomous vehicles may encounter situations where they must make ethical decisions, such as prioritizing the safety of passengers versus pedestrians. Addressing these ethical dilemmas and defining ethical standards for autonomous vehicles remains a contentious issue.
- ❖ Public Acceptance and Trust: Building public trust in autonomous vehicles is essential for their widespread adoption. High-profile accidents and cybersecurity incidents can erode trust in the technology, making it challenging to gain public acceptance (Luo et al., 2024).
- ❖ Infrastructure Compatibility: The successful deployment of cloud robotics for autonomous vehicles may require upgrades to existing infrastructure, such as roads, traffic management systems, and communication networks. Ensuring compatibility between autonomous vehicles and existing infrastructure can be a logistical challenge.
- ❖ Environmental Impact: While autonomous vehicles have the potential to reduce emissions and traffic congestion, there are concerns about the environmental impact of data centres and the energy consumption associated with cloud robotics infrastructure (Lavisse et al., 2023).

Understanding and addressing these challenges and limitations is crucial for the development and deployment of cloud robotics for autonomous vehicles. As the technology evolves, it is expected that innovative solutions and best practices will emerge to overcome these hurdles and make autonomous vehicles safer, more efficient, and more accessible.

6. CLOUD-BASED DATA PROCESSING

Cloud-based data processing is a cornerstone of cloud robotics for autonomous vehicles. It involves the collection, transmission, storage, and analysis of data in the cloud, enabling autonomous vehicles to make informed decisions, adapt to their environment, and optimize their operations (Rocha et al., 2023). This section explores the intricacies of cloud-based data processing in the context of autonomous vehicles.

- ❖ Data Collection: Autonomous vehicles are equipped with an array of sensors, including cameras, lidar, radar, ultrasonic sensors, and GPS receivers, to gather information about their surroundings. These sensors generate a continuous stream of data, capturing details such as the vehicle's location, road conditions, the presence of other vehicles, pedestrians, and traffic signals.
- ❖ Data Transmission: The collected data is transmitted to remote cloud servers through high-speed internet connections. This data transfer is typically done in real-time to provide up-to-the-minute information for decision-making. To ensure data integrity and security, robust encryption and authentication protocols are used.
- ❖ Data Storage: In the cloud, the data is stored in vast data centers, where it can be organized, archived, and made easily accessible. This storage capability is essential for maintaining historical data records, tracking vehicle performance, and facilitating forensic analysis of accidents or incidents.

- ❖ Real-Time Analysis: Cloud-based data processing systems employ advanced analytics and machine learning algorithms to process the incoming data. This analysis serves various purposes, including object detection and tracking, road condition assessment, traffic prediction, and route planning. These algorithms help the vehicle make sense of the sensory data and make informed decisions (Liu et al., 2023).
- ❖ Mapping and Localization: Cloud-based data processing often involves high-definition mapping and localization services. Accurate maps are essential for autonomous vehicles to understand their position within the environment. These maps may include details about road layouts, traffic signs, and lane markings, among other crucial information (Meng et al., 2023).
- ❖ Traffic and Weather Information: Cloud-based systems can provide real-time traffic and weather information to autonomous vehicles. This data helps vehicles navigate congested areas, avoid traffic jams, and adapt to changing weather conditions. Additionally, it aids in optimizing routes and schedules (Fantin Irudaya Raj et al., 2023).
- ❖ Remote Assistance: Cloud robotics enables remote assistance for autonomous vehicles. In cases where vehicles encounter unfamiliar or challenging situations, human operators or remote experts can provide guidance and support by accessing the data stream and video feeds from the vehicle (Raj & Balaji, 2023).
- ❖ Data Sharing: Autonomous vehicles connected to the cloud can share data, creating a network of information exchange. This cooperative approach enhances situational awareness and helps vehicles make collective decisions, ultimately contributing to safer and more efficient traffic flow.
- ❖ Software Updates: Cloud-based data processing facilitates over-the-air software updates for autonomous vehicles. Manufacturers can remotely deploy patches, improvements, and new features, ensuring vehicles remain updated with the latest algorithms and safety measures.
- ❖ Predictive Analysis: The historical data stored in the cloud can be used for predictive analysis. By examining past traffic patterns and accident data, it is possible to anticipate potential hazards and optimize routes for increased safety and efficiency (Xia et al., 2023).

Cloud-based data processing empowers autonomous vehicles with the intelligence to navigate complex environments, make informed decisions, and adapt to changing conditions. The synergy between autonomous vehicles and the cloud is not only transforming the way we travel but also paving the way for more efficient, data-driven transportation systems that have the potential to reduce congestion, lower emissions, and enhance safety. However, as with any technological innovation, challenges such as data security, latency, and connectivity must be effectively addressed to realize the full potential of cloud-based data processing for autonomous vehicles.

7. MACHINE LEARNING AND AI IN CLOUD ROBOTICS

Machine learning and artificial intelligence are central to the operation of autonomous vehicles, and their integration with cloud robotics further amplifies their impact. In this section, we will delve into the critical role that machine learning and AI play in cloud robotics for autonomous vehicles (Nakanoya et al., 2023).

- ❖ Sensory Data Processing: Autonomous vehicles are equipped with various sensors, capturing vast amounts of data about their surroundings. Machine learning and AI algorithms process this sensory data in real-time, identifying objects, detecting obstacles, and interpreting the environment. Computer vision algorithms, for example, enable the vehicle to recognize pedestrians, other vehicles, and road signs (Maheswari et al., 2023).
- ❖ Perception and Object Recognition: Machine learning models are employed for object recognition and perception. Deep learning algorithms, such as convolutional neural networks (CNNs), are capable of discerning intricate details and classifying objects in the vehicle's field of view. This capability is essential for ensuring safe navigation (Fantin Irudaya Raj & Balaji, 2022; Gampala et al., 2020).
- ❖ Localization and Mapping (SLAM): Simultaneous Localization and Mapping (SLAM) techniques utilize machine learning to create and update detailed maps of the environment while simultaneously estimating the vehicle's position within these maps. This is critical for precise navigation (Ebadi et al., 2023).
- ❖ Predictive Modeling: Machine learning models are used to predict the behaviour of other road users, such as pedestrians and other vehicles. Predictive modelling allows the autonomous vehicle to anticipate the movements of nearby objects and make proactive decisions to avoid collisions.
- ❖ Path Planning and Decision-Making: AI algorithms are responsible for path planning and decision-making. These algorithms determine the optimal route, speed, and manoeuvres for the vehicle to reach its destination while adhering to traffic rules and safety regulations.
- ❖ Adaptive Learning: Machine learning enables autonomous vehicles to adapt and improve over time. They can learn from past experiences and situations, allowing them to make better decisions in the future. Continuous learning and adaptation enhance the safety and efficiency of autonomous vehicles (Ebadi et al., 2023).
- ❖ Semantic Segmentation: Semantic segmentation is an advanced computer vision technique that categorizes each pixel in an image to identify different objects and their boundaries. This fine-grained understanding of the environment is crucial for safe navigation.
- ❖ Behaviour Prediction: AI models can predict the behaviour of other drivers and pedestrians based on historical data and real-time observations. This predictive capability allows the vehicle to make anticipatory decisions to prevent accidents (Neelakandan et al., 2022).
- ❖ Data Anomaly Detection: Machine learning algorithms can detect anomalies or irregularities in the sensory data. This is essential for identifying sensor malfunctions or unusual road conditions and triggering appropriate responses.
- ❖ Human-Machine Interaction: AI plays a role in human-vehicle interaction. Natural language processing (NLP) and speech recognition systems enable passengers to communicate with the vehicle, providing instructions and receiving information.
- ❖ Cyber-security and Intrusion Detection: Machine learning is used to detect and respond to cyber-security threats. It can identify unusual patterns in data traffic that may indicate a cyber-attack or unauthorized access to the vehicle's systems (Raj, 2023).
- ❖ Fleet Learning: In cloud robotics, machine learning models and AI can facilitate fleet learning. The collective data and experiences of a fleet of autonomous vehicles can be shared and used to improve the performance and safety of all vehicles in the network (Huo et al., 2023).

The combination of cloud robotics and machine learning/AI empowers autonomous vehicles to adapt to complex, dynamic environments. These technologies are fundamental in making autonomous vehicles more reliable, safe, and capable of delivering efficient and convenient transportation solutions. However, challenges related to data privacy, model robustness, and continuous improvement must be addressed to fully harness the potential of machine learning and AI in cloud robotics for autonomous vehicles.

8. SAFETY AND SECURITY

Safety and security are paramount in the development and deployment of autonomous vehicles leveraging cloud robotics. Ensuring the protection of passengers, and pedestrians, and the integrity of the vehicles themselves is of utmost importance (Giannaros et al., 2023). This section explores the critical safety and security considerations within this transformative technology landscape.

8.1 Safety Considerations

❖ Collision Avoidance: Autonomous vehicles must employ advanced sensor technology and real-time data processing to detect and respond to potential collisions. Safety systems, including emergency braking and evasive manoeuvres, should be in place to avoid accidents (Ashwin & Naveen Raj, 2023).

❖ Redundancy and Fail-Safe Mechanisms: Autonomous vehicles should incorporate redundancy in critical systems and employ fail-safe mechanisms. If one component or system fails, backup systems must ensure safe operation and enable the vehicle to come to a controlled stop.

❖ Testing and Simulation: Rigorous testing and simulation environments are essential to validate the safety of autonomous vehicles before they are deployed on public roads. These tests can identify and rectify potential safety issues without putting human lives at risk.

❖ Ethical Dilemmas: Autonomous vehicles may encounter situations where they must make ethical decisions, such as prioritising the safety of passengers versus pedestrians. Addressing these ethical dilemmas and defining ethical standards is a complex and evolving challenge (Dhirani et al., 2023).

❖ Data Security: Ensuring the security of sensor data and decision-making algorithms is crucial. Unauthorized access to or manipulation of data can have dire consequences for the safety of autonomous vehicles.

❖ Interoperability: As autonomous vehicle fleets grow, ensuring the safe interaction between vehicles from different manufacturers and with traditional human-driven vehicles is a significant safety challenge (Zhang et al., 2023).

8.2 Security Considerations

❖ Cybersecurity: Autonomous vehicles are susceptible to cyberattacks, including remote hacking. Ensuring the cybersecurity of both the vehicle's onboard systems and the cloud infrastructure is critical to prevent unauthorized access, data breaches, and malicious interference (Zhao, Fang, Min, Wu, Wang, & Teixeira, 2023).

❖ Data Encryption: All data transmitted between the vehicle and the cloud, as well as within the cloud infrastructure itself, must be encrypted to prevent eavesdropping and data theft.

❖ Authentication and Access Control: Access to critical systems and data should be restricted to authorized personnel only. Implementing robust authentication and access control mechanisms is vital.

- ❖ Software Updates: Ensuring the security of software updates is essential. Malicious software updates could compromise the safety and security of autonomous vehicles. A secure update mechanism is necessary to protect against this threat (Iclodean et al., 2023).
- ❖ Vulnerability Assessments: Continuous vulnerability assessments and penetration testing are required to identify and address security weaknesses in both the vehicle's systems and the cloud infrastructure.
- ❖ Regulatory Compliance: Complying with cybersecurity regulations and standards is essential. Regulatory bodies are increasingly defining cybersecurity requirements for autonomous vehicles to ensure their safety.
- ❖ Anomaly Detection: Implementing machine learning algorithms for anomaly detection can help identify unusual patterns or behaviours in the vehicle's systems or data traffic, potentially indicating a security breach.
- ❖ Privacy: Respecting passenger privacy is a significant concern. Autonomous vehicles collect a wealth of data, and appropriate measures must be in place to protect passenger privacy and ensure that data is used only for legitimate purposes (Ljubi & Groznik, 2023).

Balancing safety and security in the development and operation of autonomous vehicles is a complex task. It requires a multifaceted approach that encompasses technology, regulation, and public awareness. As autonomous vehicles become more integrated into our transportation systems, safety and security must remain at the forefront of all efforts to ensure that these vehicles bring the promised benefits while minimizing risks.

9. REAL-WORLD EXAMPLES OF CLOUD ROBOTICS IN VEHICLES

9.1. Waymo (Formerly Google's Self-Driving Car Project)

- ❖ Waymo (Hu et al., 2023), a subsidiary of Alphabet Inc., is a leader in autonomous vehicle technology. They have been utilizing cloud robotics extensively to enhance their autonomous driving capabilities.
- ❖ Waymo's fleet of autonomous vehicles collects a massive amount of data from sensors, cameras, and lidar devices. This data is processed in the cloud to improve decision-making algorithms.
- ❖ In a notable case, Waymo used cloud robotics to teach their vehicles how to handle various challenging scenarios, including complex intersections, highway merges, and adverse weather conditions. The data and knowledge from one vehicle can be shared and benefit the entire fleet, contributing to collective learning.

9.2. Tesla Autopilot and Full Self-Driving (FSD)

- ❖ Tesla, an electric vehicle manufacturer, utilizes over-the-air updates to enhance its vehicles' autonomous capabilities. These updates leverage cloud robotics for distributing new software features and improvements (Talpes et al., 2020).

- ❖ Tesla's vehicles rely on a network of sensors and cameras to enable features such as Autopilot, Navigate on Autopilot, and Full Self-Driving (FSD). Data collected from Tesla's vast fleet of vehicles is used to train and improve its autonomous driving algorithms in the cloud.
- ❖ Through software updates delivered via the cloud, Tesla has enhanced the functionality of its vehicles over time, allowing them to perform tasks like lane changes, merging onto highways, and even navigating complex city streets.

9.3. Uber Advanced Technologies Group (ATG)

- ❖ Uber's ATG (Macrae, 2022) division has been working on autonomous vehicles to revolutionize the ride-sharing industry. Cloud robotics plays a vital role in their development.
- ❖ Uber's autonomous vehicles use a combination of onboard processing and cloud-based data analysis to navigate urban environments. Cloud-based systems help enhance mapping, traffic prediction, and localization for these vehicles.
- ❖ In a high-profile case, Uber's ATG vehicles were involved in the testing and deployment of autonomous taxis in select cities. The company leveraged cloud robotics to support the real-time decision-making of its vehicles as they transported passengers.

9.4. Cruise (a Subsidiary of General Motors)

- ❖ Cruise, a self-driving technology company backed by General Motors, uses cloud robotics to power its autonomous vehicles (Cole, 2020).
- ❖ Cruise's approach to autonomous driving includes the utilization of cloud-based data analysis for training its AI algorithms. Their vehicles continuously upload data about various driving scenarios, which is used to improve the performance of the entire fleet.
- ❖ In practice, Cruise has been conducting pilot programs in which autonomous vehicles are used for ride-sharing in selected cities, relying on the cloud for real-time data processing, route planning, and safety monitoring.

These case studies showcase the practical applications of cloud robotics in autonomous vehicles across a range of companies and approaches. They highlight how cloud-based data processing, machine learning, and real-time connectivity are leveraged to enhance the capabilities, safety, and efficiency of autonomous vehicles in real-world scenarios. Additionally, they demonstrate how cloud robotics facilitates ongoing learning and improvement, making autonomous vehicles increasingly reliable and capable over time.

10. REGULATORY AND ETHICAL CONSIDERATIONS

The integration of cloud robotics in autonomous vehicles presents a myriad of regulatory and ethical challenges that must be addressed to ensure the responsible and safe deployment of this transformative technology (Hemphill, 2020; Wang et al., 2019). In this section, we delve into the key regulatory and ethical considerations.

10.1. Regulatory Considerations

- ❖ Certification and Standards: Regulatory bodies and governments worldwide are developing certification and safety standards for autonomous vehicles. These standards set the bar for vehicle safety and functionality. Compliance with these standards is a fundamental requirement for deploying autonomous vehicles.
- ❖ Data Privacy: Autonomous vehicles collect a significant amount of data about passengers and their surroundings. Regulators are focusing on data privacy and the management of this information. Ensuring that data is used ethically, with user consent and within legal boundaries, is a critical aspect of compliance (Fantin Irudaya Raj & Appadurai, 2021).
- ❖ Liability and Insurance: Determining liability in the event of accidents involving autonomous vehicles is a complex legal challenge. Regulators are working to establish guidelines for insurance and liability to ensure that victims of accidents are appropriately compensated.
- ❖ Data Sharing and Connectivity: Regulations may dictate how data is shared between vehicles and the cloud, aiming to prevent data misuse or unauthorized access. Ensuring secure and standardized data-sharing protocols is vital.
- ❖ Safety Certification: Autonomous vehicles may need to undergo safety certification processes, similar to those for traditional vehicles. This includes crash tests, safety assessments, and evaluations of autonomous systems' ability to handle various scenarios.
- ❖ Infrastructure Integration: Regulatory bodies may need to oversee the integration of autonomous vehicles with existing infrastructure. This includes traffic management systems, road markings, and communication networks to ensure safe and efficient operations.

10.2. Ethical Considerations

- ❖ Ethical Decision-Making: Autonomous vehicles can encounter scenarios where they must make ethical decisions, such as prioritising the safety of passengers versus pedestrians. Ethical frameworks must be established to guide these decisions and reflect societal values.
- ❖ Transparency and Accountability: Ensuring transparency in how autonomous vehicles make decisions is essential for building trust. Users and regulators should understand the underlying algorithms and data sources. Accountability measures are necessary in case of malfunctions or accidents (Pokam et al., 2019).
- ❖ Job Displacement: The widespread adoption of autonomous vehicles could lead to job displacement for professional drivers, such as truck drivers and taxi drivers. Addressing the economic and employment impacts of this technology is an ethical concern.
- ❖ Access and Equity: Ensuring that autonomous vehicles are accessible to all members of society is an ethical imperative. Efforts should be made to avoid exacerbating transportation disparities and ensure that autonomous vehicles benefit underserved communities.
- ❖ Data Privacy: Ethical considerations for data privacy encompass the responsible collection, storage, and use of passenger and environmental data. Users should have control over how their data is used, and it should be handled with the utmost care (Priyadarsini et al., 2020).

- ❖ Education and Awareness: Raising awareness and educating the public about the capabilities, limitations, and ethical considerations of autonomous vehicles is important for fostering a responsible and informed approach to this technology.
- ❖ Environmental Impact: Evaluating the environmental impact of autonomous vehicles, including the energy consumption of data centres and cloud infrastructure, is an ethical concern. Efforts should be made to minimize the environmental footprint of cloud robotics.

Addressing these regulatory and ethical considerations is essential to ensure the responsible and equitable deployment of cloud robotics in autonomous vehicles. Collaboration among governments, industry stakeholders, and the public is crucial to establishing a framework that promotes the safe and ethical use of this technology while maximizing its benefits.

11. FUTURE TRENDS AND DEVELOPMENTS

The evolution of cloud robotics for autonomous vehicles is poised to bring about transformative changes in transportation. As technology continues to advance and regulatory frameworks mature, several key trends and developments (Damaj et al., 2022) are expected to shape the future of this field:

- ❖ 5G and Beyond: The deployment of 5G and future-generation wireless networks will provide ultra-fast and low-latency connectivity, enabling real-time data transfer between autonomous vehicles and the cloud. This will significantly enhance the capabilities and safety of autonomous vehicles.
- ❖ Edge and Cloud Hybrid Solutions: To mitigate latency concerns, autonomous vehicles will adopt edge computing solutions, allowing local data processing and decision-making. A hybrid approach that balances cloud and edge computing will emerge as the norm, ensuring faster response times.
- ❖ Global Mapping and Localization: Collaborative efforts will lead to a more comprehensive global mapping and localization infrastructure. High-definition maps that are constantly updated will become vital for the precise navigation of autonomous vehicles worldwide.
- ❖ Enhanced Human-Machine Interaction: Future vehicles will offer advanced human-machine interaction, with natural language processing, personalized services, and improved user interfaces. Passengers will have more interactive and seamless experiences during their journeys.
- ❖ AI-Driven Fleet Management: AI and machine learning will play a central role in fleet management, optimizing routes, maintenance schedules, and energy consumption. Predictive analytics will minimize downtime and improve the overall efficiency of autonomous vehicle fleets.
- ❖ Quantum Computing: As quantum computing technology matures, it may offer breakthroughs in processing power and data analysis. Quantum computing could revolutionize the complexity and speed at which autonomous vehicles operate.
- ❖ Environmental Sustainability: Autonomous vehicles, when integrated with sustainable energy sources and efficient routing, will contribute to a reduction in carbon emissions and traffic congestion. Transportation planning will be more environmentally conscious.
- ❖ Urban Planning and Infrastructure Development: Governments and city planners will need to adapt urban infrastructure to accommodate autonomous vehicles. This includes rethinking road designs, traffic management systems, and public transportation to better integrate autonomous vehicles into urban environments.

- ❖ Regulatory Evolution: Regulatory frameworks will continue to evolve, with an emphasis on data privacy, cybersecurity, and ethical considerations. Governments will establish standardized requirements for autonomous vehicles to ensure safety and compliance.
- ❖ Industry Collaboration: Collaboration among automotive manufacturers, tech companies, and other stakeholders will become more common. Industry partnerships and open standards will enable faster development and adoption of autonomous technologies.
- ❖ Multi-Modal Transportation: Autonomous vehicles will be integrated into a broader ecosystem of transportation options, including ridesharing, public transit, and micro-mobility services. Multimodal transportation platforms will offer passengers a seamless and efficient way to get around.
- ❖ Customization and Personalization: Autonomous vehicles will offer customization options for passengers, catering to individual preferences, from entertainment choices to interior layouts. Personalized travel experiences will be a selling point for service providers.
- ❖ Safety Innovations: The ongoing quest for safer autonomous vehicles will result in the development of novel safety technologies, such as advanced sensor systems, collision avoidance mechanisms, and emergency response systems.

The future of cloud robotics for autonomous vehicles is exciting, with continued advancements in technology and an increasing focus on safety, efficiency, and sustainability. As the ecosystem evolves, it will not only revolutionize transportation but also reshape urban planning, industry practices, and the way we live and work in the connected world.

12. CONCLUSION

The fusion of cloud robotics and autonomous vehicles represents a technological revolution that is reshaping the landscape of transportation. Cloud robotics empowers autonomous vehicles with real-time data processing, machine learning, and the ability to learn and adapt, resulting in safer, more efficient, and smarter vehicles. This technology relies on an ecosystem of components, including sensors, cloud infrastructure, connectivity, and edge computing, working together to navigate the complexities of the modern world. The advantages of cloud robotics for autonomous vehicles are far-reaching. These vehicles can make split-second decisions based on real-time data, optimize routes, and contribute to eco-friendly transportation solutions. Moreover, they offer a personalized user experience and facilitate remote monitoring and support. However, the path to widespread adoption is not without its challenges. Latency, connectivity reliability, data security, and regulatory hurdles pose significant obstacles. Ensuring safety, cybersecurity, and addressing ethical concerns are also paramount. Striking a balance between innovation and responsibility is the key to a successful transition to autonomous vehicles powered by cloud robotics.

In conclusion, cloud robotics for autonomous vehicles is not just a technological advancement but a societal transformation. It has the potential to make transportation safer, more efficient, and sustainable. However, it also presents a myriad of complex issues that require continued collaboration among industry stakeholders, governments, and the public. It is crucial to navigate the challenges of cloud robotics in autonomous vehicles with responsibility to realize their full potential.

REFERENCES

Appadurai, M., & Fantin Irudaya Raj, E. (2021). Finite element analysis of lightweight robot fingers actuated by pneumatic pressure. In *Recent Advances in Manufacturing, Automation, Design and Energy Technologies: Proceedings from ICoFT 2020* (pp. 379–385). Springer Singapore.

Ashwin, S. H., & Naveen Raj, R. (2023). Deep reinforcement learning for autonomous vehicles: Lane keep and overtaking scenarios with collision avoidance. *International Journal of Information Technology : an Official Journal of Bharati Vidyapeeth's Institute of Computer Applications and Management*, *15*(7), 3541–3553. doi:10.1007/s41870-023-01412-6

Bendiab, G., Hameurlaine, A., Germanos, G., Kolokotronis, N., & Shiaeles, S. (2023). Autonomous vehicles security: Challenges and solutions using blockchain and artificial intelligence. *IEEE Transactions on Intelligent Transportation Systems*, *24*(4), 3614–3637. doi:10.1109/TITS.2023.3236274

Buerkle, A., Eaton, W., Al-Yacoub, A., Zimmer, M., Kinnell, P., Henshaw, M., Coombes, M., Chen, W.-H., & Lohse, N. (2023). Towards industrial robots as a service (IRaaS): Flexibility, usability, safety and business models. *Robotics and Computer-integrated Manufacturing*, *81*, 102484. doi:10.1016/j.rcim.2022.102484

Chai, Y. T., Che, H. S., Tan, C., Tan, W. N., Yip, S. C., & Gan, M. T. (2023). A two-stage optimization method for Vehicle to Grid coordination considering building and Electric Vehicle user expectations. *International Journal of Electrical Power & Energy Systems*, *148*, 108984. doi:10.1016/j.ijepes.2023.108984

Chen, C., Wang, C., Liu, B., He, C., Cong, L., & Wan, S. (2023). Edge intelligence empowered vehicle detection and image segmentation for autonomous vehicles. *IEEE Transactions on Intelligent Transportation Systems*, *24*(11), 13023–13034. doi:10.1109/TITS.2022.3232153

Chen, W., Yaguchi, Y., Naruse, K., Watanobe, Y., Nakamura, K., & Ogawa, J. (2018). A study of robotic cooperation in cloud robotics: Architecture and challenges. *IEEE Access : Practical Innovations, Open Solutions*, *6*, 36662–36682. doi:10.1109/ACCESS.2018.2852295

Cole, A. (2020). *Fundamental valuation of the General Motors Company: assessing the value in the midst of a pandemic* (Master's thesis).

Damaj, I. W., Yousafzai, J. K., & Mouftah, H. T. (2022). Future trends in connected and autonomous vehicles: Enabling communications and processing technologies. *IEEE Access : Practical Innovations, Open Solutions*, *10*, 42334–42345. doi:10.1109/ACCESS.2022.3168320

Darwin, S., & Fantin Irudaya Raj, E. (2023). The Emergence of Cyber-Physical System in the Context of Self-Powered Soft Robotics. *Self-Powered Cyber Physical Systems*, 57-90.

Dhirani, L. L., Mukhtiar, N., Chowdhry, B. S., & Newe, T. (2023). Ethical dilemmas and privacy issues in emerging technologies: A review. *Sensors (Basel)*, *23*(3), 1151. doi:10.3390/s23031151 PMID:36772190

Ebadi, K., Bernreiter, L., Biggie, H., Catt, G., Chang, Y., Chatterjee, A., ... Carlone, L. (2023). Present and future of SLAM in extreme environments: The DARPA subT challenge. *IEEE Transactions on Robotics*.

Fantin Irudaya Raj, E. (2022). Implementation of machine learning techniques in unmanned aerial vehicle control and its various applications. In *Computational Intelligence for Unmanned Aerial Vehicles Communication Networks* (pp. 17–33). Springer International Publishing. doi:10.1007/978-3-030-97113-7_2

Fantin Irudaya Raj, E., & Appadurai, M. (2021). The hybrid electric vehicle (HEV)—an overview. *Emerging Solutions for e-Mobility and Smart Grids: Select Proceedings of ICRES 2020*, 25-36.

Fantin Irudaya Raj, E., Appadurai, M., Chithamabara Thanu, M., & Francy Irudaya Rani, E. (2023). IoT-Based Smart Parking System for Indian Smart Cities. *Machine Intelligence, Big Data Analytics, and IoT in Image Processing: Practical Applications*, 369-398.

Fantin Irudaya Raj, E., & Balaji, M. (2021). Analysis and classification of faults in switched reluctance motors using deep learning neural networks. *Arabian Journal for Science and Engineering*, *46*(2), 1313–1332. doi:10.1007/s13369-020-05051-y

Fantin Irudaya Raj, E., & Balaji, M. (2022). Application of deep learning and machine learning in pattern recognition. In *Advance Concepts of Image Processing and Pattern Recognition: Effective Solution for Global Challenges* (pp. 63–89). Springer Singapore. doi:10.1007/978-981-16-9324-3_4

Gampala, V., Kumar, M. S., Sushama, C., & Raj, E. F. I. (2020). Deep learning based image processing approaches for image deblurring. *Materials Today: Proceedings*. Advance online publication. doi:10.1016/j.matpr.2020.11.076

Giannaros, A., Karras, A., Theodorakopoulos, L., Karras, C., Kranias, P., Schizas, N., Kalogeratos, G., & Tsolis, D. (2023). Autonomous vehicles: Sophisticated attacks, safety issues, challenges, open topics, blockchain, and future directions. *Journal of Cybersecurity and Privacy*, *3*(3), 493–543. doi:10.3390/jcp3030025

Hakak, S., Gadekallu, T. R., Maddikunta, P. K. R., Ramu, S. P., Parimala, M., De Alwis, C., & Liyanage, M. (2023). Autonomous Vehicles in 5G and beyond: A Survey. *Vehicular Communications*, *39*, 100551. doi:10.1016/j.vehcom.2022.100551

Hao, Y., Liu, J., Liu, Y., Liu, X., Meng, Z., & Xing, F. (2023). Global Visual–Inertial Localization for Autonomous Vehicles with Pre-Built Map. *Sensors (Basel)*, *23*(9), 4510. doi:10.3390/s23094510 PMID:37177714

Hemphill, T. A. (2020). Autonomous vehicles: US regulatory policy challenges. *Technology in Society*, *61*, 101232. doi:10.1016/j.techsoc.2020.101232

Hu, X., Zheng, Z., Chen, D., & Sun, J. (2023). Autonomous vehicle's impact on traffic: Empirical evidence from Waymo Open Dataset and implications from modelling. *IEEE Transactions on Intelligent Transportation Systems*, *24*(6), 6711–6724. doi:10.1109/TITS.2023.3258145

Huo, D., Sari, Y. A., Kealey, R., & Zhang, Q. (2023). Reinforcement learning-based fleet dispatching for greenhouse gas emission reduction in open-pit mining operations. *Resources, Conservation and Recycling*, *188*, 106664. doi:10.1016/j.resconrec.2022.106664

Iclodean, C., Varga, B. O., & Pfister, F. (2023). Autonomous Vehicles Technological Trends. *Electronics (Basel)*, *12*(5), 1149. doi:10.3390/electronics12051149

Kato, S., Takeuchi, E., Ishiguro, Y., Ninomiya, Y., Takeda, K., & Hamada, T. (2015). An open approach to autonomous vehicles. *IEEE Micro*, *35*(6), 60–68. doi:10.1109/MM.2015.133

Lavisse, T., Panariello, R., Perdu, F., & Zwolinski, P. (2023). Integrating an ageing model within Life Cycle Assessment to evaluate the environmental impacts of electric batteries. *Procedia CIRP*, *116*, 251–256. doi:10.1016/j.procir.2023.02.043

Liu, X., Zhu, W., Lian, X., & Xu, X. (2023). Monitoring mining surface subsidence with multi-temporal three-dimensional unmanned aerial vehicle point cloud. *Remote Sensing (Basel)*, *15*(2), 374. doi:10.3390/rs15020374

Ljubi, K., & Groznik, A. (2023). Role played by social factors and privacy concerns in autonomous vehicle adoption. *Transport Policy*, *132*, 1–15. doi:10.1016/j.tranpol.2022.12.013

Luo, C., He, M., & Xing, C. (2024). Public acceptance of autonomous vehicles in China. *International Journal of Human-Computer Interaction*, *40*(2), 315–326. doi:10.1080/10447318.2022.2115336

Macrae, C. (2022). Learning from the failure of autonomous and intelligent systems: Accidents, safety, and sociotechnical sources of risk. *Risk Analysis*, *42*(9), 1999–2025. doi:10.1111/risa.13850 PMID:34814229

Maheswari, B. U., Imambi, S. S., Hasan, D., Meenakshi, S., Pratheep, V. G., & Boopathi, S. (2023). Internet of things and machine learning-integrated smart robotics. In Global Perspectives on Robotics and Autonomous Systems: Development and Applications (pp. 240-258). IGI Global. doi:10.4018/978-1-6684-7791-5.ch010

Mello, R. C., Scheidegger, W. M., Múnera, M. C., Cifuentes, C. A., Ribeiro, M. R., & Frizera-Neto, A. (2022). The PoundCloud framework for ROS-based cloud robotics: Case studies on autonomous navigation and human–robot interaction. *Robotics and Autonomous Systems*, *150*, 103981. doi:10.1016/j.robot.2021.103981

Meng, W., Yang, Y., Zang, J., Li, H., & Lu, R. (2023). DTUAV: A novel cloud–based digital twin system for unmanned aerial vehicles. *Simulation*, *99*(1), 69–87. doi:10.1177/00375497221109575

Ming, G. (2023). Exploration of the intelligent control system of autonomous vehicles based on edge computing. *PLoS One*, *18*(2), e0281294. doi:10.1371/journal.pone.0281294 PMID:36730359

Mushtaq, A., Haq, I. U., Sarwar, M. A., Khan, A., Khalil, W., & Mughal, M. A. (2023). Multi-Agent Reinforcement Learning for Traffic Flow Management of Autonomous Vehicles. *Sensors (Basel)*, *23*(5), 2373. doi:10.3390/s23052373 PMID:36904577

Nakanoya, M., Narasimhan, S. S., Bhat, S., Anemogiannis, A., Datta, A., Katti, S., Chinchali, S., & Pavone, M. (2023). Co-design of communication and machine inference for cloud robotics. *Autonomous Robots*, *47*(5), 1–16. doi:10.1007/s10514-023-10093-w

Neelakandan, S., Beulah, J. R., Prathiba, L., Murthy, G. L. N., Irudaya Raj, E. F., & Arulkumar, N. (2022). Blockchain with deep learning-enabled secure healthcare data transmission and diagnostic model. *International Journal of Modeling, Simulation, and Scientific Computing*, *13*(04), 2241006. doi:10.1142/S1793962322410069

Othman, K. (2023). Exploring the evolution of public acceptance towards autonomous vehicles with the level of knowledge. *Innovative Infrastructure Solutions*, *8*(8), 208. doi:10.1007/s41062-023-01180-z

Pokam, R., Debernard, S., Chauvin, C., & Langlois, S. (2019). Principles of transparency for autonomous vehicles: First results of an experiment with an augmented reality human–machine interface. *Cognition Technology and Work*, *21*(4), 643–656. doi:10.1007/s10111-019-00552-9

Prasad, A. O., Mishra, P., Jain, U., Pandey, A., Sinha, A., Yadav, A. S., Kumar, R., Sharma, A., Kumar, G., Hazim Salem, K., Sharma, A., & Dixit, A. K. (2023). Design and development of software stack of an autonomous vehicle using robot operating system. *Robotics and Autonomous Systems*, *161*, 104340. doi:10.1016/j.robot.2022.104340

Priyadarsini, K., Raj, E. F. I., Begum, A. Y., & Shanmugasundaram, V. (2020). Comparing DevOps procedures from the context of a systems engineer. *Materials Today: Proceedings*. Advance online publication. doi:10.1016/j.matpr.2020.09.624

Raj, E. F. I. (2023). Clustering and Classification of Digital Forensic Data Using Machine Learning and Data Mining Approaches. In *Advancements in Cybercrime Investigation and Digital Forensics* (pp. 115–136). Apple Academic Press. doi:10.1201/9781003369479-6

Raj, E. F. I., & Balaji, M. (2023). Shape Feature Extraction Techniques for Computer Vision Applications. In *Smart Computer Vision* (pp. 81–102). Springer International Publishing. doi:10.1007/978-3-031-20541-5_4

Rana, M. M., & Hossain, K. (2023). Connected and autonomous vehicles and infrastructures: A literature review. *International Journal of Pavement Research and Technology*, *16*(2), 264–284. doi:10.1007/s42947-021-00130-1

Rani, E. F. I., Pushparaj, T. L., & Raj, E. F. I. (2022). Escalating the resolution of an urban aerial image via novel shadow amputation algorithm. *Earth Science Informatics*, *15*(2), 905–913. doi:10.1007/s12145-022-00772-0

Rocha, D., Teixeira, G., Vieira, E., Almeida, J., & Ferreira, J. (2023). A Modular In-Vehicle C-ITS Architecture for Sensor Data Collection, Vehicular Communications and Cloud Connectivity. *Sensors (Basel)*, *23*(3), 1724. doi:10.3390/s23031724 PMID:36772760

Sijini, A. C., Fantin, E., & Ranjit, L. P. (2016). Switched reluctance motor for hybrid electric vehicle. *Middle East Journal of Scientific Research*, *24*(3), 734–739.

Syamala, M., Malathi, J., Singh, V., GS, H. P., Maheswari, B. U., & Murugan, S. (2024). Cloud Solutions for Smart Parking and Traffic Control in Smart Cities. In Handbook of Research on AI and ML for Intelligent Machines and Systems (pp. 169-194). IGI Global.

Talpes, E., Sarma, D. D., Venkataramanan, G., Bannon, P., McGee, B., Floering, B., Jalote, A., Hsiong, C., Arora, S., Gorti, A., & Sachdev, G. S. (2020). Compute solution for tesla's full self-driving computer. *IEEE Micro*, *40*(2), 25–35. doi:10.1109/MM.2020.2975764

Tengilimoglu, O., Carsten, O., & Wadud, Z. (2023). Infrastructure requirements for the safe operation of automated vehicles: Opinions from experts and stakeholders. *Transport Policy*, *133*, 209–222. doi:10.1016/j.tranpol.2023.02.001

Wang, Y., Liu, Z., Zuo, Z., Li, Z., Wang, L., & Luo, X. (2019). Trajectory planning and safety assessment of autonomous vehicles based on motion prediction and model predictive control. *IEEE Transactions on Vehicular Technology*, *68*(9), 8546–8556. doi:10.1109/TVT.2019.2930684

Woo, S., Youtie, J., Ott, I., & Scheu, F. (2021). Understanding the long-term emergence of autonomous vehicles technologies. *Technological Forecasting and Social Change*, *170*, 120852. doi:10.1016/j.techfore.2021.120852

Xia, X., Meng, Z., Han, X., Li, H., Tsukiji, T., Xu, R., Zheng, Z., & Ma, J. (2023). An automated driving systems data acquisition and analytics platform. *Transportation Research Part C, Emerging Technologies*, *151*, 104120. doi:10.1016/j.trc.2023.104120

Xue, Y., Wang, C., Ding, C., Yu, B., & Cui, S. (2024). Observer-based event-triggered adaptive platooning control for autonomous vehicles with motion uncertainties. *Transportation Research Part C, Emerging Technologies*, *159*, 104462. doi:10.1016/j.trc.2023.104462

Zhang, F., Shi, W., Chen, M., Huang, W., & Liu, X. (2023). Open HD map service model: An interoperable high-Definition map data model for autonomous driving. *International Journal of Digital Earth*, *16*(1), 2089–2110. doi:10.1080/17538947.2023.2220615

Zhao, X., Fang, Y., Min, H., Wu, X., Wang, W., & Teixeira, R. (2023). Potential sources of sensor data anomalies for autonomous vehicles: An overview from road vehicle safety perspective. *Expert Systems with Applications*, 121358.

Chapter 21
Agricultural Automation Using Cloud Robotics

E. Fantin Irudaya Raj
https://orcid.org/0000-0003-2051-3383
Dr. Sivanthi Aditanar College of Engineering, India

E. Francy Irudaya Rani
https://orcid.org/0000-0003-2864-8880
Francis Xavier Engineering College, India

C. Sweetline Jenita
https://orcid.org/0009-0006-0933-4375
Dr. Sivanthi Aditanar College of Engineering, India

V. S. Winstor Jebakumar
Dr. Sivanthi Aditanar College of Engineering, India

ABSTRACT

Agriculture, as a critical sector in the global economy, faces increasing challenges such as growing demand for food, limited resources, and the need for sustainability. The integration of automation technologies and cloud robotics in agriculture has emerged as a transformative solution to address these challenges. The chapter introduces and elucidates how it leverages cloud computing and data analytics to bring efficiency, scalability, and intelligence to the agricultural landscape. Several case studies are presented to illustrate real-world applications of cloud robotics in diverse farming settings, highlighting their impact on resource management, crop yield optimization, and environmental sustainability. Furthermore, the chapter identifies the hurdles and barriers hindering the widespread adoption of cloud robotics in agriculture, including cost, connectivity, and data security concerns, and offers potential solutions to overcome these challenges.

1. INTRODUCTION

Agriculture is the cornerstone of human civilization, providing sustenance, economic stability, and livelihoods to billions of people worldwide. However, the agricultural sector faces a multitude of challenges

DOI: 10.4018/979-8-3693-1914-7.ch021

in the 21st century, including a burgeoning global population, resource constraints, climate change, and the imperative of producing more food with fewer resources (Agrimonti et al., 2021). To address these challenges and ensure the future sustainability of agriculture, technological innovations have become indispensable. One such innovation is the fusion of cloud computing and robotics in the agricultural sector, a synergy that holds the promise of revolutionizing traditional farming practices (Phasinam et al., 2022).

The concept of agricultural automation is not new. For decades, the industry has witnessed the deployment of various automation technologies, such as tractors, automated irrigation systems, and crop monitoring sensors. These technologies have undeniably increased efficiency, reduced labor costs, and improved yields. However, the integration of cloud robotics in agriculture represents a transformative leap forward (Kalyani & Collier, 2021). Figure 1 shows the automated robotic arm controlled through cloud server used in agriculture.

Cloud robotics, a relatively recent field, combines the power of robotics with cloud computing, artificial intelligence, and the Internet of Things (IoT) (Lytridis et al., 2021). It has the potential to enhance agricultural automation in numerous ways, including real-time data analysis, remote monitoring, predictive analytics, and precise decision-making. By harnessing the computing and storage capabilities of cloud-based systems, agricultural robots can access vast amounts of data, process it in real-time, and operate with a level of sophistication previously unimaginable (Afrin et al., 2021).

The present chapter aims to explore the significant role of cloud robotics in agricultural automation. We will delve into the array of automation technologies currently employed in agriculture, highlighting the advantages and limitations of these technologies. We will then transition to an in-depth analysis of cloud robotics and its relevance to agriculture, illustrating how it enhances the efficiency, scalability, and intelligence of agricultural processes. As with any transformative technology, cloud robotics in agriculture presents a set of challenges and barriers, including concerns about cost, connectivity, and data security. The present chapter will explore these challenges and propose strategies to overcome them, fostering a conducive environment for the widespread adoption of cloud robotics in agriculture.

It also examined the landscape of cloud robotics platforms and software solutions, providing insights into the options available to agricultural stakeholders. This information will enable farmers and industry participants to make informed decisions when considering the implementation of cloud robotics in their

Figure 1. Automated robotic arm controlled through cloud server

operations. Finally, the chapter will conclude with a forward-looking perspective, discussing the anticipated future trends and the evolving landscape of agricultural automation using cloud robotics. As we stand at the intersection of agriculture and cutting-edge technology, it is crucial to understand how this fusion of fields will shape the future of farming practices, resource management, and global food security.

2. AGRICULTURAL AUTOMATION TECHNOLOGIES

Agricultural automation technologies encompass a wide range of tools and systems that aim to optimize and streamline various aspects of farming. These technologies have evolved significantly over the years, enhancing productivity, reducing labor requirements, and improving overall efficiency in agricultural processes. In this section, the chapter will explore some of the key agricultural automation technologies, highlighting their features, applications, and the benefits they offer to the agricultural industry.

2.1. Robotics in Agriculture

Robotics in agriculture has made substantial strides in recent years. Agricultural robots come in various forms, including autonomous tractors, robotic arms, and even specialized drones. These robots are equipped with sensors, cameras, and AI algorithms, enabling them to perform a wide range of tasks such as planting, harvesting, weeding, and crop monitoring. The advantages of robotic automation in agriculture include (Mahmud et al., 2020):

a) Labor Efficiency: Robots can work tirelessly, reducing the need for human labor during labor-intensive tasks, b) Precision Agriculture: Robots can apply fertilizers and pesticides with high precision, reducing waste and environmental impact. c) 24/7 Operation: Robots can work day and night, improving productivity and timely execution of tasks. However, challenges associated with agricultural robotics include high initial costs, technical complexities, and the need for advanced training for operators.

2.2. Unmanned Aerial Vehicles (UAVs)

Drones, often referred to as UAVs, have gained popularity in agriculture for their ability to capture aerial imagery and collect data about crops and fields. Drones are equipped with high-resolution cameras and sensors that can monitor crop health, identify disease or pest infestations, and assess irrigation needs (del Cerro et al., 2021; Fantin Irudaya Raj, 2022). The benefits of drones in agriculture include: a)Data Collection: Drones provide a bird's-eye view of the entire field, allowing for rapid data collection and analysis, b) Timely Decision-Making: The real-time data collected by drones can aid farmers in making prompt and informed decisions, c) Cost-Effective: Drones are a cost-effective solution for large-scale crop monitoring. Nevertheless, limitations such as battery life, limited payload capacity, and regulatory constraints must be addressed when using drones in agriculture.

2.3. IoT Sensors and Precision Agriculture

The Internet of Things (IoT) has revolutionized precision agriculture by incorporating a network of sensors and data analytics to monitor and manage various aspects of farming. IoT sensors can measure soil moisture, temperature, humidity, and other environmental variables (Mangalraj, 2023). These sensors provide farmers with valuable insights, allowing them to optimize resource use, reduce waste, and increase crop yields. The advantages of IoT sensors in agriculture include: a) Data-Driven Decisions: IoT sensors provide real-time data, enabling farmers to make informed decisions. b) Resource Optimization: By monitoring environmental conditions, farmers can optimize water and fertilizer usage. c) Sustainability: Precision agriculture practices driven by IoT help reduce the environmental impact of farming. Nonetheless, the adoption of IoT technology may require investments in infrastructure and data analytics capabilities.

In the realm of agricultural automation, these technologies have significantly improved the industry's ability to adapt to the challenges posed by a growing global population and changing environmental conditions. In the following sections, we will explore how cloud robotics integrates with these technologies to unlock new possibilities for agriculture.

3. CLOUD ROBOTICS IN AGRICULTURE

Cloud robotics represents the next frontier in agricultural automation, leveraging the power of cloud computing, artificial intelligence (AI), and the Internet of Things (IoT) to revolutionize farming practices. It introduces a new paradigm in which agricultural robots can seamlessly interact with cloud-based platforms, enabling real-time data analysis, remote monitoring, and decision-making that transcend the capabilities of standalone devices (Shakya, 2020). In this section, we will delve into the concept of cloud robotics in agriculture, elucidating its relevance and the ways it enhances the efficiency, scalability, and intelligence of agricultural processes.

3.1. Understanding Cloud Robotics

Cloud robotics in agriculture combines the strengths of traditional agricultural automation technologies with cloud-based systems and advanced data analytics. At its core, it involves the connection of agricultural robots to cloud servers, where vast amounts of data can be processed in real time (Paraforos & Griepentrog, 2021). The implications of this connection are profound, as it empowers robots to access a wealth of information and insights that can be used to optimize farming operations.

Key components of cloud robotics in agriculture include:

- ❖ Cloud Computing: Cloud platforms serve as a central hub for data storage, processing, and analytics. These platforms offer the computational power needed to make sense of data generated by agricultural robots.
- ❖ Real-time Data Analysis: With cloud robotics, data can be analyzed in real time, enabling timely decision-making and adaptive responses to changing conditions.
- ❖ Connectivity: Agricultural robots are connected to the cloud via the Internet, allowing for remote monitoring and control.

3.2. The Relevance of Cloud Robotics in Agriculture

Cloud robotics is highly relevant in the context of modern agriculture. It addresses several critical challenges (Araújo et al., 2021) faced by the industry:

- ❖ Data-Intensive Farming: Modern farming generates vast amounts of data, including crop health metrics, weather conditions, and market trends. Cloud robotics can efficiently handle this data, turning it into actionable insights.
- ❖ Scalability: Cloud robotics enables scalability, allowing farmers to deploy a fleet of robots that can work collaboratively and share data, knowledge, and resources.
- ❖ Resource Optimization: By analyzing data from multiple sources, cloud robotics can optimize resource usage, such as water, fertilizer, and energy, to minimize waste and maximize productivity.

3.3. Real-Time Monitoring and Decision-Making

One of the defining features of cloud robotics in agriculture is the ability to monitor and make decisions in real time (Chouhan et al., 2021). Agricultural robots can continuously transmit data to the cloud, where AI algorithms analyze it and provide recommendations to farmers. This capability allows for: a) Immediate Responses: In the event of a pest infestation or adverse weather conditions, cloud-connected robots can trigger responses such as applying pesticides or adjusting irrigation in real time, b) Predictive Analytics: The historical and real-time data collected by robots can be used to predict crop yields, disease outbreaks, and optimal harvesting times.

4. BENEFITS OF CLOUD ROBOTICS IN AGRICULTURE

The integration of cloud robotics into the agricultural sector brings forth a multitude of advantages and benefits that significantly impact farming operations. By combining cloud computing with robotic systems, agriculture gains access to real-time data analysis, remote monitoring, and intelligent decision-making, fostering enhanced efficiency, scalability, and sustainability. The present section explores the wide-ranging benefits of using cloud robotics in agriculture. Figure 2 represents the cloud-based agriculture monitoring system and Table 1 differentiates the traditional and cloud robotics-based agriculture.

4.1. Real-Time Data Analysis

One of the primary advantages of cloud robotics in agriculture is the ability to conduct real-time data analysis. Agricultural robots equipped with sensors continuously collect data on crop health, environmental conditions, and operational performance (Kumar et al., 2021). This data is then transmitted to cloud servers, where it is processed and analyzed by advanced algorithms. The benefits of real-time data analysis include:

- ❖ Timely Decision-Making: With instant access to data insights, farmers and agronomists can make informed decisions swiftly, responding to changing conditions in real time.
- ❖ Early Problem Detection: Cloud robotics can identify issues such as disease outbreaks, pest infestations, or drought stress at an early stage, enabling proactive measures.
- ❖ Optimized Resource Use: Real-time analysis helps fine-tune resource allocation, leading to efficient water and fertilizer usage, reduced waste, and cost savings.

Figure 2. Cloud-based agriculture monitoring system

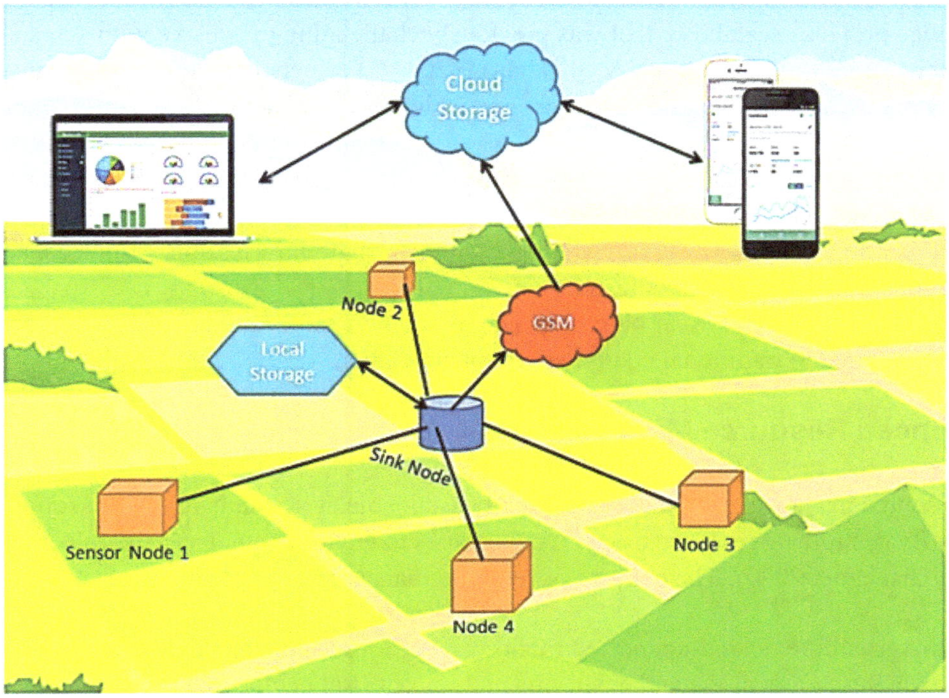

Table 1. Difference between traditional and modern cloud robotics-based agriculture

Sl. No.	Traditional Farming	Cloud Robotic-Based Modern Farming
1	Low rate of production	Higher rate of production
2	Use of compost and manure	Use chemical-based fertilizers and pesticides
3	More farm labour and use of cattle	Less farm labour and the use of tractors and robots

4.2. Remote Monitoring and Control

Cloud robotics facilitates remote monitoring and control of agricultural operations. Farmers can access the status of their robotic fleet, view real-time data, and make adjustments from anywhere with an internet connection (Rani et al., 2022; Sishodia et al., 2020). The benefits of remote monitoring and control include:

- ❖ Reduced Labor Demands: Farmers and operators can oversee multiple robots and tasks simultaneously, reducing the need for physical presence in the field.
- ❖ Accessibility: Remote access allows for continuous monitoring even during adverse weather conditions or at irregular hours.
- ❖ Global Connectivity: Farm managers and stakeholders can monitor and collaborate on operations, regardless of geographic location.

4.3. Scalability

Cloud robotics provides scalability that was previously challenging to achieve with standalone robotic systems (Deivakani et al., 2021; Dey & Shekhawat, 2021). Farmers can deploy and manage fleets of robots that work collaboratively and share data and insights. The benefits of scalability include:

- ❖ Increased Productivity: Scalable operations can cover larger areas and handle more tasks, boosting overall productivity.
- ❖ Resource Sharing: Robots within a scalable system can share information, resources, and learning, enhancing their collective intelligence.
- ❖ Cost-Efficiency: Scalability can be cost-effective, as the cost of adding additional robots is often lower compared to traditional equipment or manual labor.

4.4. Enhanced Resource Management

Efficient resource management is a cornerstone of sustainable agriculture, and cloud robotics excels in this regard. By optimizing resource usage, it contributes to environmental sustainability and economic efficiency (Zhai et al., 2020). The benefits of resource management in cloud robotics include:

- ❖ Precision Agriculture: Cloud-connected robots can precisely apply fertilizers, pesticides, and irrigation, reducing excess usage and environmental impact (Raj et al., 2022).
- ❖ Sustainable Practices: Resource optimization contributes to sustainable agricultural practices, supporting long-term soil health and reduced ecological footprints.
- ❖ Yield Improvement: By minimizing waste and promoting healthy crop growth, cloud robotics can lead to improved crop yields and overall profitability.

In summary, the benefits of cloud robotics in agriculture are far-reaching, offering real-time data analysis, remote monitoring and control, scalability, and enhanced resource management. These advantages translate into increased efficiency, sustainability, and productivity, making cloud robotics a transformative technology with the potential to revolutionize the agricultural landscape.

5. CHALLENGES AND BARRIERS

While cloud robotics in agriculture holds significant promise, its adoption is not without challenges and barriers (Kasemi et al., 2022). These hurdles must be acknowledged and addressed to ensure the successful integration of cloud robotics into farming operations. The present section delves into the key challenges and barriers that may impede the widespread adoption of this transformative technology.

5.1. Cost and Initial Investment

The initial cost of implementing cloud robotics in agriculture can be a significant barrier for many farmers, particularly small-scale operators. The expenses associated with purchasing robotic equipment, setting up cloud infrastructure, and training personnel can be substantial. Additionally, ongoing maintenance and software updates may require further financial commitment (Gil et al., 2023).

5.2. Connectivity and Infrastructure

Effective use of cloud robotics relies on stable and high-speed internet connectivity. In many rural areas, access to reliable internet service can be limited, affecting the ability to transmit data in real time and maintain consistent communication with cloud-based systems. Moreover, the development of the required infrastructure, including IoT sensor networks and cloud servers, may be challenging in remote regions (Uddin et al., 2021).

5.3. Data Security and Privacy

The vast amounts of data generated by cloud-connected agricultural robots raise concerns about data security and privacy. Farmers may be hesitant to transmit sensitive information, such as crop yield data or financial records, to external cloud servers. Ensuring the robust security of data in transit and at rest is imperative to alleviate these concerns (Ferrag et al., 2020).

5.4. Technical Complexity and Training

Working with cloud robotics requires a certain level of technical expertise and knowledge of data analytics and software management. Training personnel and farm workers to operate and troubleshoot these systems can be time-consuming and may necessitate additional resources (Zhou et al., 2020).

5.5. Integration and Compatibility

Integrating cloud robotics with existing farming infrastructure and equipment can be a complex task. Compatibility issues may arise when connecting older machinery or sensors with modern cloud-based systems. Compatibility challenges can slow down the adoption process and increase the overall cost of implementation.

5.6. Regulation and Compliance

The agricultural industry is subject to various regulations and compliance standards. Implementing cloud robotics may require adherence to additional guidelines, such as data handling regulations and safety standards for autonomous machinery. Navigating the regulatory landscape can be a barrier to widespread adoption (Jayalakshmi & Gomathi, 2020).

5.7. Reliability and Downtime

As with any technology, cloud robotics systems are not immune to technical issues and downtime. A system failure or network interruption can disrupt farming operations and result in financial losses. Ensuring the reliability and redundancy of cloud systems is a key challenge (Priyadarsini et al., 2020).

5.8. Resistance to Change

Resistance to change is a common human factor that can impede the adoption of new technologies. Some farmers may be hesitant to shift from traditional farming methods to cloud robotics, fearing the loss of control, job displacement, or uncertainty about the technology's long-term viability (Zhang et al., 2022).

Addressing these challenges and barriers is crucial for the successful adoption of cloud robotics in agriculture. Strategies to overcome these obstacles include financial incentives, infrastructure development, improved training programs, data security protocols, and advocacy for supportive policies. The next section will delve into cloud robotics platforms and software solutions, offering insights into the tools available to agricultural stakeholders as they navigate these challenges.

6. FUTURE TRENDS AND OUTLOOK

The landscape of cloud robotics in agriculture is ever-evolving, driven by technological advancements, changing agricultural practices, and the need to address global challenges. As we look to the future, several key trends and developments are expected to shape the continued integration of cloud robotics in farming operations. In this section, we will explore these future trends and provide an outlook on the evolving role of cloud robotics in agriculture.

6.1. Enhanced Data Analytics and AI

The future of cloud robotics in agriculture will see a deeper integration of data analytics and artificial intelligence. Advanced machine learning algorithms will enable robots to process and interpret data with higher accuracy, resulting in more precise decision-making. This will lead to improved crop management, pest control, and resource allocation (Fantin Irudaya Raj & Balaji, 2022).

6.2. Autonomous Systems

Autonomy will play a pivotal role in the future of cloud robotics in agriculture. Robots will become increasingly autonomous, capable of conducting complex tasks with minimal human intervention. From autonomous planting and harvesting to predictive maintenance, these systems will enable farmers to focus on higher-level management and strategic decision-making (Appadurai & Fantin Irudaya Raj, 2021).

6.3. Connectivity and 5G Technology

As global connectivity improves, particularly in rural areas, cloud robotics will benefit from faster and more reliable internet access. The deployment of 5G technology will facilitate real-time data transmission, reducing latency and enhancing the efficiency of cloud-connected robots (Tang et al., 2021).

6.4. Integration With Blockchain for Transparency

Blockchain technology will play a role in ensuring transparency and traceability in agricultural supply chains. By integrating blockchain with cloud robotics, it will be possible to track the entire journey of a crop, from planting to market, ensuring product authenticity and quality control (Neelakandan et al., 2022).

6.5. Regulatory Frameworks

The future of cloud robotics in agriculture will likely see the development of standardized regulatory frameworks to address safety, data privacy, and compliance issues. Governments and industry bodies will work together to create guidelines that support the responsible adoption of this technology.

6.6. Sustainability and Environmental Monitoring

The increasing focus on sustainability and environmental responsibility will drive cloud robotics to play a larger role in monitoring and managing ecological impact. Robots will not only optimize resource use but also contribute to sustainable farming practices and environmental conservation (Sarkar et al., 2020).

6.7. Collaboration With Other Technologies

Cloud robotics will increasingly collaborate with other emerging technologies, such as drones, satellite imaging, and IoT devices. This synergy will provide farmers with a holistic view of their operations and enable them to make data-driven decisions on a broader scale.

6.8. Supportive Policies and Incentives

Governments and agricultural organizations may introduce policies and incentives to encourage the adoption of cloud robotics in agriculture. These measures may include subsidies for technology adoption, research grants, and support for training programs.

6.9. Market Growth

The market for cloud robotics in agriculture is expected to grow significantly in the coming years. As technology becomes more affordable and accessible, an increasing number of farmers, from small-scale to large-scale operations, will incorporate cloud robotics into their farming practices (Borsellino et al., 2020).

In conclusion, the future of cloud robotics in agriculture holds immense promise. As the technology evolves and integrates with other innovations, it will continue to address the challenges of feeding a growing global population while optimizing resource use and sustainability. The outlook for cloud robotics in agriculture is one of innovation, efficiency, and a more resilient agricultural industry.

7. CONCLUSION

The journey through the realm of agricultural automation using cloud robotics has revealed a transformative landscape with remarkable potential to revolutionize farming practices. Cloud robotics, as a synergy of cloud computing, artificial intelligence, and agricultural automation technologies, offers a path forward in addressing the complex challenges that modern agriculture faces. The chapter has explored the diverse facets of cloud robotics in agriculture, from the core technologies and advantages to the practical implementation through case studies. We have witnessed how cloud robotics enhances data analysis, enables real-time monitoring, and empowers precise decision-making. This technology has redefined the boundaries of agricultural automation, allowing for scalability, resource optimization, and sustainability. Yet, as with any technological innovation, cloud robotics in agriculture is not without its challenges. The issues of cost, connectivity, data security, and regulatory compliance must be addressed to ensure its widespread adoption. The future of cloud robotics in agriculture holds promise, with trends pointing towards enhanced data analytics, increased autonomy, better connectivity, and a focus on sustainability. The collaboration of cloud robotics with other emerging technologies will further enrich the potential for innovation and growth.

REFERENCES

Afrin, M., Jin, J., Rahman, A., Rahman, A., Wan, J., & Hossain, E. (2021). Resource allocation and service provisioning in multi-agent cloud robotics: A comprehensive survey. *IEEE Communications Surveys and Tutorials*, *23*(2), 842–870. doi:10.1109/COMST.2021.3061435

Agrimonti, C., Lauro, M., & Visioli, G. (2021). Smart agriculture for food quality: Facing climate change in the 21st century. *Critical Reviews in Food Science and Nutrition*, *61*(6), 971–981. doi:10.1080/10408398.2020.1749555 PMID:32270688

Appadurai, M., & Fantin Irudaya Raj, E. (2021). Finite element analysis of lightweight robot fingers actuated by pneumatic pressure. In *Recent Advances in Manufacturing, Automation, Design and Energy Technologies: Proceedings from ICoFT 2020* (pp. 379–385). Springer Singapore.

Araújo, S. O., Peres, R. S., Barata, J., Lidon, F., & Ramalho, J. C. (2021). Characterising the agriculture 4.0 landscape—Emerging trends, challenges and opportunities. *Agronomy (Basel)*, *11*(4), 667. doi:10.3390/agronomy11040667

Borsellino, V., Schimmenti, E., & El Bilali, H. (2020). Agri-food markets towards sustainable patterns. *Sustainability (Basel)*, *12*(6), 2193. doi:10.3390/su12062193

Chouhan, A. S., Purohit, N., Annaiah, H., Saravanan, D., Raj, E. F. I., & David, D. D. S. (2021). A Real-Time Gesture Based Image Classification System with FPGAand Convolutional Neural Network. *International Journal of Modern Agriculture*, *10*(2), 2565–2576.

Deivakani, M., Kumar, S. S., Kumar, N. U., Raj, E. F. I., & Ramakrishna, V. (2021, March). VLSI implementation of discrete cosine transform approximation recursive algorithm. *Journal of Physics: Conference Series*, *1817*(1), 012017. doi:10.1088/1742-6596/1817/1/012017 del Cerro, J., Cruz Ulloa, C., Barrientos, A., & de León Rivas, J. (2021). Unmanned aerial vehicles in agriculture: A survey. *Agronomy (Basel)*, *11*(2), 203. doi:10.3390/agronomy11020203

Dey, K., & Shekhawat, U. (2021). Blockchain for sustainable e-agriculture: Literature review, architecture for data management, and implications. *Journal of Cleaner Production*, *316*, 128254. doi:10.1016/j.jclepro.2021.128254

Fantin Irudaya Raj, E. (2022). Implementation of machine learning techniques in unmanned aerial vehicle control and its various applications. In *Computational Intelligence for Unmanned Aerial Vehicles Communication Networks* (pp. 17–33). Springer International Publishing. doi:10.1007/978-3-030-97113-7_2

Fantin Irudaya Raj, E., & Balaji, M. (2022). Application of deep learning and machine learning in pattern recognition. In *Advance Concepts of Image Processing and Pattern Recognition: Effective Solution for Global Challenges* (pp. 63–89). Springer Singapore. doi:10.1007/978-981-16-9324-3_4

Ferrag, M. A., Shu, L., Yang, X., Derhab, A., & Maglaras, L. (2020). Security and privacy for green IoT-based agriculture: Review, blockchain solutions, and challenges. *IEEE Access : Practical Innovations, Open Solutions*, *8*, 32031–32053. doi:10.1109/ACCESS.2020.2973178

Gil, G., Casagrande, D. E., Cortés, L. P., & Verschae, R. (2023). Why the low adoption of robotics in the farms? Challenges for the establishment of commercial agricultural robots. *Smart Agricultural Technology*, *3*, 100069. doi:10.1016/j.atech.2022.100069

Jayalakshmi, M., & Gomathi, V. (2020). Sensor-cloud based precision agriculture approach for intelligent water management. *International Journal of Plant Production*, *14*(2), 177–186. doi:10.1007/s42106-019-00077-1

Kalyani, Y., & Collier, R. (2021). A systematic survey on the role of cloud, fog, and edge computing combination in smart agriculture. *Sensors (Basel)*, *21*(17), 5922. doi:10.3390/s21175922 PMID:34502813

Kasemi, R., Lammer, L., & Vincze, M. (2022). The gap between technology and agriculture, barrier identification and potential solution analysis. *IFAC-PapersOnLine*, *55*(39), 314–318. doi:10.1016/j.ifacol.2022.12.042

Kumar, R., Mishra, R., Gupta, H. P., & Dutta, T. (2021). Smart sensing for agriculture: Applications, advancements, and challenges. *IEEE Consumer Electronics Magazine*, *10*(4), 51–56. doi:10.1109/MCE.2021.3049623

Lytridis, C., Kaburlasos, V. G., Pachidis, T., Manios, M., Vrochidou, E., Kalampokas, T., & Chatzistamatis, S. (2021). An overview of cooperative robotics in agriculture. *Agronomy (Basel)*, *11*(9), 1818. doi:10.3390/agronomy11091818

Mahmud, M. S. A., Abidin, M. S. Z., Emmanuel, A. A., & Hasan, H. S. (2020). Robotics and automation in agriculture: Present and future applications. *Applications of Modelling and Simulation*, *4*, 130–140.

Mangalraj, A. (2023). Smart Water Management System Powered by the Internet of Things (IoT) for Real-World Applications. In Internet of Things (pp. 129-142). Chapman and Hall/CRC.

Neelakandan, S., Beulah, J. R., Prathiba, L., Murthy, G. L. N., Irudaya Raj, E. F., & Arulkumar, N. (2022). Blockchain with deep learning-enabled secure healthcare data transmission and diagnostic model. *International Journal of Modeling, Simulation, and Scientific Computing*, *13*(04), 2241006. doi:10.1142/S1793962322410069

Paraforos, D. S., & Griepentrog, H. W. (2021). Digital farming and field robotics: Internet of things, cloud computing, and big data. *Fundamentals of Agricultural and Field Robotics*, 365-385.

Phasinam, K., Kassanuk, T., Shinde, P. P., Thakar, C. M., Sharma, D. K., Mohiddin, M. K., & Rahmani, A. W. (2022). Application of IoT and cloud computing in automation of agriculture irrigation. *Journal of Food Quality*, *2022*, 1–8. doi:10.1155/2022/8285969

Priyadarsini, K., Raj, E. F. I., Begum, A. Y., & Shanmugasundaram, V. (2020). Comparing DevOps procedures from the context of a systems engineer. *Materials Today: Proceedings*. Advance online publication. doi:10.1016/j.matpr.2020.09.624

Raj, E. F. I., Appadurai, M., & Athiappan, K. (2022). Precision farming in modern agriculture. In *Smart Agriculture Automation Using Advanced Technologies: Data Analytics and Machine Learning, Cloud Architecture, Automation and IoT* (pp. 61–87). Springer Singapore.

Rani, E. F. I., Pushparaj, T. L., & Raj, E. F. I. (2022). Escalating the resolution of an urban aerial image via novel shadow amputation algorithm. *Earth Science Informatics*, *15*(2), 905–913. doi:10.1007/s12145-022-00772-0

Sarkar, S., Skalicky, M., Hossain, A., Brestic, M., Saha, S., Garai, S., Ray, K., & Brahmachari, K. (2020). Management of crop residues for improving input use efficiency and agricultural sustainability. *Sustainability (Basel)*, *12*(23), 9808. doi:10.3390/su12239808

Shakya, D. S. (2020). Survey on cloud based robotics architecture, challenges and applications. *Journal of Ubiquitous Computing and Communication Technologies*, *2*(1), 10–18. doi:10.36548/jucct.2020.1.002

Sishodia, R. P., Ray, R. L., & Singh, S. K. (2020). Applications of remote sensing in precision agriculture: A review. *Remote Sensing (Basel)*, *12*(19), 3136. doi:10.3390/rs12193136

Tang, Y., Dananjayan, S., Hou, C., Guo, Q., Luo, S., & He, Y. (2021). A survey on the 5G network and its impact on agriculture: Challenges and opportunities. *Computers and Electronics in Agriculture*, *180*, 105895. doi:10.1016/j.compag.2020.105895

Uddin, M. A., Ayaz, M., Mansour, A., Aggoune, E. H. M., Sharif, Z., & Razzak, I. (2021). Cloud-connected flying edge computing for smart agriculture. *Peer-to-Peer Networking and Applications*, *14*(6), 3405–3415. doi:10.1007/s12083-021-01191-6

Zhai, Z., Martínez, J. F., Beltran, V., & Martínez, N. L. (2020). Decision support systems for agriculture 4.0: Survey and challenges. *Computers and Electronics in Agriculture*, *170*, 105256. doi:10.1016/j.compag.2020.105256

Zhang, F., Xiao, Y., Gao, L., Ma, D., Su, R., & Yang, Q. (2022). How agricultural water use efficiency varies in China—A spatial-temporal analysis considering unexpected outputs. *Agricultural Water Management*, *260*, 107297. doi:10.1016/j.agwat.2021.107297

Zhou, K., Wen, Y., Wu, W., Ni, Z., Jin, T., & Long, X. (2020). Cloud service optimization method based on dynamic artificial ant-bee colony algorithm in agricultural equipment manufacturing. *Mathematical Problems in Engineering*, *2020*, 1–11. doi:10.1155/2020/9134695

Compilation of References

. Jain, P., Firke, P. N., Kapadnis, K. N., Patil, T. S., & Rode, S. S (2014). RF based Spy robot. *Int. Journal of Engineering Research and Applications, 4*(4), 06-09.

Aarthi, V., Visali, R., & Nagarajan, K. K. (2021). Smart Spying Robot with IR Thermal Vision. *7th International Conference on Advanced Computing and Communication Systems (ICACCS), 1*, 183-187. 10.1109/ICACCS51430.2021.9441821

Abdulrahman, Y., Arnautović, E., Parezanović, V., & Svetinovic, D. (2023). AI and Blockchain Synergy in Aerospace Engineering: An Impact Survey on Operational Efficiency and Technological Challenges. *IEEE Access : Practical Innovations, Open Solutions*, *11*, 87790–87804. doi:10.1109/ACCESS.2023.3305325

Adams, W. R., & Goodwin, L. D. (2013). Interactive and collaborative cloud-based robotics programming using RoboBlockly. In *Proceedings of the 15th international conference on Information Integration and Web-based Applications & Services* (pp. 283-292). Academic Press.

Afrin, M., Jin, J., Rahman, A., Rahman, A., Wan, J., & Hossain, E. (2021). Resource allocation and service provisioning in multi-agent cloud robotics: A comprehensive survey. *IEEE Communications Surveys and Tutorials*, *23*(2), 842–870. doi:10.1109/COMST.2021.3061435

Agapito, G., & Cannataro, M. (2023). An Overview on the Challenges and Limitations Using Cloud Computing in Healthcare Corporations. *Big Data and Cognitive Computing*, *7*(2), 68. doi:10.3390/bdcc7020068

Agrimonti, C., Lauro, M., & Visioli, G. (2021). Smart agriculture for food quality: Facing climate change in the 21st century. *Critical Reviews in Food Science and Nutrition*, *61*(6), 971–981. doi:10.1080/10408398.2020.1749555 PMID:32270688

Ahmad, A. (2024). Ethical implications of artificial intelligence in accounting: A framework for responsible ai adoption in multinational corporations in Jordan. *International Journal of Data and Network Science*, *8*(1), 401–414. doi:10.5267/j.ijdns.2023.9.014

Ahmad, F., Wysocki, R. W., Fernandez, J. J., Cohen, M. S., & Simcock, X. C. (2023). Patient perspectives on telemedicine during the COVID-19 pandemic. *Hand (New York, N.Y.)*, *18*(3), 522–526. doi:10.1177/15589447211030692 PMID:34515550

Ahmed, A. A., Abdalla, T. Y., & Abed, A. A. (2015). Path Planning of Mobile Robot by using Modified Optimized Potential Field Method. *International Journal of Computer Applications*, *113*(4), 6–10. doi:10.5120/19812-1614

Alhwarin, F., & Wang, C. (2008). Improved SIFT-Features Matching for Object Recognition. *BCS International Academic Conference 2008 -Visions of Computer Science,* 179-190. 10.14236/ewic/VOCS2008.16

Ali, O., AlAhmad, A., & Kahtan, H. (2023). A review of advanced technologies available to improve the healthcare performance during COVID-19 pandemic. *Procedia Computer Science*, *217*, 205–216. doi:10.1016/j.procs.2022.12.216 PMID:36687286

Alirezazadeh, S., & Alexandre, L. (2021). Optimal algorithm allocation for single robot cloud systems. *IEEE Transactions on Cloud Computing*.

Al-Jumaili, A. H. A., Muniyandi, R. C., Hasan, M. K., Paw, J. K. S., & Singh, M. J. (2023). Big Data Analytics Using Cloud Computing Based Frameworks for Power Management Systems: Status, Constraints, and Future Recommendations. *Sensors (Basel)*, *23*(6), 2952. doi:10.3390/s23062952 PMID:36991663

Alkhafaji, F. S. M. (2021, December). Modeling and control high speed robotic arm for industrial applications. In *2021 Global Congress on Electrical Engineering (GC-ElecEng)* (pp. 121-130). IEEE. 10.1109/GC-ElecEng52322.2021.9788417

Alosaimi, R., & Alnuem, M. (2016). Risk management frameworks for cloud computing: A critical review. *AIRCC's International Journal of Computer Science and Information Technologies*, *8*(4), 1–11. doi:10.5121/ijcsit.2016.8401

Alotaibi, E.T., Alqefari, S.S., & Koubaa, A. (2019). LSAR: Multi-UAV Collaboration for Search and Rescue Missions. *IEEE Access*, *7*, 55817–55832.

Alouffi, B., Hasnain, M., Alharbi, A., Alosaimi, W., Alyami, H., & Ayaz, M. (2021). A systematic literature review on cloud computing security: Threats and mitigation strategies. *IEEE Access : Practical Innovations, Open Solutions*, *9*, 57792–57807. doi:10.1109/ACCESS.2021.3073203

Alseddiqi, M., AlMannaei, B., Najam, O., & Al-Mofleh, A. (2022, December). The Importance of Medical Robots in Improving Healthcare Services. In *2022 International Conference on Automation, Computing and Renewable Systems (ICACRS)* (pp. 1397-1401). IEEE. 10.1109/ICACRS55517.2022.10029105

Alzubaidi, L., Fadhel, M. A., Al-Shamma, O., Zhang, J., Santamaría, J., Duan, Y., & Oleiwi, R., S. (. (2020). Towards a better understanding of transfer learning for medical imaging: A case study. *Applied Sciences (Basel, Switzerland)*, *10*(13), 4523. doi:10.3390/app10134523

Amaran, M. H., Noh, N. A. M., Rohmad, M. S., & Hashim, H. (2015). A comparison of lightweight communication protocols in robotic applications. *Procedia Computer Science*, *76*, 400–405. doi:10.1016/j.procs.2015.12.318

Andtfolk, M., Nyholm, L., Eide, H., Rauhala, A., & Fagerström, L. (2021). Attitudes toward the use of humanoid robots in healthcare—A cross-sectional study. *AI & Society*, 1–10.

Appadurai, M., & Fantin Irudaya Raj, E. (2021). Finite element analysis of lightweight robot fingers actuated by pneumatic pressure. In *Recent Advances in Manufacturing, Automation, Design and Energy Technologies: Proceedings from ICoFT 2020* (pp. 379–385). Springer Singapore.

Araújo, S. O., Peres, R. S., Barata, J., Lidon, F., & Ramalho, J. C. (2021). Characterising the agriculture 4.0 landscape—Emerging trends, challenges and opportunities. *Agronomy (Basel)*, *11*(4), 667. doi:10.3390/agronomy11040667

Arunajyothi, G. (2016). A study on cloud robotics architecture, challenges and applications. *International Journal of Computer Science and Network Security*, *16*(9), 110. doi:10.18535/ijecs/v5i9.27

Arzo, S. T., Naiga, C., Granelli, F., Bassoli, R., Devetsikiotis, M., & Fitzek, F. H. (2021). A theoretical discussion and survey of network automation for IoT: Challenges and opportunity. *IEEE Internet of Things Journal*, *8*(15), 12021–12045. doi:10.1109/JIOT.2021.3075901

Ashwin, S. H., & Naveen Raj, R. (2023). Deep reinforcement learning for autonomous vehicles: Lane keep and overtaking scenarios with collision avoidance. *International Journal of Information Technology : an Official Journal of Bharati Vidyapeeth's Institute of Computer Applications and Management*, *15*(7), 3541–3553. doi:10.1007/s41870-023-01412-6

Asiri, A., AlBishi, S., AlMadani, W., ElMetwally, A., & Househ, M. (2018). The use of telemedicine in surgical care: A systematic review. *Acta Informatica Medica*, *26*(3), 201. doi:10.5455/aim.2018.26.201-206 PMID:30515013

Avinash, K., & Kumar, A. (2023). A novel approach of unsupervised feature selection using iterative shrinking and expansion algorithm. *Journal of Interdisciplinary Mathematics*, *26*(3), 519–530. doi:10.47974/JIM-1678

Babaeizadeh, M., Bakhtiari, M., & Mohammed, A. M. (2015). Authentication methods in cloud computing: A survey. *Research Journal of Applied Sciences, Engineering and Technology*, *9*(8), 655–664. doi:10.19026/rjaset.9.1451

Bakhshipour, M., Ghadi, M. J., & Namdari, F. (2017). Swarm robotics search & rescue: A novel artificial intelligence-inspired optimisation approach. *Applied Soft Computing*, *57*, 708–726. doi:10.1016/j.asoc.2017.02.028

Balakrishnan, M., Gowthaman, S., Kumaran, S. J., & Sabhapathy, G. R. (2015). A smart spy robot charged and controlled by wireless systems. *Proceeding of International Conference on Innovations in Information, Embedded and Communication Systems (ICIIECS)*, 1-4. 10.1109/ICIIECS.2015.7193096

Bangemann, T., Karnouskos, S., Camp, R., Carlsson, O., Riedl, M., McLeod, S., ... Stluka, P. (2014). State of the art in industrial automation. *Industrial Cloud-Based Cyber-Physical Systems: The IMC-AESOP Approach*, 23-47.

Banimfreg, B. H. (2023). A comprehensive review and conceptual framework for cloud computing adoption in bioinformatics. *Healthcare Analytics*, *3*, 100190. doi:10.1016/j.health.2023.100190

Beecher, R. C. (1979). Puma: Programmable universal machine for assembly. In *Computer vision and sensor-based robots* (pp. 141–152). Springer US. doi:10.1007/978-1-4613-3027-1_8

Behar, J. A., Liu, C., Kotzen, K., Tsutsui, K., Corino, V. D., Singh, J., Pimentel, M. A. F., Warrick, P., Zaunseder, S., Andreotti, F., Sebag, D., Kopanitsa, G., McSharry, P. E., Karlen, W., Karmakar, C., & Clifford, G. D. (2020). Remote health diagnosis and monitoring in the time of COVID-19. *Physiological Measurement*, *41*(10), 10TR01. doi:10.1088/1361-6579/abba0a PMID:32947271

Bekbolatova, M., Mayer, J., Ong, C. W., & Toma, M. (2024, January). Transformative Potential of AI in Healthcare: Definitions, Applications, and Navigating the Ethical Landscape and Public Perspectives. In Healthcare (Vol. 12, No. 2, p. 125). MDPI.

Bellini, P., Nesi, P., & Pantaleo, G. (2022). IoT-enabled smart cities: A review of concepts, frameworks and key technologies. *Applied Sciences (Basel, Switzerland)*, *12*(3), 1607. doi:10.3390/app12031607

Bendiab, G., Hameurlaine, A., Germanos, G., Kolokotronis, N., & Shiaeles, S. (2023). Autonomous vehicles security: Challenges and solutions using blockchain and artificial intelligence. *IEEE Transactions on Intelligent Transportation Systems*, *24*(4), 3614–3637. doi:10.1109/TITS.2023.3236274

Berlet, M., Fuchtmann, J., Krumpholz, R., Naceri, A., Macari, D., Jähne-Schon, C., Haddadin, S., Friess, H., Feussner, H., & Wilhelm, D. (2024). Toward telemedical diagnostics—Clinical evaluation of a robotic examination system for emergency patients. *Digital Health*, *10*. doi:10.1177/20552076231225084 PMID:38205033

Bhaskar, S., Bradley, S., Sakhamuri, S., Moguilner, S., Chattu, V. K., Pandya, S., Schroeder, S., Ray, D., & Banach, M. (2020). Designing futuristic telemedicine using artificial intelligence and robotics in the COVID-19 era. *Frontiers in Public Health*, *8*, 708. doi:10.3389/fpubh.2020.556789 PMID:33224912

Bhat, M., & Kapoor, P. (2012). Application of sad algorithm in image proc processing for motion detection and simulink block sets for object tracking. *International Journal of Engineering Science & Advanced Technology*, *2*(3), 731–736.

Bhuvaneswary, N., Pravallika, S., & Jayapriya, V., & Himabindu. (2021). Night Surveillance Military Spy Robot using Raspberry Pi. *Annals of the Romanian Society for Cell Biology*, 5740–5747.

Bicchi, A., & Kumar, V. (2000, April). Robotic grasping and contact: A review. In *Proceedings 2000 ICRA. Millennium conference. IEEE international conference on robotics and automation. Symposia proceedings (Cat. No. 00CH37065)* (Vol. 1, pp. 348-353). IEEE.

Bishop, A. J., Sheng, W., Carlson, B. W., & Jones, N. F. (2023). The evolution and rise of robotic health assistants: The new human-machine frontier of geriatric home care. *(Re) designing the Continuum of Care for Older Adults: The Future of Long-Term Care Settings*, 97-121.

Blais, M.-A., & Akhloufi, M. A. (2023). *Reinforcement learning for swarm robotics: An overview of applications, algorithms and simulators*. Cognitive Robotics.

Bogue, R. (2017). Cloud robotics: A review of technologies, developments and applications. Industrial Robot. *International Journal (Toronto, Ont.)*, *44*(1), 1–5.

Bonaccorsi, M., Fiorini, L., Cavallo, F., Saffiotti, A., & Dario, P. (2016). A cloud robotics solution to improve social assistive robots for active and healthy aging. *International Journal of Social Robotics*, *8*(3), 393–408. doi:10.1007/s12369-016-0351-1

Borkar, S., Ghutke, P., Patil, W., Joshi, S., & Sorte, S. (2023, April). A Review of Pick and Place Robots for the Pharmaceutical Industry. In *2023 11th International Conference on Emerging Trends in Engineering & Technology-Signal and Information Processing (ICETET-SIP)* (pp. 1-6). IEEE. 10.1109/ICETET-SIP58143.2023.10151652

Borsellino, V., Schimmenti, E., & El Bilali, H. (2020). Agri-food markets towards sustainable patterns. *Sustainability (Basel)*, *12*(6), 2193. doi:10.3390/su12062193

Bousdekis, A., Lepenioti, K., Ntalaperas, D., Vergeti, D., Apostolou, D., & Boursinos, V. (2019). A RAMI 4.0 View of Predictive Maintenance: Software Architecture, Platform and Case Study in Steel Industry. *International Conference on Advanced Information Systems Engineering CAiSE 2019: Advanced Information Systems Engineering Workshops*, 95-106.

Bouziane, R., Terrissa, L. S., Ayad, S., & Brethé, J. F. (2022). Towards an architecture for cloud robotic services. *International Journal of Computers and Applications*, *44*(9), 863–874. doi:10.1080/1206212X.2021.1964790

Browne, C. (2011). Evolutionary game design. *IEEE Transactions on Computational Intelligence and AI in Games*, *2*, 11–21.

Buerkle, A., Eaton, W., Al-Yacoub, A., Zimmer, M., Kinnell, P., Henshaw, M., Coombes, M., Chen, W.-H., & Lohse, N. (2023). Towards industrial robots as a service (IRaaS): Flexibility, usability, safety and business models. *Robotics and Computer-integrated Manufacturing*, *81*, 102484. doi:10.1016/j.rcim.2022.102484

Butun, I., Österberg, P., & Song, H. (2019). Security of the Internet of Things: Vulnerabilities, attacks, and countermeasures. *IEEE Communications Surveys and Tutorials*, *22*(1), 616–644. doi:10.1109/COMST.2019.2953364

Cacciamani, G. E., Chen, A., Gill, I. S., & Hung, A. J. (2024). Artificial intelligence and urology: Ethical considerations for urologists and patients. *Nature Reviews. Urology*, *21*(1), 50–59. doi:10.1038/s41585-023-00796-1 PMID:37524914

Chai, Y. T., Che, H. S., Tan, C., Tan, W. N., Yip, S. C., & Gan, M. T. (2023). A two-stage optimization method for Vehicle to Grid coordination considering building and Electric Vehicle user expectations. *International Journal of Electrical Power & Energy Systems*, *148*, 108984. doi:10.1016/j.ijepes.2023.108984

Chang, C., & Lin, C. (2011). LIBSVM: A library for support vector machines. *ACM Transactions on Intelligent Systems and Technology*, *2*(3).

Chavan, M. P., Kharde, S., Jade, M., & Jadhav, A. (2021). Wireless spy robot. *International Engineering Journal For Research & Development*, *6*, 4–4.

Cheikhrouhou, O., Koubâa, A., & Zarrad, A. (2020). A cloud based disaster management system. *Journal of Sensor and Actuator Networks*, *9*(1), 6. doi:10.3390/jsan9010006

Chen, C., Wang, C., Liu, B., He, C., Cong, L., & Wan, S. (2023). Edge intelligence empowered vehicle detection and image segmentation for autonomous vehicles. *IEEE Transactions on Intelligent Transportation Systems*, *24*(11), 13023–13034. doi:10.1109/TITS.2022.3232153

Chen, H. (2021). Cloud-Enhanced Robotics in College Engineering Labs: A Case Study. *IEEE Transactions on Learning Technologies*, *14*(4), 623–638.

Chen, M., Ma, Y., Ullah, S., Cai, W., & Song, E. (2013, September). ROCHAS: Robotics and cloud-assisted healthcare system for empty nester. In *Proceedings of the 8th international conference on body area networks* (pp. 217-220). 10.4108/icst.bodynets.2013.253922

Chen, Q., & Hsu, M. (2001). Inter-enterprise collaborative business process management. *Proceedings 17th International Conference on Data Engineering*. 10.1109/ICDE.2001.914836

Chen, W. (2016). Cloud-Based Programming Environments for Educational Robotics: A Comparative Study. *IEEE Transactions on Emerging Topics in Computing*, *4*(2), 234–247.

Chen, W., Yaguchi, Y., Naruse, K., Watanobe, Y., Nakamura, K., & Ogawa, J. (2018). A study of robotic cooperation in cloud robotics: Architecture and challenges. *IEEE Access : Practical Innovations, Open Solutions*, *6*, 36662–36682. doi:10.1109/ACCESS.2018.2852295

Chen, Y. B., Yu, J. Q., Su, X. L., & Luo, G. C. (2015, January). Path Planning for Multi-UAV Formation. *Journal of Intelligent & Robotic Systems*, *77*(1), 229–246. doi:10.1007/s10846-014-0077-y

Chen, Y., Wang, F., & Liu, Y. (2020). Cybersecurity threats and defense in networked surveillance systems: Challenges and opportunities. *IEEE Access : Practical Innovations, Open Solutions*, *8*, 187096–187110.

Chen, Y., Wang, F., & Liu, Y. (2022). Ethical considerations in AI-driven surveillance: Addressing algorithmic bias and privacy concerns. *IEEE Transactions on Systems, Man, and Cybernetics. Systems*, *52*(1), 50–65.

Ch, G., Jana, S., Majji, S., Kuncha, P., & Tigadi, A. (2021). Diagnosis of COVID-19 using 3D CT scans and vaccination for COVID-19. *World Journal of Engineering*, *19*(2), 189–194. doi:10.1108/WJE-03-2021-0161

Chhabra, M. (2023). Implications of Cloud Computing for Health Care. In *Cloud-based Intelligent Informative Engineering for Society 5.0* (pp. 41–59). Chapman and Hall/CRC. doi:10.1201/9781003213895-3

Chinamuttevi, V. S. (2023). A novel RF-SMOTE model to enhance the definite apprehensions for IoT security attacks. *Journal of Discrete Mathematical Sciences and Cryptography*, *26*(3), 861–873. doi:10.47974/JDMSC-1766

Choudhary, S., & Singh, N. (2022). Analysis of security-based access control models for cloud computing. *International Journal of Cloud Applications and Computing*, *12*(1), 1–19. doi:10.4018/IJCAC.2022010104

Chouhan, A. S., Purohit, N., Annaiah, H., Saravanan, D., Raj, E. F. I., & David, D. D. S. (2021). A Real-Time Gesture Based Image Classification System with FPGAand Convolutional Neural Network. *International Journal of Modern Agriculture*, *10*(2), 2565–2576.

Chung, J.-M., Park, K., Won, T., Oh, W., & Choi, S. (2010). New protocols for future wireless systems. *2010 53rd IEEE International Midwest Symposium on Circuits and Systems*.

Chung, J., & Lee, Y. S. (2023). The evolving impact of robots on jobs. *Industrial & Labor Relations Review*, *76*(2), 290–319. doi:10.1177/00197939221137822

Cole, A. (2020). *Fundamental valuation of the General Motors Company: assessing the value in the midst of a pandemic* (Master's thesis).

Cordioli, E., Giavina-Bianchi, M., Pedrotti, C. H. S., & Podgaec, S. (2023). Brazilian Medical Survey on Telemedicine since the onset of COVID-19. *Einstein (São Paulo), 21*, eAE0428.

Correia, A. W., Pope, C. A. III, Dockery, D. W., Wang, Y., Ezzati, M., & Dominici, F. (2013). Effect of air pollution control on life expectancy in the United States: An analysis of 545 US counties for the period from 2000 to 2007. *Epidemiology (Cambridge, Mass.)*, *24*(1), 23–31. doi:10.1097/EDE.0b013e3182770237 PMID:23211349

Costas, Palaiologou, & Alifragis. (2006). Virtual and remote robotic laboratory: Comparative experimental evaluation. *IEEE Transactions on Education, 49*(3), 360-369.

Dahl, T. S., & Kamel Boulos, M. N. (2013). Robots in health and social care: A complementary technology to home care and telehealthcare? *Robotics (Basel, Switzerland)*, *3*(1), 1–21. doi:10.3390/robotics3010001

Damaj, I. W., Yousafzai, J. K., & Mouftah, H. T. (2022). Future trends in connected and autonomous vehicles: Enabling communications and processing technologies. *IEEE Access : Practical Innovations, Open Solutions*, *10*, 42334–42345. doi:10.1109/ACCESS.2022.3168320

Dang, L. M., Piran, M. J., Han, D., Min, K., & Moon, H. (2019). A survey on internet of things and cloud computing for healthcare. *Electronics (Basel)*, *8*(7), 768. doi:10.3390/electronics8070768

Darwin, S., & Fantin Irudaya Raj, E. (2023). The Emergence of Cyber-Physical System in the Context of Self-Powered Soft Robotics. *Self-Powered Cyber Physical Systems*, 57-90.

Darwin, S., & Raj, E. F. I. (2023). A detailed review on embedded based heartbeat monitoring systems. *Smart Embedded Systems and Applications*, 79.

Dawarka, V., & Bekaroo, G. (2022). Building and evaluating cloud robotic systems: A systematic review. *Robotics and Computer-integrated Manufacturing*, *73*, 102240. doi:10.1016/j.rcim.2021.102240

de Almeida, J. P. L. S., Nakashima, R. T., Neves-Jr, F., & de Arruda, L. V. R. (2019). Bio-inspired online path planner for cooperative exploration of unknown environment by a Multi-Robot System. *Robotics and Autonomous Systems*, *112*, 32–48. doi:10.1016/j.robot.2018.11.005

Deiva Ganesh, A., & Kalpana, P. (2022). Future of artificial intelligence and its influence on supply chain risk management – A systematic review. *Computers & Industrial Engineering*, *169*, 108206. doi:10.1016/j.cie.2022.108206

Deivakani, M., Kumar, S. S., Kumar, N. U., Raj, E. F. I., & Ramakrishna, V. (2021, March). VLSI implementation of discrete cosine transform approximation recursive algorithm. *Journal of Physics: Conference Series*, *1817*(1), 012017. doi:10.1088/1742-6596/1817/1/012017

del Cerro, J., Cruz Ulloa, C., Barrientos, A., & de León Rivas, J. (2021). Unmanned aerial vehicles in agriculture: A survey. *Agronomy (Basel)*, *11*(2), 203. doi:10.3390/agronomy11020203

DePold, H., Ros, R., Semsch, E., & Adya, A. (2019). Cloud Robotics in Education. In *2019 IEEE International Conference on Robotics and Automation (ICRA)* (pp. 5582-5589). IEEE.

Dey, K., & Shekhawat, U. (2021). Blockchain for sustainable e-agriculture: Literature review, architecture for data management, and implications. *Journal of Cleaner Production*, *316*, 128254. doi:10.1016/j.jclepro.2021.128254

Dhirani, L. L., Mukhtiar, N., Chowdhry, B. S., & Newe, T. (2023). Ethical dilemmas and privacy issues in emerging technologies: A review. *Sensors (Basel)*, *23*(3), 1151. doi:10.3390/s23031151 PMID:36772190

Diftler, M. A., Permenter, F. A., Hargrave, B. K., Platt, R., Savely, R. T., Ambrose, R. O., Mehling, J. S., Abdallah, M. E., Radford, N. A., Bridgwater, L. B., Sanders, A. M., Askew, R. S., Linn, D. M., & Yamokoski, J. D. (2011). Robonaut 2 - The first humanoid robot in space. *2011 IEEE International Conference on Robotics and Automation.* 10.1109/ICRA.2011.5979830

Dittakavi, R. S. S. (2022). Evaluating the Efficiency and Limitations of Configuration Strategies in Hybrid Cloud Environments. *International Journal of Intelligent Automation and Computing*, *5*(2), 29–45.

Djellid, A., Benmeddour, F., Bahri, L., & Ghodbane, A. (2022). Novel high-gain and compact UWB microstrip antenna for WIFi, WIMAX, WLAN, X band and 5G applications. *Optical and Quantum Electronics*, *54*(7), 441. doi:10.1007/s11082-022-03731-7

Dragomir, D., Gheorghe, L., Costea, S., & Radovici, A. (2016). A survey on secure communication protocols for IoT systems. *2016 International Workshop on Secure Internet of Things (SIoT).* 10.1109/SIoT.2016.012

Dulce-Galindo, J. A., Santos, M. A., Raffo, G. V., & Pena, P. N. (2022). Distributed supervisory control for multiple robot autonomous navigation performing single-robot tasks. *Mechatronics*, *86*, 102848. doi:10.1016/j.mechatronics.2022.102848

Ebadi, K., Bernreiter, L., Biggie, H., Catt, G., Chang, Y., Chatterjee, A., ... Carlone, L. (2023). Present and future of SLAM in extreme environments: The DARPA subT challenge. *IEEE Transactions on Robotics*.

Edwards, M. (1984). Robots in industry: An overview. *Applied Ergonomics*, *15*(1), 45–53. doi:10.1016/S0003-6870(84)90121-2 PMID:15676500

Elfaki, A. O., Abduljabbar, M., Ali, L., Alnajjar, F., Mehiar, D. A., Marei, A. M., Alhmiedat, T., & Al-Jumaily, A. (2023). Revolutionizing Social Robotics: A Cloud-Based Framework for Enhancing the Intelligence and Autonomy of Social Robots. *Robotics (Basel, Switzerland)*, *12*(2), 48. doi:10.3390/robotics12020048

El-Gendy, S. (2020). IoT Based AI and its Implementations in Industries. *2020 15th International Conference on Computer Engineering and Systems (ICCES)*, 1-6. 10.1109/ICCES51560.2020.9334627

Emam, Y., Wilson, S., Hakenberg, M., Munz, U., & Egerstedt, M. (2020). A Receding Horizon Scheduling Approach for Search & Rescue Scenarios. *IFAC-PapersOnLine*, *53*(2), 3451–3456. doi:10.1016/j.ifacol.2020.12.1680

Emil Selvan, G. S. R., Jagadeesh Kannan, R., Sangapu, S. C., Shalini, S., Pandurangan, R., & Raj, E. F. I. (2023). Novel Development of Fuzzy Controller Based Multi-Agent System for Efficient Navigation of Autonomous Robots. *International Journal of Uncertainty, Fuzziness and Knowledge-based Systems*, *31*(Supp01), 17–30. doi:10.1142/S0218488523400020

Engel, Y., Etzion, O., & Feldman, Z. (2012). A basic model for proactive event- driven computing. *Proceedings of the 6th ACM International Conference on Distributed Event-based Systems*, 107-118. 10.1145/2335484.2335496

Ewert, D., Schilberg, D., & Jeschke, S. (2013). Problem based learning of object-oriented Programming with LEGO Mindstorms and leJOS. Automation. *Communication and Cybernetics in Science and Engineering*, *2011/2012*, 315–323.

Fantin Irudaya Raj, E., & Appadurai, M. (2021). The hybrid electric vehicle (HEV)—an overview. *Emerging Solutions for e-Mobility and Smart Grids: Select Proceedings of ICRES 2020*, 25-36.

Fantin Irudaya Raj, E., Appadurai, M., Chithamabara Thanu, M., & Francy Irudaya Rani, E. (2023). IoT-Based Smart Parking System for Indian Smart Cities. *Machine Intelligence, Big Data Analytics, and IoT in Image Processing: Practical Applications*, 369-398.

Fantin Irudaya Raj, E. (2022). Implementation of machine learning techniques in unmanned aerial vehicle control and its various applications. In *Computational Intelligence for Unmanned Aerial Vehicles Communication Networks* (pp. 17–33). Springer International Publishing. doi:10.1007/978-3-030-97113-7_2

Fantin Irudaya Raj, E., & Balaji, M. (2021). Analysis and classification of faults in switched reluctance motors using deep learning neural networks. *Arabian Journal for Science and Engineering*, *46*(2), 1313–1332. doi:10.1007/s13369-020-05051-y

Fantin Irudaya Raj, E., & Balaji, M. (2022). Application of deep learning and machine learning in pattern recognition. In *Advance Concepts of Image Processing and Pattern Recognition: Effective Solution for Global Challenges* (pp. 63–89). Springer Singapore. doi:10.1007/978-981-16-9324-3_4

Fdhila, W., Rinderle-Ma, S., Knuplesch, D., & Reichert, M. (2015). Change and compliance in collaborative processes. IEEE SCC, 162–169. doi:10.1109/SCC.2015.31

Felzenszwalb, P. F., Girshick, R. B., McAllester, D., & Ramanan, D. (2010). Object Detection with Discriminatively Trained Part Based Models. *IEEE Transactions on Pattern Analysis and Machine Intelligence*, *32*(9), 1–20. doi:10.1109/TPAMI.2009.167 PMID:20634557

Ferrag, M. A., Shu, L., Yang, X., Derhab, A., & Maglaras, L. (2020). Security and privacy for green IoT-based agriculture: Review, blockchain solutions, and challenges. *IEEE Access : Practical Innovations, Open Solutions*, *8*, 32031–32053. doi:10.1109/ACCESS.2020.2973178

Ferraiolo, D. F., Barkley, J. F., & Kuhn, D. R. (1999). A role-based access control model and reference implementation within a corporate intranet. *ACM Transactions on Information and System Security*, *2*(1), 34–64. doi:10.1145/300830.300834

Fiorini, L., Esposito, R., Bonaccorsi, M., Petrazzuolo, C., Saponara, F., Giannantonio, R., Petris, G. D., Dario, P., & Cavallo, F. (2017). Enabling personalised medical support for chronic disease management through a hybrid robot-cloud approach. *Autonomous Robots*, *41*(5), 1263–1276. doi:10.1007/s10514-016-9586-9

Firouzi, F., Farahani, B., Daneshmand, M., Grise, K., Song, J., Saracco, R., & Luo, A. (2021). Harnessing the power of smart and connected health to tackle COVID-19: IoT, AI, robotics, and blockchain for a better world. *IEEE Internet of Things Journal*, *8*(16), 12826–12846. doi:10.1109/JIOT.2021.3073904 PMID:35782886

Floridi, L., & Taddeo, M. (2016). What is data ethics? *Philosophical Transactions. Series A, Mathematical, Physical, and Engineering Sciences*, *374*(2083), 20160360. doi:10.1098/rsta.2016.0360 PMID:28336805

Fosch-Villaronga, E., & Millard, C. (2019). Cloud robotics law and regulation: Challenges in the governance of complex and dynamic cyber–physical ecosystems. *Robotics and Autonomous Systems*, *119*, 77–91. doi:10.1016/j.robot.2019.06.003

Francy Irudaya Rani, E., Lurthu Pushparaj, T., & Fantin Irudaya Raj, E. (2023). Adopting a Fusion Approach for Optical Amplification. *Modeling and Optimization of Optical Communication Networks*, 11-34.

Gampala, V., Kumar, M. S., Sushama, C., & Raj, E. F. I. (2020). Deep learning based image processing approaches for image deblurring. *Materials Today: Proceedings*. Advance online publication. doi:10.1016/j.matpr.2020.11.076

Gao, J. (2019). *Machine Learning Applications for Data Center Optimization*. Available at: https://static.googleusercontent.com/media /research.google.com/en//pubs/ archive/42542.pdf

Garcia, M., & Rodriguez, P. (2021). Integration of Cloud Robotics in the Curriculum: A Case Study in High School Robotics Clubs. *IEEE Transactions on Learning Technologies*, *14*(2), 289–305.

Garg, S., Sünderhauf, N., Dayoub, F., Morrison, D., Cosgun, A., Carneiro, G., ... Milford, M. (2020). Semantics for robotic mapping, perception and interaction: A survey. *Foundations and Trends® in Robotics, 8*(1–2), 1-224.

Garg, V., Arora, S., & Gupta, C. (2011). Cloud computing approaches to accelerate drug discovery value chain. *Combinatorial Chemistry & High Throughput Screening*, *14*(10), 861–871. doi:10.2174/138620711797537085 PMID:21843145

Gatti, R. R., Singh, C., Agrawal, R., & Serrao, F. J. (Eds.). (2023). *Self-Powered Cyber Physical Systems*. John Wiley & Sons. doi:10.1002/9781119842026

Ghodsian, N., Benfriha, K., Olabi, A., Gopinath, V., Arnou, A., Charrier, Q., & Helou, M. El. (2022). Toward designing an integration architecture for a mobile manipulator in production systems: Industry 4.0. *Procedia CIRP, 109*, 443–448. https://doi.org/https://doi.org/10.1016/j.procir.2022.05.276

Giannaros, A., Karras, A., Theodorakopoulos, L., Karras, C., Kranias, P., Schizas, N., Kalogeratos, G., & Tsolis, D. (2023). Autonomous vehicles: Sophisticated attacks, safety issues, challenges, open topics, blockchain, and future directions. *Journal of Cybersecurity and Privacy*, *3*(3), 493–543. doi:10.3390/jcp3030025

Gil, G., Casagrande, D. E., Cortés, L. P., & Verschae, R. (2023). Why the low adoption of robotics in the farms? Challenges for the establishment of commercial agricultural robots. *Smart Agricultural Technology*, *3*, 100069. doi:10.1016/j.atech.2022.100069

Givehchi, O., Trsek, H., & Jasperneite, J. (2013). Cloud computing for industrial automation systems — A comprehensive overview. *2013 IEEE 18th Conference on Emerging Technologies & Factory Automation (ETFA)*, 1-4. 10.1109/ETFA.2013.6648080

Gnanasekar, A. K., Deivakani, M., Bathala, N., Raj, E. F. I., & Ramakrishna, V. (2021). Novel low-noise CMOS bioamplifier for the characterization of neurodegenerative diseases. In GeNeDis 2020: Genetics and Neurodegenerative Diseases (pp. 221-226). Springer International Publishing. doi:10.1007/978-3-030-78787-5_27

Goel, P., & Singh, D. (2014, February). Efficient ABC Algorithm for Dynamic Path Planning. *International Journal of Computer Applications*, *88*(2), 15–18. doi:10.5120/15322-3634

Gonçalves, C. P. dos S. (2018). Quantum robotics, neural networks and the quantum force interpretation. SSRN *Electronic Journal*. doi:10.2139/ssrn.3244327

Gouda, M. G., & Multari, N. J. (1991). Stabilizing communication protocols. IEEE Transactions on Computers. *IEEE Transactions on Computers*, *40*(4), 448–458. doi:10.1109/12.88464

Grzybowski, A., Jin, K., & Wu, H. (2024). Challenges of artificial intelligence in medicine and dermatology. *Clinics in Dermatology*. Advance online publication. doi:10.1016/j.clindermatol.2023.12.013 PMID:38184124

Guillén, A. J., Crespo, A., Gómez, J. F., & Sanz, M. D. (2016). A framework for effective management of condition based maintenance programs in the context of industrial development of E-Maintenance strategies. *Computers in Industry*, *82*, 170–185. doi:10.1016/j.compind.2016.07.003

Gundu, S. R., Panem, C. A., & Timmapuram, A. (2020). Robotic technology-based cloud computing for improved services. *SN Computer Science*, *1*(4), 190. doi:10.1007/s42979-020-00203-1

Gunge, V. S., & Yalagi, P. S. (2016). Smart home automation: A literature review. *International Journal of Computer Applications*, *975*(8887-8891).

Gupta, S., Modgil, S., Kumar, A., Sivarajah, U., & Irani, Z. (2022). Artificial intelligence and cloud-based Collaborative Platforms for Managing Disaster, extreme weather and emergency operations. *International Journal of Production Economics, 254*. doi:10.1016/j.ijpe.2022.108642

Gupta, S. M., & Rizvi, S. W. A. (2023). Cloud Security for Healthcare Services. *Journal of Management and Service Science*, *3*(1), 1–9.

Hakak, S., Gadekallu, T. R., Maddikunta, P. K. R., Ramu, S. P., Parimala, M., De Alwis, C., & Liyanage, M. (2023). Autonomous Vehicles in 5G and beyond: A Survey. *Vehicular Communications*, *39*, 100551. doi:10.1016/j.vehcom.2022.100551

Haleem, A., Javaid, M., Singh, R. P., & Suman, R. (2022). Medical 4.0 technologies for healthcare: Features, capabilities, and applications. *Internet of Things and Cyber-Physical Systems*, *2*, 12–30. doi:10.1016/j.iotcps.2022.04.001

Hao, Y., Liu, J., Liu, Y., Liu, X., Meng, Z., & Xing, F. (2023). Global Visual–Inertial Localization for Autonomous Vehicles with Pre-Built Map. *Sensors (Basel)*, *23*(9), 4510. doi:10.3390/s23094510 PMID:37177714

Harbin, J., Gerasimou, S., Matragkas, N., Zolotas, T., Calinescu, R., & Alpizar Santana, M. (2023). Model-driven design space exploration for multi-robot systems in simulation. *Software & Systems Modeling*, *22*(5), 1665–1688. doi:10.1007/s10270-022-01041-w

Hegge, K. K., & Agrawal, R. (2023). Industrial Automated Multipurpose Robot Using WIFI. *2023 4th International Conference for Emerging Technology (INCET),* 1-8.

Hemphill, T. A. (2020). Autonomous vehicles: US regulatory policy challenges. *Technology in Society*, *61*, 101232. doi:10.1016/j.techsoc.2020.101232

Herm, L. V., Janiesch, C., Helm, A., Imgrund, F., Hofmann, A., & Winkelmann, A. (2023). A framework for implementing robotic process automation projects. *Information Systems and e-Business Management*, *21*(1), 1–35. doi:10.1007/s10257-022-00553-8

Hernández, E., Velázquez, R., Giannoccaro, N. I., & Gutíerrez, C. A. (2017, November). Kinematic analysis and workspace simulation of humanoid robot KUBO. In *2017 IEEE 37th Central America and Panama Convention (CONCAPAN XXXVII)* (pp. 1-6). IEEE.

Holland, J., Kingston, L., McCarthy, C., Armstrong, E., O'Dwyer, P., Merz, F., & McConnell, M. (2021). Service robots in the healthcare sector. *Robotics (Basel, Switzerland)*, *10*(1), 47. doi:10.3390/robotics10010047

Hribernik, K., von Stietencron, M., Bousdekis, A., Bredehorst, B., Mentzas, G., & Thoben, K. D. (2018). Towards a unified predictive maintenance system-a use case in production logistics in aeronautics. *Procedia Manufacturing*, *16*, 131–138. doi:10.1016/j.promfg.2018.10.168

Huang, Y. (2022) Cloud-Enhanced Robotics for Inclusive Education: A Case Study with Special Needs Students. *IEEE Transactions on Learning Technologies*, *15*(1), 56–71.

Hu, G., Tay, W., & Wen, Y. (2012). Cloud robotics: Architecture, challenges and applications. *IEEE Network*, *26*(3), 21–28. doi:10.1109/MNET.2012.6201212

Hunkeler, U., Truong, H. L., & Stanford-Clark, A. (2008). MQTT-S — A publish/subscribe protocol for Wireless Sensor Networks. *2008 3rd International Conference on Communication Systems Software and Middleware and Workshops (COMSWARE '08).*

Huo, D., Sari, Y. A., Kealey, R., & Zhang, Q. (2023). Reinforcement learning-based fleet dispatching for greenhouse gas emission reduction in open-pit mining operations. *Resources, Conservation and Recycling*, *188*, 106664. doi:10.1016/j.resconrec.2022.106664

Hussain, N., & Samal, A. (2024). Design and Implementation of Single to Multi-cloud Securities in Network. In N. R. Shetty, N. H. Prasad, & N. Nalini (Eds.), *Advances in Computing and Information. ERCICA 2023. Lecture Notes in Electrical Engineering* (Vol. 1104). Springer.

Hu, X., Zheng, Z., Chen, D., & Sun, J. (2023). Autonomous vehicle's impact on traffic: Empirical evidence from Waymo Open Dataset and implications from modelling. *IEEE Transactions on Intelligent Transportation Systems*, *24*(6), 6711–6724. doi:10.1109/TITS.2023.3258145

Ibrahim Khalil, M. W., & Kamel, M. A. E. (2023). Towards smart sustainable cities vision and challenges. *International Journal of Nonlinear Analysis and Applications*.

Ichnowski, J., Chen, K., Dharmarajan, K., Adebola, S., Danielczuk, M., Mayoral-Vilches, V., ... Goldberg, K. (2023, May). Fogros 2: An adaptive and extensible platform for cloud and fog robotics using ros 2. *Proceedings IEEE International Conference on Robotics and Automation*.

Iclodean, C., Varga, B. O., & Pfister, F. (2023). Autonomous Vehicles Technological Trends. *Electronics (Basel)*, *12*(5), 1149. doi:10.3390/electronics12051149

Intuitive. (n.d.). https://www.intuitive.com/en-us/patients/da-vinci-robotic-surgery/about-the-systems

Jain, I., & Singh, B. (2023). Analysing the Applications of Cloud Computing in Smart Agriculture. In Convergence of Cloud Computing, AI, and Agricultural Science (pp. 100-119). IGI Global. doi:10.4018/979-8-3693-0200-2.ch005

Jain, S., & Doriya, R. (2019). Security issues and solutions in cloud robotics: a survey. In *Next Generation Computing Technologies on Computational Intelligence: 4th International Conference, NGCT 2018, Dehradun, India, November 21–22, 2018, Revised Selected Papers 4* (pp. 64-76). Springer Singapore. 10.1007/978-981-15-1718-1_6

Jain, S., & Doriya, R. (2022). Security framework to healthcare robots for secure sharing of healthcare data from cloud. *International Journal of Information Technology : an Official Journal of Bharati Vidyapeeth's Institute of Computer Applications and Management*, *14*(5), 2429–2439. doi:10.1007/s41870-022-00997-8 PMID:35633945

Jamwal, P. K., Niyetkaliyev, A., Hussain, S., Sharma, A., & Van Vliet, P. (2023). Utilizing the intelligence edge framework for robotic upper limb rehabilitation in home. *MethodsX*, *11*, 102312. doi:10.1016/j.mex.2023.102312 PMID:37593414

Jayalakshmi, M., & Gomathi, V. (2020). Sensor-cloud based precision agriculture approach for intelligent water management. *International Journal of Plant Production*, *14*(2), 177–186. doi:10.1007/s42106-019-00077-1

Jayanthi, A., & Reddy, B. E. (2023). An intelligent cloud robotics personalized medicine system by assessment of DNA and electronic health records using fuzzy-based progressive alignment sequence (F-Pas). *Soft Computing*, 1–11. doi:10.1007/s00500-023-08359-7

Jensen, M., Schwenk, J., Gruschka, N., & Iacono, L. L. (2009, September). On technical security issues in cloud computing. *Proceedings. IEEE International Conference on Cloud Computing*, 109–116.

Jevtić, A., Andina, D., Jaimes, A., Gomez, J., & Jamshidi, M. (2010). Unmanned aerial vehicle route optimisation using ant system algorithm. *2010 5th International Conference on System of Systems Engineering*, 1–6.

Jijesh & Shivashankar. (2017). Design and implementation of automatic medicine dispensing machine. *RTEICT 2017 - 2nd IEEE International Conference on Recent Trends in Electronics, Information and Communication Technology, Proceedings*, 1962–1966.

Johnson, A., Brown, D., & Davis, C. (2021). AI-driven UGVs for perimeter surveillance: Enhancing security through autonomous patrolling. *Journal of Intelligent Robotics*, *40*(3), 312–327.

Johnson, R. (2018). Design and Implementation of a Cloud-Connected Robotics Course for Undergraduate Education. *IEEE Transactions on Robotics*, *37*(2), 189–204.

Jones, A., Smith, B., & Davis, C. (2018). Unmanned aerial vehicles for surveillance: Current capabilities and future trends. *Journal of Security Engineering*, *15*(2), 45–58.

Jumani, A. K., Siddique, W. A., & Laghari, A. A. (2023). Cloud and Machine Learning Based Solutions for Healthcare and Prevention. In *Image Based Computing for Food and Health Analytics: Requirements, Challenges, Solutions and Practices: IBCFHA* (pp. 163–192). Springer International Publishing. doi:10.1007/978-3-031-22959-6_10

K. S & Arun. (2022). Priority Queueing Model-Based IoT Middleware for Load Balancing. *2022 6th International Conference on Intelligent Computing and Control Systems (ICICCS)*, 425-430.

K. S. & Arun. (2022). Priority Queueing Model-Based IoT Middleware for Load Balancing. *2022 6th International Conference on Intelligent Computing and Control Systems (ICICCS)*, 425-430.

Kabir, H., Tham, M. L., & Chang, Y. C. (2023). Internet of robotic things for mobile robots: Concepts, technologies, challenges, applications, and future directions. *Digital Communications and Networks*, *9*(6), 1265–1290. doi:10.1016/j.dcan.2023.05.006

Kalatehjari, E. H., Hosseini, M. M., & Behrad, A. (2012). Object Recognition Based on Sift Features and a Novel Feature Matching Algorithm. *RNIS: Research Notes in Information Science*, *10*(0), 7–14. doi:10.4156/rnis.vol10.issue0.2

Kalyani, Y., & Collier, R. (2021). A systematic survey on the role of cloud, fog, and edge computing combination in smart agriculture. *Sensors (Basel)*, *21*(17), 5922. doi:10.3390/s21175922 PMID:34502813

Kaphungkui, N.K. (2015). RF based remote control for home electrical appliances. *International Journal of Innovative Research in Electrical, Electronics, Instrumentation and Control Engineering, 3*(7), 42-44.

Kasemi, R., Lammer, L., & Vincze, M. (2022). The gap between technology and agriculture, barrier identification and potential solution analysis. *IFAC-PapersOnLine*, *55*(39), 314–318. doi:10.1016/j.ifacol.2022.12.042

Kato, S., Takeuchi, E., Ishiguro, Y., Ninomiya, Y., Takeda, K., & Hamada, T. (2015). An open approach to autonomous vehicles. *IEEE Micro*, *35*(6), 60–68. doi:10.1109/MM.2015.133

Kaur, J. (2023, May). How is Robotic Process Automation Revolutionising the Way Healthcare Sector Works? In *International Conference on Information, Communication and Computing Technology* (pp. 1037-1055). Singapore: Springer Nature Singapore. 10.1007/978-981-99-5166-6_70

Kaur, J. (2024). Towards a Sustainable Triad: Uniting Energy Management Systems, Smart Cities, and Green Healthcare for a Greener Future. In Emerging Materials, Technologies, and Solutions for Energy Harvesting (pp. 258-285). IGI Global.

Kaur, J. (2023). Robotic Process Automation in Healthcare Sector. In *E3S Web of Conferences* (Vol. 391, p. 01008). EDP Sciences.

Kehoe, B., Patil, S., Abbeel, P., & Goldberg, K. (2015). A survey of research on cloud robotics and automation. *IEEE Transactions on Automation Science and Engineering, 12*(2), 398–409. doi:10.1109/TASE.2014.2376492

Keung, K. L., Lee, C. K. M., & Ji, P. (2021). Data-driven order correlation pattern and storage location assignment in robotic mobile fulfillment and process automation system. *Advanced Engineering Informatics*, *50*, 101369. doi:10.1016/j.aei.2021.101369

Khan, J., Ali, F., Ullah, F., Bacha, J., & Lee, S. (2019). Artificial Intelligence and Internet of Things (AI-IoT) Technologies in Response to COVID-19 Pandemic: A Systematic Review. *IEEE Access : Practical Innovations, Open Solutions*, *10*, 62613–62660. doi:10.1109/ACCESS.2022.3181605

Khazetdinov, A., Aleksandrov, A., Zakiev, A., Magid, E., & Hsia, K. H. (2020). RFID-based warehouse management system prototyping using a heterogeneous team of robots. *Robots in Human Life*, 263.

Kim, J., Gu, G. M., & Heo, P. (2016). Robotics for healthcare. *Biomedical Engineering: Frontier Research and Converging Technologies*, 489-509.

Kim, H., & Lee, S. (2017). Cloud Robotics in STEM Education: Challenges and Opportunities. *IEEE Transactions on Education*, *60*(1), 30–37.

Kim, S., & Choi, J. (2016). A Cloud-Based Robotics Platform for K-12 Education: Design and Evaluation. *IEEE Transactions on Education*, *59*(2), 91–98.

Kim, S., & Lee, J. (2021). Ethics of surveillance technologies: Balancing security and privacy in the age of enhanced spy robotics. *Journal of Information Ethics*, *9*(2), 178–193.

Klauschen, F., Dippel, J., Keyl, P., Jurmeister, P., Bockmayr, M., Mock, A., ... Müller, K. R. (2024). Toward explainable artificial intelligence for precision pathology. *Annual Review of Pathology: Mechanisms of Disease, 19*.

Ko, H., Keoh, S. L., & Jin, J. (2017). Privacy Protection for Cloud-Based Robotic Networks. *Security and Privacy in Cyber-Physical Systems: Foundations, Principles and Applications,* 195-222. 10.1002/9781119226079.ch10

Konishi, M., Sasaki, T., & Cai, K. (2022). Efficient Safe Control via Deep Reinforcement Learning and Supervisory Control – Case Study on Multi-Robot Warehouse Automation. *IFAC-PapersOnLine*, *55*(28), 16–21. doi:10.1016/j.ifacol.2022.10.318

Koot, M., Mes, M. R. K., & Iacob, M. E. (2021). A systematic literature review of supply chain decision making supported by the Internet of Things and Big Data Analytics. *Computers & Industrial Engineering*, *154*, 107076. doi:10.1016/j.cie.2020.107076

Kormushev, P., Calinon, S., & Caldwell, D. (2013). Reinforcement learning in robotics: Applications and real-world challenges. *Robotics (Basel, Switzerland)*, *2*(3), 122–148. doi:10.3390/robotics2030122

Koubaa, A., Allouch, A., Alajlan, M., Javed, Y., Belghith, A., & Khalgui, M. (2019). Micro Air Vehicle Link (MAVlink) in a Nutshell: A Survey. *IEEE Access : Practical Innovations, Open Solutions*, *7*, 87658–87680. doi:10.1109/ACCESS.2019.2924410

Koubaa, A., Qureshi, B., Sriti, M. F., Allouch, A., Javed, Y., Alajlan, M., Cheikhrouhou, O., Khalgui, M., & Tovar, E. (2019). Dronemap Planner: A service-oriented cloud-based management system for the Internet-of-Drones. *Ad Hoc Networks*, *86*, 46–62. doi:10.1016/j.adhoc.2018.09.013

Kranthi, M., & Tanguturi, R. C. (2023). Design of Intelligent Medical Integrity Authentication and Secure Information for Public Cloud in Hospital Administration. *2023 2nd International Conference on Edge Computing and Applications (ICECAA),* 256-261.

Krishna, B., Amuthavalli, G., & Stalin, DRaj, E. FSaravanan, D. (2021). Certain investigation of SARS-COVID-2-induced Kawasaki-like disease in Indian Youngsters. *Annals of the Romanian Society for Cell Biology*, 1167–1182.

Krishnan, A., & Swarna, S. (2020). Robotics, IoT, and AI in the Automation of Agricultural Industry: A Review. *2020 IEEE Bangalore Humanitarian Technology Conference (B-HTC)*, 1-6. 10.1109/B-HTC50970.2020.9297856

Kumar & Singhal. (2020). Design and Analysis of IoT-Based Intelligent Robot for Real-Time Monitoring and Control. *2020 International Conference on Power Electronics & IoT Applications in Renewable Energy and its Control (PARC)*, 549-552.

Kumar, A., & Jaiswal, A. (2020). A deep swarm-optimised model for leveraging industrial data analytics in cognitive manufacturing. *IEEE Transactions on Industrial Informatics*, *17*(4), 2938–2946. doi:10.1109/TII.2020.3005532

Kumar, A., Kalpana, V., & Hariram, V. (2023). Artificial Intelligence based Health Monitoring System on IoTH platform. *2023 Second International Conference on Augmented Intelligence and Sustainable Systems (ICAISS)*, 1458-1463.

Kumar, P. S., & Sharma, A. (2019). IoT Enabled Indoor Autonomous Mobile Robot using CNN and Q-Learning. *2019 IEEE International Conference on Industry 4.0, Artificial Intelligence, and Communications Technology (IAICT)*, 7-13.

Kumar, R., Mishra, R., Gupta, H. P., & Dutta, T. (2021). Smart sensing for agriculture: Applications, advancements, and challenges. *IEEE Consumer Electronics Magazine*, *10*(4), 51–56. doi:10.1109/MCE.2021.3049623

Kunduru, A. R. (2023). Cloud BPM Application (Appian) Robotic Process Automation Capabilities. *Asian Journal of Research in Computer Science*, *16*(3), 267–280. doi:10.9734/ajrcos/2023/v16i3361

Kurunathan, H., Severino, R., Koubaa, A., & Tovar, E. (2019). DynaMO—Dynamic Multisuperframe Tuning for Adaptive IEEE 802.15.4e DSME Networks. *IEEE Access, 7*, 122522–122535.

Kurunathan, H., Severino, R., Koubaa, A., & Tovar, E. (2018). IEEE 802.15. 4e in a nutshell: Survey and performance evaluation. *IEEE Communications Surveys and Tutorials*, *20*(3), 1989–2010. doi:10.1109/COMST.2018.2800898

Lara, C. L., & Wassick, J. (2023). The future of supply chain - a perspective from the process and online retail industries. *Computers & Chemical Engineering*, *179*, 108401. doi:10.1016/j.compchemeng.2023.108401

Latsou, C., Farsi, M., & Erkoyuncu, J. A. (2023). Digital twin-enabled automated anomaly detection and bottleneck identification in complex manufacturing systems using a multi-agent approach. *Journal of Manufacturing Systems*, *67*, 242–264. doi:10.1016/j.jmsy.2023.02.008

Lavisse, T., Panariello, R., Perdu, F., & Zwolinski, P. (2023). Integrating an ageing model within Life Cycle Assessment to evaluate the environmental impacts of electric batteries. *Procedia CIRP*, *116*, 251–256. doi:10.1016/j.procir.2023.02.043

Le, D. N., Van Le, C., Tromp, J. G., & Nguyen, G. N. (Eds.). (2018). Emerging technologies for health and medicine: virtual reality, augmented reality, artificial intelligence, internet of things, robotics, industry 4.0. Academic Press.

Lee, K.B., Cheon, S., & Kim, C.O. (2017). A Convolutional Neural Network for Fault Classification and Diagnosis in Semiconductor Manufacturing Processes. *IEEE Transactions on Semiconductor Manufacturing, 30*(2), 135-142.

Lee, C., & Park, H. (2021). Cloud Robotics and Gamification: Engaging Students in Programming and Problem-Solving. *IEEE Transactions on Robotics*, *38*(4), 981–995.

Lee, J., Ardakani, H. D., Yang, S., & Bagheri, B. (2015). Industrial big data analytics and cyber-physical systems for future maintenance & service innovation. *Procedia CIRP*, *38*, 3–7. doi:10.1016/j.procir.2015.08.026

Lee, J., Kao, H.-A., & Yang, S. (2014). Service innovation and smart analytics for industry 4.0 and big data environment. *Procedia CIRP*, *16*, 3–8. doi:10.1016/j.procir.2014.02.001

Lee, J., & Kim, S. (2021). Ethical considerations in the deployment of surveillance robotics: Balancing security and privacy. *Journal of Ethics in Technology*, *7*(3), 112–128.

Li, L., Bagheri, S., Goote, H., Hasan, A., & Hazard, G. (2013). *Risk adjustment of patient expenditures: A big data analytics approach.* Paper presented at the Big Data, 2013 IEEE International Conference on. 10.1109/BigData.2013.6691790

Li, J., & Wu, S. (2017). Cloud Robotics for Programming Education: A Practical Approach. *IEEE Transactions on Robotics*, *36*(4), 532–546.

Lin, C. C., & Tseng, H. Y. (2005, January). A neural network application for reliability modelling and condition-based predictive maintenance. *International Journal of Advanced Manufacturing Technology*, *25*(1–2), 174–179. doi:10.1007/s00170-003-1835-3

Li, Q., Wang, H., & Zhang, L. (2022). Sensor fusion in aerial surveillance drones: Advancements and challenges. *IEEE Sensors Journal*, *22*(4), 2150–2165.

Liu, Q., Wang, C., Li, X., & Gao, L. (2023). An improved genetic algorithm with modified critical path-based searching for integrated process planning and scheduling problem considering automated guided vehicle transportation task. *Journal of Manufacturing Systems*, *70*, 127–136. doi:10.1016/j.jmsy.2023.07.004

Liu, W. (2021). Improvement of Navigation of Mobile Robotics Based on IoT System. *2021 IEEE International Conference on Robotics, Automation and Artificial Intelligence (RAAI)*, 69-72. 10.1109/RAAI52226.2021.9507826

Liu, X., Zhu, W., Lian, X., & Xu, X. (2023). Monitoring mining surface subsidence with multi-temporal three-dimensional unmanned aerial vehicle point cloud. *Remote Sensing (Basel)*, *15*(2), 374. doi:10.3390/rs15020374

Liu, Y. (2018). Remote Learning with Cloud-Connected Robots: A Case Study in Elementary Education. *IEEE Robotics & Automation Magazine*, *25*(2), 45–53.

Liu, Y., Zhang, H., & Huang, C. (2019). A novel RGB-D SLAM algorithm based on cloud robotics. *Sensors (Basel)*, *19*(23), 5288. doi:10.3390/s19235288 PMID:31805628

Li, X., & Wang, Y. (2017). A Cloud-Based Platform for Collaborative Robotics Learning. *IEEE Transactions on Education*, *60*(4), 286–293.

Li, Z., & Millar-Bilbao, F. (2021, March). Characterising edge-cloud data transmission for patient-centric healthcare systems. In *2020 IEEE International Conference on E-health Networking, Application & Services (HEALTHCOM)* (pp. 1-6). IEEE. 10.1109/HEALTHCOM49281.2021.9398991

Ljubi, K., & Groznik, A. (2023). Role played by social factors and privacy concerns in autonomous vehicle adoption. *Transport Policy*, *132*, 1–15. doi:10.1016/j.tranpol.2022.12.013

Logeshwaran, J., Ramesh, G., & Aravindarajan, V. (2023). A Secured Database Monitoring Method to Improve Data Backup and Recovery Operations in Cloud Computing. *BOHR International Journal of Computer Science*, *2*(1), 1–7. doi:10.54646/bijcs.019

Lowe, D. G. (1999). Object Recognition from Local Scale-Invariant Features. *Proc. of the International Conference on Computer Vision*, 1-8.

Luo, C., He, M., & Xing, C. (2024). Public acceptance of autonomous vehicles in China. *International Journal of Human-Computer Interaction*, *40*(2), 315–326. doi:10.1080/10447318.2022.2115336

Luo, S., Meng, Q., Li, S., & Yu, H. (2023). Research of intent recognition in rehabilitation robots: A systematic review. *Disability and Rehabilitation. Assistive Technology*, 1–12. doi:10.1080/17483107.2023.2170477 PMID:36695473

Lytridis, C., Kaburlasos, V. G., Pachidis, T., Manios, M., Vrochidou, E., Kalampokas, T., & Chatzistamatis, S. (2021). An overview of cooperative robotics in agriculture. *Agronomy (Basel)*, *11*(9), 1818. doi:10.3390/agronomy11091818

Macrae, C. (2022). Learning from the failure of autonomous and intelligent systems: Accidents, safety, and sociotechnical sources of risk. *Risk Analysis*, *42*(9), 1999–2025. doi:10.1111/risa.13850 PMID:34814229

Madhiarasan, M. (2021). Design and development of IoT based solar powered versatile moving robot for military application. *International Journal of System Assurance Engineering and Management*, *12*(3), 437–450. doi:10.1007/s13198-021-01089-9

Magesh, G., Muthuswamy, P., & Singh, B. (2015). Use of Information Technology among school students in the State of Tamil Nadu, India. *International Journal of Applied Engineering Research: IJAER*, *10*(1), 2201–2209.

Mahajan, H. B., Rashid, A. S., Junnarkar, A. A., Uke, N., Deshpande, S. D., Futane, P. R., Alkhayyat, A., & Alhayani, B. (2023). Integration of Healthcare 4.0 and blockchain into secure cloud-based electronic health records systems. *Applied Nanoscience*, *13*(3), 2329–2342. doi:10.1007/s13204-021-02164-0 PMID:35136707

Maheswari, B. U., Imambi, S. S., Hasan, D., Meenakshi, S., Pratheep, V. G., & Boopathi, S. (2023). Internet of things and machine learning-integrated smart robotics. In Global Perspectives on Robotics and Autonomous Systems: Development and Applications (pp. 240-258). IGI Global. doi:10.4018/978-1-6684-7791-5.ch010

Mahmud, M. S. A., Abidin, M. S. Z., Emmanuel, A. A., & Hasan, H. S. (2020). Robotics and automation in agriculture: Present and future applications. *Applications of Modelling and Simulation*, *4*, 130–140.

Makhdoom, I., Abolhasan, M., Abbas, H., & Ni, W. (2019, January). Blockchain's adoption in IoT: The challenges, and a way forward. *Journal of Network and Computer Applications*, *125*(1), 251–279. doi:10.1016/j.jnca.2018.10.019

Mangalraj, A. (2023). Smart Water Management System Powered by the Internet of Things (IoT) for Real-World Applications. In Internet of Things (pp. 129-142). Chapman and Hall/CRC.

Manimala, K., Raj, E. F. I., & Jagatheswari, S. (2023). Advancements in Deep Learning Techniques for Analyzing Electronic Medical Records. In *Deep Learning for Healthcare Decision Making* (pp. 133–158). River Publishers. doi:10.1201/9781003373261-6

Marks, R. (2023). Robotics as a Form of Non-Surgical Complementary Osteoarthritis Treatment. *International Journal of Complementary and Internal Medicine*, *5*(1), 189–198.

Martinez, A., & Davis, C. (2019). Remote Control and Programming of Educational Robots Using Cloud Technologies. *IEEE Transactions on Robotics*, *36*(1), 98–112.

Matsubara, A., & Ibaraki, S. (2009). Monitoring and control of cutting forces in machining processes: A review. *International Journal of Automotive Technology*, *3*(4), 445–456. doi:10.20965/ijat.2009.p0445

Ma, Y., Zhang, Y., Wan, J., Zhang, D., & Pan, N. (2015). Robot and cloud-assisted multi-modal healthcare system. *Cluster Computing*, *18*(3), 1295–1306. doi:10.1007/s10586-015-0453-9

Mehmood, M. S., Shahid, M. R., Jamil, A., Ashraf, R., Mahmood, T., & Mehmood, A. (2019, November). A comprehensive literature review of data encryption techniques in cloud computing and IoT environment. In *2019 8th International Conference on Information and Communication Technologies (ICICT)* (pp. 54-59). IEEE. 10.1109/ICICT47744.2019.9001945

Mehta, K., Gaur, S., Maheshwari, S., Chugh, H., & Anibhushan Kumar, M. (2023, April). Big Data Analytics Cloud based Smart IoT Healthcare Network. In *2023 7th International Conference on Trends in Electronics and Informatics (ICOEI)* (pp. 437-443). IEEE.

Mell, P., & Grance, T. (2011). *The NIST Definition of Cloud Computing*. NIST Special Publication. https://csrc.nist.gov/publications/nistpubs/800-145/SP800-145.pdf

Mello, R. C., Ribeiro, M. R., & Frizera-Neto, A. (2022). *Implementing Cloud Robotics for Practical Applications: From Human-robot Interaction to Autonomous Navigation* (Vol. 152). Springer Nature.

Mello, R. C., Scheidegger, W. M., Múnera, M. C., Cifuentes, C. A., Ribeiro, M. R., & Frizera-Neto, A. (2022). The PoundCloud framework for ROS-based cloud robotics: Case studies on autonomous navigation and human–robot interaction. *Robotics and Autonomous Systems*, *150*, 103981. doi:10.1016/j.robot.2021.103981

Meng, W., Yang, Y., Zang, J., Li, H., & Lu, R. (2023). DTUAV: A novel cloud–based digital twin system for unmanned aerial vehicles. *Simulation*, *99*(1), 69–87. doi:10.1177/00375497221109575

Mester, G. (2006). Distance Learning in Robotics. *Proceedings of The Third International Conference on Informatics, Educational Technology and New Media in Education*, 239-245. 10.1109/ICL.2012.6402174

Michalos, G., Kousi, N., Makris, S., & Chryssolouris, G. (2016). Performance Assessment of Production Systems with Mobile Robots. *Procedia CIRP*, *41*, 195–200. doi:10.1016/j.procir.2015.12.097

Mijwil, M., Al-Mistarehi, A. H., Abotaleb, M., El-kenawy, E. S. M., Ibrahim, A., Abdelhamid, A. A., & Eid, M. M. (2023). From Pixels to Diagnoses: Deep Learning's Impact on Medical Image Processing-A Survey. *Wasit Journal of Computer and Mathematics Science*, *2*(3), 9–15. doi:10.31185/wjcms.178

Militano, L., Arteaga, A., Toffetti, G., & Mitton, N. (2023). The cloud-to-edge-to-IoT continuum as an enabler for search and rescue operations. *Future Internet*, *15*(2), 55. doi:10.3390/fi15020055

Milner, E., Sooriyabandara, M., & Hauert, S. (2022). Stochastic behaviours for retrieval of storage items using simulated robot swarms. *Artificial Life and Robotics*, *27*(2), 264–271. doi:10.1007/s10015-022-00749-8

Ming, G. (2023). Exploration of the intelligent control system of autonomous vehicles based on edge computing. *PLoS One*, *18*(2), e0281294. doi:10.1371/journal.pone.0281294 PMID:36730359

Mishra, A., Banerjee, S., & Singh, B. (2023). Deep Learning Techniques for Smart Agriculture Applications. In Machine Learning and Deep Learning for Smart Agriculture and Applications (pp. 1-23). IGI Global.

Mishra, H. M., Qtaishat, A., & Kumar, M. (2023). IoT Based Smart Systems using Artificial Intelligence and Machine Learning: Accessible and Intelligent Solutions. *2023 6th International Conference on Information Systems and Computer Networks (ISCON)*, 1-6.

Mistry, I., Tanwar, S., Tyagi, S., & Kumar, N. (2020). Blockchain for 5G-enabled IoT for industrial automation: A systematic review, solutions, and challenges. *Mechanical Systems and Signal Processing*, *135*, 106382. doi:10.1016/j.ymssp.2019.106382

Mittelstadt, B. D., Allo, P., Taddeo, M., Wachter, S., & Floridi, L. (2016). The ethics of algorithms: Mapping the debate. *Big Data & Society*, *3*(2). doi:10.1177/2053951716679679

Mohammed, M. A., Ibrahim, D. A., & Abdulkareem, K. H. (2023). Bio-inspired robotics enabled schemes in blockchain-fog-cloud assisted IoMT environment. *Journal of King Saud University. Computer and Information Sciences*, *35*(1), 1–12. doi:10.1016/j.jksuci.2021.11.009

Mohanarajah, G., Hunziker, D., D'Andrea, R., & Waibel, M. (2015). Rapyuta: A cloud robotics platform. *IEEE Transactions on Automation Science and Engineering, 12*(2), 481–493. doi:10.1109/TASE.2014.2329556

Mohanty, P. K., & Parhi, D. R. (2014, December). A new efficient optimal path planner for mobile robot based on Invasive Weed Optimization algorithm. *Frontiers of Mechanical Engineering*, *9*(4), 317–330. doi:10.1007/s11465-014-0304-z

Molla-Hosseini, M., Hosseini, M., Vossoughi, G., & Sharifzadeh, A. M. (2022). Robotics: A Key Driver of Industry 4.0. *Industry 4.0 Vision for Energy and Materials: Enabling Technologies and Case Studies*, 73-98.

Mondal, B., Rakshit, S., Billaha, M.A., Roy, B., & Sarkar, R. (2019). Low-Cost Wireless Data Transmission System for Industrial Applications. *Advances in Computer, Communication and Control,* 509-523.

Monostori, L., Valckenaers, P., Dolgui, A., Panetto, H., Brdys, M., & Csáji, B. C. (2014). Cooperative Control in Production and Logistics. *IFAC Proceedings Volumes, 47*(3), 4246–4265. https://doi.org/https://doi.org/10.3182/20140824-6-ZA-1003.01026

Monteiro, A. C. B., França, R. P., Arthur, R., & Iano, Y. (2021). An overview of medical Internet of Things, artificial intelligence, and cloud computing employed in health care from a modern panorama. *The Fusion of Internet of Things, Artificial Intelligence, and Cloud Computing in Health Care,* 3-23.

Muhtaroglu, F., Demir, S., Obali, M., & Girgin, C. (2013). *Business model canvas perspective on big data applications.* Paper presented at the Big Data, 2013 IEEE 10.1109/BigData.2013.6691684

Mukherjee, A. (2017). Cloud Robotics Architectures. *Challenges and Applications,TTIC*, *1*, 27–33.

Mushtaq, A., Haq, I. U., Sarwar, M. A., Khan, A., Khalil, W., & Mughal, M. A. (2023). Multi-Agent Reinforcement Learning for Traffic Flow Management of Autonomous Vehicles. *Sensors (Basel)*, *23*(5), 2373. doi:10.3390/s23052373 PMID:36904577

Naik, Y. M., Deshpande, C. M., Shah, R. R., & Kulkarni, R. R (2013). Android Controlled Spy-Robot. *International Journal of Software and Web Sciences 4*(1), 54-57.

Nakanoya, M., Narasimhan, S. S., Bhat, S., Anemogiannis, A., Datta, A., Katti, S., Chinchali, S., & Pavone, M. (2023). Co-design of communication and machine inference for cloud robotics. *Autonomous Robots*, *47*(5), 1–16. doi:10.1007/s10514-023-10093-w

Narula, A., Narula, N. K., Khanna, S., Narula, R., Narula, J., & Narula, A. (2014). Future prospects of artificial intelligence in robotics software, a healthcare perspective. *International Journal of Applied Engineering Research: IJAER*, *9*(22), 10271–10280.

Nawrat, Z., & Krawczyk, D. (2023). Robot-Based Medicine. Robots in Medicine: Mobile Robots Versus Mobile Decision, Necessity Versus Possibility and Future Challenges. *Mobile Robot: Motion Control and Path Planning*, 127-162.

Nayyar & Nagpal. (2018). Internet of Robotic Things: Driving Intelligent Robotics of Future - Concept, Architecture, Applications and Technologies. *2018 4th International Conference on Computing Sciences (ICCS)*, 151-160.

Neelakandan, S., Beulah, J. R., Prathiba, L., Murthy, G. L. N., Irudaya Raj, E. F., & Arulkumar, N. (2022). Blockchain with deep learning-enabled secure healthcare data transmission and diagnostic model. *International Journal of Modeling, Simulation, and Scientific Computing*, *13*(04), 2241006. doi:10.1142/S1793962322410069

Ness, S., Shepherd, N. J., & Xuan, T. R. (2023). Synergy Between AI and Robotics: A Comprehensive Integration. *Asian Journal of Research in Computer Science*, *16*(4), 80–94. doi:10.9734/ajrcos/2023/v16i4372

Niihara, M., Saito, N., Yukawa, C., Toyoshima, K., Oda, T., Hirota, M., & Barolli, L. (2022). A Transportation Routing Method Based on A* Algorithm and Hill Climbing for Swarm Robots in WLAN Environment. *International Conference on Broadband and Wireless Computing. Communication and Applications*, 361–368.

Ni, J., Wu, W., Shen, J., & Fan, X. (2014). An Improved VFF Approach for Robot Path Planning in Unknown and Dynamic Environments. *Mathematical Problems in Engineering*, *18*(12), 1–10. doi:10.1155/2014/461237

Oji, R. (2012). An Automatic Algorithm for Object Recognition and Detection Based on ASIFT Key points. *Signal & Image Processing International Journal (Toronto, Ont.)*, *3*(5), 29–39.

Oliff, H., Liu, Y., Kumar, M., Williams, M., & Ryan, M. (2020). Reinforcement learning for facilitating human-robot-interaction in manufacturing. *Journal of Manufacturing Systems*, *56*, 326–340. doi:10.1016/j.jmsy.2020.06.018

Oniani, S., Marques, G., Barnovi, S., Pires, I. M., & Bhoi, A. K. (2021). Artificial intelligence for internet of things and enhanced medical systems. *Bio-inspired neurocomputing*, 43-59.

Orhan, K., Gülbeş, M. M., Jadhav, A., & Jagtap, R. (2024). Medico-Legal Problems of Artificial Intelligence. In *Artificial Intelligence in Dentistry* (pp. 259–282). Springer International Publishing.

Othman, K. (2023). Exploring the evolution of public acceptance towards autonomous vehicles with the level of knowledge. *Innovative Infrastructure Solutions*, *8*(8), 208. doi:10.1007/s41062-023-01180-z

Ouedraogo, M., Mignon, S., Cholez, H., Furnell, S., & Dubois, E. (2015). Security transparency: The next frontier for security research in the cloud. *Journal of Cloud Computing (Heidelberg, Germany)*, *4*(1), 1–14. doi:10.1186/s13677-015-0037-5

Pal, D., & Joshi, S. (2023). AI, IoT and Robotics in Smart Farming: Current Applications and Future Potentials. *2023 International Conference on Sustainable Computing and Data Communication Systems (ICSCDS)*, 1096-1101. 10.1109/ICSCDS56580.2023.10105101

Pande, A., & Mishra, D. (2023). The Synergy between a Humanoid Robot and Whisper: Bridging a Gap in Education. *Electronics (Basel)*, *12*(19), 3995. doi:10.3390/electronics12193995

Pandey, N. K., Kumar, K., Saini, G., & Mishra, A. K. (2023). Security issues and challenges in cloud of things-based applications for industrial automation. *Annals of Operations Research*, 1–20. doi:10.1007/s10479-023-05285-7 PMID:37361100

Paraforos, D. S., & Griepentrog, H. W. (2021). Digital farming and field robotics: Internet of things, cloud computing, and big data. *Fundamentals of Agricultural and Field Robotics*, 365-385.

Park, S., & Kim, J. (2018). Exploring the Impact of Cloud Robotics on Programming Skills in Middle School Students. *IEEE Transactions on Learning Technologies*, *11*(3), 412–427.

Park, Y. (2019). Exploring the Use of Cloud Robotics in Informal Learning Environments: A Case Study in Science Museums. *IEEE Transactions on Learning Technologies*, *12*(3), 372–387.

Patel, A., Shah, N., Ramoliya, D., & Nayak, A. (2020, November). A detailed review of cloud security: issues, threats & attacks. In *2020 4th International conference on electronics, communication and aerospace technology (ICECA)* (pp. 758-764). IEEE. 10.1109/ICECA49313.2020.9297572

Patoliya, J., Mehta, H., & Patel, H. (2015) Arduino controlled war field spy robot using night vision wireless camera and Android application. *Proceeding of 5th Nirma University International Conference on Engineering (NUiCONE)*, 1-5 10.1109/NUICONE.2015.7449624

Pavankumar, P., & Darwante, N. K. (2022). Performance Monitoring and Dynamic Scaling Algorithm for Queue Based Internet of Things. *2022 International Conference on Innovative Computing, Intelligent Communication and Smart Electrical Systems (ICSES)*, 1-7.

Peng, Y., Dong, M., & Zuo, M. J. (2010). Current status of machine prognostics in condition-based maintenance: A review. *International Journal of Advanced Manufacturing Technology*, *50*(1–4), 297–313. doi:10.1007/s00170-009-2482-0

Phasinam, K., Kassanuk, T., Shinde, P. P., Thakar, C. M., Sharma, D. K., Mohiddin, M. K., & Rahmani, A. W. (2022). Application of IoT and cloud computing in automation of agriculture irrigation. *Journal of Food Quality*, *2022*, 1–8. doi:10.1155/2022/8285969

Pokam, R., Debernard, S., Chauvin, C., & Langlois, S. (2019). Principles of transparency for autonomous vehicles: First results of an experiment with an augmented reality human–machine interface. *Cognition Technology and Work*, *21*(4), 643–656. doi:10.1007/s10111-019-00552-9

Prabhu, Shetty, GowdaSingh, & S. (2023). Design & Implementation of Brain Controlled Interface For Healthcare Applications. *IEEE International Conference on Distributed Computing, VLSI, Electrical Circuits and Robotics (DISCOVER)*, 197-200. 10.1109/DISCOVER58830.2023.10316733

Prasad, V. S., Mahajan, Y., & Suneetha, S. (2023). A Cloud-Based UV Monitoring System for Remote Real-Time UV Exposure Tracking. *2023 4th International Conference on Smart Electronics and Communication (ICOSEC)*, 1764-1770.

Prasad, A. O., Mishra, P., Jain, U., Pandey, A., Sinha, A., Yadav, A. S., Kumar, R., Sharma, A., Kumar, G., Hazim Salem, K., Sharma, A., & Dixit, A. K. (2023). Design and development of software stack of an autonomous vehicle using robot operating system. *Robotics and Autonomous Systems*, *161*, 104340. doi:10.1016/j.robot.2022.104340

Prasad, K., & Gugan, I. (2023). Enhancing Microcomputer Edge Computing for Autonomous IoT Motion Control. *International Journal on Recent and Innovation Trends in Computing and Communication*, *11*(3), 58–67. doi:10.17762/ijritcc.v11i3.6202

Preethesh, B., D'Costa, H. A., Alva, S. R., Kanchan, V. H., & Gatti, R. R. (2023). Feasibility Study of Digital Twin in Automotive Industry—Trends and Challenges. *Self-Powered Cyber Physical Systems,* 265-281.

Priyadarsini, K., Raj, E. F. I., Begum, A. Y., & Shanmugasundaram, V. (2020). Comparing DevOps procedures from the context of a systems engineer. *Materials Today: Proceedings*. Advance online publication. doi:10.1016/j.matpr.2020.09.624

Pushparaj, T. L., Rani, E. F. I., Raj, E. F. I., & Appadurai, M. (2022). Machine-based drug design to inhibit SARS-CoV-2 virus. *Smart Health Technologies for the COVID-19 Pandemic. Internet of Medical Things Perspectives*, *42*, 295.

Qin, R., Dzombak, R., Amin, R., & Mehta, K. (2013). Reliability of a telemedicine system designed for rural Kenya. *Journal of Primary Care & Community Health*, *4*(3), 177–181. doi:10.1177/2150131912461797 PMID:23799704

Qu, H., & Simon, X. (2009, September). Real-time robot path planning based on a modified pulse-coupled neural network model. *IEEE Transactions on Neural Networks*, *20*(11), 1724–1739. doi:10.1109/TNN.2009.2029858 PMID:19775961

Qureshi, B., & Koubâa, A. (2014). Five traits of performance enhancement using cloud robotics: A survey. *Procedia Computer Science*, *37*, 220–227. doi:10.1016/j.procs.2014.08.033

Raina, D., Singh, H., Saha, S. K., Arora, C., Agarwal, A., Chandrashekhara, S. H., ... Nandi, S. (2021, November). Comprehensive telerobotic ultrasound system for abdominal imaging: Development and in-vivo feasibility study. In *2021 International Symposium on Medical Robotics (ISMR)* (pp. 1-7). IEEE. 10.1109/ISMR48346.2021.9661578

Raj, E. F. I. (2023). Clustering and Classification of Digital Forensic Data Using Machine Learning and Data Mining Approaches. In *Advancements in Cybercrime Investigation and Digital Forensics* (pp. 115–136). Apple Academic Press. doi:10.1201/9781003369479-6

Raj, E. F. I., Appadurai, M., & Athiappan, K. (2022). Precision farming in modern agriculture. In *Smart Agriculture Automation Using Advanced Technologies: Data Analytics and Machine Learning, Cloud Architecture, Automation and IoT* (pp. 61–87). Springer Singapore.

Raj, E. F. I., Appadurai, M., Thiyaharajan, D., & Pushparaj, T. L. (2024). State-Of-The-Art Technologies for Crop Health Monitoring in the Modern Precision Agriculture. In *Precision Agriculture for Sustainability* (pp. 21–39). Apple Academic Press.

Raj, E. F. I., & Balaji, M. (2023). Shape Feature Extraction Techniques for Computer Vision Applications. In *Smart Computer Vision* (pp. 81–102). Springer International Publishing. doi:10.1007/978-3-031-20541-5_4

Raj, E. F. I., Manimala, K., & Appadurai, M. (2023). Differential Protection Scheme along with Backup Blockchain System. In *Smart and Power Grid Systems–Design Challenges and Paradigms* (pp. 135–156). River Publishers. doi:10.1201/9781003339557-6

Ramesh Naidu, P., Guruprasad, N., & Dankan Gowda, V. (2021). A High-Availability and Integrity Layer for Cloud Storage, Cloud Computing Security: From Single to Multi-Clouds. *Journal of Physics: Conference Series*, *1921*(1), 012072. doi:10.1088/1742-6596/1921/1/012072

Ramesha, M., & Pai, G. N. (2020). Internet of things: Internet revolution, impact, technology road map and features. *Adv. Math. Sci. J.*, *9*(7), 4405–4414. doi:10.37418/amsj.9.7.11

Rana, M. M., & Hossain, K. (2023). Connected and autonomous vehicles and infrastructures: A literature review. *International Journal of Pavement Research and Technology*, *16*(2), 264–284. doi:10.1007/s42947-021-00130-1

Rana, M., & Bhushan, M. (2023). Machine learning and deep learning approach for medical image analysis: Diagnosis to detection. *Multimedia Tools and Applications*, *82*(17), 26731–26769. doi:10.1007/s11042-022-14305-w PMID:36588765

Rani, E. F. I., Pushparaj, T. L., & Raj, E. F. I. (2022). Escalating the resolution of an urban aerial image via novel shadow amputation algorithm. *Earth Science Informatics*, *15*(2), 905–913. doi:10.1007/s12145-022-00772-0

Rani, E. F. I., Pushparaj, T. L., Raj, E. F. I., & Appadurai, M. (2022). New approaches in machine-based image analysis for medical oncology. In *Machine Learning and Deep Learning Techniques for Medical Science* (pp. 333–359). CRC Press. doi:10.1201/9781003217497-17

Ranjan, A., & Agrawal, R. (2023). Wireless and Solar-Powered Multipurpose Robot for Industrial Automation: A Sustainable Solution. *2023 8th International Conference on Communication and Electronics Systems (ICCES)*, 146-151.

Ranjan, A., Hegge, K. K., & Agrawal, R. (2023). Wireless and Solar-Powered Multipurpose Robot for Industrial Automation: A Sustainable Solution. *2023 8th International Conference on Communication and Electronics Systems (ICCES)*, 146-151.

Rastgoo, Nakisa, & Kebangsaan. (2014). A critical evaluation of literature on robot Path planning in dynamic environment. *Journal of Theoretical and Applied Information Technology*, *70*(1), 177-185.

Raunholt, T., Rodriguez, I., Mogensen, P., & Larsen, M. (2021). Towards a 5G mobile edge cloud planner for autonomous mobile robots. *2021 IEEE 94th Vehicular Technology Conference (VTC2021-Fall).*

Ravi, R. V., Goyal, S. B., Neagu, B. C., Raboaca, M. S., & Verma, C. (2022). A Low-Cost Industrial Automation System Using IoT and Cloud Computing. *2022 International Conference and Exposition on Electrical And Power Engineering (EPE)*, 649-653. 10.1109/EPE56121.2022.9959772

Razavian, A. A., & Sun, J. (2005). Cognitive based adaptive path planning algorithm for autonomous robotic vehicles. *IEEE Proceedings SoutheastCon*, 153 – 160. 10.1109/SECON.2005.1423236

Reddy, N. S., & Ashreetha, B. (2023). Technologies for Comprehensive Information Security in the IoT. *2023 International Conference for Advancement in Technology (ICONAT)*, 1-5.

Reddy, S., & Patwal, P. P. S. (2022). Data Analytics and Cloud-Based Platform for Internet of Things Applications in Smart Cities. *2022 International Conference on Industry 4.0 Technology (I4Tech)*, 1-6.

Reitze, D. D. (2014, September). *Using commercial web services to build Automated Test Equipment cloud based applications. In 2014 IEEE AUTOTEST*. IEEE.

Resende, C. H. L., Geraldes, C. A. S., & Lima, F. R. (2021). Decision Models for Supplier Selection in Industry 4.0 Era: A Systematic Literature Review. *Procedia Manufacturing*, *55*, 492–499. doi:10.1016/j.promfg.2021.10.067

Ribeiro, M. I. (2004). Kalman and extended kalman filters: Concept, derivation and properties. *Institute for Systems and Robotics*, *43*(46), 3736–3741.

Rivero-Moreno, Y., Echevarria, S., Vidal-Valderrama, C., Stefano-Pianetti, L., Cordova-Guilarte, J., Navarro-Gonzalez, J., ... Avila, G. L. D. (2023). Robotic surgery: A comprehensive review of the literature and current trends. *Cureus*, *15*(7). Advance online publication. doi:10.7759/cureus.42370 PMID:37621804

Rocha, D., Teixeira, G., Vieira, E., Almeida, J., & Ferreira, J. (2023). A Modular In-Vehicle C-ITS Architecture for Sensor Data Collection, Vehicular Communications and Cloud Connectivity. *Sensors (Basel)*, *23*(3), 1724. doi:10.3390/s23031724 PMID:36772760

Romeo, L., Petitti, A., Marani, R., & Milella, A. (2020). Internet of robotic things in smart domains: Applications and challenges. *Sensors (Basel)*, *20*(12), 3355. doi:10.3390/s20123355 PMID:32545700

Sadybekov, A. V., & Katritch, V. (2023). Computational approaches streamlining drug discovery. *Nature*, *616*(7958), 673–685. doi:10.1038/s41586-023-05905-z PMID:37100941

Saha, O., & Dasgupta, P. (2018). A comprehensive survey of recent trends in cloud robotics architectures and applications. *Robotics (Basel, Switzerland)*, *7*(3), 47. doi:10.3390/robotics7030047

Sahoo, S. K., & Choudhury, B. B. (2023). Challenges and opportunities for enhanced patient care with mobile robots in healthcare. *Journal of Mechatronics and Artificial Intelligence in Engineering*, *4*(2), 83–103. doi:10.21595/jmai.2023.23410

Sali, S. M., & Joy, K. R. (2023). Intelligent Rover: An IoT Based Smart Surveillance Robotic Car for Military. *2023 2nd International Conference on Computational Systems and Communication (ICCSC)*, 1-6. 10.1109/ICCSC56913.2023.10143011

Sarkar, S., Skalicky, M., Hossain, A., Brestic, M., Saha, S., Garai, S., Ray, K., & Brahmachari, K. (2020). Management of crop residues for improving input use efficiency and agricultural sustainability. *Sustainability (Basel)*, *12*(23), 9808. doi:10.3390/su12239808

Sarker, S., Jamal, L., Ahmed, S. F., & Irtisam, N. (2021). Robotics and artificial intelligence in healthcare during COVID-19 pandemic: A systematic review. *Robotics and Autonomous Systems*, *146*, 103902. doi:10.1016/j.robot.2021.103902 PMID:34629751

Sathyaprakash, B. P. (2021). Surveillance Spy Robot with Wireless Night Vision Camera and Metal Detection. *Turkish Journal of Computer and Mathematics Education*, *12*(12), 2397–2402.

Saxena, A., & Chaturvedi, R. (2021). Advancement of Industrial Automation in Integration with Robotics. *2021 International Conference on Simulation, Automation & Smart Manufacturing (SASM)*, 1-6. 10.1109/SASM51857.2021.9841071

Scholz, M., Zhang, X., Kreitlein, S., & Franke, J. (2018). Decentralised Intelligence: The Key for an Energy Efficient and Sustainable Intralogistics. *Procedia Manufacturing*, *21*, 679–685. doi:10.1016/j.promfg.2018.02.171

Sendi, A. S., & Cheriet, M. (2014, March). Cloud computing: A risk assessment model. In *2014 IEEE International Conference on Cloud Engineering* (pp. 147-152). IEEE. 10.1109/IC2E.2014.17

Serpen, G., & Dou, C. (2015, March). Automated robotic parking systems: Real-time, concurrent and multi-robot path planning in dynamic environments. *Applied Intelligence*, *42*(2), 231–251. doi:10.1007/s10489-014-0598-x

Sethuraman, S., & Singh, B. (2023). Smart Agriculture Applications Using Internet of Things. In Sustainable Science and Intelligent Technologies for Societal Development (pp. 380-392). IGI Global.

Sfeir, Saad, & Saliah-Hassane. (2011). An improved Artificial Potential Field approach to real-time mobile robot path planning in an unknown environment. *IEEE International Symposium on Robotic and Sensors Environments (ROSE),* 208 – 213.

Shakya, D. S. (2020). Survey on cloud based robotics architecture, challenges and applications. *Journal of Ubiquitous Computing and Communication Technologies*, *2*(1), 10–18. doi:10.36548/jucct.2020.1.002

Sharma, K. (2024). Personalized Telemedicine Utilizing Artificial Intelligence, Robotics, and Internet of Medical Things (IOMT). In Handbook of Research on Artificial Intelligence and Soft Computing Techniques in Personalized Healthcare Services (pp. 301-323). Apple Academic Press.

Sharma, M., Tomar, A., & Hazra, A. (2024). Edge Computing for Industry 5.0: Fundamental, Applications and Research Challenges. *IEEE Internet of Things Journal*, 1. doi:10.1109/JIOT.2024.3359297

Sharma, V., Kaur, S., Mahajan, R., Chawla, S., Khatri, T., & Dhankar, M. (2015). High-Performance Surveillance and Operating Robot with Raspberry Pi. *International Journal of Control and Automation*, *8*(11), 407–414. doi:10.14257/ijca.2015.8.11.37

Shekhar, R. (2023). Securing networked image transmission using public-key cryptography and identity authentication. *Journal of Discrete Mathematical Sciences and Cryptography*, *26*(3), 779–791. doi:10.47974/JDMSC-1754

Shen, X., Frazzoli, E., Rus, D., & Ang, M. H. (2016, October). Fast joint compatibility branch and bound for feature cloud matching. In *2016 IEEE/RSJ International Conference on Intelligent Robots and Systems (IROS)* (pp. 1757-1764). IEEE. 10.1109/IROS.2016.7759281

Shetty, S. H., Shetty, S., Singh, C., & Rao, A. (2022). Supervised Machine Learning: Algorithms and Applications. *Fundamentals and Methods of Machine and Deep Learning: Algorithms, Tools and Applications*, 1-16.

Shishehgar, M., Kerr, D., & Blake, J. (2018). A systematic review of research into how robotic technology can help older people. *Smart Health (Amsterdam, Netherlands)*, *7*, 1–18. doi:10.1016/j.smhl.2018.03.002

Shruthi & Naveen. (2020). Internet of Things: A Survey on Devices, Ecosystem, Components and Communication Protocols. *2020 4th International Conference on Electronics, Communication and Aerospace Technology (ICECA)*, 611-616.

Sijini, A. C., Fantin, E., & Ranjit, L. P. (2016). Switched reluctance motor for hybrid electric vehicle. *Middle East Journal of Scientific Research*, *24*(3), 734–739.

Sikorska, J., Hodkiewicz, M., & Ma, L. (2011). *Prognostic modelling options for remaining useful life estimation by industry*. Mechanical Systems and Signal Pro. doi:10.1016/j.ymssp.2010.11.018

Sikorski, J. J., Haughton, J., Kraft, M., Street, P., & Drive, P. F. (2016, June). Blockchain technology in the chemical industry: Machine-to-machine electricity market. *Applied Energy*, *195*(178), 234–246.

Singh, B. (2016). Nurse's attitude towards computerization in private hospitals of Tamil Nadu, India. Research journal of pharmacy and technology, 9(12), 2205-2210.

Singh, G., Banga, V. K., & Yingthawornsuk, T. (2022). Artificial Intelligence and Industrial Robot. *2022 16th International Conference on Signal-Image Technology & Internet-Based Systems (SITIS)*, 622-625. 10.1109/SITIS57111.2022.00098

Singh, A., Gupta, T., & Korde, M. (2017) Bluetooth controlled spy robot. *International Conference on Information, Communication, Instrumentation and Control (ICICIC),* 1-4.

Singh, B., & Acharjya, D. P. (2020). Computational intelligence techniques for efficient delivery of healthcare. *Health and Technology*, *10*(1), 167–185. doi:10.1007/s12553-018-00280-6

Singh, B., Lopez, D., & Ramadan, R. (2023). Internet of things in Healthcare: A conventional literature review. *Health and Technology*, *13*(5), 699–719. doi:10.1007/s12553-023-00771-1

Singh, B., & Muthuswamy, P. (2013). Factors affecting the adoption of electronic health records by nurses. *World Applied Sciences Journal*, *28*(11), 1531–1535.

Singh, B., & Senthil, J. (2015). Use of information technology by nurses in private hospitals in the state of Tamil Nadu in India. *Mediterranean Journal of Social Sciences*, *6*(4), 658–666. doi:10.5901/mjss.2015.v6n4s2p658

Singla, A., & Sharma, N. (2022). IoT Group Key Management using Incremental Gaussian Mixture Model. *2022 3rd International Conference on Electronics and Sustainable Communication Systems (ICESC)*, 469-474.

Siripurapu, S., Darimireddy, N. K., Chehri, A., Sridhar, B., & Paramkusam, A. V. (2023). Technological Advancements and Elucidation Gadgets for Healthcare Applications: An Exhaustive Methodological Review-Part-I (AI, Big Data, Block Chain, Open-Source Technologies, and Cloud Computing). *Electronics (Basel)*, *12*(3), 750. doi:10.3390/electronics12030750

Sishodia, R. P., Ray, R. L., & Singh, S. K. (2020). Applications of remote sensing in precision agriculture: A review. *Remote Sensing (Basel)*, *12*(19), 3136. doi:10.3390/rs12193136

Smith, E., & Brown, D. (2020). Advancing security through autonomous surveillance: The role of machine learning algorithms. *The International Journal of Robotics Research*, *39*(4), 567–582.

Smith, E., & Brown, D. (2020). Swarm robotics for urban surveillance: Opportunities and challenges. *IEEE Robotics & Automation Magazine*, *27*(2), 75–89.

Smith, J. R., & Johnson, M. (2020). The Role of Cloud Robotics in the Education Sector: Current Status and Future Trends. *Robotics (Basel, Switzerland)*, *9*(3), 59.

Smith, J., & Johnson, A. (2019). Cloud Robotics: A Comprehensive Review. *IEEE Transactions on Robotics*, *35*(3), 657–672.

Sood, V. M., Singh, Y., Bhargava, B., & Narang, S. K. (Eds.). (2024). *Intelligent Security Solutions for Cyber-Physical Systems*. CRC Press. doi:10.1201/9781003406105

Soori, M., Arezoo, B., & Habibi, M. (2013). Dimensional and geometrical errors of three-axis CNC milling machines in a virtual machining system. *Computer-Aided Design*.

Sørensen, M., Andersen, Z. J., Nordsborg, R. B., Jensen, S. S., Lillelund, K. G., Beelen, R., Schmidt, E. B., Tjønneland, A., Overvad, K., & Raaschou-Nielsen, O. (2012). Road traffic noise and incident myocardial infarction: A prospective cohort study. *PLoS One*, *7*(6), e39283. doi:10.1371/journal.pone.0039283 PMID:22745727

Sorrentino, A., Cavallo, F., & Fiorini, L. (2020). A plug and play transparent communication layer for cloud robotics architectures. *Robotics (Basel, Switzerland)*, *9*(1), 17. doi:10.3390/robotics9010017

Sridaran, G., & Dhulekar, N. (2001). Object Recognition and Tracking in Video Sequences. *IEEE Transactions on Pattern Analysis and Machine Intelligence*, *23*(4), 1335–1342.

Srinivas, K. N. V., & Lakineni, P. K. (2023). Development of a Real-time Location Monitoring App with Emergency Alert Features for Android Devices. *2023 4th IEEE Global Conference for Advancement in Technology (GCAT),* 1-8.

Srinivas, V., & De, T. (2023). Wi-Fi Router Signal Coverage Position Prediction System using Machine Learning Algorithms. *2023 International Conference on Sustainable Computing and Smart Systems (ICSCSS),* 253-258.

Stasevych, M., & Zvarych, V. (2023). Innovative robotic technologies and artificial intelligence in pharmacy and medicine: Paving the way for the future of health care—a review. *Big Data and Cognitive Computing*, *7*(3), 147. doi:10.3390/bdcc7030147

Steele, J. P. H. (1989). An algorithm with real-time response for avoiding moving obstacles. *Systems, Man and Cybernetics Conference Proceedings, IEEE International Conference, 2*, 551-556. 10.1109/ICSMC.1989.71357

Sultan, N. (2014). Making use of cloud computing for healthcare provision: Opportunities and challenges. *International Journal of Information Management*, *34*(2), 177–184. doi:10.1016/j.ijinfomgt.2013.12.011

Sundararajan, A. (2014). Peer-to-peer businesses and the sharing (collaborative) economy: Overview, economic effects and regulatory issues. *The Power of Connection: Peer to Peer Businesses*.

Sun, L., Jiang, X., Ren, H., & Guo, Y. (2020). Edge-cloud computing and artificial intelligence in internet of medical things: Architecture, technology and application. *IEEE Access : Practical Innovations, Open Solutions*, *8*, 101079–101092. doi:10.1109/ACCESS.2020.2997831

Sun, L., Ma, J., Wang, H., Zhang, Y., & Yong, J. (2015). Cloud service description model: An extension of USDL for cloud services. *IEEE Transactions on Services Computing*, *11*(2), 354–368. doi:10.1109/TSC.2015.2474386

Swasthik, M., Pawan, C. T., Shankar, H., Harshitha, V., & Vinay, N. A. (2021). SPi-Bot: A Gesture Controlled Virtual Telepresence Robot. *International Journal of Modern Agriculture*, *10*(2), 3782–3792.

Syamala, M., Malathi, J., Singh, V., GS, H. P., Maheswari, B. U., & Murugan, S. (2024). Cloud Solutions for Smart Parking and Traffic Control in Smart Cities. In Handbook of Research on AI and ML for Intelligent Machines and Systems (pp. 169-194). IGI Global.

Talpes, E., Sarma, D. D., Venkataramanan, G., Bannon, P., McGee, B., Floering, B., Jalote, A., Hsiong, C., Arora, S., Gorti, A., & Sachdev, G. S. (2020). Compute solution for tesla's full self-driving computer. *IEEE Micro*, *40*(2), 25–35. doi:10.1109/MM.2020.2975764

Tang, Y., Dananjayan, S., Hou, C., Guo, Q., Luo, S., & He, Y. (2021). A survey on the 5G network and its impact on agriculture: Challenges and opportunities. *Computers and Electronics in Agriculture*, *180*, 105895. doi:10.1016/j.compag.2020.105895

Tao, Y., Skubic, M., Han, T., Xia, Y., & Chi, X. (2010). Performance Evaluation of SIFT-Based Descriptors for Object Recognition. *Proceedings of the International Multi Conference of engineers and computer scientist, 2,* 1453-1456.

Tardioli, D. (2014). A wireless communication protocol for distributed robotics applications. *2014 IEEE International Conference on Autonomous Robot Systems and Competitions (ICARSC).* 10.1109/ICARSC.2014.6849795

Tengilimoglu, O., Carsten, O., & Wadud, Z. (2023). Infrastructure requirements for the safe operation of automated vehicles: Opinions from experts and stakeholders. *Transport Policy*, *133*, 209–222. doi:10.1016/j.tranpol.2023.02.001

Touhid, M. T. B., Marne, M., Oskroba, T., Mirahmadi, S. A., Zhu, E., Mehrabian, A., Defersha, F., & Yang, S. (2023). Building a cloud-based digital twin for remote monitoring and control of a robotic assembly system. *International Journal of Advanced Manufacturing Technology*, *129*(9), 4045–4057. doi:10.1007/s00170-023-12611-7

Trianni, V., IJsselmuiden, J., & Haken, R. (2016). *The saga concept: swarm robotics for agricultural applications*. Technical Report. 2016. Available online: http://laral. istc. cnr. it/saga…

Tyagi, A. K. (2024). Blockchain and Artificial Intelligence for Cyber Security in the Era of Internet of Things and Industrial Internet of Things Applications. In AI and Blockchain Applications in Industrial Robotics (pp. 171-199). IGI Global.

Uddin, M. A., Ayaz, M., Mansour, A., Aggoune, E. H. M., Sharif, Z., & Razzak, I. (2021). Cloud-connected flying edge computing for smart agriculture. *Peer-to-Peer Networking and Applications*, *14*(6), 3405–3415. doi:10.1007/s12083-021-01191-6

Udendhran, R., Yamini, G., Badrinath, N., Jegathesh Amalraj, J., & Suresh, A. (2023). Enhancing representational learning for cloud robotic vision through explainable fuzzy convolutional autoencoder framework. *Soft Computing*, 1–11. doi:10.1007/s00500-023-08570-6

Uscher-Pines, L., & Mehrotra, A. (2014). Analysis of Teladoc use seems to indicate expanded access to care for patients without prior connection to a provider. *Health Affairs*, *33*(2), 258–264. doi:10.1377/hlthaff.2013.0989 PMID:24493769

Vallès-Peris, N., & Domènech, M. (2023). Caring in the in-between: A proposal to introduce responsible AI and robotics to healthcare. *AI & Society*, *38*(4), 1685–1695. doi:10.1007/s00146-021-01330-w

van der Aalst, W. (2000). Process-oriented architectures for electronic commerce and inter organizational workflow. *Information Systems*, *24*(8), 639–671. doi:10.1016/S0306-4379(00)00003-X

Varun, R. S., & Rajesh, T. (2018). Implementation of swarm intelligence in obstacle avoidance. *RTEICT 2017 - 2nd IEEE International Conference on Recent Trends in Electronics, Information and Communication Technology, Proceedings*, 525–528.

Vásárhelyi, G., Virágh, C., Somorjai, G., Nepusz, T., Eiben, A. E., & Vicsek, T. (2018). Optimised flocking of autonomous drones in confined environments. *Science Robotics*, *3*(20), eaat3536. doi:10.1126/scirobotics.aat3536 PMID:33141727

Velásquez Villagrán, N., Pesado, P., & Estevez, E. (2020). Cloud robotics for industry 4.0 - A literature review. In *Communications in Computer and Information Science* (pp. 3–15). Springer International Publishing.

Vellela, S. S., Reddy, B. V., Chaitanya, K. K., & Rao, M. V. (2023, January). An Integrated Approach to Improve E-Healthcare System using Dynamic Cloud Computing Platform. In *2023 5th International Conference on Smart Systems and Inventive Technology (ICSSIT)* (pp. 776-782). IEEE. 10.1109/ICSSIT55814.2023.10060945

Vermesan, O., Bröring, A., Tragos, E., Serrano, M., Bacciu, D., Chessa, S., & Bahr, R. (2022). Internet of robotic things–converging sensing/actuating, hyperconnectivity, artificial intelligence and IoT platforms. In *Cognitive Hyperconnected Digital Transformation* (pp. 97–155). River Publishers. doi:10.1201/9781003337584-4

Vijeth, G., & Gatti, R. R. (2023). Review on Robot Operating System. *Self-Powered Cyber Physical Systems*, 297-307.

Voisin, A., Levrat, E., Cocheteux, P., & Iung, B. (2010). Generic prognosis model for proactive maintenance decision support: Application to pre-industrial e- maintenance test bed. *Journal of Intelligent Manufacturing*, *21*(2), 177–193. doi:10.1007/s10845-008-0196-z

Wan, J., Tang, S., Shu, Z., Li, D., Wang, S., Imran, M., & Vasilakos, A. V. (2016). Software-Defined Industrial Internet of Things in the Context of Industry 4.0. *IEEE Sensors Journal, 16*(20).

Wang, Q., & Chen, L. (2020). Enhancing Educational Robotics through Cloud-Based Programming Platforms. *IEEE Transactions on Learning Technologies*, *13*(4), 789–804.

Wang, X. (2017). Remote Programming of Educational Robots via Cloud Services: A Case Study in Middle School Classrooms. *IEEE Transactions on Learning Technologies*, *10*(4), 509–524.

Wang, Y., Liu, Z., Zuo, Z., Li, Z., Wang, L., & Luo, X. (2019). Trajectory planning and safety assessment of autonomous vehicles based on motion prediction and model predictive control. *IEEE Transactions on Vehicular Technology*, *68*(9), 8546–8556. doi:10.1109/TVT.2019.2930684

Wan, J., Tang, S., Yan, H., Li, D., Wang, S., & Vasilakos, A. V. (2016). Cloud robotics: Current status and open issues. *IEEE Access : Practical Innovations, Open Solutions*. doi:10.1109/ACCESS.2016.2574979

Watson, P. (2018). Integrating Cloud Robotics into Classroom and Learning Experiences. *Journal of Computing in Higher Education*, *30*(2), 358–371.

Wielki, J. (2013). *Implementation of the big data concept in organizations-possi-bilities, impediments and challenges.* Paper presented at the Computer Science.

Wior, I., Roth, P., & Fay, A. (2012). Stability Criterion for a Single Decision-Making Node in Decentralized Material Handling Systems. *IFAC Proceedings Volumes, 45*(24), 320–327. https://doi.org/https://doi.org/10.3182/20120912-3-BG-2031.00068

Woo, S., Youtie, J., Ott, I., & Scheu, F. (2021). Understanding the long-term emergence of autonomous vehicles technologies. *Technological Forecasting and Social Change*, *170*, 120852. doi:10.1016/j.techfore.2021.120852

Wu, S.-j., Gebraeel, N., Lawley, M. A., & Yih, Y. (2007). A neural network integrated decision support system for condition-based optimal predictive maintenance policy. *Systems, Man and Cybernetics, Part A: Systems and Humans. IEEE Transactions on, 37*(2), 226–236.

Xia, X., Meng, Z., Han, X., Li, H., Tsukiji, T., Xu, R., Zheng, Z., & Ma, J. (2023). An automated driving systems data acquisition and analytics platform. *Transportation Research Part C, Emerging Technologies*, *151*, 104120. doi:10.1016/j.trc.2023.104120

Xin, N. Y., & Ling, L. Y. (2013). *How we could realize big data value*. Paper presented at the Instrumentation and Measurement, Sensor Network and Automation (IMSNA), 2013 2nd International Symposium on.

Xue, Y., Wang, C., Ding, C., Yu, B., & Cui, S. (2024). Observer-based event-triggered adaptive platooning control for autonomous vehicles with motion uncertainties. *Transportation Research Part C, Emerging Technologies*, *159*, 104462. doi:10.1016/j.trc.2023.104462

Xu, L. (2020). Cloud Robotics and its Application in Project-Based Learning for Engineering Students. *IEEE Transactions on Learning Technologies*, *13*(1), 45–58.

Yadav, S., Kaushik, A., & Sharma, S. (2021). Simplify the difficult: artificial intelligence and cloud computing in healthcare. *IoT and Cloud Computing for Societal Good*, 101-124.

Yalcin, M., & Patoglu, V. (2012, June). Kinematics and design of AssistOn-SE: A self-adjusting shoulder-elbow exoskeleton. In *2012 4th IEEE RAS & EMBS International Conference on Biomedical Robotics and Biomechatronics (BioRob)* (pp. 1579-1585). IEEE. 10.1109/BioRob.2012.6290928

Yan, H., Wan, J., Zhang, C., Tang, S., Hua, Q., & Wang, Z. (2018). Industrial Big Data Analytics for Prediction of Remaining Useful Life Based on Deep Learning, Real-Time Edge Analytics for Big Data in Internet of Things. *IEEE Access, 6*, 17190 – 17197.

Yan, J., Meng, Y., Lu, L., & Li, L. (2017). Industrial Big Data in an Industry 4.0 Environment: Challenges, Schemes, and Applications for Predictive Maintenance, Complex System Health Management based on Condition Monitoring and Test Data. *IEEE Access, 5*, 23484 – 23491.

Yang, G., Pang, Z., Deen, M. J., Dong, M., Zhang, Y. T., Lovell, N., & Rahmani, A. M. (2020). Homecare robotic systems for healthcare 4.0: Visions and enabling technologies. *IEEE Journal of Biomedical and Health Informatics*, *24*(9), 2535–2549. doi:10.1109/JBHI.2020.2990529 PMID:32340971

Yang, G., & Zhang, H. (2019). Cloud Robotics in Higher Education: A Survey of Current Practices and Future Directions. *IEEE Transactions on Education*, *62*(3), 205–220.

Yang, P., Xiong, N., & Ren, J. (2020). Data security and privacy protection for cloud storage: A survey. *IEEE Access : Practical Innovations, Open Solutions*, *8*, 131723–131740. doi:10.1109/ACCESS.2020.3009876

Yu, H., & Zhou, Z. (2021). Optimization of IoT-based artificial intelligence assisted telemedicine health analysis system. *IEEE Access : Practical Innovations, Open Solutions*, *9*, 85034–85048. doi:10.1109/ACCESS.2021.3088262

Yung, K. L., Tsang, Y. P., Wu, C. H., & Ip, W. H. (2023). An autonomous, multi-agent, IoT-empowered space logistics system for mission-critical inventory packing. *ISA Transactions*, *132*, 167–181. doi:10.1016/j.isatra.2022.10.035 PMID:36404152

Zereik, E., Bibuli, M., Mišković, N., Ridao, P., & Pascoal, A. (2018). Challenges and future trends in marine robotics. *Annual Reviews in Control*, *46*, 350–368. doi:10.1016/j.arcontrol.2018.10.002

Zhai, Z., Martínez, J. F., Beltran, V., & Martínez, N. L. (2020). Decision support systems for agriculture 4.0: Survey and challenges. *Computers and Electronics in Agriculture*, *170*, 105256. doi:10.1016/j.compag.2020.105256

Zhang, Y., Ren, S., Liu, Y., & Si, S. (2017). A big data analytics architecture for cleaner manufacturing and maintenance processes of complex products. *Journal of Cleaner Production 142*(2), 626-641.

Zhang, F., Shi, W., Chen, M., Huang, W., & Liu, X. (2023). Open HD map service model: An interoperable high-Definition map data model for autonomous driving. *International Journal of Digital Earth*, *16*(1), 2089–2110. doi:10.1080/17538947.2023.2220615

Zhang, F., Xiao, Y., Gao, L., Ma, D., Su, R., & Yang, Q. (2022). How agricultural water use efficiency varies in China—A spatial-temporal analysis considering unexpected outputs. *Agricultural Water Management*, *260*, 107297. doi:10.1016/j.agwat.2021.107297

Zhang, H., & Zhang, L. (2019, April). Cloud robotics architecture: trends and challenges. In *2019 IEEE International Conference on Service-Oriented System Engineering (SOSE)* (pp. 362-3625). IEEE. 10.1109/SOSE.2019.00061

Zhang, L., Wang, H., & Li, Q. (2019). Swarm robotics for collaborative surveillance: Challenges and opportunities. *IEEE Transactions on Robotics*, *35*(3), 647–662.

Zhang, R., Zhou, Y., Zhang, J., & Zhao, J. (2023). Cloud-integrated robotics: transforming healthcare and rehabilitation for individuals with disabilities. *Proceedings of the Indian National Science Academy*, 1-12.

Zhang, X., Xiong, G., Ai, Y., Liu, K., & Chen, L. (2023). Vehicle dynamic dispatching using curriculum-driven reinforcement learning. *Mechanical Systems and Signal Processing*, *204*, 110698. doi:10.1016/j.ymssp.2023.110698

Zhang, Y. (2018). Cloud Robotics for Online Robotics Competitions in High Schools. *IEEE Transactions on Learning Technologies*, *11*(1), 132–147.

Zhao, X., Fang, Y., Min, H., Wu, X., Wang, W., & Teixeira, R. (2023). Potential sources of sensor data anomalies for autonomous vehicles: An overview from road vehicle safety perspective. *Expert Systems with Applications*, 121358.

Zheng, Y., Shen, M., Shen, X., & Zhu, X. (2017). A cloud-based robot programming system for STEM education. In *2017 IEEE/RSJ International Conference on Intelligent Robots and Systems (IROS)* (pp. 5197-5202). IEEE.

Zhou, K., Wen, Y., Wu, W., Ni, Z., Jin, T., & Long, X. (2020). Cloud service optimization method based on dynamic artificial ant-bee colony algorithm in agricultural equipment manufacturing. *Mathematical Problems in Engineering*, *2020*, 1–11. doi:10.1155/2020/9134695

Zygouris, N. C., Striftou, A., Dadaliaris, A. N., Stamoulis, G. I., Xenakis, A. C., & Vavougios, D. (2017, April). The use of LEGO mindstorms in elementary schools. In *2017 IEEE Global Engineering Education Conference (EDUCON)* (pp. 514-516). IEEE. 10.1109/EDUCON.2017.7942895

About the Contributors

Rathishchandra Ramachandra Gatti currently works as Professor and Head of the Department of Mechanical Engineering & Robotics at Sahyadri College of Engineering & Management, India. Previously he worked as Sessional Academic at Curtin University - Australia, Lead NPI Engineer at General Electric, Product development Engineer at SVI and Automotive Design Engineer at Meritor-Automotive, and Manufacturing Engineer Trainee at Rapsri. He has 21 years of work experience with 19 years in Industrial and Academic R&D and product development. He has published 15 patents(3 Granted), 50+ peer reviewed publications and edited/authored 7 books. His research interests are self powered autonomous systems, energy harvesting, robotics and AI in physical systems.

Chandra Singh is currently working as a Assistant Professor of Department of Electronics & Communication Engineering. He pursued his Bachelors in Electronics & Communication Engineering at SSE,Mukka, M.Tech from NMAM Institute of Technology, Nitte and Pursuing PhD from VTU Belagavi. He has 4 Patents (3 Australia, 1 South Africa), 25+ peer-reviewed publications and editor of 7 research books(3 Books Published). His research interests include communication, wireless communication, IoT AI/ML, CPS, Robotics.

* * *

Masooda A. currently serves as an Assistant Professor at Sahyadri College of Engineering and Management. She holds both a Bachelor's and Master's degree from VTU University. Her academic pursuits and expertise lie in the dynamic realms of Natural Language Processing, Cloud Computing, IoT, Machine Learning, and Deep Learning.

M. Balaji, Associate Professor in the Department of Electrical and Electronics Engineering, has 19 years of teaching and research experience. He received his B.E (EEE) degree first class with distinction from Annamalai University, M.E. Power Electronics and Drives (Anna University 5th Rank) from Sri Venkateswara College of Engineering,AnnaUniversity and Ph.D in the area of Special Electrical Machines from Anna University.He has published 40 papers in international journals and around 50 papers in national and international conferences. He has coordinated a National level conference, National level Workshops and has delivered guest lectures at various conferences and workshops. He is a recognized supervisor for Anna University. He is currently guiding 4 Ph.D scholars working on Design and Control aspects of Electrical Machines, Hybrid Electric Vehicles and Power Converters. Seven scholars have completed Ph.D under his guidance. He has completed DST -SERB project titled "Design and Development of Hybrid Switched Reluctance Motor Drive",2015-18, Rs 22.9 Lakhs under Fast track scheme for young scientist. He is the co-principal investigator for the project titled "Design and Development of Permanent Magnet Assisted Reluctance Motor Drives for Pump Application funded by Science and Engineering Research Board (SERB).

Jyoti Batra is an Assistant Professor in the School of Computer Applications at Lovely Professional University, Punjab, India. Her research interests lie in areas like programming.

Sneep Bhat has completed BE in Electronics and Communication Engineering from VTU, Belagavi and M.Tech in Digital Electronics and Communication from VTU and completed Ph D in the specialization od AI based Robots under VTU Belagavi, Karnataka, India. He is having more than 20 years of teaching and 12 years of research experience. He has published more than 25 research papers in reputed journals and conference proceedings' and he has one patent in his credit and written a text book, 'Image Processing'.

Hrnakshi C. is an Assistant Professor at Sahyadri College of Engineering and Management. She received a bachelor's & Master' s degree in Computer Science from VTU University. She is interested in Data Analytics, Machine Learning and Big data Analytics research domains.

E atn Irudaya Raj is currently working as an Assistant Professor at Dr. Sivanthi Aditanar College of Engineering, Tiruchendur. He obtained his PhD from Anna University, ME in Power electronics and Drives from SSN College of Engineering, Chennai and Bachelor of Engineering in Electrical and Electronics Engineering from Dr. Sivanthi Aditanar College of Engineering, Tiruchendur. he has 9+ patents,3+ Copyrights,32+ journal papers, 16+ conference papers,26 book chapters, 1 authored book and 2 edited books. He has an h-index of 23 and an i-10 index of 38. He was recognized as an NPTEL Domain Scholar under the domain "Faculty Domain – Advanced" in October 2022, by NPTEL, IIT Madras.

E rancy Irudaya Rani has been working as an assistant professor in the Electronics and Communication Engineering Department at Francis Xavier Engineering College, Tamil Nadu, India. She has completed a major research project on an IoT-based smart wheelchair sponsored by TNSCST, Tamil Nadu, India. She has published many research articles in reputed journals and guided many ME and BE students. Her Major research work is in medical image processing, IoT, and its applications.

Dna owda V. is currently working as an Assistant Professor in the Department of Electronics and Communication Engineering at BMS Institute of Technology and Management in Bangalore. Previously, he worked as a Research Fellow at ADA DRDO and as a Software Engineer at Robert Bosch. With a total experience of 14 years, including teaching and industry, Dr. Gowda has made significant contributions to both academia and research. He has published over 80 research papers in renowned international journals and conferences. In recognition of his innovative work, Dr. Gowda has been granted six patents, including four from Indian authorities and two international patents. His research interests primarily lie in the fields of IoT and Signal Processing, where he has conducted workshops and handled industry projects. Dr. Gowda is passionate about teaching and strives to create an engaging and effective learning environment for his students. He consistently seeks opportunities to enhance his knowledge and stay updated with the latest advancements in his field.

Sandeepkumar Hegde is working as Associate Professor in the department of Computer Science and Engineering at NMAM Institute of Technology and completed his PhD from Visveswarya Technological University, Belagavi. He has more than 12 years of teaching experience and 3 years of industrial experience. He has more than 6 years of research experience. He has published more than 18 research papers in international journals and conferences. His area of interest include artificial intelligence and machine learning.

Mnimala K. has 27 years of teaching experience. She is currently working as Associate Professor in the department of Information Science and Technology, College of Engineering, Guindy, Anna University, Chennai. She has published 30 journal papers in reputed journals like Elsevier, Springer, Taylor & Francis and has published several conference papers, book chapters. She is a life time member in Professional bodies like Indian Society of Technical Education and Institution of Engineers (India). She served as a member of MCA syllabus reviewing committee in Xavier's College (Autonomous), Tirunelveli. She is also a reviewer for journals like Elsevier, IET and Springer. She has received funds from Institution of Engineers (India) for 5 student projects under her guidance. She has also received a fund of Rs.6.8 Lakhs from DRDO for her project titled "Implementation of path planning optimization algorithm for efficient navigation of UAV". Currently her project titled "Leader Follower Autonomous Flight control for UAV swarm" for 29.7 Lakhs was sanctioned by DRDO ARDB panel and is in ongoing stage. Her interests are doing projects in Machine learning, IOT, Cyber Security, Power Electronics.

Rakshith K. is currently pursuing his BE in mechanical engineering from Sahyadri College of Engineering & Management.

Jsreet Kaur is currently working as an Assistant Professor in University Business School, Chandigarh University, Mohali, Punjab. She is a post graduate (MBA-H.R) from Panjab University, Chandigarh. She has also qualified UGC NET JRF in Human Resource Management/Labour and Social Welfare and has pursued PhD in Business Management from Chandigarh University, Mohali. She has over 8 years of experience in academic and administrative assignments. She also received "Best Teacher of the Department Award" in the year 2019 and 2021 in the field of imparting quality education. Her research interests include Employee Engagement, Management of Organizational Change and Organization Development. She has published several research papers and articles in reputed international and national journals and has presented papers in various national and international conferences. She also contributed one edited book and 10 book chapters on various topics.

Mandeep Kaur is working as Assistant Professor in Department of Electronics and Communication engineering, Punjabi University Patiala. She has thirteen years of teaching experience. She has published more than 60 research papers in SCI and Scopus journal and International/National conferences. She has guided 25 M.Tech students. Her area of interest is VLSI design, VLSI DSP, optimization and machine learning.

Dvender Kumar is an Assistant Professor in the School of Computer Applications at Lovely Professional University, Punjab, India. His research interests lie in areas like wireless sensor networks and artificial intelligence, particularly focusing on optimizing energy efficiency and data transmission. Dr. Kumar is passionate about leveraging technology to enhance learning experiences and is actively involved in developing innovative approaches for computer science education.

T urthu Pushparaj received his Doctor of Philosophy Degree in Chemistry from the University of Madras, India. He is working as an assistant professor in the PG Department of Chemistry at TD-MNS College, India. He has published many research articles in reputed journals and guided many PG and Ph.D. research scholars. His major research interests are in smart MRI CAs, material science, and green engineering.

Shifan Mohammed is currently pursuing his BE in mechanical engineering from Sahyadri College of Engineering & Management.

Thangavel Murugan (Senior Member – IEEE) is serving as an Assistant Professor in the Department of Information Systems and Security, College of Information Technology, United Arab Emirates University. His academic and research specialization is Information Security, High Performance Computing, Ethical Hacking, Cyberforensics, Blockchain, Cybersecurity Intelligence and Educational Technology. He received Doctorate from Madras Institute of Technology (MIT) Campus, Anna University – Chennai, received Post Graduate degree as M.E. Computer Science and Engineering from J.J. College of Engineering and Technology, Trichy under Anna University – Chennai (University First Rank Holder & Gold Medalist) and received Bachelor's degree as B.E. Computer Science and Engineering from M.A.M College of Engineering, Trichy under Anna University – Chennai (College First Rank Holder & Gold Medalist). He presently holds 11 years of Teaching and Research experience from various academic institutions. He has published 10+ articles in International Journals, 15+ book chapters in International Publishers, 25+ in the proceedings of International Conferences and 3 in the proceedings of national conferences / seminars. He has been actively participating as reviewers in the international journals and conferences. He has attended 100+ and organized 50+ Workshops / FDPs /Contests/Industry based courses over the past years of experience. He has been a technical speaker in various Workshops/FDPs/Conferences. He has been recognized by IIT Bombay & SAP CSR as SAP Award of Excellence with a cash reward of Rs.5000/- for the best Participation in IITBombayX: FDPICT001x Use of ICT in Education for Online and Blended Learning. He shows interest in student counseling, in motivating for better career and in helping them design value-based lifestyle.

Shreeraj N. K. is currently pursuing his BE in mechanical engineering from Sahyadri College of Engineering & Management.

N on Subathira completed B.E Electrical And Electronics Engineering under Anna University. She Completed M.E Embedded Systems Under Anna University. She is Working as Junior Research Fellow under DRDO Project in Dr. Sivanthi Aditanar College of Engineering Tiruchendur. Her areas of interest are machine learning, unmanned aerial vehicles (UAV), and artificial intelligence.

Rajalaxmi Prabhu B. is working as Associate Professor in the department of Computer Science and Engineering at NMAM Institute of Technology and completed her PhD from Visveswarya Technological University, Belagavi. She has more than 11 years of teaching experience and more than 6 years of research experience. She has published more than 15 research papers in international journals and conferences. Her area of interest include artificial intelligence and machine learning, Natural language processing.

K. D. V. Prasad works as a Faculty of Research at Symbiosis Institute of Business Management, Hyderabad; Symbiosis International (Deemed University), Pune, India. Dr. Prasad holds a Master's in Computer Applications and a Master's in Software Systems from BITS, Pilani; MBA (Human Resources), IGNOU, New Delhi. Dr. Prasad possesses a Ph.D. in Business Management (Kanpur University), and PhD in Business Administration (RTM Nagpur University. He is AIMA-Certified Management Teacher and Fellow, World HR Board, Carlton Advanced Management Institute, USA; Fellowship of the Management and Business Research Council (FMBRC).

Nthya R. received the B.E. degree in Computer science and Engineering from Nandha Engineering College, Erode in 2008. She received her M.E. degree in Computer science and Engineering at SNS College of Technology, Coimbatore, India in the year 2010. She has completed her Ph.D. degree in Information and Communication Engineering from Anna University, Chennai in 2017. Currently working as an Associate Professor at the Department of Computer Science Engineering, Vivekananda College of Engineering for Women. She has over 12 years of experience in educational Institution, her research and development experience is 11 years. She has more than 10 publications in International Journals, presented papers in International conferences. Her Area of Research interests are IoT, Wireless Networks, Machine Learning, Deep learning, Artificial intelligence.

Bieda Singh received Ph.D from VIT University, Vellore, Tamilnadu, India. He is an Assistant Professor (Senior) in the School of Information Technology Engineering and Information Systems at VIT Vellore, Tamilnadu, India. His research interests include Health Informatics, Soft Computing Techniques, Artificial Intelligence, Rough Set, Knowledge Representations.

C Sweetline Jenita completed her Bachelor of Engineering in Electrical and Electronics Engineering under Anna University. She have pursued her M.E degree in Power Electronics and Drives under Anna University. Presently, she is working as Junior Research Fellow (JRF) for DRDO-ARDB project in Dr.Sivanthi Aditanar Engineering College, Tiruchendur. Her area of interest is Machine Learning, Unmanned Area Vehicle (UAV).

Sai Venkatramana Prasada G. S. is currently working as Assistant Professor in the Department of Electronics and communication Engineering in Vivekananda College of Engineering and Technology, Puttur, Karnataka, India. He completed his PhD from REVA University, Bengaluru, India in the area of VLSI. He has 14 years of teaching experience. His research area of interest is Robotics and Artificial Intelligence, Low Power, High speed VLSI design.

Index

A

C

N

O

P

R

S

T

V

W

www.ingramcontent.com/pod-product-compliance
Lightning Source LLC
LaVergne TN
LVHW080120011026
849160LV00013B/1142